W9-CHO-708

A

CYCLE

OF

CELESTIAL OBJECTS

FOR THE USE OF NAVAL, MILITARY, AND
PRIVATE ASTRONOMERS

OBSERVED, REDUCED, AND DISCUSSED

BY

CAPTAIN WILLIAM HENRY SMYTH, R.N., K.S.F., D.C.L.,

ONE OF THE BOARD OF VISITORS OF THE ROYAL OBSERVATORY;

FELLOW OF THE ROYAL, THE ANTIQUARIAN, THE ASTRONOMICAL, AND THE
GEOGRAPHICAL SOCIETIES OF LONDON; VICE-PRESIDENT OF THE
UNITED SERVICE INSTITUTION;
CORRESPONDING MEMBER OF THE INSTITUTE OF FRANCE;
HONORARY MEMBER OF THE ROYAL IRISH ACADEMY; AND OF THE
SCIENTIFIC ACADEMIES OF NAPLES, PALERMO, FLORENCE,
WASHINGTON, AND NEW YORK.

VOLUME THE SECOND.

THE BEDFORD CATALOGUE.

Published by:

Willmann–Bell, Inc.

P. O. Box 35025
Richmond, Virginia 23235   ☎ (804)
United States of America   320-7016

Publishers and Booksellers

Serving Astronomers Worldwide
Since 1973

Library of Congress Cataloging-in-Publication Data

Smyth, W. H. (William Henry), 1788-1865.
    The Bedford catalogue.

        Includes index.
        1. Astronomy--Amateurs' manuals.   I. Title.
QB63.S64   1986   523   85-29621
ISBN 0-943396-10-7

Printed in the United States of America
10 9 8 7 6 5 4 3 2 1

This edition first published by John W. Parker, London (1844)

FOREWORD

Admiral William Henry Smyth's *Cycle of Celestial Objects* has long been regarded as the patriarch of celestial observing guides, particularly the second volume, here offered, which was named *The Bedford Catalogue* after the site of Smyth's private observatory. What makes it so special is that it is the first true celestial Baedeker and not just another "cold" catalogue of mere numbers and data. Like the original Baedeker travel guidebooks of the last century, this work is full of colorful commentary on the highlights of the heavenly scene and heavily influenced several subsequent works of its type, even to the present day.

Smyth had an interesting career. He lived from 1788 to 1865 and, although born in England, had a father who was a colonial resident of New Jersey and emigrated to England as a Loyalist after the Revolutionary War; he was also a descendant of Captain John Smith, the principal founder of the first permanent English colony in North America at Jamestown, Virginia.

The early part of Smyth's career was with the Royal Navy, with which he served in the Mediterranean during the Napoleonic conflicts. During this period he visited the noted Sicilian astronomer Giuseppe Piazzi, famous for his discovery in 1801 of the first asteroid ever found—Ceres. Smyth's visit to Piazzi's Palermo observatory was decisive in turning his interests to astronomy, and after retiring from the Royal Navy in 1825, he settled in Bedford, England, and started building what was then the finest private observatory in that country. It was equipped with a 6-inch refractor, a substantial instrument back then for an astronomical hobbyist.

During the 1830's Smyth put this instrument to good use, observing a huge variety of deep-sky objects, including clusters, nebulae, double stars, and much else. He kept detailed systematic notes of these observations, and was particularly interested in double stars.

These notes were the basis of his *Cycle of Celestial Objects* which appeared in 1844 and was so well received that the Royal Astronomical Society awarded Smyth its gold medal and made him its president for one term.

The second and most famous volume of his "Cycle," *The Bedford Catalogue*, is one of those true astronomical classics whose value has not disappeared with time. Granted, much of its data must be approached with caution not only because of its age, but also because some of it—particularly double-star position values—were the result of relatively inaccurate measurements by Smyth. But such information can today be readily gleaned from any of several modern listings and catalogues.

It is in the descriptive material that Smyth is a delight. He not only describes what the user of a small telescope will see, but also includes much fascinating astronomical, mythological, and historical lore. Many of these descriptions are especially valuable for the novice and user of small telescopes of a size similar to Smyth's. All too often such people see lavish photographs of celestial objects in astronomy books which were taken with huge observatory instruments and time exposures of several hours, and are then disappointed that their modest telescope won't offer them similar views. But back in Smyth's day such lavish photographic observations were a number of decades in the future and he described what the eye sees through a telescope. Incidentally, the human eye has here some real advantages over a photograph, particularly in its ability to discern fine delicate details and contrasts over a great range of brightness, as can be found in objects such as the Orion Nebula (M42).

Note particularly his lengthy discussions of certain objects, such as the Andromeda "nebula" on pages 14-17 (see back of this page). No one knew its true nature as an enormous aggregation of stars at the time and no one yet observed or photographed it in a manner that revealed its true form, spiral arms and all. It was merely one of the most prominent "nebulae" —and that was the term then applied to virtually all non-stellar celestial smudges.

Also, look at Smyth's discussion of the famous double star Gamma Virginis on pages 275-281. Here is the first such star to have its orbit computed, done at the time he lived, and we feel his excitement in describing this major astronomical advance which represented the first application of the law of gravitation outside the solar system. Such material retains permanent historical value.

A few other cautions: Smyth lists some now-obsolete constellations such as Anse, Antinous, Argo Navis, and Taurus Poniatowskii, which are, respectively, the Goose, Young Boy Antinous, Ship, and Bull of Poniatowskii. Other groups are listed differently than they would be today, such as Clypeus Sobieskii, Piscis Australis, Pyxis Nautica, and Scorpio; today they are Scutum,

Piscis Austrinus, Pyxis, and Scorpius. The constellation Argo Navis is now broken up into Puppis, the Stern; Vela, the Sails; Carina, the Keel; and Pyxis, the Mariner's Compass. Smyth also lists the Pleiades star cluster as if it were a separate constellation, whereas it is part of Taurus, the Bull. Also, certain stellar and deep-sky object names and designations are given differently from how we would encounter them today. Finally, he only includes those constellations he could see from England, omitting a large sky area in the southern hemisphere around its celestial pole.

Part of this work's charm is its classic 19th-century style, which also found expression in some of those "talky" novels of the period. But we must watch out for a few obsolete terms here and there, such as the word "comes" for the fainter component of a double star. But such instances occur relatively infrequently.

But, most of all, we can enjoy Smyth as not only a true astronomical classic which more than a few observers over the years and decades have yearned to get their hands on, but also as a warm, friendly guide to the splendors of the sky which can be appreciated equally well in the armchair as well as at the eyepiece.

George Lovi
January 1986

The following four pages from Volume 1--*Prolegomena*--are repeated here so that the reader might know the particulars of the classification and form of registry observed in *The Bedford Catalogue.*

NOTANDA ON THE BEDFORD CATALOGUE.

LES observations des étoiles doubles sont très-délicates, et c'est un vaste et un très-fertile champ à défricher, que nous recommandons aux soins des amateurs qui voudront se rendre utiles, et faire encore autres choses que des observations banales qu'on répète par-tout.—BARON DE ZACH.

CERTAIN official duties with which I was charged, and other considerations, conspired to make me desirous of examining the heavens with better means, and greater attention, than the nature of the service I had been upon, permitted: and this inclination received its sharpest spur at the close of 1813, when I accidentally assisted Piazzi in reading some of the proof sheets of the Palermo Catalogue. No sooner, therefore, was I released from the survey of the Mediterranean Sea, which had then been under my direction, than a new task opened to my view. My first intention was, to re-examine the mean places of all those stars visible in this hemisphere, from the first to the fourth magnitudes inclusive, by a scheme of comparative operations upon the standard Greenwich points; and my working lists were made accordingly. But the high pressure just then applied to our public observatories diminished the necessity of such a task, and the acquirement of a powerful telescope in 1830, induced me to modify my plan; by observing fewer meridian objects, and entering upon a wider scrutiny of the general sidereal phenomena. The CYCLE, therefore, contains the most interesting double and multiple stars, of which the primaries are in Piazzi's Catalogue—a selection of clusters and nebulæ from the works of the two Herschels—and the *élite* of Messier's list of objects, inserted in the *Connaissance des Temps* for 1784. The contents are:

Nebulæ	-	-	- 98	Binary stars - -	20
Clusters	-	-	72	Triple stars - -	46
Stars and *comites*		-	161	Quadruple stars -	13
Double stars		-	419	Multiple stars -	21

In the following Catalogue, these objects, amounting to 850, are placed in the order of right ascension, and are thus registered:

I. The designation and synonyme; with the mean apparent place, reduced from the time of observation to the common epoch of 1840; with the respective annual precession, worked by Dr. Pearson's Tables.

II. The position and distance of the double or multiple star, and the epoch of the measurements; to these are subjoined, in parentheses, weights of the comparative value of the observations, in numbers from one to ten, the first representing nearly worthlessness, and the latter perfection. These weights are derived from a mean of those which were assigned to the estimated value of the measures before the angles and micrometer heads were read off, and when they ran pretty equal, a mere arithmetical mean was taken; but where they differed greatly, the reduction was made on an algebraic form, similar to the rule for finding the centre of gravity of a number of weights. I am indebted to Sir J. Herschel for this application of the method of least squares, and by it, the mean reduction of several nights was brought to one epoch and value.

III. This epoch is given in the year and its hundredth parts; and where it is applied to the nebulæ, it signifies the mean date when the observation was made for differentiating their mean places, for many of the principal ones were under constant examination.

IV. This preceding matter, which constitutes the substantial result of my observations, is followed by a general description of the object, with the magnitudes and colours of its components; its place in the asterism to which it belongs, and the most authentic details of its history. In this portion the following siglæ occur:

A. Argelander.	H. and S. Herschel and South.
B. Baily.	M. Messier.
Br. Brioschi.	*P.* Piazzi.
D. Dawes.	S. South.
H. Herschel I.	Σ. Struve.
H. Herschel II.	*T.* Taylor.

These letters are not to be confounded with those which particularize the component stars. Some astronomers put each group in the order of A, B, C, &c., as they enter the field of view; but I have preferred considering each as a separate and distinct system, calling the brightest A, the next B, and so on. My star A is invariably registered under Piazzi's magnitude, as I know that he took much pains on the subject; and the instrument, climate at Palermo, and skill, were equal to the task. The degree of brightness is however, as already stated, among the desiderata of practical astronomy, for the best eye is but a bad photometer. Assuming, therefore, Piazzi's magnitude for the primary of each object, I have classed the secondaries by a comparison of their light, according to a rule-of-thumb estimation of the refractive and dispersive properties of the existing state of the atmosphere. This, of course, is arbitrary; but in the absence of a more certain and conventional rule, it must suffice. What, from this practice, I term the 16th magnitude, is the *minimum visibile* of my powers both in eye and instrument, and merely indicates a minute point that can be caught by glimpses, when watching under favourable circumstances of atmosphere and telescope.

The observatory form of registry used for the double stars, is precisely that given by Sir John Herschel in the fifth volume of the Astronomical Society's *Memoirs*, p. 92; and I had the advantage of having many hundreds of copies pulled for me by Mr. Moyes, at the same time with Sir John's. In speaking of *above* and *below*, *preceding* or *following*, it is always as seen in an inverting telescope. My double-wire micrometer is one of the very best which the skill and practice of Mr. Dollond was able to produce; and it is really a charming instrument to use. Sir John recommended my trying a full red screen for the lamp, and after a fair trial of this I could never tolerate green again, which so long had seemed the best softener of the illumination. In measuring distances, the alternate + and − reading of the micrometer was practised, to get rid of the zero correction. The position, or angle, made by the line joining the two stars with the direction of diurnal rotation at the meridian, was mostly taken by placing the stars between two thick parallel lines; and was always altered many degrees backwards or forwards to prevent the eye's being biassed. The readings are reckoned from the north, in the direction *n f s p n*, (*north, following, south, preceding, north*,) a method which Sir J. Herschel proposed for adoption in 1830, as well for its convenience, as "its avoidance of the continual and most annoying mistakes" of the method introduced by his father, of reading by quadrants. The following diagram shows both forms, as used in the reversed field of a telescope:

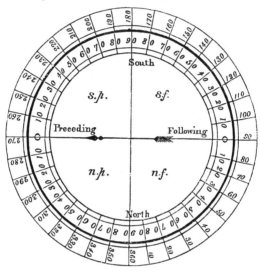

The extremely faint objects were estimated either by an annular micrometer, or a vertical bar which I had fitted to an excellent eye-piece, with a small hole in its centre. I also found Mr. Dollond's spherical crystal double-image micrometer excellent for ascertaining the position of objects too faint for illumination. The axis of motion in this sphere lies at right angles with its natural axis of formation, and when the latter is exactly parallel with the direction of the telescope, it exhibits but one image. In proportion, however, as the sphere is turned, by the milled head of the index at the side, the double image separates; and as soon as the *four* images which will be made by a double star are brought to appear as three in a line, the divided arc shows the angular position.

SYNOPTICAL TABLE OF CONTENTS.

Synonyme.	Order.	Rt. Ascension. Jan. 1, 1840.	Declination. Jan. 1, 1840.	Page

A Q U A R I U S (continued).

Synonyme.	Order.	R.A.	Dec.	Page
41 Aquarii	Double star	22ʰ 05ᵐ	S 21° 52′	513
γ Aquarii	Double star	22 13	2 11	514
53 Aquarii	Double star	22 18	17 33	517
ζ Aquarii	Binary star	22 20	0 50	517
200 P. xxii. Aquarii	Double star	22 34	9 09	521
209 P. xxii. Aquarii	Double star	22 37	10 29	522
τ¹ Aquarii	Double star	22 39	14 54	523
219 P. xxii. Aquarii	Triple star	22 39	5 03	523
ψ¹ Aquarii	Double star	23 07	9 57	531
94 Aquarii	Double star	23 10	14 20	532
69 P. xxiii. Aquarii	Double star	23 15	9 20	533
107 Aquarii	Double star	23 37	19 34	538

A Q U I L A.

Synonyme.	Order.	R.A.	Dec.	Page
2 Aquilæ	Star and comes	18 33	9 12	426
5 Aquilæ	Triple star	18 38	1 07	427
2024 H. Aquilæ	Double star	18 48	N 10 10	437
11 Aquilæ	Double star	18 51	13 25	438
263 P. xviii. Aquilæ	Double star	18 52	14 42	438
15 Aquilæ	Double star	18 56	S 4 16	440
302 P. xviii. Aquilæ	Double star	18 58	N 6 19	441
ζ Aquilæ	Gr. star and comes	18 58	13 38	441
2035 H. Aquilæ	Cluster	19 08	S 1 12	443
23 Aquilæ	Double star	19 10	N 0 48	444
28 Aquilæ	Double star	19 12	12 05	445
δ Aquilæ	Gr. star and comes	19 17	2 48	446
144 P. xix. Aquilæ	Double star	19 22	2 34	449
38 H. vi. Aquilæ	Nebula	19 24	8 54	449
241 P. xix. Aquilæ	Nebula	19 35	8 01	453
250 P. xix. Aquilæ	Double star	19 36	12 09	454
257 P. xix. Aquilæ	Double star	19 37	10 24	454
γ Aquilæ	Gr. star and comes	19 38	10 14	455
π Aquilæ	Double star	19 41	11 25	458
a Aquilæ	Star and comes	19 43	8 27	460
307 P. xix. Aquilæ	Gr. star, double	19 44	9 57	462
56 Aquilæ	Double star	19 45	S 8 59	462
57 Aquilæ	Double star	19 45	8 38	463
β Aquilæ	Gr. star and comes	19 47	N 6 00	464
2 P. xx. Aquilæ	Double star	20 02	16 27	468
43 P. xx. Aquilæ	Double star	20 06	6 06	470

A R G O   N A V I S.

Synonyme.	Order.	R.A.	Dec.	Page
149 P. vii Argo Navis	Double star	7 28	S 23 08	181
38 H. viii. Argo Navis	Double star	7 29	14 08	181
46 H. viii. Argo Navis	Cluster	7 32	16 00	184
175 P. vii. Argo Navis	Double star	7 32	26 26	185
46 M. Argo Navis	Double star	7 35	14 27	185
64 H. iv. Argo Navis	Nebula	7 35	17 50	186
93 M. Argo Navis	Cluster	7 38	23 29	188
2 Argo Navis	Double star	7 38	14 18	189
5 Argo Navis	Double star	7 40	11 48	189
37 H. vi. Argo Navis	Cluster	7 52	10 20	190
ι Argo Navis	Gr. star and comes	8 01	23 51	192
11 H. vii. Argo Navis	Cluster	8 03	12 23	194
72 P. viii. Argo Navis	Double star	8 18	23 32	197

	OBJECTS.	RT. ASCENSION.	DECLINATION.	
Synonyme.	Order.	Jan. 1, 1840.	Jan. 1, 1840.	Page

C Y G N U S (continued).

61^1 Cygni	.... Gr. binary star	20^h 59^m	.... N 37° 58′	 494
1 P. xxi. Cygni	.... Double star	21 02	.... 29 34	 497
ζ Cygni	.... Gr. star and *comes*	21 06	.... 29 34	 497
39 M. Cygni	.... Cluster	21 26	.... 47 44	 504
μ Cygni	.... Double star	21 37	.... 28 01	 508

DELPHINUS.

16 ♓. iv. Delphini	.... Nebula	20 15	.... 19 36	 477
178 P. xx. Delphini	.... Quadruple	20 23	.... 10 44	 480
103 ♓. i. Delphini	.... Cluster	20 26	.... 6 53	 481
β Delphini	.... Triple star	20 30	.... 14 02	 482
a Delphini	.... Star and *comes*	20 32	.... 15 21	 483
γ Delphini	.... Double star	20 39	.... 15 33	 487
χ Delphini	.... Double star	20 48	.... 11 57	 490

DRACO.

λ Draconis	.... Star and *comes*	11 22	.... 70 13	 253
a Draconis	.... Star and *comes*	14 00	.... 65 08	 313
219 ♓. i. Draconis	.... Nebula	15 02	.... 56 23	 335
22 M. (?) Draconis	.... Nebula	15 05	.... 57 36	 335
ι Draconis	.... Star and *comes*	15 21	.... 59 32	 344
764 ♓. ii. Draconis	.... Nebula	15 36	.... 59 52	 347
η Draconis	.... Gr. star and *comes*	16 22	.... 61 52	 365
17 Draconis	.... Triple star	16 32	.... 53 15	 367
20 Draconis	.... Double star	16 55	.... 65 17	 378
μ Draconis	.... Binary star	17 02	.... 54 41	 380
147 P. xvii. Draconis	.... Double star	17 25	.... 51 00	 391
β Draconis	.... Gr. star and *comes*	17 26	.... 52 25	 391
ν^1 Draconis	.... Double star	17 29	.... 55 18	 394
ψ^1 Draconis	.... Double star	17 44	.... 72 14	 398
γ Draconis	.... Gr. star and *comes*	17 52	.... 51 31	 400
37 ♓. iv. Draconis	.... Nebula	17 58	.... 66 38	 410
40 Draconis	.... Double star	18 12	.... 79 58	 418
39 Draconis	.... Triple star	18 21	.... 58 42	 420
χ Draconis	.... Star and *comes*	18 24	.... 72 40	 421
226 P. xviii. Draconis	.... Double star	18 44	.... 59 09	 434
o Draconis	.... Double star	18 49	.... 59 12	 437
287 P. xviii. Draconis	.... Double star	18 55	.... 58 00	 440
δ Draconis	.... Star and *comes*	19 12	.... 67 23	 445
108 P. xix. Draconis	.... Double star	19 15	.... 62 55	 446
ε Draconis	.... Double star	19 48	.... 69 52	 465
30 P. xx. Draconis	.... Double star	20 03	.... 63 14	 470

EQUULEUS.

355 P. xx. Equulei	.... Double star	20 45	.... 6 44	 489
376 P. xx. Equulei	.... Double star	20 48	.... 3 55	 490
ε Equulei (1 Fl.)	.... Triple star	20 51	.... 3 41	 490
λ Equulei (2 Fl.)	.... Double star	20 54	.... 6 33	 491
δ Equulei	.... Double star	21 06	.... 9 22	 497
β Equulei	.... Star and *comites*	21 15	.... 6 08	 501

ERIDANUS.

η Eridani	.... Star and *comes*	2 49	.... S 9 32	 73
τ^4 Eridani	.... Star and *comes*	3 12	.... 22 21	 79
98 P. iii. Eridani	.... Double star	3 29	.... N 0 04	 82
107 ♓. i. Eridani	.... Nebula	3 33	.... S 19 05	 83

Objects.		Rt. Ascension.	Declination.	
Synonyme.	Order.	Jan. 1, 1840.	Jan. 1, 1840.	Page

LYRA.

α Lyræ	.... Gr. star and *comes*	18ʰ 31ᵐ	.... N 38° 38′	 423
151 P. xviii. Lyræ	.... Double star	18 33	.... 35 55	 426
ε Lyræ	.... Multiple star	18 39	.... 39 30	 428
ζ Lyræ	.... Double star	18 39	.... 37 26	 429
ν¹ Lyræ	.... Quadruple star	18 44	.... 32 38	 432
β Lyræ	.... Gr. star and *comes*	18 44	.... 33 11	 433
57 M. Lyræ	.... Nebula	18 47	.... 32 50	 435
γ Lyræ	.... Star and *comes*	18 53	.... 32 28	 439
299 P. xviii. Lyræ	.... Double star	18 57	.... 46 43	 440
17 Lyræ	.... Double star	19 01	.... 35 15	 442
8 P. xix. Lyræ	.... Double star	19 02	.... 38 41	 442
13 P. xix. Lyræ	.... Double double st.	19 03	.... 37 39	 442
η Lyræ	.... Double star	19 08	.... 38 52	 443
56 M. Lyræ	.... Cluster	19 10	.... 29 54	 444

MONOCEROS.

58 P. vi. Monocerotis	Double star	6 11	.... 12 21	 147
8 Monocerotis	Double star	6 15	.... 4 40	 149
104 P. vi. Monocerotis	Triple star	6 19	.... 0 32	 151
10 Monocerotis	Double star	6 20	.... S 4 40	 151
11 Monocerotis	Triple star	6 21	.... 6 56	 152
2 ⬡. vii. Monocerotis	Cluster	6 23	.... N 5 03	 152
14 Monocerotis	Double star	6 26	.... 7 41	 153
15 Monocerotis	Triple star	6 32	.... 10 02	 156
31 ⬡. viii. Monocerotis	Cluster	6 40	.... S 3 00	 163
27 ⬡ vi. Monocerotis	Cluster	6 44	.... N 0 38	 165
50 M. Monocerotis	Double star	6 55	.... S 8 07	 169
33 ⬡. viii. Monocerotis	Dble. star in clus.	7 01	.... 10 22	 170
34 ⬡. viii. Monocerotis	Double star	7 07	.... 10 01	 171
116 P. vii. Monocerotis	Double star	7 20	.... 11 14	 176
52 ⬡. viii. Monocerotis	Cluster	7 26	.... 12 42	 180
29 Monocerotis	Triple star	8 01	.... 2 31	 191
22 ⬡. vi. Monocerotis	Double star	8 06	.... 5 19	 195
81 P. viii. Monocerotis....	Double star	8 20	.... 1 59	 197

OPHIUCHUS.

δ Ophiuchi	.... Gr. star and *comes*	16 06	.... 3 17	 356
ρ Ophiuchi	.... Double star	16 16	.... 23 04	 361
88 P. xvi. Ophiuchi	.... Double star	16 20	.... 7 46	 364
λ Ophiuchi	.... Binary star	16 23	.... N 2 20	 365
40 ⬡. vi. Ophiuchi	.... Cluster	16 23	.... S 12 41	 366
12 M. Ophiuchi	.... Cluster	16 39	.... 1 40	 374
19 Ophiuchi	.... Double star	16 39	.... N 2 21	 374
10 M. Ophiuchi	.... Cluster	16 49	.... S 3 52	 376
19 M. Ophiuchi	.... Cluster	16 53	.... 26 02	 377
270 P. xvi. Ophiuchi	.... Double star	16 54	.... N 8 41	 378
η Ophiuchi	.... Star and *comes*	17 01	.... S 15 31	 379
36 Ophiuchi	.... Multiple star	17 05	.... 26 21	 381
39 Ophiuchi	.... Double star	17 08	.... 24 06	 386
9 M. Ophiuchi	.... Cluster	17 10	.... 18 21	 388
94 P. xvii. Ophiuchi	.... Double star	17 17	.... N 15 45	 389
54 Ophiuchi	.... Double star	17 27	.... 13 16	 392
53 Ophiuchi	.... Double star	17 27	.... 9 42	 392
α Ophiuchi	.... Gr. star and *comes*	17 27	.... 12 41	 393
14 M. Ophiuchi	.... Cluster	17 29	.... S 3 09	 395

Objects.		Rt. Ascension.	Declination.	
Synonyme.	Order.	Jan. 1, 1840.	Jan. 1, 1840.	Page

PERSEUS.

76 M. Persei	Nebula	1ʰ 32ᵐ	N 50° 46′	40
χ Persei	Triple star	2 07	56 46	57
33 Ḥ. vi. Persei	Double star	2 08	56 24	58
227 H. Persei	Cluster	2 22	56 49	63
156 Ḥ. i. Persei	Nebula	2 30	38 21	65
34 M. Persei	Cluster	2 32	42 02	66
12 Persei	Star and pair	2 32	39 31	66
θ Persei	Triple star	2 33	48 33	67
η Persei	Double star	2 39	55 13	70
20 Persei	Double star	2 43	37 41	72
220 P. ii. Persei	Double star	2 50	51 42	74
γ Persei	Double star	2 53	52 52	75
β Persei	Star and comes	2 58	40 20	77
25 Ḥ. vi. Persei	Cluster	3 04	46 38	78
a Persei	Gr. star and comes	3 13	49 17	80
δ Persei	Star and comes	3 32	47 16	82
40 Persei	Double star	3 32	33 27	83
80 Ḥ. viii. Persei	Double star	3 37	52 10	85
ζ Persei	Quadruple star	3 44	31 24	90
43 Persei	Double star	3 45	50 13	90
ε Persei	Double star	3 47	39 32	91
60 Ḥ. vii. Persei	Group of stars	3 58	49 04	93
μ Persei	Star and comes	4 03	47 59	94
57 Persei	Double star	4 22	42 43	101
58 Persei	Star and pair	4 26	40 56	101

PISCES.

34 Piscium	Double star	0 02	10 14	4
35 Piscium	Double star	0 07	7 56	5
38 Piscium	Double star	0 09	7 59	6
42 Piscium	Double star	0 14	12 35	7
49 Piscium	Double star	0 22	15 09	8
51 Piscium	Double star	0 24	6 04	9
52 Piscium	Double star	0 24	19 25	10
55 Piscium	Double star	0 31	20 33	13
65 Piscium	Double star	0 41	26 50	20
251 P. O. Piscium	Double star	0 51	S 0 05	23
ψ¹ Piscium	Double star	0 57	N 20 37	24
σ² Piscium	Triple star	0 57	31 19	24
77 Piscium	Double star	0 58	4 03	26
4 P. i. Piscium	Double star	1 02	8 42	31
φ Piscium	Double star	1 05	23 44	31
ζ Piscium	Double star	1 05	6 44	32
85 P. i. Piscium	Star and comes	1 20	7 08	36
100 Piscium	Double star	1 26	11 44	38
123 P. i. Piscium	Double star	1 28	6 49	38
145 P. i. Piscium	Double star	1 32	24 56	39
107 Piscium	Double star	1 34	19 29	41
209 P. i. Piscium	Double star	1 47	1 03	46
a Piscium	Double star	1 54	1 59	48
430 Ḥ. ii. Piscium	Nebula	23 06	3 39	531
ι Piscium	Gr. star and comes	23 32	4 45	536
179 P. xxiii. Piscium	Double star	23 38	S 0 37	538

PISCIS AUSTRALIS.

a Piscis Australis	Gr. star and comes	22 49	30 28	524

ERRATUM.

The reader is requested to alter the following inadvertence, with his pen :
p. 371, l. 2, for 15ʰ in ℞, read 16ʰ.

THE BEDFORD CATALOGUE.

I. α ANDROMEDÆ.

Æ 0ʰ 0ᵐ 08ˢ   Prec. + 3ˢ·07

Dec. N 28° 12′·5   —— N 20″·06

Position 267°·1 (to 6) Distance 65″·9 (to 3) Epoch 1834·64

—— 266°·9 (to 8)   —— 64″·8 (to 6)   —— 1837·74

A STANDARD Greenwich star, with a minute companion. A 1, bright white; B 11, purplish. According to my adopted plan, A is assumed as a first magnitude star from Piazzi; though otherwise it would have been rated only as a second, which is the size assigned to it by Ptolemy, who probably copied it from the Catasterisms of Hipparchus. B did not escape the vigilance of Sir William Herschel, who classed the object 32 v. with these results:

Pos. 259° 23′ Dist. 55″·53 Ep. 1781·56

The next micrometrical observations, are those of the Rev. Mr. Dawes, who managed to obtain them with a five-foot achromatic, of 3¾ inches aperture, and kindly forwarded them to me.

Pos. 264° 10′ Dist. 66″·57 Ep. 1830·68

On a comparison of all the measures previous to my observations, the increase of angle and distance in this object may be charged to the *sf* movements of the large individual in Æ and Dec., with an allowance for errors of operation in so difficult a star. The amount of the proper motions has been thus valued:

P.... Æ + 0″·14   Dec. − 0″·21
B....   + 0″·19   − 0″·13
A....   + 0″·16   − 0″·15

The extensive northern constellation of which this star is now the lucida, was one of the old 48 asterisms, and its components, as optical means advanced, have been thus registered:

Ptolemy . . . 23 stars   Hevelius . . . . 47 stars
Bulialdus . . . 26   Flamsteed . . . 66
Bayer . . . 27   Bode 226

Andromeda is conspicuously figured near her father, mother, and lover, in the bonds which Aratus says she carried to Heaven; and has been also designated *Virgo devota, Mulier catenata*, and *Persea;* while Schickard, on the part of the Mosaicists, claimed her as *Abigail*. The Arabians, whose tenets prohibited their drawing the human figure, represented her as a sea-calf: but the principal star was called *Sirrah*, and *Alpherat*, from *Sirrat-al faras*, the horse's navel, it having formerly been quartered on Pegasus, whence it was taken to decorate the tresses of the lady. Ulugh Beigh calls it *Rás-al-marat al muselselah*, the head of the woman in bonds. Warm imaginations perceive a resemblance to chains, by drawing the eye from 51 and 54 of Flamsteed, on the lady's left foot, over χ between the feet, to τ on the right calf; and from Alamak on the right foot, through χ and ξ to ϕ on the left knee. Owing to the derangements which the inadvertence or ignorance of the celestial map-makers have occasioned, there is no little confusion in this particular, for Flamsteed's Nos. 51 and 54 Andromedæ, are ϕ and υ Persei, though placed exactly where Ptolemy wished them to be —on the lady's foot: so also a in this asterism has been lettered δ Pegasi by Bayer, and β has been the lucida of the Northern Fish.

Several members of this configuration have been placed among the *stellæ versatiles*, whose brightness varies; first by Mr. Pigott, in the Philosophical Transactions for 1786, and next by Sir William Herschel, in the same work for 1797. They were also reviewed carefully by Harding and Westphal.

Sirrah is useful in alignment, or the mode of finding from a few stars with which a spectator is familiar, others which are unknown to him. Thus, an imaginary line drawn from the belt of Orion, which all the world knows or ought to know, through Aries, will lead to the head of Andromeda. Certain brackish rhymes then state:

And on, from where the pinion'd maid,	From Alpherat down to Markab's beams,
Her cruel fate attends,	Let a cross line be sent,
Wide o'er the heavens his fabled form	Then will four stars, upon the horse,
Wing'd Pegasus extends.	A spacious square present.

Of this notable square, *Alpherat* and *Scheat* form the northern side, while *Markab* and *Algenib* mark the southern; and these are useful in extending the alignment to other sought objects.

II. β CASSIOPEÆ.

$\mathbb{R}$ 0ʰ 0ᵐ 42ˢ   Prec. + 3ˢ·07

Dec. N 58° 16′·3   —— N 20″·05

Position 339°·6 (w 1)   Distance 201″·0 (w 1)   Epoch 1838·65

A bright star, whose acolyte is so small that it is here rather estimated than measured. A $2\frac{1}{2}$, whitish; B $11\frac{1}{2}$, dusky. This object is called *Caph*, from *Kaff-al-Khadib*, the stained hand, a name from which a scientific friend supposes, that although now only the *lucida cathedræ*— or bright star on the couch-frame—one of the hands may have reached it in the earlier designs. But the Arabians applied the term *Kaff*, a flat hand, to the whole asterism, whose five brightest stars represented the thumb and fingers, coloured as if stained with henna, after the Oriental custom. This general name came to be fixed upon β. Mohammed al Tízíní records it as *Sanám al-nákah*, the camel's hump—for ardent fancies figured a kneeling camel of the principal stars. Ptolemy describes it as being on the female's back. Caph is considered to be variable, from the second to the fourth magnitudes; but to me it has generally appeared of the brightness above recorded. Mayer assigns to it a proper motion in $\mathit{R} = + 0''\cdot773$; and other astronomers give as the amount—

$$
\begin{array}{lll}
P.....\mathit{R} & + 0''\cdot82 & \text{Dec.} \; - 0''\cdot25 \\
B.... & + 1''\cdot01 & \quad\quad - 0''\cdot17 \\
A.... & + 0''\cdot97 & \quad\quad - 0''\cdot19 \\
\end{array}
$$

A glance from the Pole-star to Alpherat, passes through Caph, nearly in mid-distance: or a line from between γ and δ, the following stars in the wain of the Great Bear, carried over the pole, strikes upon it, at a similar distance beyond Polaris:

In yonder stars, which form a Cross,   Io, Caph precedes the whole,
A Cross more glorious than that   which decks the austral pole.

III. 147 ♄. III. ANDROMEDÆ.

R 0ʰ 1ᵐ 43ˢ   Prec.   +   3ˢ·07
Dec. N 25°   1′·2   ——   N 20″·05

Position 120°·0 (w 1)   Distance 28″·0 (w 1)   Epoch 1836·81

A double star in a coarse cluster, occupying the spot where I looked for Sir William Herschel's faint nebula. A 10, and B 11, both pale blue. It lies on the crown of Andromeda's head, and about 3° south of the star a. I saw none of the nebulosity alluded to by Sir William in his registry of 1784,—but a perceptible glow in a tolerably rich and darkened field was indicative of nebulous matter. This part was followed by three principal stars nearly in a line at almost equal distances, and each with a companion np. The third being the smallest and closest, is here estimated. Another double star follows in the upper part of the field, at about three minutes Δ R.

If this glow really be a mass of that self-shining element, or modification of matter, which Sir William Herschel has taught us to recognise as being a thin filmy substance diffused in the vast regions of space, it must be at an astounding distance from our system. Sir William only considered that he could here perceive nebulous fluid,—but neither his instrument, nor any other yet constructed, has sufficed to give it a distinct form, let alone a resolvable aspect. From every indication, and inductive analogy, the numerous small stars perceivable in the field are posited between us and the nebulosity: it must therefore be at a degree of remoteness, which cannot be expressed in language. There are some of the star-gazing class, who would consider such a shadowy pellicle as barely worth enrolment; but to the superior order of observers it is a most important object, both for scrutiny and study, in our present ignorance of the nature of this self-luminous substance. To the astronomer of a future day, it may offer a more condensed aspect.

IV. 34 PISCIUM.

$\mathbb{R}$ 0ʰ 1ᵐ 53ˢ   Prec. +   3ˢ·07

Dec. N 10° 14′·6   ——   N 20″·04

Position 165°·0 (w 2)   Distance 7″·0 (w 1)   Epoch 1838·77

A neat double star. A 6, silvery white; B 13½ pale blue; and they point to some small stars in the *sf* quadrant. This fine object, though numbered to Pisces, is astern of the leading Fish's tail, and near the wing of Pegasus; and 4° *s*—a little preceding—the bright star γ Pegasi. From the delicacy of the *comes* it is so excessively difficult to measure, that I only mark a mean of careful estimations. It was discovered by M. Struve, and is No. 5 of his grand Catalogue. The first Dorpat measures were :

Pos. 160° 08′   Dist. 8″·37   Ep. 1828·73

A slight proper motion in space is attributed to A, which, however small, has been thus valued:

P....$\mathbb{R}$ − 0″·06   Dec. − 0″·01
Br.... − 0″·03   − 0″·02
B.... + 0″·06   − 0″·05

V. 22 ANDROMEDÆ.

$\mathbb{R}$ 0ʰ 2ᵐ 2ˢ   Prec. +   3ˢ·08

Dec. N 45° 10′·9   ——   N 20″·04

Position B C 85°·1 (w 5)   Distance 4″·6 (w 3)   Epoch 1830·84

————— 84°·0 (w 9)   ——— 4″·9 (w 5)   —— 1838·92

A star leading to a distant pair. A 5, white; B 8, pale yellow, and C 9, bluish. A is in the Galaxy, between the left hand of Andromeda, and the head of Cassiopea; and it may be fished up by a line through γ and α of the latter, at three times the interval between them in distance. It is here introduced as a pointer to the charming double star B C in the *np* quadrant, on a line 351°·5, and Δ $\mathbb{R}$ = 18ˢ. It is in a fine field with several stars between the individuals. B and C form 83 ♄. ii., and were thus measured:

Pos. 84° 12′   Dist. 3″·50   Ep. 1783·16

By Sir James South, No. 381, they were:

Pos. 85° 21′   Dist. 5″·01   Ep. 1825·99

which shows the position to be stationary; and as ♄.'s distance was based on an allowance upon the apparent magnitude of the large star, no exact inference, as to change in this element, can be drawn. The orbital measures of 1838·92 are most satisfactory, and taken under very favourable conditions. The proper motions of A have been thus registered:

P....$\mathbb{R}$ − 0″·07   Dec. − 0″·02
B.... + 0″·11   − 0″·00

VI. γ PEGASI.

Æ 0ʰ 5ᵐ 0ˢ   PREC. + 3ˢ·08

DEC. N 14° 17′·7   —— N 20″·04

POSITION 300°·2 (ω 1)   DISTANCE 181″·0 (ω 1)   EPOCH 1835·07

A standard Greenwich star, with a distant companion. A 2½ white; B 13, pale blue, with a small *comes* in the *sp*,—a line from B carried through A, leads to two small stars in the *sf* quadrant. In Tycho Brahé's catalogue this is erroneously placed in the constellation Pisces; but it is on the extreme of the wing of Pegasus, whence it obtained the name of Algenib, from the Arabic *Jenáh-al-faras*, the horse's wing. A comparison of the distance between this star and Regulus, by ancient and modern astronomers, shows Ptolemy to be out − 12′ 18″. γ Pegasi has a slight proper motion through space, the amount of which has been thus variously given:

$$P....Æ - 0''·03 \qquad Dec. - 0''·09$$
$$B.... + 0''·01 \qquad\qquad - 0''·01$$
$$A.... + 0''·04 \qquad\qquad - 0''·01$$

To find this star by alignment, lead a line from the Pleiades through Aries, or look about 14 degrees south of Alpherat, where it will be identified by its lustre. With α Andromedæ it forms the twenty-seventh of the Mansions of the Moon, under the designation of *Alfargu*, from the Arabian *Al farigh al-muäkhker*, the hindmost loiterer.

VII. 35 PISCIUM.

Æ 0ʰ 6ᵐ 44ˢ   PREC. + 3ˢ·07

DEC. N 7° 55′·9   —— N 20″·04

POSITION 150°·1 (ω 8)   DISTANCE 11″·6 (ω 7)   EPOCH 1832·04

—— 149°·5 (ω 7)   —— 11″·9 (ω 7)   —— 1837·89

A neat double star. A 6, pale white; B 8, violet tint. This fine object is 62 ♓ III., who describes it as being in "lino austrino" of the constellation; but by Sir J. Lubbock's map, it is on the south tip of the tail of the preceding Fish. A line from α Andromedæ through γ Pegasi, extended about 6° to the south, strikes upon 35 Piscium. It stood when first classed thus:

Pos. 148° 54′ Dist. 12″·50 Ep. 1782·68

which, compared with the recent measures of Σ. H. S. and myself, embracing an interval of fifty-six years, leaves no doubt as to its fixity. From the weights attached to my MS. observations, I place great dependance on the results here given, at both the above epochs.

VIII.   38 PISCIUM.

$\mathbb{R}$ 0^h 9^m 9^s Prec. +   3^s·08

Dec. N 7° 59′·2   —— N 20″·03

Position 236°·4 (w 6)   Distance 4″·5 (w 4)   Epoch 1830·96

——— 235°·9 (w 8)   ——— 4″·8 (w 4)   —— 1837·89

A very neat double star on the following tip of the preceding Fish's tail, following the former object nearly on the parallel. A 7½, light yellow: B 8, flushed white. This elegant pair is 50 H. ii., and it was thought to be binary by its discoverer, from the following measurements:

Pos. 244° 57′   Dist. 4″·00   Ep. 1782·68
　　 235° 27′   4″·00   1802·67

These observations appeared to give a retrograde motion, which subsequent astronomers have not confirmed. My own observations are entitled to weight from their great coincidence. Those sage astrologers who dubbed Pisces a most malignant sign, ought to have contemplated this beautiful object; had this been done, every notion of stellar unpropitiousness and malevolence must have vanished.

Sig. Carlo Brioschi attributed a larger amount of proper motion to this star, than appears to be borne out by the comparison of standard epochs; the deductions most worthy of attention are:

P....$\mathbb{R}$ − 0″·16   Dec. + 0″·05
B....　 + 0″·08   + 0″·11
T....　 + 0″·30   − 0″·02

IX.   ı CETI.

$\mathbb{R}$ 0^h 11^m 17^s Prec. +   3^s·06

Dec. S 9° 42′·7   —— N 20″·02

Position 12°·0 (w 1)   Distance 45″·0 (w 1)   Epoch 1838·82

A wide double star on the north extreme of the tail; whence it was called *Dheneb Kaïtos shemâli*, the northern branch of the Whale's tail. A 4, bright yellow; B 15, deep blue. This is an excessively difficult object, being only discernible after long attention, and by occasionally averting the eye to another part of the field in view. The position and distance are therefore only the result of cautious estimation. It is No. 1953 of H.'s Fifth Series, where the companion is rated as of the 12th magnitude. There is a small star near the vertical, in the *sp* quadrant. The proper motion of A has been deduced by reference to standard Catalogues, from which it was thus valued:

Fiamsteed . . $\mathbb{R}$ + 0″·161   Dec. + 0″·08
Roemer . . .   + 0″·118   − 0″·05
Bradley . . .   − 0″·114   − 0″·06

and by Mr. Baily's recent deductions, its amount is, in $\mathbb{R}$ + 0″·02, and in Dec. $-0''\cdot05$. The object may be found by a line carried through a Andromedæ and γ Pegasi, and extended to about $24°$ south of the latter, where it will be seen as the north-eastern apex of a nearly equilateral triangle formed by ι, η, and β Ceti.

X. 42 PISCIUM.

$$\text{Æ} \quad 0^h \; 14^m \; 9^s \qquad \text{Prec.} \; + \; 3^s\cdot09$$
$$\text{Dec. N } 12° \; 35'\cdot6 \qquad \text{———} \quad \text{N } 20''\cdot01$$

Position $341°\cdot5$ (w 4) Distance $35''\cdot0$ (w 1) Epoch $1833\cdot95$

A delicate double star following γ Pegasi at about $2\frac{1}{2}°$ in the sf quarter. A 7, topaz yellow; B 13, emerald green. It is in the boundary, but not in the figure of the Fishes; and though not close, has an elegant aspect from the strong contrast of its colours in so barren a field of view. As the small star defies illumination, the position is approached by the rock-crystal micrometer, and the distance is cautiously inferred. This object is Σ.'s No. 27, and was thus first measured:

Pos. $344° \; 54'$ Dist. $31''\cdot63$ Ep. $1828\cdot76$

The star A has an appreciable movement in space; but it is so slight a quantity, that until observations can be rendered perfect, its exact amount will be debateable. The values have been thus given:

$$P....\text{Æ} - 0''\cdot06 \qquad \text{Dec.} + 0''\cdot04$$
$$Br.... \quad - 0''\cdot01 \qquad\qquad + 0''\cdot04$$
$$B.... \quad + 0''\cdot15 \qquad\qquad + 0''\cdot04$$

XI. 22 H. CASSIOPEÆ.

$$\text{Æ} \quad 0^h \; 18^m \; 10^s \qquad \text{Prec.} \; + \; 3^s\cdot36$$
$$\text{Dec. N } 70° \; 30'\cdot3 \qquad \text{———} \quad \text{N } 19''\cdot98$$

Mean Epoch of the Observation .... $1838\cdot90$

A large and straggling group of small stars, between the Lady's footstool and the knee of Cepheus; a line from the γ of one asterism to the γ of the other, and $\frac{1}{3}$ the distance from that of Cepheus, hits 22 H. The place here given is that of a coarse double star, the components of which are of the $8\frac{1}{2}$ and 11 magnitudes, both greyish, in the following portion of the mass; and which is, in a manner, insulated. It was first registered by Sir John Herschel, and described as a very loose but pretty rich cluster. On the whole, this object offers a fair test for trying the light and defining power of a telescope.

XII.   12 CETI.

$\mathcal{R}$ 0^h 21^m 53^s PREC. $+$ 3^s·06

DEC. S 4° 50′·6   —— N 19″·95

POSITION A B 180°·8 (*w* 1)   DISTANCE   6″·5 (*w* 1) ⎫
—— A C 113°·4 (*w* 2)   ————   201″·0 (*w* 1) ⎬ EPOCH 1837·89
                                       ⎭

A triple star, or rather, a double one with a distant companion, above half-way in a line shot from γ Pegasi to β Ceti. A 6, topaz yellow; B 15, bright blue; C 11, dusky,—other telescopic stars in the field. This beautiful, but most difficult test object, is No. 322 of H.'s Second Series. It lies between the Whale's tail and the Southern Fish, nearly mid-distance of two stars to the *np* and *sf*, but trending towards the parallel,—the following individual being of the 11th magnitude, and the largest. B is only discernible by the closest attention under favour-ing circumstances, though when once caught, is tolerably well seen; the detail here given is therefore a mere estimation. Piazzi remarks: " Probably 11^a of Mayer, the sign of the declination being wrong, as in the 9^a, neither of which are found in a northern sky."

The principal star of this object, has had the following very minute amount of proper motions attributed to it:

P....$\mathcal{R}$ $+$ 0″·15   Dec. $-$ 0″·05
B.... $+$ 0″·05 $-$ 0″·01

XIII.   49 PISCIUM.

$\mathcal{R}$ 0^h 22^m 29^s PREC. $+$ 3^s·10

DEC. N 15° 9′·2   —— N 19″·95

POSITION 109°·5 (*w* 2)   DISTANCE 15″·0 (*w* 1)   EPOCH 1835·87

A delicate double star, *nf* γ Pegasi about 4°. A 7, silvery white; and B 13, cerulean blue. Though quartered in Pisces this very deli-cate object is actually between the wing of Pegasus, and the right hand of Andromeda: and this, though a minor one, is among the many errors calling for a reform of the constellations. It is followed nearly on the parallel by a yellow star of the 7·8 magnitude, which must be the one alluded to by Piazzi, in Note O. 92, though the distance is nearer 50 than 30 seconds. The companion to A is so minute as to vanish under the slightest illumination, the details are therefore merely esti-mated,—but with great care. It was discovered by Σ., whose measures with the great Dorpat telescope are:

Pos. 107° 42′ Dist. 13″·26 Ep. 1828·74

XIV. 28 H. CASSIOPEÆ.

Æ 0^h 24^m 5^s Prec. + 3^s·34
Dec. N 62° 23'·9   —— N 19"·93

Position 228°·0 (w 1)   Distance 6"·5 (w 1)   Epoch 1837·97

A neat double star in a cluster. A 10, and B 11, both pale grey. They are near the centre of an elegant and rich, but somewhat straggling field of stars; and being too small to admit of light, their position and distance are only estimated. The vicinity is splendidly strewed with stars from the 10th to the 15th sizes, of which the most clustering part is about 8' or 9' in extent. It is closely *nf* of κ in the throne of Cassiopea, a beautiful individual of a bright yellow colour, and 4th magnitude.

XV. 51 PISCIUM.

Æ 0^h 24^m 9^s Prec. + 3^s·08
Dec. N 6° 04'·3   —— N 19"·93

Position 82°·5 (w 9)   Distance 27"·6 (w 6)   Epoch 1835·91

A fine double star in a line about one-third the distance from γ Pegasi to η Ceti. A 6½, pearl white; B 9, lilac tint. This is in the centre of that part of the *kheït*, or ribbon, of the sign Pisces, which is near the tail of the preceding fish; and my observations for position and distance are highly satisfactory. It is 70 Ħ. iv.; and a comparison with his first measures would indicate a sensible increase of distance between the two stars, as well as a retrocession of the orbital angle: but from the concordance of recent measures, it may still be questioned whether the position has not been stationary. Indeed, when we recollect that Ħ. first classed these interesting objects, and made the instruments wherewith to grapple their details, we can only admire how well they stand such rigorous comparisons. The previous points of departure are these:

	Pos:	Dist.	Ep.
Ħ.	89° 24'	22"·48	1782·68
H. and S.	82° 49'	25"·08	1823·87
Σ.	82° 35'	27"·42	1833·20

The proper motions of 51 Piscium, have been thus valued:

P....Æ − 0"·10   Dec. + 0"·10
B....   + 0"·05   + 0"·07

XVI. 52 PISCIUM.

Æ   0ʰ 24ᵐ 13ˢ   Prec. +   3ˢ·12
Dec. N 19° 24′·8   —— N 19″·93

Position 311°·0 (w 1)   Distance 25″·0 (w 1)   Epoch 1836·92

A double star, between Andromeda's right arm and the back of Pegasus; and nearly midway of a line from γ Pegasi to ζ Andromedæ. A 6, fine yellow; B 14, deep blue. This is a neat and most delicate object, whose position and distance are carefully estimated. It is followed at about 12ˢ, by a pale star of the 12th magnitude. The object was first discovered by H., and is No. 1982 of his Fifth Series. Piazzi has remarked in his note on 50 Piscium, " cujus declinatio 19° 11′ (*Bradley*) omnino non invenitur:" and Mr. Baily has shown, in his edition of the British Catalogue, that No. 50 does not exist, but that this star, No. 52, was the individual observed and registered by Flamsteed in 1692. The amount of spacial proper motions assigned to it is:

P....Æ + 0″·24   Dec. − 0″·06
B....   + 0″·18   − 0″·03

XVII. 113 P. O. CETI.

Æ   0ʰ 26ᵐ 20ˢ   Prec. +   3ˢ·05
Dec. S   5° 25′·8   —— N 19″·91

Position 44°·6 (w 8)   Distance 19″·6 (w 8)   Epoch 1832·87

A neat double star, between the whale's tail and the *chétil* of Pisces, on a line striking from β Ceti through α Andromedæ, and about one-third of the distance. A 7, cream yellow; and B 9, smalt blue. This is a fine object, being nearly mid-way between two stars, one in the *sf* quadrant, and the smallest in the *np*. It is No. 39 of Σ.'s grand Catalogue, where the mean of his observations gives:

Pos. 45° 27′   Dist. 20″·09   Ep. 1830·24

My own measures were taken under favourable circumstances, and may therefore be deemed good. Piazzi tells us that in Flamsteed's asterism, a companion of the 9th magnitude follows this star, by 11ˢ·2 of time, which he could not find. There is, however, at about the same distance, and to the north, a star of this character, although rather smaller. Is the follower, then, variable?

XVIII. π ANDROMEDÆ.

Æ   0ʰ 28ᵐ 21ˢ   Prec. +   3ˢ·17
Dec. N 32° 50'·3   —— N 19"·89

Position 173°·9 (w 7)   Distance 35"·6 (w 7)   Epoch 1832·90

A fine double star, between the shoulders of the chained Lady. A 4½, fine white; B 9, blue,—and they point to two small stars at a distance in the *sf* quadrant. A line carried from a Pegasi to a Andromedæ, and extended 6° beyond, strikes upon π. Ptolemy was *right* in stating it to be upon the shoulder; but as the figures of the constellations are drawn on many modern celestial maps, with the fronts towards the observer, π then appears to be situate on the breast. It is well described by Piazzi: "*duplex*," ait, "*comes* 10ᵃ *magnitudinis sequitur* 0ˢ·6 *temporis circiter* ½ *min. ad austrum.*" This object was first registered as a double star by ⩑., and is No. 17 v.; but though he pronounced his distance of 34·20" to be inaccurate, it must have been pretty near the truth, for subsequent measures indicate about the same. H. and S. obtained results thus:

Pos. 175° 26' Dist. 35"·59 Ep. 1821·88

The spacial movement attributed to π, seems to be diminishing under more rigorous treatment; at present it is thus registered:

P....Æ + 0"·30   Dec. + 0"·10
Br.... + 0"·27   + 0"·06
B.... + 0"·06   + 0"·02

XIX. δ ANDROMEDÆ

Æ   0ʰ 30ᵐ 47ˢ   Prec. +   3ˢ·17
Dec. N 29° 59'·1   —— N 19"·86

Position 208°·3 (w 2)   Distance 122"·0 (w 2)   Epoch 1833·54

A bright star with a telescopic companion to the *sp*. A 3, orange; B 11½, dusky; with three small stars in the southern part of the field. It is on the right shoulder of Andromeda, though the old Catalogues term it "clarior in sinistrâ scapulâ," as an accepted interpretation of the contested ἐν τῷ μεταφρένῳ of Ptolemy; and it is found by a line from γ passed over β, and carried about 7° beyond the latter. Some of the recent investigators of those aberrations from the common laws of precession which have been detected in so many stars, have not included this in their lists; yet its proper motions amount to

P....Æ + 0"·35   Dec. − 0"·09
B.... + 0"·20   − 0"·11

XX. α CASSIOPEÆ

Æ 0^h 31^m 29^s Prec. + 3^s·34

Dec. N 55° 39'·5   —— N 19"·87

Position 278°·4 (w 3)   Distance 96"·9 (w 3)   Epoch 1831·86

A standard Greenwich star, with a companion.   A 3, pale rose-tint; B 10½, smalt blue.   This object is in the right breast of *Dhát-al-Kursa,* the Lady of the Throne; and it also obtained the names of Lucida Cassiopea, and Schedir; the last being probably a corruption of *Al-sadr,* the breast, by the framers of the Alphonsine Tables.   The Arabians having no passion for delineating the human form, made a dog of Cepheus, and its female of Cassiopea, retaining the *Sedes Regia,* or throne of the latter; but the *Canis fœmina* meant no disrespect.   The double star forms 18 Ḥ. v., and was thus registered :

Pos. 275° 26'   Dist. 56"·17   Ep. 1781·97

which, compared with mine, does not show a greater difference of angle than might be expected from the proper motion of an object whose proximity is accidental.   But the difference in distance is so remarkable, that it must be imputed to instrumental error, rather than that the acolyte is describing an ellipse round its primary.   This acolyte must be the star alluded to by Piazzi, in Note O. 139, though he calls it the 9th magnitude, for the Δ Dec. 6"·78 makes " parumper ad boream."   The large star is one of the insulated class, and has a slight proper motion, the amount of which is thus registered:

P....Æ − 0"·05   Dec. − 0"·07
B. ..   + 0"·15   − 0"·03
A....   + 0"·10   − 0"·04

Cassiopea, one of the ancient 48 asterisms, formed by five bright stars disposed something like an M, is a well-known circumpolar constellation, next to her husband Cepheus, and on the opposite side of the pole to the Great Bear.   The earlier Arabians considered the whole as a large hand, of which the bright stars constituted the finger-points, and in which was even included the nebulous group in the left hand of Perseus. Bayer thought the Hebrews called this asterism Ebn Ezra—"*ein seltsamer Missgriff,*" exclaims Ideler: but Bayer had also seen in Juvenal's *cathedra mollis,* an allusion to Cassiopea's chair.   It has been recorded as Al Thuraiyá, the many, a name more exclusively appropriated by the Pleiades: and it was once also called Carion, or the Laconian Key.   The enumeration of its members increases, of course, as optical science advances, and successive ages have yielded these numbers:

Ptolemy   . . . 13 stars   Hevelius . . . . 38 stars
Ulugh Beigh . . 13   Flamsteed . . . 55
Tycho Brahé . . 46   Bode 134

There has been much idle discussion as to the orthography of this lady's name, whether it should be written Cassiopea, after the Latins, or Cassiepea from the Greek Κασσιέπεια; and the result has left either to the writer's choice—*utrum horum,* &c.   In the early illustrations to Hyginus, she is bound to her throne, or rather to a seat with a sort of gibbet-back, very much like the scaffold called *i tre pezzi di legno* by the Italians. Thus secured she cannot fall out in going round head downwards, pursuant to sentence.

Sir William Herschel has tabulated the comparative lustre of the stars in this constellated group, and the statement will be found in the 86th volume of the Philosophical Transactions.

The distance of Shedir from Regulus, by Ptolemy, differs from that of modern astronomers by − 23′ 37″; and the difference of latitude between it and α Trianguli is − 3′ 33″, on similar comparison.

Shedir is actually a variable star, though its usual aspect is that of a sharp, clear, third degree. H̩.'s observations in 1796, make α and γ of the 3rd magnitude, and β 3·2 in lustre of the stars in Cassiopea; and its brightness is marked γ in Ptolemy. Certainly when H. called my attention thereto, it was smaller than β or γ of the same constellation,— "That the fluctuation in splendour of this star," writes H., "should have heretofore escaped notice, is not extraordinary, since the difference between its greatest and least brightness can hardly be estimated at so much as half a magnitude." Its period is stated at about 200 days; but in July, 1839, it was positively brightening and better defined than the other two. To find this star from the northward, project a ray from Alioth, through the Pole-star, and it will pass through the middle of Cassiopea, at nearly an equal distance on the other side of the pole. The circumvolution is well marked; when Ursa is at its lowest position below the pole, Cassiopea is near the zenith, and *vice versa*. If Shedir is required from the southward, resort to the galley rhyme :

> From *alpha* Ceti, to the east   of Al'mak, towering rise,
> You'll mark on Cassiopea's breast,   where Shedir decks the skies.

XXI.   55 PISCIUM.

Æ   0ʰ 31ᵐ 31ˢ   Prec. +   3ˢ·14
Dec. N 20° 33′·6   —— N 19″·85

Position 193°·7 (*w* 6)   Distance 5″·9 (*w* 3)   Epoch 1833·83

A neat double star, ascertained by a line through δ and ε Andromedæ, and 7° beyond the latter. A 6, orange; B 9, deep blue,—the colours in good contrast, and therefore forming a rich specimen of opposed hues. This very beautiful object is between the head and right arm of Andromeda; and was found to be double by the vigilant Dorpat astronomer, being his No. 46, thus measured :

Pos. 192° 73′ Dist. 6″·37 Ep. 1830·22

The interval of time which has elapsed between the two epochs of observation, is too brief for drawing any conclusions upon the differences observable in the results.

XXII. 18 Ḥ. V. ANDROMEDÆ.

ℛ   0ʰ 31ᵐ 42ˢ   PREC. + 3ˢ·23
DEC. N 40° 48'·6   —— N 19°·85

MEAN EPOCH OF THE OBSERVATION . . . . 1836·66

A large faintish nebula of an oval form, with its major axis extending north and south. It is between the left arm and robes of Andromeda, a little to the *np* of 31 Messier; and was discovered by Miss Herschel in 1783, with a Newtonian 2-foot sweeper. It lies between two sets of stars, consisting of four each, and each disposed like the figure 7, the preceding group being the smallest; besides other telescopic stars to the south. This mysterious apparition was registered by Ḥ. as 30' long and 12' broad, but only half that size by his son; and there was a faint suspicion of a nucleus. This doubt must stand over for the present,— for whatever was a matter of uncertainty in the 20-foot reflector, would have no chance of definition in my instrument. It was carefully differentiated with β Andromedæ.

XXIII. 146 P. O. CETI.

ℛ   0ʰ 32ᵐ 34ˢ   PREC. + 3ˢ·05
DEC. S 5° 13'·8   —— N 19"·84

POSITION 289°·9 (*w* 3)   DISTANCE 57"·9 (*w* 1)   EPOCH 1837·87

A wide double star to the north of the Whale's tail, over which an imaginary line from η Ceti to α Pegasi passes, at near 9° from the former. A 6½, pale topaz; B 9, violet-tint,—several other stars in the distant parts of the field. This object, though coarse, is pleasing, from its contrasted colours: it was seen and thus described by Piazzi: "Aliæ 9ˣ magnitudinis 3"·2 temporis præcedit, ½ min. ad boream." The principal star has a very appreciable amount of proper motions, which have been thus registered:

P....ℛ – 0"·40   Dec. + 0"·35
B.... – 0"·20   + 0"·11

XXIV. 31 M. ANDROMEDÆ.

ℛ   0ʰ 34ᵐ 5ˢ   PREC. + 3ˢ·24
DEC. N 40° 23'·6   —— N 19"·82

MEAN EPOCH OF THE OBSERVATION . . . . 1833·70

An overpowering nebula, with a companion about 25' in the south vertical. It is of an oval shape, light, brightening towards the *sf* edge of the general mass, and of a milky irresolvable nebulosity; but though described " in cingulo Andromedæ," is between the robes and left arm of the Lady, and certainly below the girdle. There are numerous telescopic stars around; and three minute ones are involved in the glow, but which can have no connection with it, and are doubtless between our system and the nebulosity. The axis of direction trends *sp* and *nf;* and it may be caught by a good eye, on a very fine night, by running a fancied line from Alamak to Mirak, and from thence carrying a rectangular glance to a distance of about $6\frac{1}{2}°$. It can also be struck upon by a ray from *γ* in the mouth of Cetus, over Sheratan in the head of Aries, and through Mirak, or *β* Andromedæ, to $6\frac{1}{2}°$ beyond.

This is the oldest known nebula; for though it attracted but little notice till the seventeenth century, it was seen, at least, as far back as 905 A.D. Simon Marius re-discovered it,—if such a term can be applied to

an object seen with the naked eye: in his rare work—*De Mundo Joviali* — that astronomer acquaints us, that he first examined it with a telescope on the 15th Dec. 1612; he was astonished at the singularity of the phenomenon, but expressly says, that he leaves to others to judge whether it was a new discovery or not. It was therefore by an oversight, that Halley ascribes the discovery, in 1661, to Bulialdus (*Ismaël Boulliaud*); who himself mentions its being known as *Nebulosa in cingulo Andromedæ*, and that it had been noticed 150 years before, by an expert though anonymous astronomer. The tenuity of its boundary offering no definition for exact comparison, has made the several attempts to figure it so conflicting as to mislead. Marius describes it as resembling the diluted light of the flame of a candle seen through horn, —

Halley mentions that it emits a radiant beam,—Cassini calls it *à peu-près triangulaire*,—Le Gentil considered it round for some years, then oval, but always of an uniform light in all its parts,—while Messier represents it as resembling two cones, or pyramids of light, opposed by their bases. From such statements, Boulliaud and Kircher thought this wonderful object appeared and disappeared, like Mira; and Le Gentil had no doubt of its undergoing changes in form. But probably this discordance is a consequence of the means employed. Le Gentil, by his paper of 1749, seems to have used telescopes of various sizes, in order to see it very clearly—"*non seulement pour servir à la reconnoître, mais encore pour voir si dans la suite elle ne seroit point sujette à quelque variation, soit dans la figure, soit dans la position;*" yet fifteen years afterwards Messier differs from him, by assigning a greater brilliance to the centre than to the edges, which latter accords better with my views of it, than do our apparent mean places. It is, however, remarkable that Messier examined this giant nebula with a $4\frac{1}{2}$ foot Newtonian, and then turned the instrument upon γ Andromedæ—"qui en étoit fort près"—to compare its light with that of the star, on a beautiful night of August, 1764; but he makes no mention of the duplicity, or contrasted colours, of that lovely star.

Sir William Herschel, the Præses of all the examiners into the construction of the heavens, gave this phenomenon a rigid scrutiny, and concluded it to be the nearest of all the great nebulæ. "The brightest part of it," he says, "approaches to the resolvable nebulosity, and begins to show a faint red colour; which, from many observations on the magnitude and colour of nebulæ, I believe to be an indication that its distance in the coloured part does not exceed 2000 times the distance of Sirius." Does not exceed that distance! That is, so far from us, that light, which is endowed with the swiftest degree of motion yet known, flying along at the rate of 190,000 miles in a second of time, or nearly twelve millions of miles in a minute, would require upwards of 6000 years to traverse the awful interval: as to that type of terrestrial velocity, so often cited, the cannon-ball, with its 500-miles-an-hour pace, it would have no chance of passing the same space under nine or ten thousand millions of years. What an overwhelming idea does such an astonishing conclusion give of the All-wise and Omnipotent Intelligence!

Halley considered the light of this object as depending quite on a particular cause. In reality, he says, the spot is "nothing else but the light coming from an extraordinary great space in the ether, through which a lucid medium is diffused that shines with its own proper lustre." Other philosophers have advanced similar opinions, or at least, opinions not remotely different; and there is still a wide field for conjecture and speculation. The causes and arrangement of so astonishing a mass of nebulous matter, if not quite inscrutable, are still so unapproachable that it will probably occupy ages to detect them; but we must hesitate in the conclusion of a contemporaneous lecturer, of its being composed of the united lustre of a vast system of stars.

The companion was discovered in November, 1749, by Le Gentil, and was described by him as being about an eighth of the size of the principal one; he adds, "*elle m'a paru exactement de la même densité que l'ancienne.*" The light is certainly more feeble than here assigned. Messier—whose No. 32 it is—observed it closely in 1764, and remarked, that no change had taken place since the time of its being first recorded. In form it is nearly circular. The powerful telescope of Lord Rosse*

---

* This telescope is a reflector of three feet in diameter, of performance hitherto unequalled. It was executed by the Earl of Rosse, under a rare union of skill, assiduity, perseverance, and munificence. The years of application required to accomplish this, have not worn his Lordship's zeal and spirit; like a giant refreshed, he has returned to his task, and is now occupied upon a metallic disc of no less than six feet in diameter. Should the figure of this prove as perfect as the present one, we may soon over-leap what many absurdly look upon as the boundaries of the creation.

has been applied to this, after finding that no actual re-solution in the large nebulæ could be seen, though its edge had stellar symptoms; and it proved to be clearly resolvable into stars—the which directly interferes with Le Gentil's remark.

XXV. 78 ꙰. VIII. CASSIOPEÆ.

Æ   0ʰ 34ᵐ 15ˢ   Prec. + 3ˢ·43
Dec. N 60° 54′·7   —— N 19″·82

Position 70°·0 (w 1)   Distance 12″·0 (w 1)   Epoch 1835·68

A small double star, in a loose cluster of about thirty of the 9th and 10th magnitudes, occupying all the field; but there being no star-dust, or nebulosity intermixed, the firmament appears unusually dark between them. The most conspicuous object is the double star here carefully estimated, of which A is of the 8½ magnitude, and B of the 11th, both pale. It is No. 1046 of H.'s Fourth Series. This cluster, which is on the seat of Cassiopea, and exactly half-way between γ and κ, was discovered by Miss Herschel in 1784; and described by ꙰. as " taking up 15 or 20 minutes."

XXVI. β CETI.

Æ   0ʰ 35ᵐ 34ˢ   Prec. + 3ˢ·00
Dec. S 18° 51′·9   —— N 19″·81

Position 221°·5 (w 1)   Distance 542″·0 (w 1)   Epoch 1836·72

A Greenwich star of the second grade, with a distant companion in the *sp* quadrant. A 2½, yellow; B 12, pale blue,—and there is a 9th magnitude star in the *sf*, following by about a minute of time. The principal star has a proper motion assigned to it, to the following value:

P....Æ + 0″·21   Dec. + 0″·07
B....   + 0″·16   + 0″·05
A....   + 0″·21   + 0″·03

This star is in the south branch of the Whale's tail, whence it obtained the name of *Dheneb Kaïtós jenúbi*. But it has been more widely noticed as Diphda,—from *Difda' al tháni*, or the second frog, pertaining to an original Arabian constellation, of which the first frog was *Difda al auwel*, the same with Fom-al-hút, or *a* Piscis Australis. From various compa-risons of their lustre, β Ceti is certainly larger than *a*; they were both registered γ, or 3rd magnitude, by Ptolemy: but it seems to have been increasing in brightness. A fancied line between Fom-al-hút and Menkab passes over β Ceti, in about mid-distance.

XXVII. 175 P. O. ANDROMEDÆ.

Æ   0ʰ 37ᵐ 50ˢ   Prec. + 3ˢ·20

Dec. N 30° 04′·2   —— N 19″·77

Position 235°·8 (w 9)   Distance 46″·4 (w 8)   Epoch 1836·12

A wide double star, following δ on the Lady's right shoulder, by
about a degree and a half. Both individuals are of the 8th magnitude,
and pale yellow. B is Piazzi's No. 176; and the pair—which is not
a fine one—constitutes 123 I,l. v.; and the register of the latter may be
thus ranged in juxtaposition with deductions from the former:

$$\text{I}_0^\text{l}. \quad \text{Pos. } 237° \ 36' \quad \text{Dist. } 45''{\cdot}02 \quad \text{Ep. } 1783{\cdot}02$$
$$\text{P.} \qquad 236° \ 00' \qquad\quad 44''{\cdot}3 \qquad\quad 1800{\cdot}00$$

It was afterwards measured by H. and S., as 142 Bode's Andromedæ:

$$\text{Pos. } 236° \ 0' \quad \text{Dist. } 46''{\cdot}46 \quad \text{Ep. } 1821{\cdot}95$$

whence, on a comparison with my own observations, it must have
remained stationary; the very slight decrease of angle being as imputable
to instrumental error as to motion.

XXVIII. 181 P. O. CASSIOPEÆ.

Æ   0ʰ 38ᵐ 58ˢ   Prec. + 3ˢ·34

Dec. N 50° 34′·2   —— N 19″·75

Position 147°·2 (w 6)   Distance 2″·3 (w 4)   Epoch 1832·87

—— 146°·8 (w 8)   —— 2″·4 (w 5)   —— 1836·94

A close double star, between Andromeda's knee and the head of
Cassiopea, just following a line projected from κ through a, and carried
5° beyond the latter. A 7½, flushed white; B 9, white. This excellent
object formed 40 I,l. I.; and as he saw them " very unequal " in 1782,
and " difficult to be seen," the small star may be variable; the redness
he imputes, was probably owing to causes already alluded to.   I,l. marked
his observation as " very exact " in his manuscript, a slow change of
position, amounting to 0°·17 per annum direct, was therefore inferred by
his son; but subsequent observations do not bear this out. My own
measures, being remarkably coincident, are very satisfactory, and compared with the following, prove the companionship of these stars to be
merely optical:

I,l.	Pos. 140° 30′	Dist. 2″·0	Ep. 1782·66
S.	147° 35′	2″·57	1825·14
D.	148° 10′	2″·32	1830·78
Σ.	144° 58′	2″·19	1832·33

XXIX. η CASSIOPEÆ.

Æ	0^h 39^m 27^s	Prec. + 3^s·42
Dec.	N 56° 57′·9	—— N 19″·75

Position	87°·8 (w 5)	Distance	9″·8 (w 3)	Epoch	1830·91
——	88°·2 (w 3)	——	10″·2 (w 2)	——	1831·84
——	88°·9 (w 3)	——	9″·9 (w 3)	——	1833·70
——	90°·3 (w 7)	——	9″·8 (w 5)	——	1834·77
——	90°·9 (w 6)	——	9″·7 (w 6)	——	1835·20
——	92°·0 (w 5)	——	9″·4 (w 6)	——	1836·81
——	95°·8 (w 8)	——	9″·1 (w 5)	——	1843·19

A neat binary star, in the cestus of the seated Lady, forming the apex of a right-angled triangle with α and β. A 4, pale white; and B 7½, purple. This superb physical object was discovered to be double by ♓., who thus registered it:

Pos. 62° 04′   Dist. 11″·27   Ep. 1779·63

By ♓.'s re-examination in 1803, a rapid angular velocity was detected; and from thence to my own epochs, the following observations have formed points of great interest in their results:

Σ.	Pos. 80° 12′	Dist. 10″·80	Ep. 1819·80
H. and S.	82° 04′	8″·79	1821·90
S.	83° 05′	9″·90	1825·78
H.	86° 39′	10″·38	1829·43
D.	88° 40′	9″·74	1832·87

Sir John Herschel, in his discussion of 1831, said he would not then decide, on account of the uncertain determination of the distances, whether the motion thus established was orbital or parallactic. But, as he added, that the small star, in all probability, would be on the parallel, or in the act of changing quadrants from *nf* to *sf* in the beginning of the year 1835, I carefully watched, both before and after, and saw the prediction verified. These double-star orbits are really among the most interesting subjects which modern research has to occupy itself about; and their investigation offers a beautiful field for the *amateur* astronomer. The lapse of forty years after ♓.'s measure gives a mean velocity of + 0°·45 per annum, and the twenty-three years since elapsed + 0°·70, while the distance may be regarded as but little altered, from which a period of about 700 years is deduced.

This object is thus described by Piazzi: "Duplex. Comes 9·10ᵐ magnitudinis in eodem parallelo 1″ temporis sequitur;" and its proper motion through space is registered by

P....	Æ	+ 1″·78	Dec.	— 0″·72
B....		+ 2″·03		— 0″·48
A....		+ 1″·97		— 0″·50

XXX. 1 ♅. V. CETI.

Æ 0ʰ 39ᵐ 45ˢ   Prec. + 2ˢ·96
Dec. S 26° 10′·1   —— N 19″·74

Mean Epoch of the Observation . . . . 1836·74

A long narrow nebula, preceding the clumsy stern-frame of Cetus, but close to the boundary assigned to Apparatus Sculptoris. It is of a pale milky tint, and trends *sp* and *nf* with its brightest portion towards the south. There are several small stars in the field, of which the nearest preceding is of the 9th magnitude, and reddish. A line drawn from the 8th-magnitude star in the *np* quadrant, to the 8th in the *nf*, will be parallel to the axis of the nebula, which—owing probably to the inferiority of means—I could not make out to be of the extreme length figured by H. This singular object was discovered by Miss Herschel, in 1783: and I differentiated it with β Ceti in 1836. A line shot from α Andromedæ through β Ceti, and carried about 7° to the south, where Fom-al-hút will appear nearly at right angles with it, marks the site of the nebula.

XXXI. 65 PISCIUM.

Æ 0ʰ 41ᵐ 18ˢ   Prec. + 3ˢ·19
Dec. N 26° 50′·3   —— N 19″·72

Position 298°·2 (w 6)   Distance 4″·2 (w 4)   Epoch 1830·97
——— 298°·5 (w 8)   ——— 4″·5 (w 5)   —— 1838·17

A close double star, which, though classed in Pisces, is placed by the map artists on the right arm of Andromeda; where it may be struck upon about half-way between π and η. A 6, and B 7, both pale yellow. This fine object was discovered by ♅., who registered it thus:

Pos. 300° 57′ Dist. 4″·00 ± Ep. 1783·15

He again measured its position in 1802, when the results seemed to warrant the assumption of a slow retrograde orbital motion; but this is not confirmed by the later observations. Piazzi, who made A the companion, merely records it, "Duplex. Comes sequitur ad Austrum:" but it has been closely examined by H. S. Σ. and D. By assembling the observations in one view, H. arrived at the conclusion that the decrease might be 0°·117 per annum; and supposing the star to revolve uniformly in a circle, its period would be 3077 years. My measures, however, drawn through a similar comparison, yield only − 0°·06 per annum, and infer an *annus magnus* of a much longer period.

While these sheets are in the press, it is announced that Flamsteed's next star, 66 Piscium, is detected double in the gigantic Poulkova refractor; the components being 6th and 7th magnitudes, and half a second of space asunder.

XXXII. 36 ANDROMEDÆ.

Æ   0ʰ 46ᵐ 24ˢ   Prec. + 3ˢ·18

Dec. N 22° 45'·7   —— N 19"·64

Position 315°·7 (w 6) Distance 1"·1 (w 2) Epoch 1835·92
—— 318°·5 (w 8)   —— 1"·1 (w 2) —— 1839·77
—— 322°·9 (w 9)   —— 1"·0 (w 4) —— 1843·12

A very close double star, a miniature of η Coronæ, in the Lady's right elbow, and closely *np* of η. A 6, bright orange; B 7, yellow. This beautiful golden pair is very difficult, being designated by Σ. *aureæ vicinissimæ;* and he has, since my first observations were made, published these results:

Pos. 307°·80   Dist. 0"·847   Ep. 1832·14
320°·47   0"·937   1836·90

But the earliest measures I met with are those of H., from a comparison with which I am led to infer, that there is a decided direct orbital motion; the registered measure to which he has attached the greatest weight, being

Pos. 307° 04'   Dist. 0"·90   Ep. 1830·78

As my observations of position were very satisfactorily made, I am rather surprised to find so great a difference from that of Σ.'s last: and even omitting that, the results are rather anomalous, varying the velocity from 0°·57 to 2°·08 per annum.

My own observations offer a *mezzo termine,* since the measures were pretty tolerable. During the last operations, the stars were mostly in contact, but at times fairly separated; so that my estimation of distance by diameters of the discs, varied from 0"·75 to 1"·25.

As this was an object which demanded every assistance under high powers—and most of my positions were made with an eye-piece magnifying 600 times—I resorted to the recommendation of Sir J. Herschel, before alluded to, of applying a central paper disc to improve the separating power; and, in this instance, I think it was an advantage.

XXXIII. γ CASSIOPEÆ.

Æ   0ʰ 47ᵐ 05ˢ   Prec. + 3ˢ·54

Dec. N 59° 50'·8   —— N 19"·62

Position 347°·8 (w 1) Distance 350" (w 1) Epoch 1837·68

A bright star with a distant telescopic companion. A 3, brilliant white; B 13, blue. This fine star is on the right hip of Cassiopea, and the following part of the field has a scatter of small stars from 10th to 13th magnitudes, so as almost to make a cluster. γ is suspected of variability: and H. wrote to me from Slough, in October 1838, that it was then decidedly the chief star in that constellation. On the 28th of April following, he again addressed me on the subject, saying, "In a former letter I requested you to notice α Cassiopeæ, as then less than γ. It was so, and continued so for some time; but it soon regained its ascendancy. On the 21st instant I again got a positive observation—and α was then decidedly smaller than β or γ. I feel now quite assured not only of the change, but of its periodical recurrence." This is decisive as to α, and makes me rather regret that γ was rejected as a candidate for the Greenwich list, at the reform of 1830.

The reductions of the various meridional observations of γ, shew some slight aberrations from the general laws of precession, and therefore a proper motion in space has been assigned, of which the most authentic values are:

$$P.... \text{Æ} \quad 0''{\cdot}00 \quad \text{Dec.} + 0''{\cdot}04$$
$$B.... \quad + 0''{\cdot}03 \quad + 0''{\cdot}05$$
$$B.... \quad + 0''{\cdot}02 \quad + 0''{\cdot}02$$

To find this star by alignment, project a glance from Alioth, the inner individual of the Greater Bear's tail, through the Pole-star, and at nearly a similar distance beyond, it will meet with γ Cassiopeæ.

XXXIV.   μ ANDROMEDÆ.

$$\text{Æ} \quad 0^h\ 47^m\ 53^s \qquad \text{Prec.} + \quad 3^s{\cdot}28$$
$$\text{Dec. N } 37°\ 37'{\cdot}8 \qquad \text{——— N } 19''{\cdot}61$$

Position 115°·0 (w 1)   Distance 45″ (w 1)   Epoch 1833·88

A most delicate double star. A 4, bright white; B 16, dusky grey. On the Lady's back, and just below the girdle; and it may be found by carrying a line from γ Pegasi through δ Andromedæ, and extending it 8° or 9° beyond, to the north-west. Registered by H. as No. 1057 of his 20-foot sweeps: an object of extreme difficulty, and merely enrolled here to verify the power of the instrument; the small star, though repeatedly sought for, being only caught sight of on November 17th, when the large star was hidden behind a bar, and its place was estimated as above. Following this on the parallel, at about 20ᵐ, is a very neat double star, of the 8th and 11th magnitudes, and about 12″ apart: this is Σ. No. 104, and might have been considered an object of some delicacy, but tried after μ appears quite *staring*, and its colours, pale yellow, and green, are very decided. The proper motion of μ Andromedæ is valued as follows:

$$P.... \text{Æ} + 1''{\cdot}20 \quad \text{Dec.} + 0''{\cdot}04$$
$$B.... \quad + 0''{\cdot}21 \quad + 0''{\cdot}07$$
$$A.... \quad + 0''{\cdot}26 \quad + 0''{\cdot}07$$

Since the above was written, the Rev. Mr. Challis measured this object, as my request, with the Northumberland equatoreal, and obtained these results:

Pos. 110° 28′   Dist. 49″·19   Ep. 1842·67

XXXV. 251 P. O. PISCIUM.

Æ 0ʰ 51ᵐ 12ˢ   Prec. + 3ˢ·07

Dec. S 0° 4′·9 —— N 19‴·55

Position 299°·8 (w 8) Distance 18″·4 (w 6) Epoch 1832·98

——— 301°·8 (w 8) ——— 18″·5 (w 5) ——— 1838·03

A neat double star, bearing both illumination and high magnifying powers. A 8, pale orange; and B 9, clear blue. This, though assigned to Pisces in the Palermo Catalogue, belongs to the Whale, being in the space between the tail of Cetus and the ribbon of Pisces, at about one-third the distance of β Ceti from β Andromedæ, and nearly in the line; being one of the *amorphotæ*, of which an asterism to be called Testudo was proposed. Piazzi records this object double: "Duplex. Comes 9ᵃ magnitudinis præcedit 1″ temporis parumper ad boream;" but the first micrometrical measures I met with, are those of Sir James South:

Pos. 296° 27′ Dist. 18″·87 Ep. 1825·17

From which, when compared with my own, I inferred a sensible direct orbital motion = 0°·4 per annum; and this has been since confirmed by the Dorpat Catalogue. There is, however, on rigorous comparison, a slight proper motion in space assigned to A—of which probably B partakes—to the following amount:

B.... Æ + 0″·03   Dec. − 0″·28
T....   + 0″·01   − 0″·32

XXXVI. 26 CETI.

Æ 0ʰ 55ᵐ 35ˢ   Prec. + 3ˢ·07

Dec. N 0° 30′·5 —— N 19‴·46

Position 252°·6 (w 9) Distance 16″·4 (w 7) Epoch 1833·86

A neat double star, close to the foregoing, in the vacant space between the Whale's back and the ribbon of Pisces; being exactly in mid-distance between γ Pegasi and ζ Ceti. A 6½, pale topaz; B 9½, lilac tint. This fine object is 83 H. IV., whose measures on its first registry were:

Pos. 255° 24′ Dist. 17″·03 Ep. 1782·75

It was next examined by H. and S., thus:

Pos. 255° 21′ Dist. 15″·756 Ep. 1821·87

whence there would appear to be no material change in upwards of half a century; so that the present conclusion is, that the connexion is merely optical. The distance between them, therefore, on the assumption of a scale by their respective magnitudes, must be wonderfully vast.

XXXVII. ψ^1 PISCIUM.

Æ 0ʰ 57ᵐ 07ˢ   Prec. + 3ˢ·19
Dec. N 20° 36′·9   —— N 19″·43

Position 160°·4 (w 8)   Distance 30″·2 (w 8)   Epoch 1833·97

A fine double star, both $5\frac{1}{2}$ magnitude, and silvery white; on the
dorsal fin of the Northern Fish, with a very small star following; and
about one-third the distance of a line drawn from α Andromedæ to
γ Ceti. An easy object for a moderate telescope, B being ψ^3, or
Piazzi's No. 276, Hora 0. It is 9 Ⴙ. iv., and was thus registered:

Pos. 170° 00′   Dist. 27″·50   Ep. 1779·83

H. and S. thought the distance might have increased, while the position
had retrograded, their measures being:

Pos. 161° 02′   Dist. 30″·34   Ep. 1822·38

but my observations tend to show *fixity*, which has been confirmed by
the Dorpat Catalogue. The proper motions in space are thus valued:

P.... Æ + 0″·30   Dec. — 0″·10
B....   + 0″·15   — 0″·02

XXXVIII. σ^2 PISCIUM.

Æ 0ʰ 57ᵐ 24ˢ   Prec. + 3ˢ·27
Dec. N 31° 19′·5   —— N 19″·42

Position A B 293°·5 (w 4)   Distance 56″·0 (w 2) ⎫
—————— A C 235°·0 (w 3)   —————— 140″·0 (w 1) ⎬ Epoch 1832·81
                                                    ⎭

A coarse triple star, just above the snout of the Northern Fish, where
a line carried from α Andromedæ through δ, and 5° beyond, will hit to
the south of it. A 6, deep yellow; B $10\frac{1}{2}$, blue; C 11, ruddy. This is
a poor object, and merely examined because H. entertained some doubts
of its identity, when S. No. 393, had been measured. The results of
16 Ⴙ. v. are registered "pretty accurate;" and the whole of the observations stand thus:

Ⴙ. Pos. 285° 28′   Dist. 48″·13   Ep. 1780·59
S.   291° 08′   90″·±   1824·94

As there appears to be some error here, I examined the spot closely
for another *comes* in the direction pointed by A and B; and prevailed
upon the Rev. Mr. Challis to do the same with the great Northumberland equatoreal, in 1842. We are both satisfied that there is no other
measurable star than those I observed in 1832; the only other direct
companion in view being a bluish 9th-magnitude individual, near the
parallel of the *sp* quadrant, distant about 7′ in space. No safe conclusion can be deduced from the discrepancies observable in the position

of A and B, as the object is most delicate,—still an inference of binarity is deducible from a comparison of the registered epochs of H̲. and myself, at the rate of $0°·15$ per annum, in a *nf* direction, indicating a highly elongated ellipse, with a period of upwards of 2000 years.

XXXIX. μ CASSIOPEÆ.

Æ 0ʰ 57ᵐ 23ˢ   Prec. + 3ˢ·53
Dec. N 54° 08'·1   —— N 19"·42

Position A B 35°·0 (*w* 1)   Distance 50"·0 (*w* 1)
——— A C 157°·0 (*w* 1)   ——— 276"·0 (*w* 1)   Epoch 1832·71

A coarse triple star in the Lady's right elbow, whence, conjointly with θ, the Arabians termed it Marfak. A 5½, deep yellow; B 14, pale blue, with a minute comes *sf*; and C 11, bluish. There are several small stars in the field, and I was assured by an astronomical friend that one existed *very close* to A; but, on reference, both Mr. Challis with the Northumberland telescope, and Mr. Dawes with an eight-inch object glass, make a diagram similar to my own. My friend therefore must have been mistaken.

This star is in the British Catalogue; but Mr. Baily says he cannot find a perfect observation of it by Flamsteed. It has, however, so rapid a course through space, that it should be constantly watched, as its displacement by proper motion is the largest yet detected among stars not closely double, and having no obvious peculiarity. This is a statement of the several values:

P.... Æ + 5"·70   Dec. − 1"·65
B....   + 5"·82   − 1"·55
A....   + 5"·80   − 1"·55

Now, supposing this star to be about the same distance from the earth as analogy assigns to its magnitude, its hourly motion cannot be less than 125,000 miles, under the operation of forces incomprehensible to the human mind. Indeed, even from this remote speck, its consequences are observable; for this wonderful movement of A leads it in a course which forms an angle of 107° from its present vertical; so that in about six years B—unless participating in its journey through space— will become due south of A, telescopically speaking.

Just 18' south of μ is a star which, though of the 6th magnitude, is not in Piazzi. It is followed nearly on the parallel, about 11ˢ off, by a 9th magnitude, and both are remarkable from being red, of a decided but not deep tint. This object may have had something to do with the mistakes of Flamsteed respecting μ, alluded to by Mr. Baily. To find this star by alignment, draw a line from β through α, and extend it as far again as the distance between those two, and μ will be seen just above, with θ following it.

XL.   77 PISCIUM.

$\mathbb{R}$ 0ʰ 57ᵐ 32ˢ   Prec. +   3ˢ·09
Dec. N   4° 03′·3   —— N 19″·42

Position 82°·5 (w 8)   Distance 32″·1 (w 8)   Epoch 1835·88

A fine double star, in the centre of the *kheït*, or ribbon, connecting the two Fishes; and it may be found at rather less than a third of the distance from η Ceti towards β Andromedæ. A 7½, white; B 8, pale lilac. These are 280 and 281 of Piazzi, Hora 0,—and nearly in the middle of the line of stars running from the Whale's crest, which Vitruvius, lib. ix., assures us the Greeks named Hermedone; the which, saith the French commentator, meaneth, *les délices de Mercure;* but according to B. Baldus, *De Verborum Vitruvianorum, &c.*, it is merely a knot in the ribbon of Pisces. When ℍ. registered them 68 ɪᴠ. they were thus:

Pos. 85° 12′   Dist. 29″·60   Ep. 1782·69

About forty years afterwards, they were re-examined by H. and S.:

Pos. 82° 40′   Dist. 32″·07   Ep. 1821·91

Though as ℍ. had said, "in both measures the weather too windy for accuracy," no rigid deductions were to be drawn from the differences: by my observations fourteen years after those of H. and S., and under satisfactory conditions, they appear to be stationary with respect to each other. But A is accused of having a proper motion in space, very slight, but to the following amount:

P.... $\mathbb{R}$ + 0″·07   Dec. − 0″·02
B....   + 0″·06   − 0″·11
T....   − 0″·02   − 0″·21

XLI.   64 ℍ. VIII.   CASSIOPEÆ.

$\mathbb{R}$ 0ʰ 58ᵐ 19ˢ   Prec. +   3ˢ·67
Dec. N 60° 44′·0   —— N 19″·40

Mean Epoch of the Observation . . . . 1837·82

A lucid but loose cluster of small stars—principally 9th to 14th magnitudes, preceded by a 6th—on the robe below the right hip of Cassiopea; and it will be caught up, at about one-fourth of the distance, on a line from γ towards ε. It was discovered by Miss Herschel, in 1783, and described by her brother as a "forming cluster of pretty compressed stars." It may therefore be of interest in a future day, on which account it ought to be rigorously and mathematically figured. Indeed, rigidly accurate drawings are among the desiderata of sidereal astronomy. The mean apparent place is differentiated from γ Cassiopeæ.

XLII. η CETI.

Æ 1ʰ 0ᵐ 32ˢ   PREC. + 3ˢ·00
DEC. S 11° 01′·7   —— N 19″·35

POSITION 310°·7 (w 1)   DISTANCE 239″·0 (w 1)   EPOCH 1838·93

A bright star with a companion, in a barren field. A 3½, yellow;
B 10, livid,—only two other distant stars in view, one of which is in
the *sf*, and the other in the *nf* quadrant. It is on the monster's flank,
towards the tail, as implied by *Dheneb-al-Jenúbi*; where it has been
mistaken for the Rana Secunda of the Arabs; but which is β Ceti. As
this star is useful in the neighbouring alignments, it may be identified
by being exactly at right angles with a line shot from Fom-al-hút and
carried 8° beyond β Ceti: and it is on the same vertical with β Andromedæ. A is only marked as of ε magnitude in Ptolemy; while Tycho
Brahé and Flamsteed make it 3. Had I not adopted Piazzi's magnitudes
for my initial star, I should certainly have put this in the 4th degree. Can
it be variable? Its value of proper motion in space has been stated thus:

P.... Æ + 0″·28   Dec. − 0″·10
B....   + 0″·26   − 0″·12
A....   + 0″·23   − 0″·13

XLIII. β ANDROMEDÆ.

Æ 1ʰ 0ᵐ 47ˢ   PREC. + 3ˢ·32
DEC. N 34° 46′·3   —— N 19″·34

POSITION 299°·0 (w 1)   DISTANCE 225″·0 (w 1)   EPOCH 1839·54

A bright star with a distant telescopic companion. A 2, fine yellow;
B 12, pale blue,—and there are several small stars in the field, of which
two form a coarse pair in the *sp* quadrant. The delineations of the
Northern Fish and the body of Andromeda here create much confusion;
as the Arabs named β Andromedæ, *Jenb-al-muselselah*, or the chained
woman's side, and also *Batn-al-hút*, or the fish's belly. This star, once
in the Fish's head, is now placed on the Lady's right hip, over the
Northern Fish's mouth, whence it was called Mirach, from the mantle or
apron round her; but it became the *Miræ* of the Alphonsine Tables, which
term was substituted—on Scaliger's suggestion—by *Mízár*, girdle; an
amendment, however, that confounded it with ζ Ursæ Majoris. There
has been some difference of opinion as to its comparative brilliance. It
is certainly rather dim for the above rating, and Ptolemy enrolled it as
γ only in lustre; but Ulugh Beigh, and all the moderns, have ranked it
of the second magnitude.

This star was of importance, as forming the twenty-eighth and
last Lunar Mansion, called *Al Rishà*, the cord, because the vertical
bight of the Fish's *kheït* formed its boundary. The famous *Manázil-al-*

Ḳamar, i. e. Lunar Mansions, constituted a supposed broad circle, in Oriental astronomy, divided into twenty-eight unequal parts, corresponding with the moon's course, and therefore called the abodes of the moon. This was not a bad arrangement for a certain class of gazers, since the luminary was observed to be in or near one or other of these parts, or constellations, every night. Though tampered with by astrologers, these Lunar Mansions were probably the earliest step in ancient astronomy.

An imaginary line drawn from a Ceti, through the two stars in the head of Aries, will strike upon Mirach; or it will be at a right angle north of the line carried from β Pegasi to a Andromedæ, and extended as far again westward: or in the directions of the poet:

From Markab run a line beneath th' imprison'd Lady's head,
And over *delta* on her back to Mirach 'twill be led.

Proper motions in space are assigned to Mirach, of which, from careful comparisons, these are the best values:

$$P.... \text{ R} + 0''{\cdot}35 \qquad \text{Dec.} - 0''{\cdot}10$$
$$B.... \quad + 0''{\cdot}29 \qquad\qquad - 0''{\cdot}07$$
$$A.... \quad + 0''{\cdot}21 \qquad\qquad - 0''{\cdot}08$$

XLIV. a URSÆ MINORIS.

R   1ʰ 2ᵐ 10ˢ   Prec. + 16ˢ·47
Dec. N 88° 27'·4   —— N 19''·31

Position 209°·9 (w 6)   Distance 18''·4 (w 6)   Epoch 1830·78
——— 210°·1 (w 9)   ——— 18''·6 (w 9)   —— 1838·16

A standard Greenwich star, at the tip of the Little Bear's tail, with a companion in the *sp* quadrant. A 2½, topaz yellow; B 9½, pale white. A is Polaris, and from its perpetual apparition in this hemisphere, the most practically useful star in the heavens, whether to the astronomer or the seaman; and the want of such a constant reference at the opposite pole is severely felt. Piazzi devoted much labour to obtain all the conditions of this remarkable star, and prudently concluded that, in consequence of the great and inconstant precession in the immediate vicinity of the pole, it is difficult to separate the proper motions in space from that element: it was also narrowly watched for the detection of parallax, from 1802 to 1804, at each season, in January, July, April, and October, and it was deemed that an absolute quantity of 1''·31 was fairly deduced. It was first classed double by ♄., being his 1, ɪᴠ.; and the mean of his observations for 1779, 1781, and 1782, give:

Pos. 202° 58'   Dist. 18''·47   Ep. 1781·50

But the correction due to precession being necessary from the variation of an angle of position so near the pole, to reduce the observations from one date to another, I followed H.'s method; and the amount, −2° 39', for the forty-nine years between the above epoch and my first measures, brings ♄.'s position to 205° 37'. M. Struve made a series of interesting observations on this object from 1814 to 1819, not only to ascertain the proper motion of both components, but also to deduce the annual parallax and the aberration of light, and find whether very small stars give the latter different from the large. But he found so small a quantity for parallax, ($-0''\cdot32$) that if it was not owing to inevitable error of observation, it is at least what it ought to be. As to aberration, he found a constant of $20''\cdot112$, indicated by Polaris and its acolyte, and by other stars $= 20''\cdot300$.

H. and S. were the next examiners of this interesting star, and from the result of not fewer than a hundred measures, obtained this general mean:

Pos. 208° 49'   Dist. 18''·70   Ep. 1823·06

We may therefore, on the whole, presume that these stars are unchanged. A is marked 2·3 magnitude from the rule I have adopted, otherwise it is not even a very bright third size. It was ranked γ by Ptolemy, and Copernicus adopted it; but Tycho elevated it to the 2nd magnitude; and Kepler who, in the Rudolphine Tables, speaks of it as *vulgo Polaris*, rates it the same. At present it is only 1° 32' from the polar point, and by its northerly precession in declination will gradually approach to within 26' 30'' of it. This proximity to the actual pole will occur in A.D. 2095, but will not recur for 12,860 years. The period of the revolution of the celestial equinoctial pole about the pole of the ecliptic, is nearly 26,000 years; the north celestial pole therefore will be, about 13,000 years hence, nearly 49° from the present polar star.

The alignment rule for finding this star, is so well known that it scarcely demands repeating: yet it may be as well to remind the reader, that an imaginary line through the two well-known pointers, *a* and *β Ursæ Majoris*, nearly passes over it; and once found, it will not readily be mistaken, or forgotten, since, to the naked eye, it appears always in the same place.

In the alignment of the heavens, it may assist rough estimations to assume the distance between the Pointers at 5°, and that between the Pointers and Polaris at about 30°, which, though not the

true distances, will serve as a gazing scale. The diagram shows the direction, not the distance, of the Pole-star. From Polaris, lines of direction may be led to most of the great stars around. Hence the poetaster:

<blockquote>
The ever watchful Kokab <i>guards</i>,   while Dubhe points the Pole;
The Pole at rest, sees Heaven's bright host   unwearied round him roll.
</blockquote>

The use of the Pole-star in navigation is recorded to have been introduced by Thales; but as it was very anciently called Phœnice, and that philosopher also resided in Phœnicia, it was probably derived from the mariners of that nation, and has ever since been the "lode-star"

of seamen. Aratus mentions it as a sure sea-guide, or beacon; saying,
—*in voce Germanici*—

> Certior est Cynosura tamen sulcantibus æquor.

Dryden has happily described the infancy of navigation:

> Rude as their ships were navigated then,
> No useful compass, or meridian known;
> Coasting they kept the land within their ken,
> And knew no north but when the Pole-star shone.

Among our own seamen, the *Stella Maris*, or Pole-Star, and its companions, have immemorially been under requisition. Recorde tells us, in the *Castle of Knowledge*, nearly 300 years ago, that navigators used two pointers in Ursa,—" which many do call the Shafte, and other do name the Guardas, after the Spanish tonge." Richard Eden, in 1584, published his *Arte of Navigation*, and therein gave rules for the " starres," among which are special directions for the two called the Guards, in the mouth of the " horne," as the figure was called. See β Ursæ Minoris. In the *Safegard of Saylers*, 1619, are detailed rules for finding the hour of the night, by the " guardes:" and the Bears generally were regarded as rustic time-pieces, whence Shakspeare, in the Gadshill affair, makes the carrier exclaim, " An't be not four by the day, I'll be hanged: Charles's wain is over the new chimney, and yet our horse not packed!" As to the Little Bear, the whole animal is swung round by the tail every twenty-four hours: whence the general name for the pole was *Kotb*, which means the spindle or pin fixed in the under-stone of a mill, around which the stars typifying the upper stone turn. Hence, also, the *Ludentes*, or Dancers, of old.

I more than once attempted to fix the place of a little star, called *Blucher*, by some of the savans, which precession will have now brought within 2′ of the pole. But being only of the 10th magnitude it is a difficult object to touch in Æ, and there is a wide companion still smaller. A nebula, like a dull star, is perceivable near it, and is H. 250, *Polarissima;* so called from its proximity to the pole. Insignificant, however, as is this little star with its warlike name, it is as much the pole point of the zodiac, as α Draconis is that of the Ecliptic.

Arctos minor, or the Lesser Bear, is not mentioned by Hesiod or Homer, therefore was probably not yet admitted among the constellations in that shape: indeed, *Cynosura* was more likely to have been represented by a dog. Jacob Bryant, dreaming of Philistines, considers the word as having been borrowed by the Greeks from *Cahen Ourah*. Thales is reported to have formed it, from perceiving the seven principal stars make a similar figure to the well-known wain of the Great Bear; but reversed with respect to each other: whence Aratus assures us that both the Bears—the *magna minorque feræ* of Ovid—were called ἅμαξα, or waggons, by the Greeks. But instead of the obtuse-angled projection of the Great Bear's stern, the Lesser Bear's tail curves gradually till it reaches the Pole-star. It is, however, a perplexing asterism, from the number of hours of Æ it extends over, and its components have been thus registered:

Ptolemy . . .	8 stars	Hevelius	12 stars
Tycho Brahé . .	20	Flamsteed . . .	24
Kepler	21	Bode	75

It appears that Ursa minor was a favourite constellation among the Arabians, who called the pole-star Jedi, or the Kid; and *Al Kaukab-al-shemáli*, the Northern star, an appellation originally given to β, which in Ptolemy's time was nearer to the pole than a. On the Cufico-Arabic globe, described by Assemani, the asterism is written *Al Dubb-al-ashgar:* and in the Alphonsine Tables it is corruptly termed *Alrucaba*, which term has been discussed by Grotius, Hyde, and Ideler, as grounded in Hebrew, Chaldaic, or Arabic. We are told that the pole was also termed *Al Ḳiblah*, because of the obligation in Mahometan prayer to know which way the head is. To find the kiblah in an unknown place, they looked to Polaris and could thereby readily *orientize* themselves. To this necessity we are considered to·be indebted for the astronomy of the Abbaside Caliphs.

XLV.   4 P. I. PISCIUM.

ℛ 1ʰ 2ᵐ 31ˢ   Prec. + 3ˢ·12
Dec. N 8° 42′·0   —— N 19 ·30

Position 291°·0 (w 1)   Distance 35″·0 (w 1)   Epoch 1836·89

A very delicate double star. A 8, white; B 14, pale blue; and the two point upon a third star at a distance in the *np* quadrant. This object is in the space between the two Fishes, nearly in mid-distance between η Andromedæ and θ Ceti, where A is the apex of a triangle formed with two other stars, one to the *sp* and the other to the *np*. It is No. 634 of H.'s Second Series, and with its minute companion forms a very delicate, though wide, object. From the small star's bearing no illumination, the angle and distance are mere estimations.

XLVI.   φ PISCIUM.

ℛ 1ʰ 5ᵐ 4ˢ   Prec. + 3ˢ·23
Dec. N 23° 44′·1   —— N 19″·24

Position 226°·5 (w 2)   Distance 9″·0 (w 1)   Epoch 1834·79

A pretty close double star. A 6, orange; and B 13, flushed. This beautiful object is on the ventral fin of the Northern Fish, at a little more than half the distance from γ Pegasi towards a Trianguli; and though marked "objectum subtile" by Σ., it is steadily seen through my telescope. But it is singular that Piazzi says of it, "Duplex. Comes in eodem verticali, admodum exigua et ad austrum." He certainly could hardly have seen it double with his instrument, as it now is; but the acolyte may be variable. Σ.'s epoch is registered:

Pos. 227°·52   Dist. 7″·98   Ep. 1832·06

XLVII. ζ PISCIUM.

Æ 1ʰ 5ᵐ 21ˢ   Prec. +   3ˢ·51
Dec. N 6° 43'·7   —— N 19"·23

Position 63°·7 (w 9)   Distance 23"·3 (w 9)   Epoch 1834·88
——— 63°·8 (w 9)   ——— 23"·4 (w 9)   ——— 1839·05

A neat double star. A 6, silver white; B 8, pale grey. This fine
and easy object was classed 8 Ḥ. iv., and it is on a bend of the band
which joins the two Fishes; it constitutes Piazzi's Nos. 16 and 17, of
Hora 1, and by a reduction of his apparent places, the following somewhat vague comparison is obtained:

Ḥ. Pos. 67° 23'   Dist. 22"·19   Ep. 1779·80
P.   63° 20'   19"·60   1800

whence, after an interval of sixty years, my observations point out but
little change, and even that little is chargeable to errors of observation:
every honest and careful statement, however, is increasing in value with
the improvement of instruments. The large star may be variable.
Ptolemy calls it δ in lustre, and he is followed by Ulugh Beigh, Tycho
Brahé, and Hevelius. Mr. Baily says, " This star is stated, in the British
Catalogue, to be of the 4th magnitude; but in the original observations
it is nowhere stated to be more than the 5th; and in one place it is
marked as the 6th, but afterwards altered to the 5th, which I have
retained." A slight proper motion in space is assigned, of these values:

P. .. Æ − 0"·01   Dec. + 0"·03
B....   + 0"·19   − 0"·06
A....   + 0"·17   − 0"·09
T....   + 0"·02   + 0"·04

ζ Piscium slightly precedes an occult line drawn between β Ceti and
α Trianguli, and is nearly in the mid-distance.

XLVIII. 37 CETI.

Æ 1ʰ 6ᵐ 19ˢ   Prec. +   3ˢ·01
Dec. S 8° 47'·1   —— N 19"·21

Position AB 332°·3 (w 9)   Distance 50"·6 (w 8)⎫
——— CD 341°·1 (w 6)   ——— 20"·5 (w 4)⎭   Epoch 1838·87

A wide quadruple star. A 6, white; B 7½, light blue; C 8, yellow;
and D 10, violet. This fine, though coarse, object is on the monster's
tail joint, over η to the nf, and preceding θ by a little more than 2°.
Of the components, the larger pair are 24 Ḥ. v., of which the first
register was:

Pos. 332° 36'   Dist. 45"·15   Ep. 1783·65

The results obtained from Piazzi's mean places, and those from the micrometrical measures of H. and S., are:

P.	Pos. 332° 18′	Dist. 48″·00	Ep. 1800
H. and S.	332° 27′	50″·78	1823·79

whence, compared with my own measures, taken under the finest circumstances, it appears that the position has remained unaltered,—and the difference of distance may be imputed to some oversight in the readings. 37 Ceti has, moreover, a sensible proper motion, and in giving the registered amount of it, I must add, that my meridional reductions countenance its existence; it is according to

$$
\begin{aligned}
B.... & \text{ Æ} + 0''·11 & \text{Dec.} + 0''·33 \\
A.... & + 0''·06 & + 0''·29 \\
T.... & + 0''·27 & + 0''·34
\end{aligned}
$$

A line drawn through A B, points to a fine double star rather low down in the *np* quadrant, and there are several other stars in the field; a pretty bright one following at $a \triangle \text{Æ} = 26^s$. The second set, observed by me, or C D, form a miniature of the first pair, and are 77 ♅. IV. They precede A by about 32^s, and are 15′ to the north of it; they were thus, when first registered:

Pos. 333° 24′ Dist. 19″·10 Ep. 1782·73

and Sir James South, No. 396, found it:

Pos. 337° 34′ Dist. 19″·89 Ep. 1825·30

whence we may conclude, that no sensible change has occurred in the distance in 52 years, but that there may be a slow direct motion in the orbital angle.

XLIX. 42 ♅. VII. CASSIOPEÆ.

Æ 1^h 9^m 10^s PREC. + 3^s·69
DEC. N 57° 56′·9   —— N 19″·14

POSITION 315°·0 (*w* 1) DISTANCE 15″·0 (*w* 1) EPOCH 1837·66

A minute double star, in a cluster between the Lady's right knee and her elbow. A 9, and B 10, both white. This brilliant aigrette-shaped group of large and small stars, was discovered by ♅. on the 18th of October, 1787, and is No. 97 of H.'s Catalogue. In the centre is the fine double star before us, the position and distance of which are very carefully estimated. There is a star of the 7th magnitude at the *sf* verge of the field, which is very bright and white. The cluster is close to ϕ, and though differentiated from that star, was also observed with the meridian instruments.

The euphonist may be reminded, that the apparently barbarous Gallicism " differentiated," which occurs so frequently, is a technical expression for the manner of fixing the object's place.

L. 35 CASSIOPEÆ.

Æ 1ʰ 10ᵐ 28ˢ   Prec. + 3ˢ·89
Dec. N 63° 49'·0   —— N 19"·10

Position 352°·5 (w 8)   Distance 49"·7 (w 9)   Epoch 1830·39

A wide double star, on the lower part of the Lady's drapery, and the following of four stars describing a lozenge; which may be fished up by carrying a line from δ towards the Pole-star, and intersecting it at rather less than 5°. A 7, white; B 9, flushed,—and there are two other brightish stars in the field. In case the claim of this object to being Flamsteed's 35 should be disputed, it may be said that B is Piazzi's No. 39, and A is No. 40, Hora 1; and they constitute also 81 Ħ. v., which, when first registered, was:

Pos. 355° 12' Dist. 42"·58 Ep. 1782·56

By Sir James South, No. 397, it was:

Pos. 352° 53' Dist. 50"·36 Ep. 1824·84 |

whence there seems to be hardly any appreciable change in the position, though there is a greater variation in the distance than would be expected on an object which bears the lamp so well, and admits of such easy measurement. If there were no instrumental errors, a retrograde orbital movement might be inferred: but a reduction of Piazzi's mean places would imply a slight direct motion, his angle for 1800 being = 349°.

LI. 42 CETI.

Æ 1ʰ 11ᵐ 38ˢ   Prec. + 3ˢ·06
Dec. S 1° 21'·0   —— N 19"·07

Position 332°·8 (w 9)   Distance 1"·2 (w 4)   Epoch 1834·84

A close double star, in the space between the Whale's back, and the *kheïl*, or band of Pisces, about 10° north of η Ceti, on the line towards α Trianguli. A 6, bright white; B 8, white. A beautiful object, but very difficult to measure in distance: it is No. 113 of Σ., who terms the components *vicinæ*; and as his last measures were

Pos. 334°·30 Dist. 1"·177 Ep. 1836·91

it seems to have a direct angular movement, to the amount of about 0°·700 per annum: but this requires verification. A proper motion in space is moreover attributed to the star A, to the following values:

P.... Æ + 0"·06   Dec. + 0"·05
B....   + 0"·06   — 0"·02

LII. ψ CASSIOPEÆ.

Æ. 1ʰ 14ᵐ 42ˢ   Prec. + 4ˢ·09
Dec. N 67° 17′·5   —— N 18″·99

Position AB 102°·1 (w 9) Distance 31″·9 (w 9)⎫
—— BC 252°·6 (w 2)   —— 2″·0 (w 1)⎬ Epoch 1836·28

A fine triple star, close to the lower part of the Lady's throne, and in a line between Polaris and δ Cassiopeæ, at rather less than a third of the distance from the latter. A 4½, orange tint; B 9, blue; C 11, reddish. This object was first seen triple by Σ., and forms No. 117 of his grand Catalogue, where it shows:

Pos. AB 101°·77 Dist. 32″·22⎫ Ep. 1831·04
BC 253°·32   3″·01⎭

The two first formed 83 Ħ. v., and were thus measured by him:

Pos. 102° 12′ Dist. 33″·41 Ep. 1782·66

H. and S., who overlooked C, found AB =

Pos. 101° 19′ Dist. 33″·35 Ep. 1822·90

which, compared with my measures of A and B, show that these individuals are only optically double, having experienced no change in position or distance. Whether B and C are bodies physically connected, remains for future observers to determine. They form a delicate test. The large star has certainly a proper motion, though unnoticed by Piazzi and Argelander. Mr. Baily has given me as the quantity in Æ + 0″·43, and Dec. + 0″·04; and my reductions, though not delicate enough to decide, countenance the amount.

LIII. 124 H. CASSIOPEÆ.

Æ. 1ʰ 18ᵐ 51ˢ   Prec. + 3ˢ·89
Dec. N 61° 27′·8   —— N 18″·87

Mean Epoch of the Observation . . . . 1835·65

An open cluster, on the Lady's leg, and nearly in mid-distance from ε towards γ. It is a gathering of large and small stars, with glimpses of star-dust of considerable extent, and irregular figure; but a few of the principal individuals assume a form not unlike that of an hour-glass. There is no particular compression or condensation of the stars to suggest the existence of a central force; yet the group is sufficiently separated to indicate its forming a peculiar system of its own. The mean apparent place was carefully differentiated from δ Cassiopeæ.

LIV.   85 P. I. PISCIUM.

$\mathcal{R}$ 1^h 20^m 00^s Prec. +   3^s·13
Dec. N   7° 07'·8   —— N 18"·84

Position 98°·7 (w 6)   Distance 68"·3 (w 6)   Epoch 1836·99

A star with a distant companion, in the space between the Fishes, in front of the Ram; and nearly half-way from η Ceti towards α Trianguli. A 7, yellow; B 8½, pale blue, being Piazzi's No. 87. It is a coarse object, in a poor field, and was first micrometrically measured by Sir J. South, who numbered it 398, and obtained these results:

Pos. 98° 17'   Dist. 69"·75   Ep. 1825·00

which are nearly identical with those which I afterwards found. The mean apparent places must have been well settled by Piazzi, since the reductions bring out the following conditions for 1800, viz.: Position 97°, and Distance 67"·60.

LV.   103 M. CASSIOPEÆ.

$\mathcal{R}$ 1^h 22^m 41^s Prec. +   3^s·88
Dec. N 59° 51'·6   —— N 18"·75

Position 140°·9 (w 8)   Distance 14"·4 (w 4)   Epoch 1832·66

A neat double star in a cluster, on Cassiopea's knee, about a degree to the *nf* of δ. A 7, straw coloured; B 9, dusky blue. This is a fan-shaped group, diverging from a sharp star in the *nf* quadrant. The cluster is brilliant from the splash of a score of its largest members, the four principal ones of which are from the 7th to the 9th magnitude; and under the largest, in the *sf*, is a red star of the 8th magnitude, which must be that mentioned by H., No. 126 of his Catalogue of 1833.

My attention was first drawn to this object, by seeing it among Σ.'s *acervi;* but I soon found that it was also the 103 which Messier describes so vaguely, as being between δ and ε Cassiopeæ, whereas it is pretty close to δ, on the Lady's knee.

LVI.   100 H. I. CETI.

$\mathcal{R}$ 1^h 23^m 20^s Prec. +   2^s·99
Dec. S   7° 41'·8   —— N 18"·73

Mean Epoch of the Observation . . . . 1833·71

A tolerably bright round nebula, of a pearly tint, just above the Whale's back; discovered and registered by Sir William Herschel, in

September, 1785. The field is very interesting, for nearly south of the little nebula is a neat double star, the components of which are of the 9th and 11th magnitudes, the latter in the *sf* quadrant; and there are three other telescopic stars on the northern verge. A line from the pair above to the minute star below, would fall just before the nebula.

LVII.   33 M. TRIANGULI.

$\mathbb{R}$ 1h 24m 51s   Prec. +   3s·35
Dec. N 29° 51'·3   —— N 18"·69

Mean Epoch of the Observation . . . . 1838·74

A large and distinct, but faint pale white nebula, in the precincts of Triangulum, between it and the head of the Northern Fish; with a bright star a little *np*, and five others following at a distance, between which and the object, there is an indistinct gleam of *mere nebulous matter*. It was discovered by Messier in 1764; and to H.I. had a mottled aspect under his seven-foot reflector, in 1783: but afterwards applying a larger telescope, he resolved it into stars—"the smallest points imaginable." By a method of turning the space-penetrating power of his instrument into a gradually increasing series of gauging powers, he considered the profundity of this cluster must be of the 344th order: *i. e.*, 344 times the distance of Sirius from the earth. It is No. 131 of II.'s Catalogue of 1833; and the above place is obtained by differentiation from *a* Trianguli, from which it is about 4°, and just north of a line run from that star to δ Andromedæ.

LVIII.   40 CASSIOPEÆ.

$\mathbb{R}$ 1h 25m 52s   Prec. +   4s·59
Dec. N 72° 13'·3   —— N 18"·65

Position 240°·5 (*w* 2)   Distance 42"·0 (*w* 1)   Epoch 1834·95

A double star between the feet of Cassiopea and Cepheus, where a line from δ carried a little east of ⲯ, and about 5° further, will strike it. A 6, yellow; and B 12, pale blue. This is a delicate though rather wide object; and is one of the principal members of · *Custos Messium*, an asterism placed by Lalande between Rangifer and Cassiopea, in poorish punning compliment to his friend Messier, the " comet-ferret." The first register I find of it, is No. 2054 of H.'s Fifth Series. It is in a poor field, but about 10' or 11' to the *sp* is Σ.'s curious nebula No. 2; and nearly following it, about 4s, is a pair of minute stars lying across the parallel, about 10" apart.

The star A has had a small motion in space assigned to it, of which the following are the most authentic values:

P.... Æ − 0″·16   Dec. − 0″·15
B....   − 0″·12   + 0″·02
T....   − 0″·21   + 0″·13

LIX. 100 PISCIUM.

Æ 1ʰ 26ᵐ 22ˢ   PREC. + 3ˢ·17
DEC. N 11° 44′·3   ——— N 18″·64

POSITION 78°·9 (w 7) DISTANCE 15″·9 (w 7) EPOCH 1833·86

A neat double star. A 7, white; B 8, pale grey. This fine object is on the ribbon under the tail of the Northern Fish, and forms Piazzi's Nos. 111 and 112 of Hora 1; a line from α Arietis through γ, the first and third of the Ram's head, hits upon 100 Piscium, at about 8° southwest of γ. It is, as H. and S. have remarked, a miniature of 77 Piscium, but not so faint in my telescope as to render the measures at all difficult. It was classed by IꞮ. 131, IV., and thus measured:

Pos. 85° 00′ Dist. 15″·88 Ep. 1783·59

It was then rigidly examined by H. and S., who made it:

Pos. 80° 25′ Dist. 16″·01 Ep. 1821·91

showing a slow np sf, or retrograde orbital change, which appears to be confirmed by my measurements. This movement is also countenanced by Piazzi's mean apparent places, from which we find that, in 1800, B was at an angle from A = 83°; so that, under all the conditions, there is presumptive evidence of an angular retrocession = − 0°·120 per annum. Ulterior measures will be very desirable, since the slow progress indicates so long a period for an active revolution of the satellite, that if the orbit be circular, 3000 years may be estimated.

LX. 123 P. I. PISCIUM.

Æ 1ʰ 27ᵐ 41ˢ   PREC. + 3ˢ·13
DEC. N 6° 49′·5   ——— N 18″·59

POSITION 19°·8 (w 7) DISTANCE 1″·5 (w 3) EPOCH 1832·86
——— 26°·9 (w 9)   ——— 1″·4 (w 4)   ——— 1843·10

An interesting close double star, in the space between the two Fishes, and the meander of the ribbon, about 15° from β Arietis, or nearly half-way on a line from that star towards η Ceti. A 6½, yellowish; B 8, pale white,—a third star following at some distance; probably Piazzi's "sequitur alia 6ᵃ magnitudinis 8′ circiter ad Boream." This beautiful object is No. 92 of IꞮ.'s List of 145, and was registered by him in

October, 1792. From his remark that the stars were less than half a diameter asunder, they must be widening; and in 1801 he found the angle of position to be only about 10°. *Σ.* entered it among the *vicinæ* on his First Class, and his results are :

Pos. 20° 00′   Dist. 1″·467   Ep. 1830·23

We may hence infer that this attendant *comes* has, during the last forty-two years, been performing a small north-east portion of its orbit; although its great proximity to the primary renders the circle so small as to greatly increase the necessity of long periods for obtaining satisfactory conclusions.   To M. Savary,—who has the merit of having first determined the elliptical elements of the orbit of a binary star from observation,—we are indebted for a very ingenious suggestion, by which the dimensions of such orbit may be known.   This method depends on the fact that light does not move instantaneously, but with a certain definite velocity, so that a specific time elapses between the moment when the ray leaves a luminous body and that when it enters our eye.   To apply this, in the present case, would be difficult indeed, since its accuracy must depend upon knowing the position of the orbit with regard to the eye; and noting exactly when the errant star is at the two opposite points of its orbit.   Could these conditions be exact, the result is readily amenable to calculation; for admitting one of these components to revolve round the other in an orbit which is nearly parallel to our line of vision, it is evident that the one half of its orbit will be nearer us than the other, and that at the most distant point of its course the star will be removed from us to a distance nearly equal to the whole diameter of its orbit further than when at the point which is nearest the earth.   Hence Savary's rule: Observe the apparent times occupied by any revolving star, in going through the two halves of its orbit; and half the difference of these times will be the period in which light passes through the diameter of its orbit.

The principle of this is beautiful; and were it reducible to practice, the dimensions of such orbit could be approximated,—for as the velocity of light is a known quantity, the diameter may therefore be computed in miles if desired.   But in addition to the obstacles already cited, the long interval of time which must intervene before such an observation can be completed, is a serious difficulty; as well as deciding upon the respective brilliance of the binary components.

LXI.   145 P. I. PISCIUM.

Æ   1ʰ 32ᵐ 23ˢ   Prec. +   3ˢ·31
Dec. N 24° 56′·1   —— N 18″·44

Position 29°·2 (*w* 2)   Distance 12″·5 (*w* 1)   Epoch 1836·87

A neat double star in a barren field over the horn of Aries, which is readily found by carrying a line from β Arietis, the middle star in the Ram's head, to β Andromedæ, at somewhat less than one-third of the distance: a thwart line from β through α Trianguli passes just to the south of it, $5\frac{1}{2}°$ from α. A $6\frac{1}{2}$, cream yellow; B 13, blue,—and there is a small blue star near the vertical of the np quadrant. This very delicate object is No. 145 of Σ., who thus measured it:

Pos. 30° 30′ Dist. 11″·30 Ep. 1830·06

LXII. 76 M. PERSEI.

ÆR 1ʰ 32ᵐ 16ˢ   PREC. + 3ˢ·71
DEC. N 50° 46′·5   —— N 18″·44

POSITION 217°·0 (w 1) DISTANCE 45″·0 (w 1) EPOCH 1837·79

An oval pearly white nebula, nearly half-way between γ Andromedæ and δ Cassiopeæ; close to the toe of Andromeda, though figured in the precincts of Perseus. It trends north and south, with two stars preceding by 11ˢ and 50ˢ, and two following nearly on the parallel, by 19ˢ and 36ˢ; and just np of it is the double star above registered, of which A is 9, white; and B 14, dusky. When first discovered, Méchain considered it as a mass of nebulosity; but Messier thought it was a compressed cluster; and ℍ. that it was an irresolvable double nebula. It has an intensely rich vicinity, and with its companions, was closely watched in my observatory, as a gauge of light, during the total eclipse of the moon, on the 13th of October, 1837, being remarkably well seen during the darkness, and gradually fading as the moon emerged. In 1842, I consulted Mr. Challis upon the definition of this nebula in the great Northumberland equatoreal, and he replied: "I looked at the nebula, as you desired, and thought it had a spangled appearance. The resolution, however, was very doubtful."

LXIII. 46 ℍ. VII. CASSIOPEÆ.

ÆR 1ʰ 33ᵐ 05ˢ   PREC. + 4ˢ·02
DEC. N 61° 04′·9   —— N 18″·41

MEAN EPOCH OF THE OBSERVATION . . . . 1835·74

A cluster of stars from the 10th to the 14th magnitudes, just below the Lady's right knee; and nearly in mid-distance between δ and ϵ. It is somewhat of a triangular shape, and about 2′ or 3′ in diameter; the hypotenuse is well defined by the three brightest stars in the field, of which the central one is orange-coloured, and of the $8\frac{1}{2}$ magnitude, perhaps Sir William Herschel's "ruddy": from analogy it is between us and the components of the cluster. This object was discovered and registered in November, 1787.

LXIV. 146 H. CASSIOPEÆ.

℞ 1ʰ 33ᵐ 24ˢ   PREC. + 3ˢ·83
DEC. N 55° 04′·1   —— N 18″·40

POSITION 120°·0 (w 1)   DISTANCE 10″·0 (w 1)   EPOCH 1835·74

A double star, A 8, pale white; B 12, dusky. In a loose cluster, between the weapon of Perseus and the elbow of Cassiopea, one-third the distance from α of the latter to α of the former constellation: and it may be fished up by carrying a line from κ Cassiopeæ through γ to double the distance beyond. It was discovered by H., and consists of a gathering of small stars, of 10th to 13th magnitude, divided into two distinct groups; one *sf* of A, and the other *nf*. It is a neat but difficult double star, whose angle of position and distance from each other, were very carefully estimated. These observations were made during a vivid and strongly coloured Aurora Borealis.

LXV. 107 PISCIUM.

℞ 1ʰ 33ᵐ 50ˢ   PREC. + 3ˢ·26
DEC. N 19° 29′·4   —— N 18″·39

POSITION 318°·3 (w 2)   DISTANCE 55″·0 (w 1)   Epoch 1837·03

A double star, just before the horn of Aries, where it is the apex of an isosceles triangle, of which β and γ Arietis form the base. A 5½, pale yellow; and B 14, dusky. This is a very delicate but wide object in a barren field, with a minute star in the *nf* towards the vertical. It is No. 2071 of H.'s Fifth Series. The *comes* is so minute that light is inadmissible; the position is therefore gained by the spherical crystal micrometer, and the distance carefully estimated. The following proper motion through space has been assigned to A:

$$P.... \quad ℞ - 0″·32 \qquad Dec. - 0″·57$$
$$B.... \quad - 0″·27 \qquad - 0″·66$$
$$A.... \quad - 0″·30 \qquad - 0″·67$$

LXVI. 31 ♌. VI. CASSIOPEÆ.

℞ 1ʰ 35ᵐ 17ˢ   PREC. + 4ˢ·02
DEC. N 60° 26′·1   —— N 18″·34

POSITION 70°·2 (w 2)   DISTANCE 8″·0 (w 1)   EPOCH 1833·70

A neat double star, in a cluster near the Lady's right knee; and it may be found by drawing a line from α through δ, and carrying it about

$2\frac{1}{2}°$ further. A 9, and B $10\frac{1}{2}$, both bluish. This object is in an elegant field of large and small stars, from a certain degree of brilliance down to infinitesimal points; but without any disposition to form, except that the larger members incline towards a parallelogram in which there are several coarse pairs. In the *sp* quadrant of this cluster, is a fine ruby star of the 8th magnitude.

LXVII.　179 P. I. ARIETIS.

Æ	$1^h 41^m 19^s$	Prec. +	$3^s·29$
Dec.	N 21° 28'·7	——	N 18"·12

Position 171°·1 (*w* 6)　Distance 2"·6 (*w* 3)　Epoch 1831·98
—— 169°·9 (*w* 8)　——　2"·4 (*w* 5)　—— 1836·11

A close double star, on the Ram's horn, about $1\frac{1}{2}°$ from β on a line towards β Andromedæ. A 6, topaz yellow; B 8, smalt blue. This fine object was classed by H̶. as 73, I., and was thus registered:
Pos. 167° 24'　Dist. 3"·0±　Ep. 1782·98
It was then measured by S. as Bode's 304 Piscium :
Pos. 172° 26'　Dist. 3"·378　Ep. 1823·98
which encouraged an opinion of a motion in orbit $= + 0°·123$ per annum. But my observations, taken with rigorous care, do not confirm this, any more than do those of Σ., which have since arrived in this country. Nor do I think the stars are approaching each other, for the measures are as correspondent as, under all the conditions of the case, can be expected. The early distances of H̶., estimated by diameters of the stellar discs, were but approximations, since, exclusive of that important element, the magnifying power, it would alter according to atmospheric and other circumstances at the time of observation. As an example of the method, we may here give the remarks for this star's distance: "With 227, about $\frac{3}{4}$ diameter of L; with 460, full $1\frac{1}{4}$, or about $1\frac{1}{2}$ of L, when best seen."

LXVIII.　191 P. I. CETI.

Æ	$1^h 43^m 33^s$	Prec. +	$3^s·17$
Dec.	N 10° 01'·0	——	N 18"·03

Position 194°·1 (*w* 8)　Distance 3"·6 (*w* 6)　Epoch 1834·99

A close double star. A $7\frac{1}{2}$, and B 8, both lucid white. This beautiful object, though catalogued of the Whale, is on the fore leg of Aries; with a distant telescopic star near the vertical *sf*, and another near the parallel,—but the field is otherwise barren. It will readily be fished up, by drawing an imaginary line from a Trianguli through γ Arietis, and carrying it about $8\frac{1}{2}°$ to the southward, or nearly as far beyond. It is Σ.'s First Class, No. 178, and was first measured thus:

Pos. 192° 48′   Dist. 3″·12   Ep. 1825·81

LXIX. ζ CETI.

$\textrm{Æ}$ 1ʰ 43ᵐ 34ˢ   Prec. + 2ˢ·95
Dec. S 11° 07′·6   —— N 18″·03

Position 40°·4 (w 3)   Distance 165″·0 (w 1)   Epoch 1835·87

A bright star, with a distant companion, in a poor field. A 3, topaz yellow; B 9, white, with a small star to the *nf*. This object is in the midst of the Whale's body, whence it was called *batn Kaïtós*, the belly of Cetus, by the Arabian astronomers. It is on the line from θ towards π, and about one third of the distance from the former; and a ray carried from υ through ζ will stretch out to β Andromedæ. A has a slight movement in space, but the value and direction are differently stated:

P....	$\textrm{Æ}$ − 0″·16	Dec. + 0″·11
Br...	+ 0″·11	+ 0″·14
B....	+ 0″·06	− 0″·12

LXX. 55 ANDROMEDÆ.

$\textrm{Æ}$ 1ʰ 43ᵐ 42ˢ   Prec. + 3ˢ·56
Dec. N 39° 56′·2   —— N 18″·03

Position 350°·0 (w 1)   Distance 25″·0 (w 1)   Epoch 1832·95

A most delicate double star, on the Lady's right leg, about 3° from γ Andromedæ, a little south of the line from that star towards β. A 5½, yellow; B 16, bluish. This is No. 1094 of H.'s Fourth Series, and designated by him "a fine specimen of a nebulous star." It is singular that it was marked nebulous by Flamsteed, in the British Catalogue, perhaps in consequence of some small stars near it. It sometimes had a *burred* aspect to my gaze, and the companion was only caught by intense attention, and then only by evanescent glimpses, being a *minimum visibile* for my telescope: its position and distance are therefore only estimated. Is the intense blue which some of these mere points of light present, an optical illusion? Pigott suspected A of variability.

The acolyte of this object being of the last degree of faintness, it was necessary to apply that singular method of obtaining a view, viz. to avert the eye, and direct it to another part of the field. H. accounts for the success of this stratagem, by supposing the lateral portions of the retina to be less exhausted than the central ones.

LXXI. α TRIANGULI.

ℛ 1ʰ 43ᵐ 58ˢ   PREC. + 3ˢ·38
DEC. N 28° 47′·8   —— N 18″·02

POSITION 179°·0 (w 1)   DISTANCE 110″·0 (w 1)   EPOCH 1834·67

A bright star with a telescopic companion. A 3½, yellow; B 11, lilac. This object is at the preceding angle of Deltōton, and, except a 10th-magnitude star near the *sf* vertical beyond B, is in a barren field. It is the apex of a large oblique-angled triangle, the base of which is formed to the *np* and *nf* by β and γ Andromedæ: and it is also 6° beyond *a* Arietis, on a line brought from γ Ceti. It was named by the Arabians *Rás al Mothallath*, or "Caput Trianguli;" and has had this amount of proper motion through space assigned to it:

$$P.... ℛ + 0″·04 \quad Dec. - 0″·36$$
$$B.... \quad + 0″·06 \quad\quad - 0″·21$$
$$A.... \quad + 0″·03 \quad\quad - 0″·23$$

Though small, Triangulum, Deltōton, or Trigōnus, is one of the ancient 48 asterisms, and is supposed to have derived its name from the Egyptian Delta; but others insist that the Triangle alludes to Trinacria, or Sicily; an island favoured by Ceres, and whence her planet was lately revealed to Piazzi. The members have been thus numbered:

Ptolemy	. . . 4 stars	Flamsteed	. . . 16 stars
Bayer	 5	Piazzi	 25
Hevelius	. . . 9	Bode	 33

Several very old illustrations delineate Deltoton as an equilateral triangle, with a star at each angle—"in unoquoque angulo unum;" but it has latterly been drawn as a scalene figure. Anciently there was only a single triangle, but Hevelius took three other stars between it and the head of Aries, to form *Triangulum minus*: the figure, however, is discontinued. A line drawn from Sheratan to Al'mak passes the *lucida Trianguli*, at about one-third of the distance.

LXXII. γ ARIETIS.

ℛ 1ʰ 44ᵐ 45ˢ   PREC. + 3ˢ·27
DEC. N 18° 30′·5   —— N 17″·99

POSITION 360°·0 (w 7)   DISTANCE 9″·1 (w 6)   EPOCH 1833·07
—— 359°·8 (w 9)   —— 8″·8 (w 9)   —— 1837·93

A neat double star, the duplicity of which was discovered by Dr. Hook, in 1664. A 4½, bright white; and B 5, pale grey. This fine object is formed by Piazzi's Nos. 196 and 197, Hora 1, and is placed at the lower bend of the Ram's horn, where it precedes β and *a*; it is followed in the *nf*, but nearly on the parallel, and about 3′·5 distant, by the 9th-magnitude star which that astronomer describes. A and B are

9 Ḥ. III., and these are the results of the earliest recognition of them as a pair; with reductions from the Palermo Catalogue:

Ḥ. Pos. 356° 05′   Dist. 10″·17   Ep. 1779·68
P.   360° 00′   8″·90   1800

From observations made by Ḥ. twenty-three years afterwards, he concluded the orbital angle to have increased; but the subsequent measures of H. S. Σ. D. and myself, indicate little or no change. It is certainly a beautiful pair, in a powerful instrument. "What would Cassini say," demands Ḥ., "if he were to view the first star of Aries, which appeared to him as split in two, through a telescope that will show η Corona Borealis, and h Draconis to be double stars?"

The larger member of this compound has these proper motions in space:

P.... Æ + 0″·15   Dec. − 0″·12
B....   + 0″·10   + 0″·09

Dr. Hook mentions that the telescope shows some stars, which appear single, to consist of two or more, so close, that to the naked eye both the images falling upon one single filament of the *tunica retinæ*, make but one impression upon the brain. "Of this kind," he continues, "the most remarkable is the star in the left horn of Aries, which, whilst I was observing the comet which appeared in the year 1664, and followed till he passed by this star, I took notice that it consisted of two small stars very near together: a like instance to which I have not else met with in all the heaven." There are some thousands now detected!

γ Arietis has been called the first star in Aries, because it was once the nearest to the equinoctial point: it was named Mesartim, owing to an erroneous deduction by Bayer from the word *Sartai*, a corruption of *Al Sharataïn*, which is the next star.

LXXIII.   β ARIETIS.

Æ   1ʰ 45ᵐ 49ˢ   Prec. +   3ˢ·28
Dec. N 20° 01′·4   ——— N 17″·94

Position 198°·8 (w 1)   Difference Æ = 7ˢ·4 (w 1)   Epoch 1835·99

A bright star with a distant companion, in a barren field near the tip of the Ram's horn, being the middle one of the three stars known as the Ram's head. A 3, pearly white; B 11, dusky, a still smaller star in the *sf* quadrant. This object was named *Sheratan* or *Sharatain*, the dual of *sharat*, a sign, signifying γ and β, the two bright stars in the head of the Ram; with an interval between them and *a*, says Kazwíní, of two *kaus*, by eye-measurement: said *kaus* being used as synonymous with the astronomical ell of 2°. An imaginary line from the Pleiades to Markab, passes between them in the mid-distance of that line. They formed the first Lunar Mansion, if Kazwíní is preferred to Dr. Sedillot. A proper motion is given to A of the following value:

P.... Æ + 0″·14   Dec. − 0″·23
B....   + 0″·12   − 0″·11

LXXIV. 56 ANDROMEDÆ.

Æ 1ʰ 46ᵐ 40ˢ   Prec. + 3ˢ·51
Dec. N 37° 27′·9   —— N 17″·91

Position 302°·4 (w 9)   Distance 176″·2 (w 7)   Epoch 1834·13

A pair of stars between the Triangle and the Lady's right knee, both of the 6th magnitude, and both yellow, being Piazzi's admirably observed Nos. 203 and 204, Hora 1. These stars are suspected of physical connection, principally on the ground that their identity of movement in space implies their union in some vast system. Their proper motions have lately been rigidly inquired into, and the following results registered:

$$B. \text{ No. } 203 \; Æ \; + \; 0''·06 \qquad Dec. \; - \; 0''·01$$
$$204 \qquad + \; 0''·24 \qquad\qquad + \; 0''·04$$
$$A. \text{ No. } 203 \; Æ \; + \; 0''·03 \qquad Dec. \; - \; 0''·02$$
$$204 \qquad + \; 0''·21 \qquad\qquad + \; 0''·01$$

This object is readily identified, by carrying a line from π Andromedæ through β, and extending it about 10° beyond the latter; and it is also nearly in mid-distance between Al'mak and Mothallath.

LXXV. 209 P. I. PISCIUM.

Æ 1ʰ 47ᵐ 38ˢ   Prec. + 3ˢ·08
Dec. N 1° 03′·2   —— N 17″·87

Position 62°·9 (w 6)   Distance 1″·5 (w 3)   Epoch 1833·83

A close double star, on the *sf* extreme of the Fishes' *kheït*, or ribbon; and it lies on a line shot from ζ Piscium to α Arietis, at about a third of the distance. A 7, silvery white; and B 7½, white. Piazzi notes No. 209 as being 84 Ḥl. v., but he evidently alludes to the distant companion in the *np* quadrant. This very fine object, resembling η Coronæ, is Σ.'s No. 186, and registered " vicinæ" in the Catalogue of 1827; it has since been thus measured at Dorpat:

Pos. 64°·72   Dist. 1″·232   Ep. 1831·12

LXXVI. λ ARIETIS.

Æ 1ʰ 49ᵐ 02ˢ   Prec. + 3ˢ·32
Dec. N 22° 48′·8   —— N 17″·82

Position 45°·6 (w 7)   Distance 36″·9 (w 7)   Epoch 1830·96

A fine double star, on the root of the Ram's horn; pointed at by a line through γ and β, and is the apex of an oblique triangle of which

a and β form the base. A $5\frac{1}{2}$, yellowish white; B 8, blue. This optical object was first registered by Ḥ. thus:

Pos. 48° 00′   Dist. 36″·61   Ep. 1779·83

which, compared with the subsequent measures, shews no greater change, than the nature of observation, and the amount of proper motions chargeable on A, would lead us to expect; they are to this amount:

$$P.... \; \mathbb{R} - 0''·06 \quad \text{Dec.} - 0''·03$$
$$Br... \quad - 0''·11 \quad\quad\quad - 0''·05$$
$$B.... \quad - 0''·08 \quad\quad\quad - 0''·00$$

LXXVII. 112 Ḥ. I. ARIETIS.

$$\mathbb{R} \; 1^h \; 50^m \; 34^s \quad \text{Prec.} + 3^s·27$$
$$\text{Dec. N } 18° \; 13′·6 \quad \text{——— N } 17'''·76$$

MEAN EPOCH OF THE OBSERVATION . . . . 1836·79

A round nebula, closely following γ on the neck of the Ram, where it may be fished for on a line carried from a Trianguli $4\frac{1}{2}°$ below λ Arietis. It is large and pale, but brightens in the centre. Ḥ. classed it in November, 1785, and considered it "not easily resolvable;" but still H., No. 181, distinguished it through a thick cloud. It lies among some small stars, the most conspicuous of which form a curve across the south part of the field. Its place was carefully differentiated from γ Arietis.

LXXVIII. 222 P. I. ARIETIS.

$$\mathbb{R} \; 1^h \; 50^m \; 43^s \quad \text{Prec.} + 3^s·30$$
$$\text{Dec. N } 20° \; 16′·7 \quad \text{——— N } 17'''·75$$

Position AB 53°·0 (w 1)	Distance 2″·5 (w 1)	
——— AC 165°·0 (w 2)	——— 40″·0 (w 1)	Epoch 1834·99
——— AD 359°·2 (w 2)	——— 165″·0 (w 2)	

A quadruple star, in mid-distance between a and γ, under the Ram's ear, lying nearly at right angles with the vertical. A 6, topaz yellow; B 15, deep blue; C 10, lilac; and D 9, pale blue. This is an exquisite object, of which the three southern members form No. 196 of Σ.'s grand Catalogue, under these measures:

Pos. AB 53°·53   Dist. 2″·370 } Ep. 1832·42
    AC 167°·38   39″·460

This group is most difficult to observe, and the results are rather estimations than measures, particularly those of A B. Still, under every disadvantage, it forms an admirable test to try the light and distinctness of a telescope.

LXXIX. 227 P. I. CETI.

$\mathcal{R}$ 1ʰ 51ᵐ 49ˢ   Prec. + 3ˢ·14

Dec. N 6° 08′·4   —— N 17‴·70

Position BC 113°·9 (w 5)   Distance 43″·6 (w 3)   Epoch 1834·87

A star pointing to a distant pair, close under the Ram's fore-foot; half-way on a line projected from ζ Ceti to α Arietis. A 7½, yellowish; B 8, light blue; C 9½, violet. This is a coarse object, of which A is assumed as a pointer to B, a star which bears from it 183°·7, with Δ $\mathcal{R}$ = 4ˢ·1; and is the apex of a scalene triangle, formed with another star there, the shortest side of which is here measured as B C.

LXXX. ε TRIANGULI.

$\mathcal{R}$ 1ʰ 53ᵐ 38ˢ   Prec. + 3ˢ·47

Dec. N 32° 30′·5   —— N 17‴·63

Position 110°·0 (w 1)   Distance 5″·0 (w 1)   Epoch 1835·71

A most delicate double star, on the np limb of Deltōton; and a line projected from γ Andromedæ to pass between α and λ Arietis, will pass over it in about mid-distance. A 5½, bright yellow; B 15, dusky. This object was marked "difficilis" in the Dorpat Catalogue of 1827. It lies diagonally between two small stars, one of which, 10th magnitude, precedes it by 9ˢ·0; and the other, a deep orange-coloured 8th magnitude, follows by 14ˢ, with a little neighbour 2ˢ farther off. Proper motions in space, but under conflicting values, have been assigned to ε; thus:

	$\mathcal{R}$	Dec.
P....	− 0″·08	− 0″·10
Br...	− 0″·10	+ 0″·06
B....	+ 0″·14	0‴·00
T....	+ 0″·24	+ 0″·06

LXXXI. α PISCIUM.

$\mathcal{R}$ 1ʰ 53ᵐ 46ˢ   Prec. + 3ˢ·09

Dec. N 1° 59′·3   —— N 17‴·63

Position 334°·7 (w 8)   Distance 3″·6 (w 9)   Epoch 1834·92

———— 333°·4 (w 9)   ———— 3″·8 (w 9)   ———— 1838·87

A close double star, at the sf extreme of the ribbon of the Fishes, where it is readily identified, by carrying a line from β Ceti over θ, and rather better than as far again to the north-west,—it is also in mid-distance between v Ceti and a Arietis. A 5, pale green; B 6, blue. This splendid object is 12 ⅓. II., and was thus registered:

Pos. 337° 23′ Dist. 5″·12 Ep. 1779·80

As that astronomer obtained an angle of 330° in position, twenty-three years afterwards, he was led to suppose a retrograde motion: all the subsequent observations, however, of H. S. D. Σ. and myself, prove the fixity of these stars. But there is an appreciable movement in space,—perhaps common to both,—of which the values for A have been thus estimated:

$$\begin{array}{llll} P.... & \mathbb{R} - 0''·06 & \text{Dec.} & - 0''·09 \\ B.... & + 0''·14 & & + 0''·01 \\ T.... & + 0''·02 & & + 0''·01 \end{array}$$

Pisces is one of the old forty-eight asterisms, and the twelfth or last sign of the old zodiac. The constellation consists of two Fishes linked by a ribbon, or string, attached to their tails, and divided by Hevelius into *linum boreum* and *linum austrinum;* they occupy a large space in the firmament, the one being under the wing of Pegasus, and the other under the right arm of Andromeda, in the position described by Manilius:

Dissimile est illis iter, in contraria versis.

The conspicuous rectangular figure in Pegasus is a guide to the position of these two Fishes; the line of a Andromedæ and γ Pegasi being parallel to the body of one Fish, and that of γ and a Pegasi to the body of the other. The equinoctial colure now passes through this " watery trigon," which was not a favourite sign with astrologers; indeed, Mr. John Gadbury,—albeit it was notorious that, under domi-nance of " ye Fysches," it was good to " wed a wyfe, and to trete frendys," —says, " I know Pisces to be a dull, treacherous, phlegmatic sign." Yet this visionary φιλομάθηματικος was consulted on mundane affairs by the Parliament of England! The star a—the Syndesmos of the Greeks —has been called *Okda,* from the ʾ*Okdah al Khaïʿaïn,* or " knot of the two threads," of the Arabian *savans;* and the component members of Pisces have been thus stated:

Ptolemy	. . . 38 stars	Hevelius	. . . . 39 stars
Tycho Brahé	. 36	Flamsteed	. . . 113
Bayer	 37	Bode	 257

Eratosthenes considered that this asterism symbolized the Syrian Derceto, and it has therefore been represented with a woman's head on a huge fish's body. The Scholiast on Aratus says, that the Northern Fish was figured with a swallow's head, and called Χελιδόνιας; while the two collectively were called *Gemini Pisces,* to distinguish them from the Southern Fish. Hence, also, the *Al-semakataïn* of the Arabians; but the term *Echiguen* in the Arabo-latin Almagest has defied com-mentators, unless Ideler's notion of its being a corruption of Ἰχθύες be accepted. The same astronomers attributed great influence over man to such planets as happened to be in this sign: and we learn from the translation of the *Ysagogicus* of Alchabitus, 1485, that in such case, Saturn had full dominion over *humeros, et brachia, et collū;* Venus presided over *collū et dorsum,* and Jupiter over *cor et caput.*

LXXXII. γ ANDROMEDÆ.

Æ. 1ʰ 54ᵐ 06ˢ   Prec. + 3ˢ·63
Dec. N 41° 33′·6   —— N 17″·61

Position A B 62°·3 (w 5)   Distance 10″·3 (к 7)   Epoch 1830·91
——— A B 62°·9 (w 9)   ——— 10″·6 (к 9)   ——— 1837·80
——— A B 61°·6 (w 9)   ——— 11″·0 (к 9)⎫
——— B C 120°·0 (w 1)   ———   0″·5 (к 1)⎭ ——— 1843·33

A splendid double star, on the right ankle of Andromeda. A 3½, orange colour; B 5½, emerald green; and of these colours I feel pretty positive, although the high authority of Ḥ. and Σ. has pronounced them to be yellow and blue. This beautiful object was seen to be double by C. Mayer. in 1778: it is No. 5 Ịḥ. iii., and was thus measured when first classed:

Pos. 70° 23′   Dist. 9″·25   Ep. 1779·65

Between this period, and my attack, the following results were recorded:

H. and S. Pos. 64° 46′   Dist. 10″·91   Ep. 1821·91
Σ.   62° 26′   10″·33   1830·02

whence there would appear to have been a slight motion *npsf,* or retrograde. But even without the excellent comparison of recent observers, my own results are sufficient to establish the relative fixity of these stars; although this is against the opinion, that high coloured stars possess the greatest velocity. A is remarkable as forming, with a star in the head, and another in the belt, an almost direct line, across the parallel, from east to west; it is called Al'mak, from the Arabic *Al-'Anák-at-ard,* the "badger," or *caracal* of Buffon. Scaliger's clever notion, that Al'mak, as the star at the foot, was derived from *Al-mauk* —cothurnus—is now given up.

This star is readily found, by drawing an imaginary line through the three stars of Orion's belt, and thence over the Pleiades; or, a ray from Thuban to the Pole-star, at about two-thirds the distance beyond, leads to it. Should Orion's neighbourhood be obscured, an occult line carried through β and α, the two brightest stars of Cassiopea, and extended to rather more than five times their distance from each other, will strike upon Al'mak, after passing the star upon Andromeda's left ankle. It has a slight proper motion assigned, which my meridian observations, albeit the determination of so delicate an element is beyond their object, do not confirm; this is the registered amount:

P.... Æ + 0″·26   Dec. — 0″·11
B....   + 0″·06   — 0″·04
A....   + 0″·03   — 0″·05

Since the above was written, Mr. Baily put into my hand a letter which he had received from M. Struve, in October, 1842, announcing the unlooked for tidings that he had detected γ Andromedæ to be triple, and that the companion is composed of two stars of equal size, separated by an interval of less than 0″·5. I lost no time in notifying this to my friend Mr. Dawes, who, as well as myself, had so repeatedly gazed at this, merely as a double star. On the 1st of November, he informed me that he charged Mr. Bishop's refractor with an excellent single lens magnifying 520 times, and when the star was best defined, became satisfied of an elongation *sf* and *np* in the companion, making it look like a dumpyish egg. By the measures he obtained, the angle of position was 125° 48′, and the distance of the centres was estimated at 0″·4. I also received a letter from the Rev. J. Challis, under date of December 9, 1842, after his attacking it with the Northumberland equatoreal, at my request, of which the following extract is most interesting. "I looked at γ Andromedæ the first opportunity after receiving your note, and was surprised to find that I could easily recognise the small star as being double. I cannot say that I saw the components separated, but there was a decided elongation, and several measures which I took of position agreed well with each other. The distance is certainly not more than 0″·5. My impression was, that the components are not equal."

When I repaired to Hartwell, in February, 1843, I was baffled in my attempts to examine this object in the evening twilight. But on returning thither in the spring I was enabled to catch some fine early views of it. On the 1st of May, the morning atmosphere was perfectly diaphonous, and I teased γ under various powers from 118 to 600, until I fairly saw that the *comes* was not round, but elongated, in a direction *np* and *sf* to the amount above estimated. It was, however, so slightly oval, that, but for M. Struve's unexpected announcement, I must assuredly have overlooked it.

LXXXIII. 10 ARIETIS.

Æ 1ʰ 54ᵐ 35ˢ   Prec. + 3ˢ·37
Dec. N 25° 09′·7   —— N 17″·59

Position 26°·8 (to 6)   Distance 2″·2 (to 4)   Epoch 1838·66

A close double star, over the Ram's head, nearly in mid-distance between α Trianguli and α Arietis, and it has several followers exactly on the parallel. A 6½, yellow; B 8½, pale grey. This is one of Σ.'s discoveries, No. 208 of the Dorpat Catalogue; and so beautiful an object, that H. calls it a miniature of ε Bootis. It has been well looked to by astrometers, and the several results for comparison are:

	Pos.	Dist.	Ep.
H.	25° 28′	2″·13	1830·79
Σ.	25° 17′	1″·98	1833·05
D.	27° 50′	2″· ±	1833·36

Though in our present knowledge of these stars, there appears to be a relative fixity, proper motions in space are attributed to the large one, of which the several amounts are thus given:

	Æ	Dec.
P....	0″·00	+ 0″·20
B ...	+ 0″·20	− 0″·03
T....	+ 0″·02	− 0″·22

E 2

LXXXIV.   61 CETI.

Æ   1ʰ 55ᵐ 37ˢ   Prec. +   3ˢ·06
Dec. S 1° 06′·5   —— N 17‴·55

Position A B 188°·8 (ω 2)   Distance 39″·0 (ω 2)
——— C D 249°·8 (ω 8)   ——— 4″·6 (ω 6) } Epoch 1834·88

A pair of double stars, at the back of the Whale's head, and 3° to the south, a little following, of α Piscium; a line from τ Ceti through ζ, carried nearly double the distance, hits 61. A 7, pearly white; B 11, greenish; C 7, white; and D 8½, blue. A and B form 102 H. v., which was thus first registered:

Pos. 193° 39′   Dist. 37″·89   Ep. 1782·78

which would imply a slight change in the orbital curve, but that the difficulty of the measures must be taken into consideration.

Near the following parallel, at a distance of 4ᵐ 57ˢ, on the angular line = 102° 33′, is the beautiful double star CD, which proved to be Σ.'s 218, of which the measures previous to mine were:

S.   Pos. 250° 29′   Dist. 4″·96   Ep. 1824·92
Σ.   250° 00′   Dist. 4″·78   1832·36

LXXXV.   α ARIETIS.

Æ   1ʰ 58ᵐ 10ˢ   Prec. +   3ˢ·34
Dec. N 22° 42′·2   —— N 17‴·44

Position 107°·3 (ω 1)   Difference of Æ 19ˢ·5 (ω 1)   Epoch 1835·10

A standard Greenwich star, on the Ram's *os frontis*. A 3, yellow; B 11, purple. The large star is followed by three small ones, forming a line across the parallel, of which the middle individual is B. The primary has a sensible proper motion in space, and the following amount has been registered:

P.... Æ + 0″·20   Dec. − 0″·20
B....   + 0″·24   − 0″·12
A....   + 0″·22   − 0″·14

Though this constellation only possesses stars of more interest than magnitude, it opened the astronomical year, 2000 years ago, as *Princeps signorum* and *Ductor exercitûs zodaici;* and bore the office for a similar period. The charge is now resigned to Pisces, for Aries has passed 30° to the eastward of the point where the equinoctial is intersected by the ecliptic, or *Via Solis.* This is owing to the precession of the equinoxes, which apparent motion of the zodiac arises from a slow vibration of the earth's axis, occasioned by planetary attraction.

Thomas Hood, the Fellow of Trinity College, Cambridge, who published directions for using the celestial globe, in 1590,—and who considered the Triangle as only placed in the heavens in order that the head of Aries might be better known—thus speaks of its first star:

" *Scholar.* Why is that same starre placed so farre from the head of Aries? me thinketh it were good to keepe the figure and the signe together.

" *Master.* That cannot be; for the starres moving continually from the west towards the east, cannot keepe one and the same distance from the vernall equinoctiall point, but are carried on forward continually, so that the starres which are now in the signe *Aries* will be hereafter in *Taurus,* and from thence will come into *Gemini,*" &c.

Aries indicates the golden fleece of the adventurous crew of the Argo, albeit a stir has been made to identify him as Abraham's ram; and he is recognisable by three stars crossing the head obliquely. Hevelius refers those who wish to be familiar with the different appellations of signs and stars, to the works of Blævius, Alsted, Ricciolus, Goldemayer, Bartschius, and Cellarius; and says he selected the name mostly used—" non attento, quòd Aries nonnunquam etiam Vervex, Chrysomalus, Jupiter Ammon, Krios, Aribib, Elhemal, et Elhamel, &c. vocetur." The Ram has long been figured in his present attitude, for Manilius accurately describes him as advancing stern foremost, with his legs bent under.

> First Aries, glorious in his golden wool,
> Looks back, and wonders at the mighty bull.

And it is so represented upon the Farnese globe: yet the erudite Bishop of Avranches insists, that the ancients made him running and looking towards the west, or before him. The star under discussion was called *Hamal,* by the Arabs, *i.e.* a sheep. It was also named *Al nátih,* the butt-er, by those who considered the first Lunar Mansion as formed by α and β Arietis. A line made to pass between the Pleiades and Hyades, from Alpherat, will pick up Hamal in the mid-distance, and pass through the Ram's flank; it may also be identified by the brackish rhymes:

> From Ras Mothallath shoot a ray,　　in a south-following line,
> And where expand huge Cetus' jaws,　to *gamma* let it join;
> One-fourth the distance thus express'd　from Triangle to Whale,
> (If thus can such odd fish be termed,)　will strike upon Hamal.

The stars of Aries have been thus enumerated:

Ptolemy	. . . . 18 stars	Hevelius	. . . . 27 stars
Tycho Brahé	. . 21	Flamsteed	. . . 66
Kepler	 23	Bode	 148

An attempt has been made to bring forward a numismatic evidence of the Roman regard for this sign, by citing a coin of Domitian bearing the Ram as *Princeps Juventutis.* But I am not aware that such a symbol was ever struck; though the aureus with that legend, and the Amalthean *goat* in a garland, is common enough. Astrologers distinguished it as an equinoctial, cardinal, and diurnal sign—the day-house of Mars—of the fiery triplicity—and what not.

LXXXVI.   14 ARIETIS.

 R   2ʰ 00ᵐ 19ˢ   Prec. + 3ˢ·38
Dec. N 25° 10′·8   —— N 17″·34

Position AB 43°·5 (w 4)   Distance 82″·6 (w 2) ⎫
—————— AC 278°·6 (w 2)   —————— 106″·5 (w 2) ⎭   Epoch 1833·92

A wide triple star, between the head of Aries and the base of Triangulum; being the centre of the group mentioned as headed by 10 Arietis, above described. A 5½, white; B 10½, blue; and C 9, lilac. A is the apex of a scalene triangle, with B in the *nf* and C in the *np* quadrant, in a field otherwise barren; but immediately followed on the parallel of A, by 16 Arietis.   A and C of these three stars are registered as 69 ♓. vi., thus:

Pos. 281° 15′   Dist. 89″·47   Ep. 1781·99

Sir James South next measured it, and found

Pos. 277° 58′   Dist. 105″·258   Ep. 1823·97

which great discrepancy, H. is inclined to attribute to the construction of his father's micrometer at that early period; the effect of which is to throw great uncertainty on the earlier distances of all the wide stars: "fortunately," adds he, "these are the least replete with interest." This object was evidently not scrutinized with rigour, since B was overlooked by former observers. Piazzi, in referring to C, says, "In eodem paral. 8″ temp. circiter, præcedit alia 9ᵅ mag."

LXXXVII.   59 ANDROMEDÆ.

R   2ʰ 1ᵐ 16ˢ   Prec. + 3ˢ·60
Dec. N 38° 16′·9   —— N 17″·30

Position 34°·7 (w 9)   Distance 16″·3 (w 9)   Epoch 1835·11

A neat double star, between the right foot of Andromeda and the Triangle; which may be readily identified by running a line from *a*, close under *β*, till it arrives nearly at right angles with *γ*. A 6, bluish white; B 7½, pale violet. Piazzi saw and recorded this as double, remarking, *præcedens observata*; and it is No. 129 ♓. iv., whose measures were:

Pos. 34° 51′   Dist. 15″·25   Ep. 1783·48

whence all the recent observations, on an average interval of fifty-two years, establish its fixity. Yet it is curious, that though all the results tread closely on each other's heels, and the objects present no peculiar difficulty, H. and S. found a difference of 3° 23′ in their measures, which they considered the maximum error to which an angle can be held liable.

LXXXVIII.   55 CASSIOPEÆ.

$\mathrm{\mathcal{R}}$ 2$^\mathrm{h}$ 2$^\mathrm{m}$ 00$^\mathrm{s}$ Prec. +   4$^\mathrm{s}$·57
Dec. N 65° 46′·2   —— N 17″·27

Position AB   24°·0 (*w* 1)   Difference $\mathcal{R} =$ 1$^\mathrm{s}$·8 (*w* 1) ⎫ Epoch 1834·63
—— AC 120°·5 (*w* 1)   ————   23$^\mathrm{c}$·8 (*w* 1) ⎭

A star with two distant companions, in the Lady's right foot; or, including an orange-coloured 8th-magnitude star near the north vertical, a quadruple object. A 6, yellowish; B 11, and C 9, both greyish. It can be identified by passing an imaginary line from *a* through the two *v*'s, and carrying it about three times that distance to the north-east. This was merely looked at because it is entered at 34 ⩍. ɪ., where it is also named as the *ι* Ptolemæi of Bayer's Map; but the latter star being ranked as one of the 4th magnitude by Ptolemy, it was soon seen that there existed a mistake in the identity; and that the elegant triple star discovered by ⩍. must be 72 P. ɪɪ.

To the northward of this object, in the open space under the *scabellum*, a new star suddenly burst forth in full splendour, in November, 1572; and the locality ought to be diligently watched. This discovery appears to have been made by Schuler at Wittemburg, on the 6th of August; but the star was seen by Hainzel at Augsburg on the 7th, and by Cornelius Gemma on the 9th of November. Correspondence was, however, in those days, pretty heavily clogged, so that each was probably a discovery independent of the other. Thus it happened to Tycho Brahé, who was astonished at the apparition, when returning to supper from his crucibles, on the 11th of November: and as it was the only change which had been known to take place in the appearance of the heavens since the revival of learning in Europe, it excited the utmost attention. This star fortunately made its apparition when astronomy was sufficiently cultivated for it to be watched with precision; and being in the circumpolar region, it was constantly in view. "By a strange instinct," says one of its historians, "by a strange instinct of Providence were those admirable instruments made and erected by Tycho, a little before the appearing of this starre, as if either the starre had stayed for his tooles, or he had foreseene the birth of that starre;" but still stranger was the instinct which made the same Tycho ashamed of publishing his observations on it, considering it "a disgrace for a nobleman either to study such subjects, or to communicate them to the public."

The stranger twinkled strongly, so that its aspect was precisely that of a star, having none of the distinctive marks of a comet: it was at first white, then yellow, afterwards reddish, and finally bluish, which led the great La Place to the strange and unsatisfactory analogy of a body under the action of fire. It grew rapidly, until it surpassed Sirius in brilliancy, being brighter than Jupiter when in perigee; and as it was even visible in the day-time, Cornelius Gemma concluded its lustre to be scarcely less than that of Venus. The maximum magnitude was of short duration, and it diminished by degrees till March, 1574, when it entirely vanished from view, and has not been since seen. During its apparition it continued to hold the same position with respect to the other stars of the constellation; and as Tycho Brahé was unable to ascertain that it had any sensible parallax, he justly concluded that its place was beyond the planetary bounds. He was the forerunner of the theory of the transformation of nebulæ into stars, in supposing that it was produced by a condensation of the celestial matter collected in the Milky Way; and he inserted it in the Catalogue appended to the *Rudolphine Tables*, as of the 6th magnitude, and No. 46 of the asterism,

Lon. ♉ 6° 54'   Lat. 53° 45'   Nova, anni 1572

Tycho Brahé, Kepler, Beza, Maurolycus, and other exact spectators, wrote dissertations upon it; but to all the reasonings as to why it had not been seen before, Reisacher's answer is perhaps the best: "God knows." The astute Dr. Dee started the idea that it moved alternately towards, and from the earth, in a direct line. His brethren tried this phenomenon by their tools, and found that it came in with the *fiery trigon*, or that in which Saturn, Jupiter, and Mars, are in the three fiery signs—Aries, Leo, and Sagittarius; an event which occurs only every 800 years. Tycho had some heterodox notions as to its origin; and Ricciolus, no friend to astrology, admitted that it was saluted by all the planets *before it was extinguished*. This remark, added to that of La Place, made Mrs. Somerville say, "It is impossible to imagine anything more tremendous than a conflagration that could be visible at such a distance;" but in a conversation upon the topic with her, I found this intelligent lady not at all inclined to grant, that so vast a combustion was within the precincts of probability. Keill conjectured it to have a period of 150 years, but as it did not return to the time, the notion was started that it might have different degrees of lustre at different times. Here, however, the mind must pause; and in our ignorance, no reasoning upon such a wonderful body can be deemed wild, except that of annihilation.

As there are vague impressions that similar stars appeared in 945 and 1264, Sir J. Herschel thinks it possible another such appearance may take place in 1872, or thereabouts. Telescopes will then be applied, owing to the want of which, in 1572, it could not be ascertained whether the stranger had any sensible diameter.

LXXXIX.   TRIANGULI.

Æ   2ʰ 3ᵐ 06ˢ   Prec. +   3ˢ·45

Dec. N 29° 33'·0   —— N 17"·22

Position 78°·1 (*w* 5)	Distance 3"·6 (*w* 3)	Epoch 1830·91
—— 77°·9 (*w* 4)	—— 3"·3 (*w* 3)	—— 1834·17
—— 78°·8 (*w* 9)	—— 3"·5 (*w* 6)	—— 1838·99

A fine close double star, under the base of the Triangle, and 4½° south of β, on a line leading through β to γ Andromedæ. A 5½, topaz yellow; B 7, green. This exquisite object is 34 Ḧ. II., by whom it was likened to a Herculis, "but smaller, and not so bright." From the existence of an angular difference of 7° 39′ between the position of Ḧ. and that of H. and S., an orbital change was suspected; but all the recent measures are against it:

Ḧ.	Pos. 85° 37′	Dist. 3″·5 ±	Ep. 1781·77
H. and S.	77° 58′	3″·88	1821·94
Σ.	77° 50′	3″·598	1830·97
D.	78° 58′	3″·63	1832·94

XC. 66 CETI.

ℛ 2ʰ 04ᵐ 37ˢ   Prec. + 3ˢ·03
Dec. S 3° 08′·6   —— N 17‴·15

Position 229°·6 (w 9)   Distance 15″·4 (w 9)   Epoch 1837·89

A neat double star, on the neck of Cetus, nearly in mid-distance between γ and θ. A 7, pale yellow; B 8½, sapphire blue. This interesting object is formed by Piazzi's Nos. 17 and 18. It was registered 25 Ḧ. IV.; but as there is a probability of direct orbital motion, or else a greater proper motion of the principal star, it is to be regretted that Ḧ. left no decided measures in 1780, those given by H. from the MS. being estimations. When H. and S. attacked it, it was found:

Pos. 226° 05′   Dist. 16″·173   Ep. 1822·90

Σ. measured it with great care, and obtained:

Pos. 228° 55′   Dist. 15″·540   Ep. 1832·67

The proper motion of A through space, has been thus valued:

P....	ℛ + 0″·81	Dec.	− 0″·05
B....	+ 0″·54		− 0″·04
A....	+ 0″·39		− 0″·08

B. must also be physically connected, and partake of this occult movement, for a reduction of Piazzi's mean apparent places gives an ∠ = 231°·5 and distance = 19″·3 for 1800, which, as he says both these stars are in motion, would by this time have altered the angle to 233°, and the distance to 18″·2.

XCI. χ PERSEI.

ℛ 2ʰ 06ᵐ 52ˢ   Prec. + 4ˢ·14
Dec. N 56° 46′·3   —— N 17‴·14

Position AB 354°·3 (w 2)   Distance 65″·0 (w 1) ⎫
    —— AC 43°·2 (w 5)   —— 122″·0 (w 3) ⎬ Epoch 1835·82
  ⎭

A coarse triple star, in the weapon hand of Perseus. A 6½, yellow; B 12, bluish; C 9½, greenish. An imaginary line projected from δ

Cassiopeæ to a Persei, will pass just below χ, at one-third of the distance. This is a multiple object, among rich fields, where the intermixture of greater and less individuals renders the vicinity very favourable for testing the light and definition of a telescope. It was classed by Ḥ. No. 19, vi., but no results given; A and C were thus measured by Sir James South, but he did not perceive B:

Pos. 43° 30′   Dist. 124″·533   Ep. 1824·99

XCII.   33 Ḥ. VI. PERSEI.

Æ   2ʰ 07ᵐ 58ˢ   Prec. +   4ˢ·13

Dec. N 56° 24′·4   —— N 17″·00

Position 321°·9 (w 2)   Distance 9″·5 (w 1)   Epoch 1836·78

A delicate double star, in the glorious cluster of Perseus's weapon hand. A 8, white; B 10, pale grey. This brilliant mass of stars, from 7th to 15th magnitudes, fills the whole field of view, and emits a peculiarly

splendid light. In the centre is a coronet, or rather ellipse, of small stars, above an 8th-magnitude one, which, with its *np comes*, is here measured. The 7th-magnitude star which follows, is handsome from the blackness of the space immediately around it. A line from the lucida, or Algenib, carried to δ Cassiopeæ, passes over this brilliant assemblage, at two-thirds the distance. Sir William Herschel considered it a protuberant part of the Milky Way, in which it is situated; and analogy indicates that it is comparatively near.

This is followed by another gorgeous group of stars, from the 7th to the 15th magnitudes, at about 3ᵐ, and nearly on the parallel. It is 34 Ḥ. vi. The components gather most towards the centre, but there is little disposition to form; the sprinkle, however, is in a direction parallel to the equator. One of the central individuals is of a fine ruby colour, and a 7th-magnitude in the *nf* is of a pale garnet tint; with two sparkling but minute triplets south of it. These two clusters are quite distinct, though the outliers of each may be brought into the same field under rather high powers; and, on the best nights, the groups and light are truly admirable, affording together one of the most brilliant telescopic objects in the heavens. It is impossible to contemplate them and not infer, that there are other laws of aggregation than those which obtain among the more scattered and insulated stars.

XCIII. 38 P. II. TRIANGULI.

Æ. 2ʰ 08ᵐ 09ˢ   Prec. + 3ˢ·45

Dec. N 28° 00′·2   —— N 16″·99

Position 209°·1 (w 9)   Distance 14″·1 (w 9)   Epoch 1834·92

A fine double star, between the Ram and the Triangle, where it will be found by shooting a line from *a* Andromedæ through *a* Trianguli, and extending it nearly 5° beyond: it is also near the mid-distance between γ Andromedæ and γ Ceti. A 8½, and B 9, both silvery white. This object, formed by Piazzi's Nos. 38 and 39, was classed 40 Ḥ. iv., in 1781; no measures of position were then taken, but the distance was found to be 17″·31, "pretty accurate." H. and S. found it thus:

Pos. 208° 56′   Dist. 14″·35   Ep. 1821·96

which is too similar to the result given by my measures thirteen years afterwards, to confirm the retrograde motion inferred from a reduction of Piazzi's mean places for 1800, which give 211°·5 for the angle, and 16″·8 as the distance.

XCIV. ο CETI.

Æ. 2ʰ 11ᵐ 16ˢ   Prec. + 3ˢ·02

Dec. S 3° 42′·3   —— N 16″·85

Position 88°·9 (w 8)   Distance 116″·0 (w 4)   Epoch 1831·03

A flushed yellow variable star, with a distant companion. A, recorded in extremes varying from 2 to 7, and from thence to invisible; B 10, pale lilac. This very extraordinary object is in the middle of the Whale's neck, and well known as *Mira*; the epithet "wonderful" being given on account of its remarkable variation in brilliance, first noticed, in 1596, by David Fabricius. Forty-two years afterwards it was observed by Phocylides Holward, and treated of in his πανσέληνος as a new star, in a right line with *lucida mandibulæ Ceti*. Bailly thus relates the circumstance: "En 1638, Holward revit l'étoile de la Baleine, et à peu-près au même lieu où elle avait été apperçue par Fabricius. Il ignoroit sa première apparition, il la perdit lorsqu'elle se cacha dans les rayons du soleil; et lorsque cet astre, en s'avançant dans l'écliptique, eût rendu visibles les étoiles de la Baleine, Holward ne retrouva plus son étoile, quoiqu'il la cherchât avec soin; mais il dût être étonné de la revoir tout-à-coup le 7 Novembre, 1639. On la vit les années 1644, 45, 46, 47, 48, avec des alternatives de disparition et de renaissance, telles qu'on ne la vit jamais une année de suite. Hévélius la suivit constamment en 1648 et en 1660." Since this time it has been found pretty regular in its periods, except in the four years 1672 to 1676, during which time Hevelius could not perceive it, though it was a particular object of his attention. Bullialdus determined its periodical time, from bright through all its gradations to bright again, to be 333 days, and Cassini made the same period to be 334. Halley mentions that it was found to appear and disappear periodically; and that its period is "precisely enough, seven revolutions in six years, though it returns not always with the same lustre. Nor is it," he adds, " ever totally extinguished, but may at all times be seen with a six-foot tube." This was singular in its kind, till that in Collo Cygni was discovered; and the attention it excited among astronomers is detailed in the *Historiola Miræ Stellæ.* ♓., however, has decidedly shown that it actually does become invisible, and he estimated that its period is 331 days, 10 hours, and 19 minutes; and a comparison of the observations of Pigott, Goodricke, Wurm, Westphal, Argelander, and Bianchi, give its recent periods of maximum brightness between September 30 and October 7.

From some discrepancies of observation between A and B, ♓. surmised a rapid change in the distance, and that the position had changed from the southern to the northern side of the parallel; but from my measures and comparisons, I am inclined to think there has been little or no movement beyond what may be ascribed to the proper motions of A in space, the values of which have been thus sifted:

$$P.... \; \mathbb{R} - 0''\cdot18 \quad \text{Dec.} - 0''\cdot15$$
$$B.... \quad\; - 0''\cdot05 \quad\quad\quad - 0''\cdot23$$

A word upon this. Piazzi's mean places of Nos. 56 and 57, Hora II., make A precede B by $104''\cdot4$, with a $\triangle$ in declination of only $0''\cdot6$, consequently the angle may be safely assumed as $= 90°$. The proper motion of A in $\mathbb{R}$ being marked *minus* would by this time increase the distance from B to $111''\cdot4$, and its movement in declination during the same interval being to the south, would reduce the angle to $87°\cdot5$; which is in fair agreement with my micrometrical measures.

Count de Hahn thought he saw another companion, but I could not detect it. We are also told that Mira alters its colour with its magnitude, yet it was always reddish when viewed in my telescope.

A fancied line led from Castor through Aldebaran, till it meets another shot from Al'mak by Hamal, will point out the place of this mutable body to the casual gazer, by whom, however, it is rarely picked up. It is exactly in the direction, and half-way between γ and ζ Ceti.

XCV.   19 ♓. V. ANDROMEDÆ.

$\mathbb{R}$ 2ʰ 12ᵐ 35ˢ	Prec. + 3ˢ·72
Dec. N 41° 36'·1	—— N 16''·78

Mean Epoch of the Observation . . . . 1836·78

An elongated nebula, on the Lady's right foot, where a line from Algol to Al'mak passes under it, at about two-thirds of the distance.

This wonderful object was most indistinctly seen, though watched with a set attention on a glorious night, with the telescope in the highest possible order: yet it was discovered by Miss Herschel in August, 1783, with a Newtonian sweeper of only twenty-seven inches focal length, charged with a magnifying power of 30. Sir William Herschel describes it as having a black division or chink, in the middle; and in my telescope it is certainly brighter at the edges than along the central part. Sir John Herschel, whose No. 218 it is, has given a beautiful drawing of its aspect in the twenty-foot reflector, and concludes that it is a flat ring, of enormous dimensions, seen very obliquely. It consists, probably, of myriads of solar systems at a most astounding distance from ours, and affords a distinct lesson that we must not limit the bounds of the universe by the limits of our senses. The adjoined sketch gives a slight notion of it.

XCVI. 61 P. II. ANDROMEDÆ.

$\mathcal{R}$ 2ʰ 12ᵐ 54ˢ   PREC. + 3ˢ·70
DEC. N 40° 40'·1   —— N 16"·77

POSITION 355°·0 (w 2)   DISTANCE 50"·0 (w 2)   EPOCH 1836·69

A wide double star, closely following the preceding object, and about one-third the distance from Al'mak to Algol. A 7, yellow; B 11, pale lilac. This object was examined on seeing Piazzi's note of *proxime* to an 8th-magnitude star closely following the north vertical at about 8' distance, which I am satisfied is No. 62 of his Catalogue; although its mean place from A is found to be at an angle of 3° 5' and a distance of 283". Between these two there are three other small stars, which the Palermo telescope overlooked; the nearest of which is here measured. When Mr. Dawes procured his doubly-refracting micrometer, he kindly measured this pair for me, as follows:

Pos. 355° 11'   Dist. 49"·0   Ep. 1842·76

XCVII. 72 P. II. CASSIOPEÆ.

$\mathcal{R}$ 2ʰ 15ᵐ 58ˢ   PREC. + 4ˢ·80
DEC. N 66° 40'·7   —— N 16"·62

POSITION AB 274°·2 (w 7)   DISTANCE 2"·1 (w 4) ⎫
———— AC 107°·1 (w 9)   ———— 7"·5 (w 9,) ⎬ EPOCH 1834·83

A beautiful triple star, under the Lady's right foot, and more than mid-distance between α Persei and γ Cephei. A 4½, pale yellow;

B 7, lilac; and C 9, fine blue; the individuals running nearly in a line, with the colours well contrasted.   There has been a little confusion as to the identity of this object, H. having entered it as 55 Cassiopeæ, and others calling it ι; but it is quite clear that it is as above named, and the 292 of the British Catalogue; where, as a note shows, it gave Mr. Baily some trouble.   When H. first enrolled it, he overlooked B, and marked A C No. 3, IV., thus:

Pos. 100° 37′   Dist. 7″·50   Ep. 1779·63

but he afterwards caught up the star B, which became No. 34, I., under these measures:

Pos. 290° 30′   Dist. 1″·50   Ep. 1782·45

Such results, and the not having seen B at first, would induce a supposition of great motion; but H. suggests that an error exists in the position of A·C, and that an observation of 1802 is more to be trusted. This measure = 18° 57′ sf, or 108° 57′, shews that there is little or no change in the elapsed time.

XCVIII.   23 H. IV. CETI.

Æ   2ʰ 19ᵐ 25ˢ   Prec. +   3ˢ·04
Dec. S 1° 51′·6   —— N 16″·45

Mean Epoch of the Observation . . . . 1836·78

A planetary nebula, in the middle of the Whale's neck, discovered by H. in January, 1785.   It is round, bluish white, and pale, but very distinct, and brightening towards the centre.   This object is situated equatorially between two very small stars; and four larger, due north, form the letter L.   It was differentiated with γ Ceti, from which it is about 7°, on the line leading upon ζ.

XCIX.   93 P. II. TRIANGULI.

Æ   2ʰ 19ᵐ 26ˢ   Prec. +   3ˢ·50
Dec. N 29° 12′·5   —— N 16″·45

Position 342°·1 (w 4)   Distance 2″·3 (w 2)   Epoch 1834·11

A close double star, between the Fly and the Triangle; being the northern of a small trapezium of telescopic stars, lying in the direction of a line carried from γ Arietis through a—the first and third stars in the Ram's head—and extended about two-thirds farther.   A 6½, yellow; B 10, grey.   This exquisite and difficult object was dicovered by H. in October, 1781, being his No. 21 of the First Class; but he only estimated a position and distance, which, however, approximate so nearly, that no motion can be assumed from the recent measures. Those of the Dorpat astronomer are:

Pos. 340°·40   Dist. 1″·903   Ep. 1832·36

The larger star is 13 Trianguli of Flamsteed, to which slight proper motions have been thus assigned:

P....	ℛ − 0″·04	Dec.	− 0″·02
B....	− 0″·03		+ 0″·05
T....	− 0″·02		− 0″·01

While this was in the press, I learned that Professor Struve had discovered, with the gigantic refractor at Poulkova, that the neighbouring star 89 P. II., was double; the components being of the $7\frac{1}{2}$ and $8\frac{1}{2}$ magnitudes, and $1\frac{1}{2}″$ apart.

C. 96 P. II. ARIETIS.

ℛ 2ʰ 21ᵐ 20ˢ   PREC. + 3ˢ·42
DEC. N 24° 31′·4   —— N 16″·35

POSITION 182°·0 (w 2)   DISTANCE 12″·5 (w 1)   EPOCH 1832·06

A very delicate double star, over the Ram's back, and nearly in mid-distance between the Pleiades and α Andromedæ, where it is intercepted by a line from β Arietis—the middle star of the Ram's head—to the lucida of Musca. A $6\frac{1}{2}$, pearl white; and B 14, blue. This fine object is in a poor field, with a 9th-magnitude star in the *nf* quadrant, between which and A, a little following, is another of the 13th. It is No. 12 of ℍ.'s last Catalogue, and 221 of Σ.'s; but so difficult, that with my means accurate measures are impracticable.

CI. 227 H. PERSEI.

ℛ 2ʰ 22ᵐ 03ˢ   PREC. + 4″·26
DEC. N 56° 48′·8   —— N 16″·32

EPOCH OF THE OBSERVATION . . . . 1838·78

An irregular but pretty rich cluster, on the weapon arm of Perseus, and in a fine vicinity. It was discovered by H., and consists of individuals from the 9th to the 15th magnitudes, preceded by some largish outliers, one of which is of a red tinge. The mean apparent place depends, by differentiation, on α Persei; and it may be fished up about 10° *np* that star, nearly in the mid-distance of a line shot from Polaris to the west of Algol. A 7th magnitude in the *np* quadrant is the avant-courier of this field, and three of the 11th magnitude form so correct a line in the *nf* as to attract attention.

CII.   ν CETI.

$\mathbb{R}$ 2ʰ 27ᵐ 29ˢ   Prec. + 3ˢ·14
Dec. N 4° 53'·5   —— N 16'·04

Position 85°·0 (w 1)   Distance 6"·0 (w 1)   Epoch 1833·88

A double star, in the Whale's eye, about 3° from γ Ceti, slightly preceding a line from γ towards β Arietis, the middle star of the Ram's head. A 4½, pale yellow; B 15, blue. This very delicate object is one of those marked by Σ. "difficilis;" and not without reason, for the *comes* can only be seen by glimpses, on ardent gazing; and its details are therefore mere estimations. It is followed exactly on the parallel, Δ $\mathbb{R}$=25ˢ, by a dusky star of the 11th magnitude. Σ.'s results are:

Pos. 83°·30 Dist. 7"·725 Ep. 1831·92

The proper motions in space attributed to A are:

P.... $\mathbb{R}$ − 0"·19   Dec. − 0"·13
B....   − 0"·08   − 0"·03

CIII.   30 ARIETIS.

$\mathbb{R}$ 2ʰ 27ᵐ 44ˢ   Prec. + 3ˢ·42
Dec. N 23° 56'·9   —— N 16"·02

Position 273°·0 (w 7)   Distance 38"·3 (w 5)   Epoch 1837·80

A fine double star, over the Ram's back, in the line from α Trianguli towards Aldebaran, and about one-fourth of the distance; it is also pointed out by a ray leading from γ Pegasi to α Arietis, and carried about 7° beyond. A 6, topaz yellow; B 7, pale grey; and Piazzi's No. 128 is on the following parallel, 44" off. The object forms 49 ℍ. v., being registered in 1781·79, under a distance of 31"·10, but without an angle of position. H. and S. measured it thus:

Pos. 272° 26' Dist. 38"·44 Ep. 1821·99

which made them assume that the distance had increased. Still the observations of Bradley do not warrant such a conjecture, because they give a Δ Dec.= 42"·8; while those of Piazzi yield 48"·0. But as ℍ. marked his result "inaccurate," and my determinations show no alteration in a lapse of sixteen years, the fixity of this star may be presumed upon.

30 Arietis is the most southern of a group of about a dozen double stars, spread over the adjoining portions of the three constellations— Aries, Musca, and Triangulum—with extensive patches of dark and blank space between them.

CIV. 156 ♅. I. PERSEI.

Æ   2ʰ 30ᵐ 25ˢ   Prec. + 3ˢ·71
Dec. N 38° 21′·3   —— N 15″·88

Mean Epoch of the Observation . . . . 1837·76

An elongated *lenticular* nebula, *sp* the head of Medusa, and pointed
out by a line led from the Hyades through
the Pleiades, and carried twice their distance further. This curious body was discovered by ♅. in October, 1786. Though
pale, it is very distinct in my instrument,
and elliptical, not—as the twenty-foot reflector defines it—lenticular; an appearance owing, perhaps, to its being a vast
ring lying obliquely to our line of vision.
It trends *nf* and *sp*, and is accompanied by
many small stars, of which the nearest is a

10th-magnitude, due south. Differentiated with β Persei for a mean
apparent place, and diagrammed as here represented, whereby a notion
of its form is afforded.

CV. 33 ARIETIS.

Æ   2ʰ 31ᵐ 20ˢ   Prec. + 3ˢ·47
Dec. N 26° 22′·2   —— N 15″·83

Position 00°·2 (*w* 9)   Distance 28″·5 (*w* 9)   Epoch 1832·12

A fine double star, over the back of Aries, but in the space assigned
to Musca; lying nearly mid-way between the Pleiades and β Andromedæ.
A 6, pale topaz; B 9, light blue. This easy object is 5 ♅. iv., and was
thus registered:

Pos. 2° 46′ Dist. 25″·53 Ep. 1779·74

Forty-three years afterwards, H. and S. found it:

Pos. 1° 40′ Dist. 29″·185 Ep. 1822·08

whence, compared with my own very satisfactory results, there may
exist a slow motion in orbit. But the inference that the stars are
receding from each other, cannot be supported, as ♅. marked his
distance " inaccurate." The proper motions of A are:

P.... Æ + 0″·21   Dec. − 0″·09
B....   + 0″·14   + 0″·01

CVI.   34 M. PERSEI.

Æ 2ʰ 31ᵐ 46ˢ   Prec. +   3ˢ·81
Dec. N 42° 02′·7   —— N 15″·81

Position 250°·8 (w 2)   Distance 14″·0 (w 2)   Epoch 1837·78

A double star in a cluster, between the right foot of Andromeda and the head of Medusa; where a line carried from Polaris between ε Cassiopeæ and α Persei to within about 2° of the parallel of Algol, will meet it. A and B, 8th magnitudes, and both white. It is in a scattered but elegant group of stars from the 8th to the 13th degree of brightness, on a dark ground, and several of them form into coarse pairs. This was first seen and registered by Messier, in 1764, as a "mass of small stars;" and in 1783 was resolved by Sir W. Herschel with a seven-foot reflector: with the twenty-foot he made it "a coarse cluster of large stars of different sizes." By the method he had applied to fathom the galaxy, he concluded the profundity of this object not to exceed the 144th order.

CVII.   12 PERSEI.

Æ 2ʰ 32ᵐ 11ˢ   Prec. +   3ˢ·75
Dec. N 39° 30′·8   —— N 15″·79

Position BC 209°·8 (w 8)   Distance 22″·9 (w 9)   Epoch 1833·85

A pointer to a double star in the *nf* quadrant, with Δ Æ 10ˢ, preceding the head of Medusa; and a line led through κ and γ Cassiopeæ passes over it between Algol and Al'mak. A 6, yellow; B 7½, pale blue; C 8, lilac. This object is 64 ⵁ. iv., and was thus measured:

Pos. 212° 03′   Dist. 21″·98   Ep. 1782·30

It was then re-examined by S., No. 416, with these results:

Pos. 209° 30′   Dist. 22″·88   Ep. 1823·99

which are so singularly coincident with those I obtained ten years afterwards, that its fixity may be held to be proved; and it must be rated as an optical object, or one which is casually juxtaposed in the heavens. Still the components must be within a distance of each other which imagination may compass, since their relative brightness is apparently so nearly the same. A bright star, distant upwards of 4′ in the *sp* quarter, is Sir James South's C; and still nearer to A is a pair of 10th magnitudes, on the parallel with each other.

CVIII. 84 CETI.

Æ 2ʰ 33ᵐ 02ˢ   Prec. + 3ˢ·05
Dec. S 1° 22'·7   —— N 15"·74

Position 334°·5 (w 2)   Distance 5"·0 (w 1)   Epoch 1833·97

A very delicate double star on the Whale's under-jaw; between a and ζ Ceti. A 6, pale yellow; B 14, lilac, with several minute stars in the field. This beautiful object was discovered by Σ., and is No. 295 of the Dorpat Catalogue, where its measures are:

Pos. 334° 62'   Dist. 4"·855   Ep. 1831·90

No inferences can be drawn on comparing our epochs, since my observations, from the difficulty of the object, are little better than estimations.

CIX. θ PERSEI.

Æ 2ʰ 33ᵐ 8ˢ   Prec. + 4ˢ·01
Dec. N 48° 32'·9   —— N 15"·72

Position AB 293°·1 (w 2)   Distance 15"·0 (w 1)⎫
——— AC 219°·0 (w 1)   ——— 27"·0 (w 1)⎬ Epoch 1833·65
                                     ⎭

A triple star in a rich field on the Hero's right shoulder; and nearly in mid-distance between Algol and δ on the knee of Cassiopea. A 4, yellow; B 13, violet; C 11, grey. Iᴴ. mentions a third star within 1' towards the south; but both he and Σ. measured the object only as double, with these results:

Iᴴ.   Pos. 290°·0   Dist. 13"·52   Ep. 1782·64
Σ.   294°·60   15"·40   1832·20

A very sensible proper motion in space has been registered to A, and the following astronomers assign as values:

P.... Æ + 0"·67   Dec. – 0"·10
B....   + 0"·56   – 0"·10
A....   + 0"·52   – 0"·12

CX. 77 M. CETI.

Æ 2ʰ 34ᵐ 30ˢ   Prec. + 3ˢ·06
Dec. S 0° 41'·2   —— N 15"·56

Mean Epoch of the Observation .... 1836·78

A round stellar nebula, near δ in the Whale's lower jaw, and about 2¼° from γ on the line towards ε, or s. by w. This was first classed

F 2

by M. in 1780 as a mass of stars containing nebulosity. It is small, bright, and exactly in a line with three small stars, one preceding and two following, of which the nearest and largest is a 9th-magnitude to the *sf*. There are other minute companions in the field; and the place is differentiated from γ Ceti.

This object is wonderfully distant and insulated, with presumptive evidence of intrinsic density in its aggregation; and bearing indication of the existence of a central force, residing either in a central body or in the centre of gravity of the whole system. Sir William Herschel, after repeatedly examining it, says,—"From the observations of the large ten-feet telescope, which has a gauging power of 75·82, we may conclude that the profundity of the nearest part is at least of the 910th order." That is, 910 times as far off as the stars of the first magnitude!

CXI. γ CETI.

Æ	2^h 35^m 01^s	PREC. +	3^s·11
DEC. N	2° 33'·5	—— N	15"·63

POSITION	289°·0 (*w* 4)	DISTANCE	2"·6 (*w* 3)	EPOCH	1831·85
——	286°·8 (*w* 8)	——	2"·7 (*w* 5)	——	1835·89
——	288°·8 (*w* 7)	——	2"·8 (*w* 4)	——	1838·92
——	285°·7 (*w* 8)	——	2"·6 (*w* 5)	——	1843·16

A second-grade Greenwich star, in the Whale's mouth, and closely double. A 3, pale yellow; B 7, lucid blue, the colours finely contrasted. With my instrument this beautiful object is certainly not so very difficult to measure, as the above discordancies would indicate; so—leaving the unravelment of the anomaly to future inquiry—on comparing my results with those of Σ. H. and D., I consider its *fixity* established. The amount of proper motion through space assigned to A, has been thus valued:

P.... Æ − 0"·35   Dec. − 0"·20
B.... − 0"·11   − 0"·16
A.... − 0"·13   − 0"·17

The Arabian astronomers applied the name of *Kaff-al-jidhmà*, the maimed hand, to a group of stars forming the Whale's head; and which, though limited by Ideler to α, δ, λ, μ, and ξ Ceti, has been latterly applied exclusively to γ. There are vestiges shewing that the Orientals had a large asterism here in very early times, probably before Cepheus and the Ethiopian plague were thought of. A line from β Andromedæ through β Arietis, the centre of the three stars in the Ram's head, points nearly upon γ Ceti at about 25° beyond, or as far again; and it is nearly in mid-distance between Algenib and Rigel.

CXII. 160 P. II. TRIANGULI.

$\mathcal{R}$ 2ʰ 35ᵐ 08ˢ   PREC. + 3ˢ·53
DEC. N 28° 46′·9   —— N 15″·63

POSITION 297° 8 (w 6) DISTANCE 2″·9 (w 4) EPOCH 1831·88

A close double star, very near the wing of Musca, and forming the apex of a nearly equilateral triangle with the two brightest in that insect. A 8, B 8½, both cream-white. This lovely object is in a barren field, and by no means of easy measurement. The earliest micrometric results I meet with are those of Sir James South, No. 418:

Pos. 294° 17′ Dist. 2″·903 Ep. 1825·78

CXIII. μ CETI.

$\mathcal{R}$ 2ʰ 36ᵐ 17ˢ   PREC. + 3ˢ·21
DEC. N 9° 26′·1   —— N 15″·56

POSITION 1°·0 (w 1) DIFFERENCE $\mathcal{R}$ = 1ˢ·5 (w 1) EPOCH 1836·78

A star with a companion, both of which are upon the hind hoof of Aries, though assigned to Cetus: it is on the line, and nearly a third of the distance from γ to π Ceti, beyond the former star, where it forms a nearly equilateral triangle with α and γ. A 4, pale orange; B 11, cinereous. A sensible proper motion through space has been registered to A, and the following astronomers assign as values:

$$P.... \mathcal{R} + 0''\cdot06 \quad \text{Dec.} + 0''\cdot20$$
$$B.... \quad + 0''\cdot30 \quad\quad - 0''\cdot05$$
$$A.... \quad + 0''\cdot27 \quad\quad - 0''\cdot05$$

CXIV. 64 ♓. I. CETI.

$\mathcal{R}$ 2ʰ 38ᵐ 08ˢ   PREC. + 2ˢ·95
DEC. S 8° 15′·1   —— N 15″·46

MEAN EPOCH OF THE OBSERVATION . . . . 1836·78

An oval nebula, on the strange pectoral fringe of the Whale's neck, at rather more than a quarter of the distance from ε to α. It is pale though distinct, and brightens towards the centre. This object was discovered by ♓. in January, 1785, and is No. 264 of his son's Catalogue. Its place was carefully differentiated with η Eridani; but to insure the identity of so delicate a spot, it needs only be raised a little in the inverted field, when three equidistant stars of the 8th magnitude will appear on the northern verge. There are several very small stars in the field, under a moderate power.

CXV. η PERSEI.

Ⅸ 2ʰ 39ᵐ 04ˢ   Prec. + 4ˢ·30
Dec. N 55° 13′·5   —— N 15″·41

Position 301°·1 (w 5)   Distance 28″·1 (w 5)   Epoch 1830·89
—— 300°·4 (w 8)   —— 28″·4 (w 8)   —— 1838·78

A fine double star, on the Hero's head, and nearly in mid-distance of a line projected from α Persei, and carried between δ and ε Cassiopeæ. A 5, orange; B 8½, smalt blue; the colours in clear contrast. This is a very neat object, of which the pair was measured for comparison with former results, otherwise there are three small stars which offer closer secondaries. There are distinctly nine stars in the group, of which the principal, as Professor Barlow has remarked, having three small stars nearly in a line on one side, and one on the other, forms a miniature representation of Jupiter and his satellites. The leader of this family precedes A a little below the parallel, or north of it, in the inverted field, with a Δ Ⅸ = 15ˢ. When A and B were first classed as 4 ♓. ɪᴠ., the measures were:

Pos. 290° 05′   Dist. 26″·00±   Ep. 1779·72

H. and S., No. 34, then found it

Pos. 299° 53′   Dist. 28″·96   Ep. 1821·94

whence an angular change of + 0·25 per annum was suspected; but which is not confirmed by my results.

This star is 179 P. ɪɪ., or 9 Hevelius; and was under no small confusion as to its identity in the British Catalogue, until Mr. Baily's correctives duly installed it 15 Persei η, No. 348. And he adds: "In the British Catalogue it is stated to be of the 6th magnitude: but in Halley's edition it is called the 4th. On consulting the *original* entry in the MS. book, I find it is there also noted as the 4th; which I have here adopted." It is certainly bright for Piazzi's rating, but I see no reason for altering it, since it is less lustrous than θ or γ Persei, its neighbours of the 4th magnitude.

CXVI. π ARIETIS.

Ⅸ 2ʰ 40ᵐ 22ˢ   Prec. + 3ˢ·33
Dec. N 16° 47′·8   —— N 15″·33

Position AB 121°·6 (w 5)   Distance 3″·1 (w 3) ⎫
—— AC 109°·9 (w 4)   —— 25″·0 (w 1) ⎬ Epoch 1835·89
                                                   ⎭

A neat triple star, on the haunch of Aries, closely on the line and about one-third the distance from β Arietis, the middle star of the Ram's head, and Aldebaran. A 5, pale yellow; B 8½, flushed; and C 11, dusky. This superb trio was discovered by ♅. in October, 1782, who describes them as lying in a line 109° 19', and pointing towards a fourth star in the *sf* quadrant. Now A and B certainly do point exactly to the star D, but C is quite out of the line, yet ♅.'s estimated distance of "25" or 26" " appears correct. He says, the smaller stars of 64 ɪ., are "both mere points," neither of which can be seen "except with considerable and long-continued attention;" but they are comparatively so easy in my instrument, that they may have become brighter. It is remarkable that a MS. remark of ♅., adduced by his son, describes C as "easier to be perceived" than B. If this was the case in 1782, and the three stars were then in a line with D, the object merits close watching, both for motion and variability. The Rev. Mr. Dawes saw it triple in his excel-lent five-foot telescope, thus:

$$\left. \begin{array}{l} \text{Pos. A B } 125° \ 19' \quad \text{Dist. } 3''\cdot17 \\ \qquad\text{AC } 111° \ 15' \quad \text{``}impossible\text{''} \end{array} \right\} \ \text{Ep. } 1836\cdot78$$

A slight movement in space is attributed to the large star, amounting to these values:

$$
\begin{array}{llll}
P.... & \cancel{R} + 0''\cdot03 & \text{Dec.} - 0''\cdot11 \\
B.... & \quad + 0''\cdot05 & \qquad - 0''\cdot04
\end{array}
$$

CXVII. 41 ARIETIS.

$$
\begin{array}{llll}
\cancel{R} & 2^h\ 40^m\ 34^s & \text{Prec.} + & 3^s\cdot50 \\
\text{Dec. N } 26° & 35'\cdot9 & \text{—— N } & 15'''\cdot32
\end{array}
$$

Position AB 250°·0 (w 2)	Distance 15"·0 (w 1)	
——— AC 196°·2 (w 5)	——— 38"·0 (w 3)	Epoch 1834·10
——— AD 225°·0 (w 5)	——— 124"·8 (w 5)	

A coarse quadruple star, in the south wing of the Fly, and forming its lucida. A 3, white; B 13, deep blue; C 11, lurid; D 9, pale grey. Of these two members A and C form 116 ♅. v. under these measures:

Pos. 188° 12'   Dist. 39"·37   Ep. 1782·98

while A and D constitute 5 ♅. 6, thus measured by H. and S.:

Pos. 226° 36'   Dist. 127"·55   Ep. 1821·96

But since Sir William Herschel registered this object as 35 Arietis, and another error of identity occurred, it should be stated that A is 186 P. ɪɪ., the *Muscæ secundæ* of the Palermo Catalogue; and to which is assigned the following degree of proper motion, viz.:

$$
\begin{array}{llll}
P.... & \cancel{R} + 0''\cdot20 & \text{Dec.} - 0''\cdot14 \\
B.... & \quad + 0''\cdot09 & \qquad - 0''\cdot10
\end{array}
$$

My friend the Rev. Mr. Dawes re-examined this object at my request in October, 1842, with results very similar to my own,—whence the difference between ♅. and myself in the angle of A and C must be only accidental. Piazzi's note 186 to Hora II., mentions a couple of distant companions—*utraque* 9^{*æ*} *magnit.*—which, though too far off to be very interesting, may be thus reduced:

$$1 \text{ Pos. } 275°\cdot6 \quad \text{Dist. } 94'' \atop 2 \qquad 213°\cdot5 \qquad 146''} \text{ Ep. } 1800$$

Musca Borealis is a little asterism to the north-east of the Ram's head, and is known by three stars of the 3rd and 4th magnitudes. It seems to have been composed from *informes* by Bartschius, the scientific son-in-law of Kepler; for which reason, perhaps, it was afterwards retained by Hevelius, though reluctantly. To identify the object here treated, let a line from Aldebaran be passed under the Pleiades and meet another carried from *a* Andromedæ over *a* Trianguli; it will pass Lucida Muscæ in the mid-distance.

CXVIII. γ^1 FORNACIS.

Æ 2^h 42^m 46^s Prec. $+$ $2^s\cdot66$
Dec. S $25°$ $13'\cdot3$ —— N $15''\cdot20$

Position $171°\cdot0$ (w 2)   Distance $45''\cdot0$ (w 1)   Epoch $1837\cdot94$

A wide but delicate double star. A 6, pale white; B 12, light blue. This object is followed, a little north of the parallel Δ Æ $= 25^s$, by a neat star of the 8th magnitude. As B vanishes under illumination its position is taken by the spherical rock-crystal micrometer, and the distance estimated.

This star is close under the Whale's paw, but in the precincts given to *Fornax Chemica* by La Caille. Bode, in altering the type of this asterism, says, " J'ai tracé au lieu de ces instrumens, la délinéation d'une des expériences de l'immortel Lavoisier." A line from *a* Ceti through η Eridani, carried a little more than as far again to the south, strikes γ Fornacis.

CXIX.   20 PERSEI.

Æ 2^h 43^m 37^s Prec. $+$ $3^s\cdot74$
Dec. N $37°$ $40'\cdot9$ —— N $15''\cdot15$

Position $236°\cdot5$ (w 9)   Distance $13''\cdot9$ (w 7)   Epoch $1832\cdot04$

A double star *sp* the *larva*, or mask of Medusa; which may be found by carrying a line from Mirphak through Algol and about 3° to the south; and a perpendicular to that point will cut 20 Persei at nearly $1\frac{1}{2}°$ to the eastward of it. A $6\frac{1}{2}$, pale white; B 10, sky-blue. This is a neat test object, being 60 I⨍. III.; and supposing with H., that Sir William by error noted down the quadrant as *sf* instead of *sp*, the first measures are:

Pos. 239° 30'   Dist. 14''·30   Ep. 1782·64

This supposition was confirmed by S., No. 420:

Pos. 236° 10′   Dist. 13″·88   Ep. 1824·91

whence, with the further corroboration of my observations, it is shown that there has been little or no change in an interval of half a century. Since this conclusion was arrived at, Σ.'s grand Catalogue has been received, and, on scrutiny, all the conditions of this case are therein placed beyond doubt.

A sensible proper motion in space is attributed to the leader of this compound, the values and directions of which are thus given:

P....	Æ + 0″·18	Dec. − 0″·09
Br...	+ 0″·15	− 0″·10
B....	+ 0″·09	− 0″·07
T....	+ 0″·12	− 0″·09

CXX. 191 P. II. CEPHEI.

Æ 2ʰ 45ᵐ 09ˢ   Prec. + 7ˢ·51

Dec. N 78° 46′·6 —— N 15″·06

Position 225°·8 (w 6)   Distance 5″·2 (w 2)   Epoch 1834·91

A double star, with two telescopic companions at a little distance. A 6, orange; B 10½, smalt blue. This charming object was discovered by Σ., and is No. 320 of the Dorpat Catalogue, where the measures are:

Pos. 226° 97′   Dist. 4″·427   Ep. 1831·60

This star is in a strange corner of the following boundary of Cepheus, but in the part where le Monnier squeezed in an asterism between the Ethiopian monarch and the Camelopard, to commemorate his operations in Lapland, in 1736, under the name of Rangifer. It is about 10° from the Pole-star, on a line leading from thence to Algol.

CXXI. η ERIDANI.

Æ 2ʰ 48ᵐ 38ˢ   Prec. + 2ˢ·92

Dec. S 9° 32′·4 —— N 14″·86

Position 55°·7 (w 1)   Difference Æ = 17ˢ·5 (w 1)   Epoch 1833·90

A star with a distant companion, pointed out by a south-east line from γ Pegasi through o Ceti, when it can be seen, and carried about 12° beyond; where it forms an oblique-angled triangle with ε and π Ceti. A 3, pale yellow, and not of a brightness corresponding to its rated magnitude; B 10, cinereous, being the nearest of a triangle of three small stars in the nf quadrant. This object is pretty close to the Whale's chest in the reach or bend of the River; and is known under the name of *Az-ha*, by an easily-made error of transcription from the Arabic word *Udh-ha*, a little nest; it being among the stars which were anciently called the Ostriches. The proper motion of A has been stated by

$$P.... \text{Æ} + 0''\cdot16 \quad \text{Dec.} - 0''\cdot30$$
$$B.... \quad + 0''\cdot12 \quad\quad - 0''\cdot22$$
$$A.... \quad + 0''\cdot09 \quad\quad - 0''\cdot23$$

CXXII. 220 P. II. PERSEI.

Æ 2ʰ 49ᵐ 30ˢ   Prec. + 4ˢ·21

Dec. N 51° 42'·6   —— N 14''·81

Position 84°·9 (w 8)   Distance 12''·4 (w 6)   Epoch 1835·10

——— 85°·5 (w 9)   ——— 12''·5 (w 7)   ——— 1843·18

A neat double star, on the nape of the Hero's neck, slightly preceding a line carried from Algol to Polaris, at one-third of the distance.   A 6, silvery white; B 8, sapphire blue.   This fine object escaped the eagle-sight of Ⱨ.; but it was marked " duplex" by Piazzi; and S., No. 422, thus measured it:

Pos. 85° 20'   Dist. 12''·96   Ep. 1823·97

The pair forms No. 331 of the Dorpat Catalogue, where the observations of Σ. confirm the fixity shown in the above results.   But though Piazzi made note of the *comes* following to the north, there can be little doubt of his having observed and entered it as No. 222 of the Palermo Catalogue, since a careful reduction yields these results:

As *comes.*   Pos. 83°·5   Dist. 15''⎫
As No. 222.   79 ·2   12 ⎭ Ep. 1800

CXXIII. ε ARIETIS.

Æ 2ʰ 50ᵐ 04ˢ   Prec. + 3ˢ·41

Dec. N 20° 41'·8   —— N 14''·77

Position 193°·5 (w 3)   Distance 0''·5 (w 1)   Epoch 1835·77

——— 195°·7 (w 4)   ——— 0''·8 (w 1)   ——— 1839·25

——— 199°·6 (w 7)   ——— 0''·9 (w 2)   ——— 1843·18

A very close double star, at the root of the tail.   A 5, pale yellow; and B 6½, whitish.   This is Σ.'s No. 333, of the order " pervicinæ;" he says that it is perhaps the closest of all his double stars—" inter omnes nostras fortasse vicinissima."   And H., writing to me in 1831, asks, " Have you tried ε Arietis?   My twenty-foot, with power 480, has fairly separated it.   I do not say it will always do so.   I should like to see the telescope that will."   It must, however, be widening, for

I have divorced them myself latterly, though they generally hung in contact at my earlier attempts. A third star in the *sp* is so nearly in a line with the angle of position, that I was enabled to make use of it in my measures. The distances are mere estimations. The Rev. W. R. Dawes first saw this object double, in my observatory.

In his Catalogue of 1827, Professor Struve merely registered ε Arietis as being *in contactu;* yet he subsequently paid such attention to it, that he suspected the components of variability, to the small amount of between 0 and 0·5 of his scale. This affords internal evidence that he watched them closely; but from their not appearing on the list of *Mensuræ Secundæ*, in 1837, he probably did not perceive their binarity. Still he affords a very valuable starting point, in the following mean:

Pos. 188° 87′ Dist. 0″·547 Ep. 1830·16

An increase of angle had, however, become so apparent to me in 1839, despite of light measures, that ε Arietis was among my strictest *agenda* on repairing to Hartwell to make some re-measurements; when I found it considerably altered, and easier to manage. From the results then obtained, I can with some confidence state, that the acolyte has a direct orbital motion. From the present state of the data, this motion may be inferred to amount to about 0°·85 per annum, so that its revolution may be made in four centuries at most. If we may place dependence on the observations, as to the slight increase of distance, it will probably still widen for a few years longer, until the satellite shall have doubled the southern point of its course, which now seems to be on an ellipse shooting out from ε in the micrometric direction of 210°, with a major-axis about thrice the length of its minor.

ε Arietis, as well as δ, was named *Botein*, from *al-botaïn*, the little belly; as forming with δ and ρ³ the second chamber of the *Manázil al Kamar*, or Lunar Mansions, which, says Kazwíni, is placed in three dark stars which form an atháfí, a trivet or tripod. It is readily found, mid-way on a line drawn between the Pleiades and Hamal; and a ray shot from γ Pegasi between β and γ Arietis, in the Ram's head, and led as far again, strikes ε.

CXXIV. γ PERSEI.

Æ 2ʰ 53ᵐ 14ˢ   Prec. + 4ˢ·28

Dec. N 52° 52′·4   —— N 14″·59

Position 226°·0 (w 1) Distance 55″·0 (w 1) Epoch 1837·65

A wide and unequal double star, on the Hero's left shoulder, where it forms the northern apex of a nearly equilateral triangle with α and θ Persei. A 4, flushed white; B 14, clear blue. This is No. 2170 of H.'s Fifth Series, and a line through them leads nearly over a 10th-magnitude star in the *nf* quadrant, the angle of which with A is = 51°·3, and △ Æ = 18ˢ·4; but I am quite at fault respecting a companion recorded by Piazzi in note No. 234, Hora II., in these terms: "Duplex: altera præcedit 0"·2 temporis, 8" ad Boream." A movement in space is assigned to the principal, of which the amount has been thus stated:

$$P.... \mathcal{R} - 0''·15 \quad Dec. - 0''·05$$
$$B.... \quad + 0''·12 \quad\quad 0''·00$$

CXXV. a CETI.

$$\mathcal{R} \quad 2^h 53^m 55^s \quad\quad Prec. + 3^s·13$$
$$Dec. N \; 3° 27'·5 \quad\quad —— N 14''·55$$

Position 258°·5 (w 2)   Difference $\mathcal{R} = 29^s·6$ (w 2)   Epoch 1833·85

A standard Greenwich star, in front of the lower jaw, with a distant companion. A 2½, bright orange; B 10, pale grey. This is a curious object under a moderate power, on account of a decided blue star in the field, north of it, of the 5½ magnitude. The leader is assigned proper motions in space, to the following amount:

$$P.... \mathcal{R} - 0''·08 \quad Dec. - 0''·15$$
$$B.... \quad + 0''·03 \quad\quad - 0''·10$$

a Ceti is numbered among the insulated stars, and is called *Menkab*, corrupted from *Al minkhir*, the nose or snout, a name which the Arabians applied, with greater propriety, to λ; but though *rated* of equal magnitude with β Ceti, it is not now so large. A line from Pollux by Aldebaran, carried nearly as far again, brings the eye to a Ceti; which star, with Hamal, forms the lower points of a gigantic W, of which Aldebaran, the Pleiades, and Algol make the upper portion. The poet says:

To know the bright star in the Whale,   the lower jaw which decks,
From fair Capella send a glance   through Pleiad's beauteous specks;
And bear in mind this cluster fine,   so admirably seen,
From Cetus' head to th' Charioteer,   lies just half-way between.

The figure of this asterism, a veritable *monstrum marinum*, with its long legs, ears, proboscis, missile tongue, and carnivorous jaws, ought rather to have retained the name 'Ορφὸς, Pristix, as given by Hyginus, than Κῆτος, Cetus, whose un-whale-like appendages did not escape the lash of Butler:

Yes, 'tis clear
'Tis Saturn; but what makes him there?
He's got between the Dragon's tail
And further leg behind o' th' Whale.

Stanislaus Lubienietzki, in his *Theatrum Cometicum*, 1667, attempted to lop off some of these redundancies; but in fishifying the animal he has given him so capacious a mouth and throat, that a Munchausen's ship might well have sailed in. Indeed, the leading distinctions of a whale seem to have been overlooked by all the celestial delineators.

Although this *fish*, as shown in Morell's edition of *Aratus*, 1559, is a very queer-looking creature, yet it is better drawn than some later figures, in that it has no legs; but as it is so furnished in the MS. of Cicero's translation of *Aratus* in the British Museum, it is evident that the monster is connected with the tale of Andromeda. Cetus, is, how-ever, the most extensive constellation of the firmament, occupying the large space to the south under Pisces and Aries; and it was one of the standard old 48 asterisms. Its constituents have been thus numbered:

Ptolemy	. . . 22 stars		Bayer	. . . 27 stars
Copernicus	. . 22		Hevelius	. . . 46
Tycho Brahé	. . 21		Flamsteed	. . 97
Kepler	. . . 25		Bode	 301

The Arabians appear to have given great attention to this fish. From a fancied resemblance of the stars of the head to the *Kaff-al-Khadib* in Cassiopea, they designated a, δ, λ, μ, and ξ, *Kaff-al-jidhmà*, the maimed hand; the five stars on the body—η, θ, τ, ζ and υ—were *al-na'ámát*, the ostriches; and ϕ 1, 2, 3, and 4, which are nearly in a straight line across the tail, were called *al-nidhám*, the necklace.

CXXVI.   52 ARIETIS.

ℛ 2ʰ 56ᵐ 05ˢ   PREC. + 3ˢ·49

DEC. N 24° 37′·7   —— N 14″·41

POSITION AB 265°·7 (w 4)   DISTANCE   0″·8 (w 2) ⎫
—— AC 355°·0 (w 1)   ——   5″·0 (w 1) ⎬ EPOCH 1835·88
—— AD 85°·0 (w 1)   ——   105″·0 (w 1) ⎭

A quadruple group, between the Ram's tail and the Fly, followed nearly on the parallel by a small dusky star. Three of these form Σ.'s No. 346, and the whole are of most difficult measurement. A is 6½, bright white; B 7, pale blue; C 15, blue; and D 13, lilac: the details of the latter two being, of course, mere estimations. It may be picked up by running the eye from 41 Arietis—the lucida of Musca—towards the Pleiades, in which direction it lies, at about one-third of the distance. A line led from γ Pegasi between α and β, in the head of the Ram, and carried nearly as far again, also hits this object.

CXXVII.   β PERSEI.

ℛ 2ʰ 57ᵐ 46ˢ   PREC. + 3ˢ·86

DEC. N 40° 20′·0   —— N 14″·37

POSITION 195°·0 (w 2)   DISTANCE 55″·0 (w 1)   EPOCH 1835·63

A variable star, in the forehead of the *larva* of Medusa, with a companion in the *sp* quadrant, and two others *np*. A 2 to 4, whitish; B 11, purple. This star is generally known as *Algol*, a variation of *Al-ghúl*, the monster or demon; so harshly depicted in Hevelius's map of this asterism. This ominous name was mightily noted in Astrology, and to its influence poor Padre Vitalis, in his dismal Jeremiad, attributed the then unhappy lot of the kingdom of Naples. Dr. Sedillot writes *rhól* for *ghúl*, but he adopted so singular a mode of expressing Arabic words, that it is not always easy to find out what he meant to denote.

Algol is the most remarkable of the periodic stars, since the increase and decrease in its changes of light occupy but seven or eight hours; and this is best observed at the recurrence of the diminished light, because when brightest it is the more difficult to determine, from its varying in brilliance at different times. The most feeble light lasts about eighteen minutes, from the examination of which, Argelander concludes that the period of Algol is not quite constant. The first who observed these variations was Montanari; and in 1694 Maraldi ascertained that it changed from the 2nd to the 4th magnitude. The period has been thus determined:

By Wurm 2^d 20^h 48^m $58^s\cdot7$
By Goodricke 2^d 20^h 48^m $56^s\cdot0$

It varies from the 2nd to the 4th size in three hours and a half, and back again to the 2nd in the same time, and so remains for the rest of the period, retaining its brightness. These singular appearances are accounted for, by supposing the body to revolve on an axis, having parts of its surface not luminous; and ⩍. observes that such stars, besides a rotatory motion on their axes, may also have other movements—"Such as nutations or changes in the inclination of their axes; which added to bodies much flattened by quick rotatory motions, or surrounded by rings like Saturn, will easily account for many new phenomena that may offer themselves to our extended views."

To find Algol by alignment, project a ray from Orion's belt through Aldebaran, and carrying it something more than double the distance, it will hit the head of Medusa; or, lisping in numbers,

Thus belt of Hero, eye of Bull,   so surely mark the place
Where Algol shines, 'bove three faint stars,   in fell Medusa's face.

These same stars collectively, were formerly called the Gorgons, the præses of which has had a movement in space attributed to it, to the following amount:

P.... $R +'0''\cdot11$ Dec. $- 0''\cdot04$
B.... $+ 0''\cdot03$ $+ 0''\cdot02$

CXXVIII.   25 ⩍. VI. PERSEI.

R $3^h 04^m 01^s$ Prec. $+ 4^s\cdot09$
Dec. N $46° 37'\cdot9$ —— N $13''\cdot92$

MEAN EPOCH OF THE OBSERVATION . . . . $1836\cdot76$

A very extensive and compressed cluster, on the right side of Perseus, in a rich portion of the galaxy; and it has a gathering spot, about 4' in diameter, where the star-dust glows among the minute points of light. This elegant sprinkle was registered, in December, 1786, by ⩍., who says, "the large stars are arranged in lines like interwoven letters." It is

II.'s No. 290, of the Catalogue of 1833: I derive its mean apparent place by differentiation from α Persei, from which it lies at an angle $= 9°$ with the vertical, at a distance of nearly $8°$ in space.

CXXIX.   94 CETI.

Æ	3^h 04^m 38^s	Prec. +	$3^s\cdot10$
Dec. S	$1°$ $47'\cdot9$	——	N $13''\cdot88$

Position $260°\cdot0$ (w l)   Distance $5''\cdot0$ (w l)   Epoch $1836\cdot75$

A most delicate double star, on the tip of the cameleon-like tongue with which the celestial Whale is often figured; and it will be struck by a line thrown from α Arietis, in the Ram's head, through α Ceti, and carried about $6\frac{1}{2}°$ beyond. A $5\frac{1}{2}$, pale cream-colour; B 16, dusky. This object was discovered by Sir John Herschel, and is No. 663 of his Third Series of Twenty-foot Sweeps, where the acolyte is registered of the 19th magnitude. But as, after several toilful trials under the best circumstances, I caught a view which, though most evanescent, and under an averted eye, was sufficient to catch a guess by, I have assigned its brightness at the point which is fixed upon as the *minimum visibile* of my telescope. It must, however, be esteemed among the *intensiva* of faintness, and has been repeatedly sought in vain, with the same instrument. Reasoning from the analogies presented by optical space-penetrating power, this acolyte, if not physically connected with A, must be almost inconceivably beyond it in the vast profundity of those remote regions which may be but the beginning of the Universe: such argument suggests the possible distance to be somewhere between 700 and 800 times that of Sirius from us.

CXXX. τ^4 ERIDANI.

Æ	3^h 12^m 24^s	Prec. +	$2^s\cdot66$
Dec. S	$22°$ $20'\cdot6$	——	N $13''\cdot39$

Position $240°\cdot8$ (w l)   Distance $150''\cdot0$ (w l)   Epoch $1836\cdot90$

A bright star with a distant companion, in the second *reach* of the River; being one of no fewer than nine stars designated by the letter τ in Bayer's Map of Eridanus. A $3\frac{1}{2}$, light orange; B 11, greyish. This object is in a barren field, and the large star seems overrated, since it appeared more than once diminished to nearly a 5th-magnitude; but the lowness of its position renders the case doubtful, from variable refraction. A lengthy line projected from α Arietis through γ Ceti, and from thence carried by η Eridani to $16°$ beyond, will strike upon τ^4.

CXXXI. α PERSEI.

Æ　3ʰ 12ᵐ 55ˢ　　Prec. + 4ˢ·23
Dec. N 49° 17′·2　　—— N 13″·36

Position 206°·0 (*w* 1)　Distance 75″·0 (*w* 1)　Epoch 1837·64

A standard Greenwich star, with a companion, in a rich galaxy field. A 2½, brilliant lilac; B 9, cinereous; they are followed at a little distance by many small stars. It is now placed in the Hero's left side, but as it was called *Mirfak*, the elbow, or more fully *Al mirfak al thureyyá*, the elbow of the Pleiades, to distinguish it from the other elbow, the figure may have once been differently situated: still its other Oriental name, *Jenb Bersháwush*, signifies the side of Perseus (*Bersháwush Περσεὺς*), the *p* being lost in Arabic. Chrysococca calls it Πλευρὰ Περσάους; and it must be from the word *Jenb* that it was corrupted to Algenib, the style and title of γ Pegasi, an equivoque complained of by Sir J. Herschel.

A fancied line projected from the Pole-star to the Pleiades, passes through the left knee of Perseus, and points in the mid-distance. to Mirfak; which is also gained by a line from Castor to Capella onwards; or by that which the rhymester points out:

A ray from Algol to the Pole　　with accuracy guide,
Near, but behind it, Mirfak shines　in Perseus' manly side.

The following quantity of proper motion through space has, on reference to epochal observations, been assigned to Mirfak:

P.... Æ − 0″·21　Dec. − 0″·02
B....　+ 0″·11　　　 − 0″·04
A....　+ 0″·07　　　 − 0″·05

Perseus, whose mythological story is too well known to require repetition, is one of the old 48 asterisms, and is placed in a very brilliant part of the Via Lactea, nearly opposite to the three stars forming the tail of Ursa Major, on the other side of the pole, and directly north of the Pleiades. In the ancient MS. of Cicero's *Aratus*, in the British Museum, the drawings of which as my late friend, Mr. W. G. Ottley, went far to prove, were executed before the age of Constantine, Perseus is represented with no other drapery than a light scarf, holding the head of Medusa in his left hand, and a singular hooked and pointed weapon in the right. The number of his stars may really be called infinite, on gazing with a powerful telescope; but of his constituents, whose mean apparent places are tabulated, the numbers run:

Ptolemy　.　.　. 29 stars　　Hevelius　.　.　.　. 46 stars
Tycho Brahé　.　. 33　　　Flamsteed　.　.　. ` 59
Bayer　.　.　.　. 38　　　Bode　.　.　.　.　. 196

This asterism, in the Arabo-Latin Almagest, is designated *Cheleab*, which Grotius refers to *kelb*, a dog, but others to *kullúb*, the harpago, or hooked weapon in the Hero's hand*. It was also called *Hámil rás-*

---

* Some of our crusaders must have imported this word; for it is assuredly a better etymon for "club," than the *cluppa* and *kluppel* cited by Dr. Johnson.

al-ghúl, the bearer of the demon's head, or, as the Germans say, *Träger des Medusenkopfs:* but it was the *Caco-dæmon* of astrologers. Such was its style and title for many ages; but in the new uranography of Schickard, the ethnic *Perseus cum capite Medusæ* was supplanted by David with the head of Goliath, as had been proposed by Novidius.

CXXXII. 46 P. III. ARIETIS.

Æ $3^h\ 14^m\ 06^s$ PREC. $+ 3^{s}\text{·}44$

DEC. N $20°\ 23'\text{·}7$ —— N $13''\text{·}28$

POSITION $87°\text{·}6$ (w 3)   DISTANCE $0''\text{·}8$ (w 1)   EPOCH $1834\text{·}19$

A close double star, just following the tail of Aries, at about one-third of the distance between δ Arietis and the Pleiades. A 8, pearl white; B 9, yellow. This exquisitely delicate object is in a line with two distant stars of the 10th magnitude in the *sp* quadrant, and there is another small one in the *nf*: it is Σ.'s No. 381, " oblongam suspicor," and was thus first measured by its discoverer:

Pos. 93°·7 Dist. 0''·75 Ep. 1827·16

The elongation is not immediately detected; and the focus was slightly distorted to examine the outline of the spurious disc. This, if well managed, is often of great use on such occasions. This star forms an angle of 28° from τ^2 Arietis, 45 P. Hora III., with a distance of 16'; but though so distant from each other, it is singular that there appears to be a slight movement in space, partaken by both, in quantities of similar amount and denomination.

CXXXIII. 7 TAURI.

Æ $3^h\ 24^m\ 58^s$ PREC. $+ 3^{s}\text{·}53$

DEC. N $23°\ 55'\text{·}4$ —— N $12''\text{·}55$

POSITION A B $265°\text{·}0$ (w 2)   DISTANCE $0''\text{·}7$ (w 1)   ⎫

————— A C $61°\text{·}9$ (w 5)   ———— $21''\text{·}8$ (w 3)   ⎬ EPOCH $1833\text{·}21$
                                    ⎭

A triple star, on the back of Taurus, about 3° to the *np* of the Pleiades. A 6, white; B $6\frac{1}{2}$, pale yellow; C 11, bluish. This is a fine and very difficult object, being Σ.'s No. 412, " vicinissimæ." A and C point to a *comes* in the *nf* quadrant, and constituted 88 H̶. IV., but Sir William did not observe that A was double. It may have opened since. His measures of A and C were

Pos. 66° 45' Dist. 19''·833 Ep. 1783·77

It was then re-examined by H. and S., No. 363, with these results:

Pos. 56° 06' Dist. 21''·055 Ep. 1821·97

and Σ., who first made it a triple object, thus registered it:

$$\text{A B Pos. } 269°·92 \quad \text{Dist. } 0''·692 \quad \text{Ep. } 1830·38$$
$$\text{A C} \qquad 63°·02 \qquad 22''·407 \qquad 1830·92$$

Now the first two epochs exhibited so great an orbital change, in less than forty years, as to excite much attention; but the accordance of those of Σ. and myself, indicate some error of observation or entry. In this conclusion, however, Σ.'s angle for 1821·95, in the *Dorpat Observations*, is rejected; since it must be deemed rather an essay than a conclusive measurement.

CXXXIV. 98 P. III. ERIDANI.

$$\text{Æ } 3^h 28^m 35^s \qquad \text{Prec. } + 3^s·07$$
$$\text{Dec. N } 0° 03'·7 \qquad \text{—— N } 12''·30$$

Position 231°·8 (*w 9*) Distance 5''·9 (*w 6*) Epoch 1834·93

A delicate double star, on a line with a Ceti and Rigel, and nearly one-third the distance: β and a Tauri also point upon it. A 6½, yellow; and B 9, pale blue. This is 45 H. III., who by measures in 1781·83, made the position angle = 234° 27'; but H. informs us, that by a MS. note he finds it declared, that the observation is too small by 6° or 8°. The first measures, therefore, for future reference, must be those of S., No. 431:

$$\text{Pos. } 225° 12' \quad \text{Dist. } 5''·812 \quad \text{Ep. } 1824·02$$

This object is between the Bull's chest and the northern branch of the Eridanus, in the part where the Abbé Hell (who also placed Herschel's telescope among the celestials) squeezed in his *Harpa Georgii*, to compliment a sovereign of these realms; having filched from Eridanus about thirty or forty stars, some of the 4th magnitude, for the purpose.

CXXXV. δ PERSEI.

$$\text{Æ } 3^h 31^m 33^s \qquad \text{Prec. } + 4^s·22$$
$$\text{Dec. N } 47° 16'·2 \qquad \text{—— N } 12''·10$$

Position 315°·0 (*w 2*) Distance 140''·0 (*w 1*) Epoch 1833·74

A bright star, with a companion, on the Warrior's hip, and is about 3° to the south-west of a Persei, as pointed by a line led from a Cephei, through the lucida of Perseus. A 3½, flake white; B 11, pale blue; the two pointing towards a 9th-magnitude star at a distance in the *np* quadrant. The vicinity is very rich. A has had a slight proper motion assigned by

$$
\begin{array}{lll}
P.... & \text{Æ} + 0''·04 & \text{Dec. } - 0''·10 \\
B.... & \phantom{\text{Æ}} + 0''·09 & \phantom{\text{Dec. }} - 0''·03 \\
A.... & \phantom{\text{Æ}} + 0''·05 & \phantom{\text{Dec. }} - 0''·05 \\
\end{array}
$$

CXXXVI. 40 PERSEI.

Æ 3ʰ 32ᵐ 14ˢ   Prec. + 3ˢ·77

Dec. N 33° 26′·8   —— N 12″·05

Position 238°·2 (w 5)   Distance 20″·6 (w 2)   Epoch 1834·92

A delicate double star, on the wing of the Hero's right ankle, if one of the *talaria* may be so called. A 6, pale white; B 10, ash-coloured. This is 39 ♃. III., being registered in September, 1781, but without measures, except one estimated distance of about 15″. We are, therefore, indebted to *Σ.* for the first micrometrical observation, as follows:

Pos. 237°·0   Dist. 19″·77   Ep. 1828·15

The identity of this star has created some little confusion, since several astronomers, among whom are even Flamsteed and Piazzi, have designated 38 and 40 Persei under the letters *o*¹ and *o*²: but Mr. Baily has shown, in his edition of the British Catalogue, that 40 Persei is the "parvula supra *o*" in the *Historia Cœlestis;* and that 38 Persei is the Greek 'ο μικρὸν, while the other is the English *o* of Bayer. It may be found by running a line from Algol a little to the westward of the Pleiades, and it will pass over 40 Persei at something more than half the distance. It is also struck by a ray carried from the cluster in the sword of Orion, over Aldebaran, and extended a little more than as far again.

CXXXVII. 107 ♃. I. ERIDANI.

Æ 3ʰ 33ᵐ 02ˢ   Prec. + 2ˢ·70

Dec. S 19° 04′·8   —— N 12″·00

Mean Epoch of the Observation .... 1837·85

A milky-white nebula, between the two northern *reaches* of the River; it is pale, but distinct, round, and bright in the centre. It lies nearly midway between and preceding two distant stars, *sf* and *nf*, the three forming an obtuse-angled triangle; there are only a few glimpse stars besides in the field. Now there is, on close gazing, such strong internal evidence of the nebula's being inconceivably beyond those specks of light, that, small as it appears, the mind is lost in considering its probable magnitude and distance. This object was registered by ♃. in October, 1785, and I differentiated it with γ¹ Eridani; to fish it up, run an imaginary line from the coarse double star Keid (40 Eridani) through γ, and extend it exactly as far again as the distance between those two points.

CXXXVIII. 19 PLEIADUM.

$\mathbb{R}$ 3^h 35^m 41^s Prec. $+$ $3^s\cdot55$
Dec. N $23°$ $57'\cdot7$ —— N $11''\cdot80$

Position $335°\cdot0$ (*w* 2)   Distance $45''\cdot0$ (*w* 1)   Epoch $1835\cdot01$

A delicate double star, in the cluster on the shoulder of Taurus.
A 5, lucid white; and B 10, violet tint. This object, though wide, is
fine; being Taygeta, a leading one of the seven sisters, whose name
appears to have been of some weight in Sparta. This has been considered as the Brood Hen's head, the slight movement of which, in space,
has been thus valued:

$$P....\mathbb{R} + 0''\cdot03 \quad \text{Dec.} - 0''\cdot07$$
$$B.... \quad + 0''\cdot06 \quad \quad - 0''\cdot02$$

In this group, Celeno and Electra, Nos. 129 and 130 P. iii., appear to
be affected with proper motions, similar in denomination though not
in amount: yet they are 10' apart, on an angle $= 353°$.

CXXXIX. 15 PLEIADUM.

$\mathbb{R}$ 3^h 36^m 23^s Prec. $+$ $3^s\cdot52$
Dec. N $22°$ $38'\cdot6$ —— N $11''\cdot76$

Position $342°\cdot0$ (*w* 2)   Distance $5''\cdot0$ (*w* 1)   Epoch $1835\cdot03$

A double star, in the cluster on the Bull's shoulder. A 8, bright
white; B 14, fine blue. This most delicate object was discovered by Σ.,
and entered as No. 444 of the great Dorpat Catalogue, with these measures:

Pos. $338°\cdot97$ Dist. $3''\cdot277$ Ep. $1832\cdot34$

As the same astronomer got an angle of $334°\cdot4$ in 1831, it may be
well for those who have means equal to the task, to watch this pair
closely. But any discrepancies at present must be referred to the
extreme difficulty of the stars.

CXL. 23 PLEIADUM.

$\mathbb{R}$ 3^h 36^m 51^s Prec. $+$ $3^s\cdot54$
Dec. N $23°$ $26'\cdot8$ —— N $11''\cdot72$

Position B C $149°\cdot9$ (*w* 2)   Distance $32''\cdot5$ (*w* 4)   Epoch $1830\cdot96$

A bright star on the Bull's shoulder, pointing to the small pair in
the south of the field, which were measured and proved to be S.'s

No. 437. A 5, silvery white; B 8, purple; and C 9, pale blue. This object—Merope—is in a fine neighbourhood, as viewed under a moderate magnifying power, being near the middle of the Pleiades. S. thus registers it:

Pos. 150° 17′ Dist. 34″·566 Ep. 1823·99

Proper motions in space are imputed to Merope, which are noted to these opposing values:

P....Æ − 0″·11   Dec. + 0″·04
B....   + 0″·09   − 0″·03

CXLI. 80 ♅. VIII. PERSEI.

Æ 3ʰ 37ᵐ 20ˢ   PREC. + 4ˢ·46
DEC. N 52° 09′·9 ——— N 11″·66

POSITION 255°·0 (w 2) DISTANCE 9″·5 (w 2) EPOCH 1836·79

A delicate double star in a cluster over the Hero's left thigh, and about one-third of the distance between γ Persei and Capella. A 8, light yellow; B 11, pale violet. The large individual is placed equatorially between two small stars, and the secondary advances into the *sp* quadrant, forming a fine object. It was first registered by ♅. in December, 1788; being on the following boundary of Perseus, on a wavy branch of the Galaxy; and was described as containing one large star, but without notice of the pair here measured, which seem to have been first detected by Σ. No. 446, whose measures were:

Pos. 252°·70 Dist. 8″·545 Ep. 1830·74

CXLII. η TAURI.

Æ 3ʰ 37ᵐ 59ˢ   PREC. + 3ˢ·55
DEC. N 23° 36′·3 ——— N 11″·65

POSITION 289°·2 (w 9) DISTANCE 115″·6 (w 9) EPOCH 1836·97

Alcyone, a Greenwich star, with a distant companion, in the midst of the Pleiades, called by the Arabians *Jauza*, the wall-nut, and *Neyyir*, bright, or lucida of the Pleiades. A 3, greenish yellow; B 7, pale white. Piazzi marked this "duplex," but the *comes* could only be 151 P. III.; and a reduction from his mean apparent places, and the micrometrical measures of Sir J. South, afford these results:

P. Pos. 288° 00′ Dist. 122″·50 Ep. 1800
S.   288° 42′   116″·40   1824

which, considering that A is chargeable with a small proper motion both in Æ and declination, is very consistent with more recent observations.

The other two small stars in the same, or *np* quadrant, form the " binæ ad boream" mentioned 150 P. III.; and were also measured by S. The proper motions alluded to, are thus valued:

$$P....\mathbb{R} - 0''{\cdot}04 \qquad Dec. - 0''{\cdot}09$$
$$B.... \quad + 0''{\cdot}06 \qquad\qquad - 0''{\cdot}05$$

This star has usually been considered as the one described under the 32nd of Taurus, in Ptolemy, and there marked ε in brightness. But Mr. Baily says, " I do not think this star can be η Tauri, on account of its magnitude: yet it is singular that the brightest star in the Pleiades should not have been noticed by Ptolemy*."

The Pleiades constitute a celebrated group of stars, or miniature constellation, on the shoulder of Taurus; their popular influences have been said and sung for many ages. Hesiod mentions them as the Seven Virgins, " of Atlas born;" and in the ancient MS. of Cicero's *Aratus*, in the British Museum, they are finely represented by female heads, inscribed Merope, Alcyone, Celæno, Electra, Taygeta, Asterope, and Maia, under the general title *Athlantides*,—while the illustrations to Julius Firmicus in 1497, represent them as well-grown women. The moral may be, that Atlas himself first rigidly observed these stars, and named them after his daughters. But various are the appellations under which they have been known. Theon likened them to a bunch of grapes; Aratus says they were called ἐπτάποροι; Manilius clusters them as *glomerabile sidus;* the Arabs said they were *Ath-thurayya*, or the little ones; the French designate them *poussinière;* the Germans, *gluck-henne;* the Italians knew them as *le gallinelle;* the Spaniards term them the *cabrillas*, or little nanny-goats, which is the key of the Duke's query to Sancho; and several *schools* called them the *brood-hen*, under the representation of a hen and chickens. There has also been much discussion as to the number of the individuals in the group, some of the ancients having advanced that there were seven, and others resolving to count only six, in the spirit of Ovid's oft-cited

Quœ *septem* dici, *sex* tamen esse solent.

The "lost Pleiad" is, however, rather a poetical than an exact expression, for in moonless nights I never had any difficulty in counting seven stars in the so-called Hexastron, with the naked eye; and indeed this is nothing to boast of, for many people may enumerate even more, though few will equal *Mœstlinus*, the discoverer of the new star of 1604, who, as Kepler avers, could distinctly see fourteen stars in the Pleiades, without any glasses. Still, if we admit the influence of variability at long periods, the seven in number may have occasionally been more distinct; so that while Homer and Attalus speak of six of them, Hipparchus and Aratus may properly mention seven. But they have a singularly brilliant light for their magnitudes, whence the unassisted eye

---

* The occultations of this star, and *h* Pleiadum, by the dark limb of the Moon, were well observed on the 19th March, 1839, by my excellent friend Lord Chief Justice Tindal; who thus elegantly occupied the evening of a tedious assize-day at Bedford. The observations were made with the 8½-foot equatoreal, charged with an eye-piece magnifying ninety-three times.

becomes dazzled. The ancients allotted to them only seven stars; but in modern catalogues, their numbers have run thus:

Kepler	. . . . 32 stars	Hook	. . . . . 78 stars
Galileo	 36	Jeaurat	 103
De la Hire	. . . 64	F. de Rheita	. . 188

And the zealous amateur may be assured, that there are yet many recruits for him who will undertake an exact chart of them, the which is still a desideratum, the cluster being directly in the Moon's path, and therefore the site of abundance of occultations. This part formerly constituted the third Lunar Mansion; and is so generally known, that its alineation need hardly be pointed out; yet it may be added, that an imaginary line through the wain of the Great Bear, passing Capella, leads to the Pleiades; or, from the southward, a line from Sirius, carried over Orion's belt, meets them.

An interest in the Pleiades is strongly excited by Job's beautiful allusion to God's power, in the ninth chapter of his book. We are held to deal largely in chronology when, by reducing the *occasus matutinus* of these stars—twenty-five days after the autumnal equinox—to this time, we find that 2480 years have elapsed since the days of Thales; but here we have recorded evidence of their being well noticed 3362 years ago! Look also to the thirty-eighth chapter, where, in convincing Job of ignorance and imbecility, the Omnipotent demands,

> Canst thou bind the sweet influences of the Pleiades, or loose the bands of Orion?
> Canst thou bring forth Mazzaroth in his season? or canst thou guide Arcturus with his sons?
> Knowest thou the ordinances of Heaven? canst thou set the dominion thereof in the earth?

Now this splendid passage, I am assured, is more correctly rendered thus:

> Canst thou shut up the delightful teemings of Chimah?
> Or the contractions of Chesil canst thou open?
> Canst thou draw forth Mazzaroth in his season?
> Or Aish and his sons canst thou guide?

In this very early description of the cardinal constellations, *Chimah* denotes Taurus, with the Pleiades; *Chesil* is Scorpio; *Mazzaroth* is Sirius, in the "chambers of the south;" and Aish the Greater Bear, the Hebrew word signifying a *bier*, which was shaped by the four well-known bright stars, while the three forming the tail were considered as the children attending a funeral. St. Augustin, in his annotations on the above passage, assures us that under the Pleiades and Orion, God comprehends all the rest of the stars, by a figure of speech, putting a part for the whole; and the argument is,—The all-powerful Deity regulates the seasons, and no mortal can intermeddle with them, or presume to scan the ordinances of Heaven.

This beautiful group of stars also attracted very early attention in Greece; and Hesiod, in the opening of the second book of *Works and Days*, has a truly astronomical passage upon the Pleiades, nearly 1000 years B.C. It is thus rendered by Cooke:

> There is a time when forty days they lie,
> And forty nights, conceal'd from human eye,
> But in the course of the revolving year,
> When the swain sharps the scythe, again appear.

Among the classical ancients the heliacal rising of the Seven Stars was esteemed the most favourable season for setting out on a voyage, though rain and storms were frequently then prevalent, whence Ideler thinks they merit the appellation of *Schiffahrts-gestirn.* Some savans tell us, that from the custom of letting fly a pigeon on the occasion, for auspices, they were named the Pleiades, or doves: others say the designation is derived from πλεῖν, to sail; while another class insist that it is derived from πλέος, full, from the genial bearings of the asterism. Thus etymologists dock and stretch words, and limbs of words, after a Procrustean fashion, to suit their own theories, a practice by which they fall into many a trap, even more fatal than that which assumed the Mount Sier of Ezekiel for Monsieur over-the-way. Of this system of convertible terms and changeable terminations, which form the etymological battery, a notable exposé occurs in Townsend's scourging of Sir W. Drummond; from which we may instance the group in question, *Succoth Benoth,* or Pleiades, on the back of " *Tur, Tor, Tau,*" whence is derived *Turris,* Τορσις, Ταρσος, Τυρσος, Ταυρος, and *Taurus,*" the Bull. By the way, Aldebaran was called *Taliyu-n-nejm,* as following or driving the Pleiades: can this have engendered the *tally-ho* of earthly chases? I have elsewhere remarked, what a capital hit a sharp wit might make between *Almack's* famous ball-room, and the beautiful double star *Al'mak,* which being on Andromeda's right foot may be assumed to symbolize dancing.

It may also be mentioned, that the night star to which Mahomet devotes the eighty-sixth chapter of the Koran, has been said to refer to the Pleiades; but I see no reason for restricting to any shining object, so vague an epithet as the "star of piercing brightness," which appeared by night. A more legitimate reason for supposing it to allude to this group rather than to *al-tárik,* the morning star, or *al-thákib,* Saturn the piercer, is the allusion to its bringing back the rain, in the Sura quoted: at least, so the commentator fancifully applies it.

CXLIII.   30 TAURI.

Æ 3ʰ 39ᵐ 30ˢ   Prec. + 3ˢ·27
Dec. N 10° 38'·8   —— N 11"·53

Position 58°·2 (to 8)   Distance 9"·2 (to 4)   Epoch 1833·85
—— 58°·5 (to 8)   —— 9"·0 (to 5)   —— 1839·90

A delicate double star, on the left shoulder-blade of Taurus, indicated by a line drawn from ζ Tauri, in the south horn, under Aldebaran, and continued as far again. A 6, pale emerald; B 10, purple. This elegant but difficult object is 66 Ḥ. iii., and the measures recorded are :

Pos. 72° 45' Dist. 11"·27 Ep. 1782·69

These results, as compared with the above, would have excited attention, but that H. has shown, from the existence of some error, that " no conclusion respecting the motion or rest of this star can be formed."

Between the time of Iμ.'s observations and my own, I find the following astrometrical details:

$$\text{S. Pos. } 58° \ 46' \quad \text{Dist. } 9''·867 \quad \text{Ep. } 1824·98$$
$$\Sigma. \quad\quad 57° \ 54' \quad\quad\quad 8''·897 \quad\quad\quad 1830·71$$

A small amount of proper motion must not be overlooked:

$$P....\ Æ + 0''·13 \quad \text{Dec.} - 0''·12$$
$$B.... \quad\ + 0''·03 \quad\quad\quad - 0''·05$$

CXLIV. 27 PLEIADUM.

$$Æ \quad 3^h \ 39^m \ 39^s \quad\quad \text{Prec.} + \quad 3^s·55$$
$$\text{Dec. N } 23° \ 33 ·6 \quad\quad \text{——} \quad N \ 11''·52$$

Position $238°·2$ (w 2)   Difference $Æ = 11^s·5$ (w 2)   Epoch $1832·96$

A bright star with a distant companion, bringing up the rear of the Pleiades. A 5, intense white; B 9, pale blue. Here the principal star is Atlas, which is marked in Σ.'s Catalogue of 1827 "fortasse cuneus;" I was therefore induced to give it a rigid examination, at various times, under my fullest powers,—but always made the disc perfectly round. On the arrival, therefore, of the grand Dorpat Catalogue, in 1837, I was not at all surprised to find that in 1836·74, Σ. gazing at this star with a power of 800, records, "Stella simplex in optima nocte." Yet, as the same excellent astronomer had undoubtedly seen it double, with a visible line between the two individuals, it should be closely watched. These were his measures:

$$\text{Pos. } 107°·5 \quad \text{Dist. } 0''·79 \quad \text{Ep. } 1827·16$$

Now as the motions of the heavenly bodies afford the most obvious instance of unlimited power, the object before us assumes the highest interest. From the conditions here stated, Atlas pursues its course with a rapid and restless activity in a circular orbital progression, performed in a plane nearly parallel to our line of vision. The revolution must consequently occupy a period so comparatively short, that imagination is confounded at its probable velocity.

A slight movement of Atlas in space is thus valued:

$$P....\ Æ + 0''·02 \quad \text{Dec.} + 0''·03$$
$$B.... \quad\ + 0''·05 \quad\quad\quad - 0''·05$$

Since the above was written, I find that Professor Struve, in examining the neighbouring star 165 P. III., with the giant refractor of 14·9 inches aperture, detected it to be double, the components being of the 8th and 10th magnitudes, and 10″ apart. Struve's 10th-magnitude will, however, be a task for most refractors. He also records η Pleiadum as being double, in the Poulkova Catalogue, 4th and 7th magnitudes, and the Sixth Class of distance.

CXLV. ζ PERSEI.

Æ   3ʰ 44ᵐ 05ˢ   Prec. + 3ˢ·74
Dec. N 31° 24′·2   —— N 11″·20

Position AB 206°·6 (*w* 6) Distance 13″·2 (*w* 4)
———— AC 198°·1 (*w* 3)   ———— 82″·9 (*w* 3)   Epoch 1832·19
———— AD 185°·0 (*w* 1)   ———— 121″·0 (*w* 1)

A delicate quadruple star, in the Hero's right foot; and about $7\frac{1}{2}°$ north—slightly following—of the Pleiades. A $3\frac{1}{2}$, flushed white; B 10, smalt blue; C 12, ash coloured; D 11, blue. This is an elegant group, to which H., No. 337 of Second Series, adds a fifth star of the 17th magnitude, at 25″ distance in the *np* quadrant. The object gave some trouble, since Ḥ.'s register 96 vi. is only for three individuals; but as I cannot think so neat and near a star as B could escape him while observing the others, I am inclined to attribute the derangement to some accidental oversight in taking his distances, and that the original " treble" he noted were A, B, and C; whose details would then be:

Pos. AB 203° 24′ Dist. *caret*
     AC 195° 00′   71″·43   Ep. 1782·66

An estimated notice of the star D, during a pressure of work, may have occasioned all the discrepancies; we therefore now turn to the results of S., No. 441:

Pos. AB 204° 58′ Dist. 13″·30
     AC 198° 46′   84″·38   Ep. 1825·00
     AD 184° 33′   119″·07

The difference of angle between A and B, may be owing rather to difficulty of observation than change.

CXLVI. 43 PERSEI.

Æ   3ʰ 44ᵐ 44ˢ   Prec. + 4ˢ·40
Dec. N 50° 13′·4   —— N 11″·16

Position 32°·0 (*w* 1) Distance 75″·0 (*w* 1) Epoch 1830·73

A wide and delicate double star, over the Hero's left thigh in the Galaxy; where it follows Mirphak by about 8°, a little to the north. A $6\frac{1}{2}$, white; B 10, red. This object is 41 Ḥ. v., enrolled in September, 1781, but no measures given. There are some companions in the field, of which a 9th and 11th in the *sf* quadrant form a coarse pair, at a Δ Æ = 50ˢ·5. The principal pair constitute No. 440 of S. A is said to be variable, by Taylor of Madras; and a proper motion is assigned to it thus:

P....Æ + 0″·21   Dec. — 0″·15
B....   + 0″·20   — 0″·16
A....   + 0″·16   — 0″·17

CXLVII. 32 ERIDANI.

Æ $3^h\ 46^m\ 16^s$ Prec. $+\ 3^s\!\cdot\!14$
Dec. S $3°\ 25'\!\cdot\!9$ —— N $11''\!\cdot\!04$

Position $347°\!\cdot\!4$ (w 5) Distance $7''\!\cdot\!3$ (w 4) Epoch $1831\!\cdot\!91$
—— $346°\!\cdot\!8$ (w 9) —— $6''\!\cdot\!8$ (w 9) —— $1838\!\cdot\!09$
—— $346°\!\cdot\!5$ (w 8) —— $6''\!\cdot\!6$ (w 7) —— $1843\!\cdot\!16$

A very neat double star, between the chest of Taurus and the River; and a line carried from γ Eridani to the following part of the Pleiades, passes it at rather better than a quarter of the distance. A 5, topaz yellow; B 7, sea-green; the colours in brilliant contrast. This fine object is 36 ♓ II., and was thus measured by its discoverer:

Pos. $343°\ 23'$ Dist. $4''\!\cdot\!33$ Ep. $1781\!\cdot\!81$

between which and my measures were these examinations:

	Pos.	Dist.	Ep.
H. and S.	$349°\ 01'$	$8''\!\cdot\!08$	$1821\!\cdot\!90$
Σ.	$349°\ 45'$	$6''\!\cdot\!75$	$1825\!\cdot\!00$
D.	$347°\ 14'$	$7''\!\cdot\!02$	$1830\!\cdot\!82$

From all which we may conclude that there has been little or no change, though the star is sufficiently easy for the results to have been more coincident.

CXLVIII. ε PERSEI.

Æ $3^h\ 47^m\ 08^s$ Prec. $+\ 3^s\!\cdot\!99$
Dec. N $39°\ 32'\!\cdot\!4$ —— N $10''\!\cdot\!98$

Position $9°\!\cdot\!1$ (w 6) Distance $8''\!\cdot\!4$ (w 6) Epoch $1832\!\cdot\!83$

A neat double star, under the right knee of Perseus; where it will be struck by a line led from the Pleiades due north through ζ Persei, and continued a little more than as far again: i. e. about 16° on the whole. A 3½, pale white; B 9, lilac. This fine and delicate object is 22 ♓ II., and was thus measured at its discovery:

Pos. $8°\ 32'$ Dist. $8''\!\cdot\!00$ Ep. $1780\!\cdot\!59$

The subsequent observations of H. and S., Σ. D. and myself, confirm its fixity. There is a third star in the sf, about 90″ distant.

As the components of ε Persei were not too faint to bear a trifling loss of light, I successfully employed a method of separating them which was suggested to me by Sir John Herschel, viz. a central paper disc, of two inches diameter, on the object-glass.

CXLIX. γ^1 ERIDANI.

AR. 3ʰ 50ᵐ 34ˢ   PREC. +   2ˢ·79

DEC. S 13° 58′·0   ——— N 10″·73

POSITION 286°·0 (w 1)   DIFFERENCE AR = 30ˢ (w 1)   EPOCH 1836·89

A Greenwich star, with a distant companion, preceding the bunch of τ's with which Bayer's map is disfigured; to be readily identified by shooting a ray from Procyon through the cluster in Orion's sword, and extending it nearly as far again to the eastward, or by a like process with Capella and the Hyades. A 2½, yellow; B 10, pale grey. It is in the south part of the upper reach of the River; and there is a third star, of the 11th magnitude, in the *sp* quadrant. The leader has a distinct movement through space, to which the following values are assigned:

$$P.... \text{ } AR + 0''·16 \quad \text{Dec.} - 0''·11$$
$$B.... \quad + 0''·09 \quad \quad - 0''·10$$

γ^1 Eridani is called Zaurak, from the *Neyyir-al-Zaurak*, or bright star of the boat, of the Arabians: and being at the flexure of the River, as well as large and bright, seems to be the one alluded to by Hipparchus, *Patav. Uranolog.*, as that which the equinoctial colure passed through in the time of Eudoxus. The same colure, however, could not have cotemporaneously passed through the right hand of Perseus.

CL. 213 P. III. TAURI.

AR. 3ʰ 51ᵐ 27ˢ   PREC. +   3ˢ·47

DEC. N 22° 44′·7   ——— N 10″·66

POSITION AB 128°·1 (w 9)   DISTANCE 7″·2 (w 9) ⎫
———————— BC 240°·0 (w 2)   ————— 60″·0 (w 1) ⎬ EPOCH 1835·12
⎭

A delicate triple star, in the neck of the Bull, at about one-third of the distance from the Pleiades towards the Hyades, and slightly to the north of the line drawn between them. A 7½, white; B 8, grey; C 12, blue. This neat object was discovered by Σ., and is No. 479 of the Dorpat Catalogue: and it may, at some distant period, prove to be a system of more intricate combination, than the present results on short epochs promise. Though registered "triplex," the distant individual, C, appears to have escaped S.'s telescope, whose measures of A and B, the first I meet with, are thus:

Pos. 127° 41′ Dist. 7″·208 Ep. 1823·98

CLI. 53 ⬡. IV. CAMELOPARDI.

Æ 3ʰ 53ᵐ 29ˢ   Prec. + 5ˢ·07
Dec. N 60° 23′·5   —— N 10″·51

Mean Epoch of the Observation . . . . 1837·79

A bright planetary nebula, of a bluish white tint, about 60″ in diameter, on the hind flank of the Camelopard. It is in a rich field of small stars, and was first registered by ⬡. in November, 1787, as an object whose light was uniform and definition abrupt. It is a curious body; and was watched under the total lunar eclipse of the 13th of October, 1837, being well seen during that shadowy obscurity which an Italian would call *un tenebroso orrore.*

Closely following the north vertical of this object, and about half a degree from it, is a beautiful and brilliant field of stars, the compact portion of which is 47 ⬡. VII., discovered in 1787. Many of the components of this group are in pairs, the brightest of which is a neat double star, both of the 7th magnitude, and decidedly red.

CLII. 60 ⬡. VII. PERSEI.

Æ 3ʰ 58ᵐ 11ˢ   Prec. + 4ˢ·40
Dec. N 49° 04′·5   —— N 10″·16

Mean Epoch of the Observation . . . . 1837·74

A pretty compressed oval group of small stars in the left knee of Perseus, nearly mid-way between λ and μ, in the space extending from Mirphak to Capella. It is a well marked object, with a crown of larger ones around, somewhat in the form of the letter D. It was first registered by ⬡. in December, 1790, and is in a very rich vicinity of splashy groups of stars, one of which to the *nf*, is magnificently radiated, and formed like a badge of knighthood. This figure will identify 60 ⬡. VII., an object which—however insignificant and dim a blot it may appear—is a myriad of worlds, for a powerful instrument reveals even thousands of stars in it: and various late operations show, that we have not yet arrived at our maximum of optical prowess. But J. Harris, F.R.S., tells us, even in 1729, that he does not "think our telescopes will be much farther improved!"

CLIII. 69 ♅. IV. TAURI.

Æ 3ʰ 59ᵐ 06ˢ   Prec. +   3ˢ·74
Dec. N 30° 20′·5   —— N 10″·02

MEAN EPOCH OF THE OBSERVATION . . . . 1837·68

A nebulous star over the Bull's neck, about one quarter the distance of a line between the following portion of the Pleiades and Capella. In the large reflectors this object presents an extraordinary aspect, but with my telescope looks only burred. It was first registered by ♅. in November, 1790, under this announcement: " A most singular phenomenon; a star 8th magnitude, with a faint luminous atmosphere of a circular form, about 3′ in diameter. The star is perfectly in the centre, and the atmosphere is so diluted, faint, and equal throughout, that there can be no surmise of its consisting of stars, nor can there be a doubt of the evident connection between the atmosphere and the star." From this wonderful aspect ♅. draws the following consequences. Granting the connection between the star and the surrounding nebulosity, if it consist of stars very remote which give the nebulous appearance, the central star, which is visible, must be immensely greater than the rest; or if the central star be no bigger than common, how extremely small and compressed must be those other luminous points which occasion the nebulosity? As, by the former supposition, the luminous central point must far exceed the standard of what we call a star, so, in the latter, the shining matter about the centre will be much too small to come under the same denomination; we therefore either have a central body which is not a star, or a star which is involved in a shining fluid, of a nature totally unknown to us. ♅. maintained at first, that all nebulæ were stellar masses; but it will be obvious to those who have studied Sir William's condensation system, after the palinody of 1791, that he adopted the last opinion on further experience. This luminous matter seems more fit to produce a star by its condensation, than to depend on the star for its existence; but, after all, it may be, that the star happens to fall in a line with the centre of the nebula, so as to be connected optically but not physically. See 19 ♅. vi., Æ 15ʰ 8ᵐ.

CLIV. μ PERSEI.

Æ 4ʰ 03ᵐ 10ˢ   Prec. + 4ˢ·36
Dec. N 47° 59′·7   —— N 9″·78

POSITION 230°·5 (w 2)   DISTANCE 92″·7 (w 2)   EPOCH 1832·10

A star with a distant companion, on the left knee of Perseus, and nearly in mid-distance between Mirphak and Capella. A 4½, greenish yellow; B 10, pale blue; the two pointing to some others at a distance in the *sp* quadrant. It was recorded by ♅. in August, 1780, but without other measures than an estimated distance; but his son applied the micrometer, and gained this result:

Pos. 231° 12′ Dist. 91″·56 Ep. 1821·94

Proper motions in space have been detected in μ Persei, of which the following rigorous comparisons afford the best results:

$$P....\text{Æ} + 0''{\cdot}09 \quad \text{Dec.} - 0''{\cdot}18$$
$$B.... \quad\quad + 0''{\cdot}03 \quad\quad\quad - 0''{\cdot}06$$

While this was in the press, I learned that Professor Struve had, with the giant refractor of Poulkova—14·93 inches aperture—detected a most minute acolyte within 12″ distance from A.

CLV. 39 ERIDANI.

Æ 4ʰ 06ᵐ 48ˢ Prec. + 2ˢ·85
Dec. S 10° 39′·4 —— N 9″·51

Position 154°·0 (*w* 4) Distance 7″·1 (*w* 2) Epoch 1832·07

A delicate double star, under the *nf* bend of the river, at one-fifth of the line which the eye carries from γ Eridani to γ Orionis, or nearly 6° from the former, where it is so insulated as to be readily identified. A 5, full yellow; B 11, deep blue—and nearly points to an 11th-magnitude in the *sf* quadrant. This elegant object was discovered by ♅. in January, 1785, but not having taken any measures he did not publish it till he gave the 145 New Double Stars in 1821. By Σ.'s observations it was thus:

Pos. 152° 12′ Dist. 6″·28 Ep. 1833·14

A movement is attributed to 39 Eridani, to the following amount:

$$P.... \text{Æ} + 0''{\cdot}07 \quad \text{Dec.} - 0''{\cdot}17$$
$$Br... \quad\quad + 0''{\cdot}10 \quad\quad\quad - 0''{\cdot}12$$
$$B.... \quad\quad + 0''{\cdot}03 \quad\quad\quad - 0''{\cdot}05$$

CLVI. 26 ♅. IV. ERIDANI.

Æ 4ʰ 06ᵐ 50ˢ Prec. + 2ˢ·79
Dec. S 13° 09′·1 —— N 9″·50

Mean Epoch of the Observation . . . . 1837·90

A planetary nebula under the *nf* bend of the River, about 4½° from γ Eridani in the direction of Rigel. A splendid though not very conspicuous object, of a greyish white colour; it is somewhat like a large star out of focus, with a planetary aspect. ♅., who observed it on the 1st of February, 1784, remarked that it was slightly elliptical, with an ill-defined disc; and concluded it might probably be a very compressed

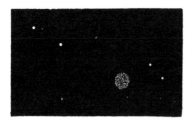

cluster of stars at an immense dis-tance. The limited aperture of my telescope only permits the object to appear in a spherical form; but the conjectural disclosure is the same. The place was carefully differentiated with 39 Eridani; and there are several telescopic stars in the field, of which two of the 8th magnitude in the *sp* quadrant, point exactly upon it, as in the annexed diagram, where the nebula is shown under its best aspect, highly magnified.

CLVII.   40 ERIDANI.

$\mathcal{R}$ 4ʰ 7ᵐ 56ˢ   Prec. + 2ˢ·90
Dec. S 7° 54′·5   —— N 9″·42

Position 107°·6 (w 8)   Distance 83″·9 (w 6)   Epoch 1837·09

A very coarse double star, in the *nf* reach of the flexuous River, designated *Keïd*, from the Arabic *al-Kaïd*, the egg-shells; being rather better than a degree to the *sf* of *o* Eridani, or *Beïd*, the egg, so called from its whiteness, and forming, with the stars around, *Az-ha-l-na'ám*, the ostrich's nest. A 5, orange colour; and B 9½, sky blue; other stars follow in the field. This object is remarkable for its amount of proper motion, being, as far as I yet know, second only to 61 Cygni, and therefore an object of very considerable interest*. It is No. 80 ♄. ɪɪ., and was thus measured:

Pos. 107° 53′   Dist. 81″·78   Ep. 1783·00

which determinations, compared with those of *Σ*. S. and myself, in a period of fifty-four years, manifest the physical connexion of these stars, since their relative position has scarcely changed a second, although the greater individual has performed so large a proper movement as to amount nearly to 250″ to the s.w. The values of the proper motions assigned to A are these:

P....$\mathcal{R}$ – 2″·21   Dec. – 3″·60
B....   – 2″·16   – 3″·45
A....   – 2″·19   – 3″·45

* While this sheet is actually under revise, the indefatigable Professor Bessel writes me—"A series of observations about the annual parallax of a ꜱᴇᴄᴏɴᴅ star (No. 1830 of Groombridge, *the proper motion of which over-rates that of* 61 *Cygni*,) which is now going on here, has been interrupted by bad weather. This star is less favourably situated, though its north polar distance is nearly the same as that of the other." Reducing this remarkable star from Groombridge's Catalogue, to the epoch of this Cycle, it will give for $\mathcal{R}$ 11ʰ 44ᵐ, and for Declination 38° 52′ north. The proper motions are stated to be in $\mathcal{R}$ = + 5″·167 and in Dec. – 5″·699.

CLVIII. φ TAURI.

Æ 4ʰ 10ᵐ 31ˢ   PREC. + 3ˢ·67
DEC. N 26° 57′·8   —— N 9″·21

POSITION 241°·8 (*w* 7)   DISTANCE 55″·9 (*w* 7)   EPOCH 1832·86

A wide double star, in the upper part of the Bull's neck; within the mid-distance from Aldebaran towards ε Persei. A 6, light red; B 8½, cerulean blue. This object is 13 ♓. v., found in 1779; but the register only records a distance of 55″·62, which is marked "inaccurate," albeit it proves to be so neat an estimation. It is No. 40 of H. and S., who gave these measures:

Pos. 240° 27′ Dist. 56″·84 Ep. 1821·95

The larger component has a proper movement in space, of which the registered value is:

P.... Æ + 0″·08   Dec. − 0″·10
B....   − 0″·05   − 0″·04

CLIX. γ TAURI.

Æ 4ʰ 10ᵐ 41ˢ   PREC. + 3ˢ·39
DEC. N 15° 14′·1   —— N 9″·20

POSITION 291°·0 (*w* 1)   DIFFERENCE Æ = 17″·8 (*w* 1)   EPOCH 1835·17

A bright star with a distant telescopic companion, in the Bull's nostril. A 3½, fine yellow; B 11, pale blue, preceded by another small star in the *sp* quadrant. This is *Hyadum primus*, or the leader of the Hyades, which, as the name implies, was esteemed a showery group; whence the *pluviasque Hyadas* of Virgil, and the *moist daughters* of Spenser. The family of Atlas was mentioned at η Tauri, but the Hyades were considered to be another batch of his daughters; though some, to lessen his burthen, dubbed them the *Dodonides*, or nurses of Bacchus. The ancients were not agreed as to their number, for while Thales merely reckoned the two eyes, α and ε, Euripides counted three, and Hesiod five. Though the identity of this star must be pretty well established, it may be stated, that it lies about one-third of the distance from the Pleiades to the cluster in Orion's sword. But we learn from the poetaster that this direction is almost needless; for

Among those gorgeous hosts aloft   so gloriously shown,
The Hyades, and Pleiades,   to all who seek are known.

Pliny gives the name Palilicium to the Hyades, while others have made it proper to Aldebaran, because they rose heretofore at Rome, on the feast day of Pales; and Ovid lumps them together as Sidus Hyantis. The group was also called Y-psilon—the Pythagorean symbol of human life—from its shape; and from thence the Roman V, α and ε being the extremes, and γ the angular point. From a notion, either that the same letter resembles a pig's jaws, or that Aldebaran with the Hyades were like a sow with her litter, the Latins designated them *Suculæ*. Cicero, however, thinks the name a corruption, from having mistaken the Greek word ὕες, *pigs*, for ὕειν, *to rain*. It must not be forgotten that γ Tauri has a very appreciable proper motion in the awful void, the amount of which is thus severally given:

$$
\begin{array}{lll}
P.... & \mathcal{R} + 0''\cdot14 & \text{Dec.} - 0''\cdot09 \\
Br... & + 0''\cdot25 & - 0''\cdot10 \\
B.... & + 0''\cdot18 & - 0''\cdot02 \\
\end{array}
$$

CLX. χ TAURI.

$$
\begin{array}{ll}
\mathcal{R} \quad 4^h\ 12^m\ 51^s & \text{Prec.} + 3^s\cdot63 \\
\text{Dec. N } 25°\ 14'\cdot7 & \text{——— N } 9''\cdot04 \\
\end{array}
$$

Position 25°·1 *(w 8)*   Distance 19''·3 *(w 8)*   Epoch 1831·93

A neat double star, at the back of the Bull's ear; where with υ it forms what the Arabians termed *Al Kelbeïn*, or the two Dogs. A 6, white; B 8, pale sky-blue. The alignment of χ is not difficult: a ray being shot from Castor through β Tauri, the tip of the Bull's northern horn, and extended about 15° further, towards the Pleiades, strikes upon it: a line from Rigel through Aldebaran, carried half that distance beyond the Bull's eye, also hits it. This object is 10 ℍ. iv., discovered in 1779, but not measured. The point of departure is therefore the epoch of H. and S., whose results are:

Pos. 23° 56'   Dist. 19''·96   Ep. 1822·90

But the star being rather difficult under illumination, the difference of position offers nothing to calculate upon. The leader's proper motion has been thus registered:

$$
\begin{array}{lll}
P....\mathcal{R} - 0''\cdot05 & \text{Dec.} - 0''\cdot23 \\
B..... + 0''\cdot08 & - 0''\cdot04 \\
\end{array}
$$

CLXI. 62 TAURI.

$$
\begin{array}{ll}
\mathcal{R} \quad 4^h\ 14^m\ 21^s & \text{Prec.} + 3^s\cdot60 \\
\text{Dec. N } 23°\ 55'\cdot4 & \text{——— N } 8''\cdot92 \\
\end{array}
$$

Position 290°·0 *(w 9)*   Distance 28''·6 *(w 8)*   Epoch 1835·98

A neat double star, on the tip of the Bull's left ear, at rather more than one-third of the distance from the Pleiades to ζ. A 7, silver white; B 8½, purple; and there are several small stars in the field. This is a fair object for a moderate telescope; being 109 ℍ. iv., thus measured when first registered:

Pos. 291° 12'   Dist. 28''·08   Ep. 1782·90

On comparing these data with the observations of H. and S. in 1821, of Σ. ten years afterwards, and my own at the epoch stated, there seems to be no appreciable change during an interval of fifty-three years; though Piazzi's remark—"Duplex. Comes 8ᵃ magnit. præcedit 1″·8 temporis, 10″ circiter ad Boream"—interposes a mystification. Meantime it has been slightly affected with proper motion, the amount of which is thus severally given:

$$P....\!R + 0''\cdot26 \quad \text{Dec.} - 0''\cdot11$$
$$B.... \quad\; + 0''\cdot06 \quad\quad\quad - 0''\cdot01$$

CLXII. 1 CAMELOPARDI.

Æ 4ʰ 19ᵐ 23ˢ   Prec. + 4ˢ·71
Dec. N 53° 33'·3   —— N 8″·52

Position 307°·9 (w 9)   Distance 10″·4 (w 9)   Epoch 1838·09

A neat double star, between the animal's hind hoofs; and nearly in mid-distance between a Persei, and δ on the head of Auriga. A 7½, white; B 8½, sapphire blue. The object consists of Piazzi's 83 and 84 of Hora IV.; and though the process of obtaining a position and distance from the observed Æ and Dec. can hardly be expected to give a result absolutely exact, the Palermo observations merit grave consideration, and are therefore placed as the earliest epoch, to the conclusions of the principal astrometers, as thus shown:

	Pos.	Dist.	Ep.
P.	299° 00'	12″·00	1800·00
H. and S.	306° 26'	10″·45	1822·05
Σ.	307° 05'	10″·13	1830·57
D.	307° 10'	10″·68	1832·90

CLXIII. θ^1 TAURI.

Æ 4ʰ 19ᵐ 26ˢ   Prec. + 3ˢ·41
Dec. N 15° 36'·0   —— N 8″·52

Position 166°·7 (w 9)   Distance 336″·8 (w 9)   Epoch 1834·13

A wide pair of stars on the Bull's face, where it forms the southern vertex of a small triangle with Aldebaran and the Hyades. A 5, pearly white; B, which is θ^2, 5½, yellowish. From an apparent identity in the values and signs of proper motions in space, the components of this object are suspected of being in physical connexion; and imagination is confounded at the probable period of the *magnus annus*, should the idea ultimately prove correct, for its curve defies human calculation. In order to aid future inquiry, the above mean apparent place and micrometrical measurements were made under the greatest caution and care, for the amount and direction of proper motion stands thus:

$$
\begin{array}{llll}
P. & \begin{cases} \theta^1 \, \mathbb{R} + 0''\cdot13 & \text{Dec.} & -0''\cdot10 \\ \theta^2 \phantom{\mathbb{R}} + 0''\cdot10 & & -0''\cdot05 \end{cases} \\
B. & \begin{cases} \theta^1 \phantom{\mathbb{R}} + 0''\cdot05 & & -0''\cdot01 \\ \theta^2 \phantom{\mathbb{R}} + 0''\cdot18 & & -0''\cdot01 \end{cases} \\
A. & \begin{cases} \theta^1 \phantom{\mathbb{R}} + 0''\cdot08 & & -0''\cdot02 \\ \theta^2 \phantom{\mathbb{R}} + 0''\cdot11 & & -0''\cdot07 \end{cases}
\end{array}
$$

Mr. Baily thinks it probable, that Ptolemy observed θ^1 and θ^2 as one star, and of course a *fixed* star; but from the quality of the capital now accumulating for posterity, it is probable that a few hundred years hence will find the list of *inerrantes*, or stars not wandering, a term so general down to our own day, very greatly restricted. And while on this, I am glad to produce another proof from Piazzi's *Præcipuarum stellarum* INERRANTIUM, of the excellence of the Palermo observations; for a reduction of the mean places, together with an application of the proper motions, affords the two following periods of comparison:

Pos. 166°·0   Dist. 340″·0   Ep. 1800
   167°·0   333″·0   1840

CLXIV.   217 ♅. I. AURIGÆ.

$\mathbb{R}$ 4^h 19^m 43^s Prec. + 3^s·91

Dec. N 34° 54′·9   —— N 8″·49

MEAN EPOCH OF THE OBSERVATION . . . . 1836·77

A round pale nebula, between the legs of Perseus and Auriga, of a slight cream-colour. It is 315 of H.'s Catalogue; and is so faint that probably I should have overlooked it, but for his having described its place so exactly, as "inclosed among six stars." It was discovered by ♅. in December, 1788, who remarked that "it stands nearly in the centre of a trapezium." Its approach is announced by a star of the 8th magnitude, in the *np* quadrant; and it lies nearly in mid-distance between Capella and the Pleiades, at about 12° from the latter.

CLXV.   80 TAURI.

$\mathbb{R}$ 4^h 21^m 01^s Prec. + 3^s·40

Dec. N 15° 17′·0   —— N 8″·39

POSITION 11°·0 (*w* 2)   DISTANCE 1″·4 (*w* 2)   EPOCH 1837·22

—— 13°·9 (*w* 5)   —— 1″·6 (*w* 3)   —— 1839·16

—— 15°·2 (*w* 8)   —— 1″·8 (*w* 4)   —— 1843·11

A close double star, on the Bull's face, and about 1½° south-west of Aldebaran. A 6, yellow; and B 8½, dusky. This beautiful object is

Σ.'s No. 554, and, as one of his *vicinæ*, is of no very easy measurement. The mean of his observations affords this result:

Pos. 12° 55′   Dist. 1″·74   Ep. 1831·18

Although I had strong doubts of my own angle of position for 1837, which is a mean of some very varying ones, still on viewing the results of Σ. with my weights for those of 1839, I could not but infer a small orbital movement; and this suspicion is greatly strengthened by my last measures at Hartwell.

CLXVI. 57 PERSEI.

$\mathbb{R}$ 4ʰ 22ᵐ 10ˢ   Prec. + 4ˢ·19
Dec. N 42° 42′·9   —— N 8″·30

Position 199°·8 (*w* 5)   Distance 110″·3 (*w* 6)   Epoch 1833·08

A wide double star, in the left ancle of Perseus, with several small stars in the field, of which a remarkable one of the 11th magnitude is to the *np* of A, and seems to have escaped the eye of ⩜. A and B are both of the 8th magnitude, and white; being Piazzi's 101 and 104 of Hora IV. The object is 99 ⩜. VI., and with a reduction from the Palermo Catalogue may be thus registered:

⩜.   Pos. 198° 09′   Dist. 96″·42   Ep. 1783·66
P.   198° 30′   116″·00   1800·00

It was next measured by H. and S., with these results:

Pos. 198° 52′   Dist. 110″·19   Ep. 1821·91

whence an extraordinary change of distance was shown. Subsequent measures, however, indicate some error in ⩜.'s register. It will be found nearly in mid-distance of a line run from *a* Persei to β Tauri, at about 10° from the former.

CLXVII. 58 PERSEI.

$\mathbb{R}$ 4ʰ 25ᵐ 37ˢ   Prec. + 4ˢ·13
Dec. N 40° 55′·8   —— N 8″·02

Position BC 30°·3 (*w* 5)   Distance 11″·6 (*w* 5)   Epoch 1838·21
———— 29°·8 (*w* 7)   ———— 11″·8 (*w* 5)   ———— 1843·18

A star on the left heel of Perseus; it is an insulated object, assumed as a pointer to the distant pair in the *sf* quadrant, with $\triangle \mathbb{R} = 4$ˢ. A 5½, orange tint; B 7½, greenish; and C 9, lilac. A line led to the north-west from the preceding star of Orion's belt through Bellatrix, and 35° further, strikes upon it; and it precedes the mid-distance of a ray shot from a Persei to β Tauri. The double star is 65 Ḥ. III., and there is another couple in the *sp* part of the field, of the 10th and 11th magnitudes, at about the same distance from A as the other pair. The results of the measures of B and C have been very accordant in distance; and also in angle, during the last twenty years, being thus:

	Pos.	Dist.	Ep.
Ḥ.	41° 06′	11″·36	1782·69
H and S.	30° 00′	12″·46	1821·97
Σ.	29° 50′	11″·71	1828·72

CLXVIII. a TAURI.

Æ.	4ʰ 26ᵐ 44ˢ	Prec.	+ 3ˢ·42
Dec.	N 16° 10′·9	——	N 7″·93

Position 35°·9 (*w* 6)   Distance 107″·9 (*w* 4)   Epoch 1836·98

A standard Greenwich star, with a telescopic companion, in the southern eye of Taurus. A 1, pale rose-tint; B 12, sky blue; a magnitude assigned on deliberate comparison, for I was surprised on readily seeing it with my 5-foot telescope of 3¾ in. aperture, but the Rev. W. R. Dawes has since shown me a diagram which he made of it in November, 1828, with a 3½-foot telescope, of 2¾ inches aperture, and a negative eye-piece magnifying 200 times. This wide object is 66 Ḥ. VI., and was thus first registered:

Pos. 37° 02′   Dist. 87″·79   Ep. 1781·97

whence it is clear that the position has undergone no appreciable change, the large star having a minute retrograde proper motion. The distance may have increased, but such an inference cannot be drawn with certainty, as the difference very probably combines instrumental error with amount of proper motion. Aldebaran is readily found by the eye, from being exactly between Bellatrix and the Pleiades. The stars in Orion's belt also point nearly in its direction; and it is moreover easily distinguished by its red colour. The rich appearance of its vicinity has been thus eulogized by the brackish poet:

In lustrous dignity aloft,	see *alpha* Tauri shine,
The splendid zone he decorates	attests the power divine:
For mark around what glitt'ring orbs	attract the wandering eye,
You'll soon confess no other star	has such attendants nigh.

It has a slight proper motion in space, of which the following amount has been estimated:

		Dec.	
P....	Æ + 0″·04		− 0″·21
B....	+ 0″·12		− 0″·15
A....	+ 0″·08		− 0″·17

Taurus is now the second in the zodiacal march, though *only* 4000 years ago he led the celestial signs, and continued to be their leader for 2000 years. The principal star is *Al-debarán*, the hindmost, because he drives the Pleiades, whence the name of *Stella dominatrix*, and *Táliyu-l-nejm*, were also applied; but it was most popularly known among the Arabians, with whom it was no favourite, as *'aïn-al-thaur*, the bull's eye, though it was placed at a little distance from the animal's head in the ancient configurations. (See *Hyades*.) Tycho considered it to be 125 times the size of our earth, while Ricciolus worked it up to 2810 times that magnitude; such unwarranted conclusions, however, are mere dreams; give us but the parallax, and the mass will soon follow. It is a red star, and I have repeatedly seen it apparently projected on the disc of the moon, even to an amount of nearly three seconds of time, at the instant of immersion, when occulted by that body, as related in the *Memoirs of the Astronomical Society*. This phenomenon seems to be owing to the greater proportionate refrangibility of the white lunar light, than that of the red light of the star, elevating her apparent disc at the time and point of contact.

All these suppositions, however, are purely arbitrary, as other stars are liable to a similar affection; and notwithstanding that the call of the Astronomical Society for observations of the occultations of Aldebaran for 1829 and 1830, was zealously responded to from various parts of Europe, nothing satisfactory was elicited. Of six observers at the Greenwich Observatory, five distinctly saw the projection on the lunar limb; and the majority of corresponding astronomers saw the star either projected or hanging on the moon's edge: but there were several practical men who saw nothing remarkable. The fact, however, of the singular phenomenon is admitted, but subject to much diversity of opinion as to its cause; for it cannot be traced either to the character of the telescope employed, of the observer, or of the weather during the observation. To those who have not the *Memoirs of the Astronomical Society* at hand, an extract from one of my reports may be illustrative:

"October 15th, 1829. I saw Aldebaran approach the bright limb of the Moon very steadily; but, from the haze, no alteration in the redness of its colour was perceptible. It kept the same steady line *to about ¾ of a minute inside the lunar disc*, where it remained, as precisely as I could estimate, two seconds and a quarter, when it suddenly vanished. In this there could be no mistake, because I clearly saw the bright line of the Moon *outside* the star, as did also Dr. Lee, who was with me. The emersion took place without anything remarkable: the dark limb not visible. Telescope 5-foot achromatic, 3¾ aperture, power 78; adjusted on the star." Dr. Lee was watching with a smaller instrument.

H. measured the apparent diameter of this *oculus Tauri* as $1''\cdot50$; and others have attempted a substantive measurement. Its ruddy aspect has long been noted, and old Leonard Digges, in his *Prognostication Everlasting*, 1555, pronounces that it is "ever a meate rodde." Indeed, all late observers agree in its redness; but Virgil wrote

> Candidus auratis aperit quum cornibus annum
> Taurus—

which golden horns must rather refer to β and ζ, the two bright stars on the tips, than to the "horns of triumph" of the Scholiast.

To account for this constellation's comprising only half the animal, the mythologists have it, that as he personates the bull which swam away with Europa, his flanks are immersed in the billows. This is very much like the Dutch effigies behind a tree; but it does not well explain why Taurus, Pegasus, or Equuleus, are deprived of their hinder parts.

Ovid, indeed, throws a doubt upon the gender of this sign, by making it the transformation of Io,—but in either case it is still the *munus amoris*, in which the heathens delighted. The classical astronomers are, however, very weak in their mythological derivations and zodiacal origins. In the rare zodiac gold-muhrs struck by Jehángír Sháh in 1618, Taurus is represented as a complete though spiritless animal, with the gibbous hump common to Indian oxen: but on the silver rupees of the same monarch, the half animal is drawn in a bold butting attitude, exactly as described by Manilius. Yet Aratus must have seen that of Eudoxus differently placed, for he puts the Pleiades in the knees. Some of the Romans represented the animal as whole; since both Vitruvius and Pliny speak of *cauda Tauri* as being formed by the Pleiades, to the derogation of those young ladies. But the Arabians retained it merely as a section, calling *o*, or Flamsteed's No. 1, the first star in *Al Khaṭ*, the slash, or section.

Taurus is one of the old 48 constellations, and contained the Fourth Mansion of the Moon. As one of the earthy triplicity, it was held to refer to the season for cultivating fields, in allusion to which the manuscript Almanac of 1386 says, that " whoso is born in yat syne schal have grace in bestis." Novidius recognised in Taurus the ox that stood with the ass by the manger, at the blessed Nativity: " but," saith Hood, " whether there were any oxe there, or no, I know not how he will prove it." It is a very rich asterism, and its components have been thus tabulated:

Ptolemy	. . . 44 stars	Hevelius	. . . . 51 stars
Ulugh Beigh	. . 43	Bullialdus	 52
Tycho Brahé	. . 43	Flamsteed	. . . 141
Bayer	 48	Bode	 394

CLXIX.   88 TAURI.

ℛ   4ʰ 26ᵐ 53ˢ   Prec. + 3ˢ·23
Dec. N   9° 49′·6   —— N 7″·92

Position 300°·4 (*w* 6)   Distance 68″·5 (*w* 3)   Epoch 1832·93

A star with a distant companion, in the right fore-leg of Taurus, being about 6° below Aldebaran, where it forms the vertex of an acute-angled triangle with that star and Bellatrix. A 5, bluish white; and B 8½, cerulean blue. Some minute stars follow A, and there is one of the 9th magnitude in the *np* quadrant. B is No. 127 of Piazzi's Hora IV., a deduction from whose mean places are given below; and the object forms 31 ℍ. vi., discovered and registered in September, 1780; but it was not micrometrically measured till the operations of H. and S. The comparison of the previous results to my own are thus:

	Pos.	Dist.	Ep.
P.	303° 24′	64″·30	1800·00
H. and S.	293° 59′	69″·45	1822·88

A discussion of all the observations leads to the inference—should they be tolerably correct in a metric sense—that the satellite will have reached the western limit of its orbit in about a century and a half; so that after the year of our Lord 2000, its distance from A will begin to decrease, since it will commence the southern half of its orbit.

CLXX. 2 CAMELOPARDI.

$\unicode{x00C6}$ 4^h 27^m 18^s Prec. + 4^s·71

Dec. N 53° 09′·0   —— N 7″·89

Position 307°·9 (w 2)   Distance 1″·9 (w 2)   Epoch 1834·49
——— 308°·7 (w 3)   ——— 1″·7 (w 3)   ——— 1836·28

A close double star, between the animal's hind hoofs. A 5½, yellow; B 7½, pale blue. An attentive observer will pick it up by casting a line from Polaris between Capella and Algol, leading it about 9° from the former; and it will be intersected by another line, drawn from a Persei to δ in the head of Auriga. This exquisite object was discovered by Σ., No. 566, who recorded it " vicinæ;" but it is certainly wider and easier of measurement than those usually so classed by him. It may, however, be increasing its distance, albeit the mean of all my observations in 1836, afford no direct testimony of the fact. These are the results of the Dorpat observations:

Pos. 311° 40′ Dist. 1″·585 Ep. 1829·79

CLXXI. τ TAURI.

$\unicode{x00C6}$ 4^h 32^m 39^s Prec. + 3^s·58

Dec. N 22° 38′·7   —— N 7″·48

Position 209°·8 (w 8)   Distance 61″·6 (w 3)   Epoch 1831·96

A star with a companion, in a barren field, at the root of the Bull's left horn; and about 6° north of Aldebaran, on a line leading from that star to Capella. A 5, bluish white, and B 8, lilac. This object is 7 ♓. vi., discovered in October, 1779; and registered with a " pretty accurate" distance of 61″ 25‴, but no angle of position appears to have been taken. Piazzi then observed the mean places, from which we obtain the following data to compare with the micrometrical measures of Sir James South, viz.:

P. Pos. 210° 00′ Dist. 56‴·60 Ep. 1800·00
S.   211° 32′   62″·82   1824·00

On the whole, weighing the different methods and instruments employed, the fixity of these stars may be held to be here established.

CLXXII. 55 ERIDANI.

Æ 4ʰ 35ᵐ 54ˢ   Pʀᴇᴄ. + 2ˢ·87
Dᴇᴄ. S 9° 05′·9   —— N 7″·19

Pᴏsɪᴛɪᴏɴ 318°·5 (w 9) Dɪsᴛᴀɴᴄᴇ 10″·2 (w 9) Eᴘᴏᴄʜ 1832·12

A neat double star, under the *nf* extreme of the River, and close to the four vertical stars of the 4th magnitude, which Kirch classed in 1688 as *Sceptrum Brandenburgicum:* and which was revived a century afterwards by Bode. A and B are both 7½, and yellowish white; being Nos. 172 and 173 of Piazzi's Hora IV., and they constitute 99 Ḥ. ɪɪɪ., being thus first registered:

Pos. 314° 09′   Dist. 9″·16   Ep. 1783·08

It was measured by H. and S., with these results:

Pos. 318° 20′   Dist. 10″·51   Ep. 1821·97

From these data, and a position of Σ. in 1820 = 322° 01′, a slight orbital increase of angle was presumed, in a direction *sf np*, or direct; but this is not confirmed by the later observations. It may be found by the out-of-door observer, by running a line from Pollux through Betelgeuze, and leading it rather more than half as far again into the south-west, where it forms the vertex of an isosceles triangle, the base of which is formed by Rigel and β Eridani.

CLXXIII. 7 CAMELOPARDI.

Æ 4ʰ 44ᵐ 28ˢ   Pʀᴇᴄ. + 4ˢ·77
Dᴇᴄ. N 53° 29′·3   —— N 6″·48

Pᴏsɪᴛɪᴏɴ 239°·2 (w 3) Dɪsᴛᴀɴᴄᴇ 27″·0 (w 2) Eᴘᴏᴄʜ 1838·71

A delicate and very difficult double star, on the animal's hind hoof; and about two-thirds of the distance from α Persei towards δ in the head of Auriga. A 5, white; B 13, orange; and they point to a third star in the *sp* quadrant, of the 12th magnitude. It is No. 610 of Σ., whose measures gave:

Pos. 238° 32′   Dist. 25″·647   Ep. 1831·57

There was apparent evidence of a considerable amount of proper motion in this star, but it has almost disappeared before the accurate observations, and discriminating comparisons, which have latterly plied it; these are the assigned values:

P.... Æ + 0″·14   Dec. − 0″·03
B....   + 0″·03   0″·00

Camelopardus is a large but indifferent constellation of recent formation, occupying the vast *sporadic* space between the Pole-star, Perseus, and Auriga. It was introduced by Bartschius, on his four-foot globe, and is said to have been reluctantly retained by Hevelius; who, though he did prefer plain to telescopic sights, yielded only to Flamsteed in diligence and accuracy of observation, among the astronomers of his day: at all events the Camelopard is not among the new constellations which Hevelius is offering to Urania and her choice staff, in the elaborate frontispiece to his *Atlas*. The animal is fairly delineated in his map, although its very existence was, even after that epoch, questioned. He assigned it 32 stars, which Flamsteed increased to 58, and Bode to 211.

CLXXIV. ω AURIGÆ.

Æ 4ʰ 48ᵐ 24ˢ   PREC. + 4ˢ·05
DEC. N 37° 38′·5   —— N 6″·16

POSITION 351°·8 (*to* 6)   DISTANCE 6″·7 (*to* 5)   EPOCH 1831·97
——— 352°·6 (*to* 7)   ——— 7″·0 (*to* 5)   ——— 1833·88

A neat double star, preceding the hip of Auriga; and about one-third down a line passed from Capella to the Hyades. A 5, pale red; B 9, light blue. Though this fine object is well defined, from the disparity of size in the pair, it is not at all of the easiest measurement: yet owing to the pains taken by astrometers, there are few double stars of which the results are more coincident:

	Pos.	Dist.	Ep.
Ḥ.	352° 37′	5″·50 ±	1779·83
H. and S.	352° 01′	7″·89	1822·90
Σ.	351° 56′	6″·46	1828·75
H.	353° 02′	6″·79	1830·96

The star *nf* this object, 5 Aurigæ, has been found to be a most delicate double star by Professor Struve—with the giant Poulkova refractor, of 14·93 inches aperture—the components being 6th and 10th magnitudes, and 1½″ apart: it should be remarked, however, that the acolyte is Σ.'s 10th, which in some cases cannot be estimated at less than my 15th.

CLXXV. 62 ERIDANI.

Æ 4ʰ 48ᵐ 32ˢ   PREC. + 2ˢ·95
DEC. S 5° 25′·8   —— N 6″·15

POSITION 73°·6 (*to* 5)   DISTANCE 63″·8 (*to* 4)   EPOCH 1831·93

A wide double star, in the centre of the *nf* end of the River. A 6, white; B 8, lilac; a third star in the *np*, of the 10th magnitude, makes

A the apex of a scalene triangle. From a comparison of the following measures:

Ⅱ.   Pos. 74° 51'   Dist. 60"·43   Ep. 1783·04
H. and S.   74° 44'   65"·86   1821·97

the distance was considered as having increased: but my observations do not support that opinion. An imaginary line led from Mintaka, the third star in Orion's belt, close over β Eridani, touches No. 62, at about 3° west of the said β.

CLXXVI. 258 P. IV. ORIONIS.

Æ   4ʰ 49ᵐ 48ˢ   Prec. + 3ˢ·10
Dec. N 1° 25'·4   —— N 6"·04

Position 180°·4 (w 8)   Distance 2"·4 (w 5)   Epoch 1833·92

A fine double star, just preceding Orion's right knee; and at rather more than a third of the distance from Rigel to Aldebaran, where it is intersected by a line passed from Pollux through Bellatrix. A 8½, white; B 9, pale grey. This exquisite object is 63 Ⅱ. I.; and having had a retrograde angular motion of 0°·269 per annum assigned to it, was very carefully attended to. My measures, though they exhibit a slight change, do not countenance the amount mentioned; since, supposing Ⅱ.'s to form a standard for its epoch, it would only be about 5° for upwards of half a century. Since this was discussed the great Dorpat Catalogue has arrived, by which a confirmation of other measures is obtained. The whole previous results are:

Ⅱ.   Pos. 174° 51'   Dist. 2"·0 ±   Ep. 1782·85
S.   173° 49'   2"·56   1825·04
Σ.   179° 54'   2"·64   1832·09

CLXXVII. 257 P. IV. TAURI.

Æ   4ʰ 49ᵐ 55ˢ   Prec. + 3ˢ·39
Dec. N 14° 17'·6   —— N 6"·04

Position AB 303°·8 (w 3)   Distance 38"·9 (w 3) ⎫
——— AC 88°·3 (w 2)   ——— 70"·0 (w 2) ⎭   Epoch 1831·95

A wide triple star, between the Bull's ear and Orion's arm; and nearly one-third of the distance from Aldebaran towards Betelgeuze, where it is also shewn by a line carried from Sirius through Bellatrix, and extended about 10° beyond. A 7, white; B, which is No. 255 of Piazzi, 8, cerulean blue; and C 10, purple, with a minute star following it. This is a pretty though coarse object, forming a neat arc; and A and B were measured by H. and S. as 26 Bode Orionis, with results which may be thus placed with the conclusions obtained from the Æs and Decs. of the Palermitan Catalogue:

P.   Pos. 301° 00′   Dist. 44″·30   Ep. 1800·00
H. and S.   304° 25′   38″·48   1822·09

A slight movement in space is attributed to A of the following value:

P.... Æ + 0″·05   Dec. − 0″·01
B....   − 0″·09   − 0″·03

CLXXVIII. 278 P. IV. ORIONIS.

Æ 4ʰ 53ᵐ 43ˢ   Prec. + 3ˢ·10
Dec. N 1° 22′·2   —— N 5″·72

Position 49°·3 (w 8)   Distance 13″·7   Epoch 1833·92

A neat double star, on Orion's right knee. A 8½, silvery white; B 9, pale blue. These are Piazzi's No. 278 and 279 of Hora IV., where the mean places are given; but they were first micrometrically measured by Sir James South, with these results:

Pos. 48° 18′   Dist. 14″·43   Ep. 1824·97

which, compared with what I obtained, promise no great motion. He who has no equatoreal instrument, may fish up this object by carrying an imaginary line from Sirius over the cluster in Orion's sword, and about 10° beyond: or it may be sought near the mid-distance between Aldebaran and Arneb, the *lucida* of Lepus.

CLXXIX. 269 P. IV. CAMELOPARDI.

Æ 4ʰ 56ᵐ 19ˢ   Prec. + 9ˢ·70
Dec. N 79° 01′·8   —— N 5″·50

Position 348°·8 (w 5)   Distance 34″·1 (w 3)   Epoch 1833·16
——— 349°·1 (w 8)   ——— 33″·8 (w 8)   ——— 1836·25

A fine double star, at the lower part of the back of the animal's neck. A 5½, light yellow; and B 9, pale blue; while in the *np* quadrant, about 2′ distant, is the little star mentioned by Piazzi, "2′ ad Boream, 2″ temporis præcedit, alia 10ᵃ magn." This object is 19 Hevelius, the No. 634 of the Dorpat Catalogue; but the first measures I meet with are those of Sir James South:

Pos. 346° 23′   Dist. 37″·01   Ep. 1825·10

whence, by graphic comparison, a slight direct orbital motion, perhaps nearly + 0°·3 in annual amount, is implied, as well as a small decrease of distance, say − 0″·3; for S.'s observations appear to be a mean of very satisfactory observations, and I am able to place considerable weight on my own. On these grounds a highly elongated elliptic orbit is to be inferred, and a period of not less than 1000 or 1200 years. The question has since been further illustrated by the arrival of Σ.'s results:

Pos. 348°·57   Dist. 34‴·042   Ep. 1834·15

To find this object look about 10° on a line carried from Polaris between Menkalinan and Capella, a and β Aurigæ: a line from Alwaid carried through the Pole-star, also reaches it at the same distance beyond.

CLXXX.   61 ℍ. VIII. AURIGÆ.

Æ   4h 57m 11s   Prec. + 4s·04
Dec. N 37° 08'·4   —— N 5″·43

Position 220°·6 (w 8)   Distance 1″·8 (w 5)   Epoch 1832·25

A loose cluster, on the lower garment of Auriga; where a line from Betelgeuze passed over the stars ζ and β Tauri, the tips of the Bull's horns, hits it at 10° beyond. A 7, topaz; B 8, amethyst. This object is a bright though freely spread band of stars, from 8th to 13th magnitudes, having four brighter ones in a curve, of which the leader is double; and there are three other pairs. It was registered by ℍ. in January, 1787, and is No. 344 of his son's Catalogue of 1830. Just to the north of the parallel, it is followed by the beautiful double star Σ. 644, which, from its aspect, being more likely to prove an optical object than the rest, was carefully measured, as above. The determinations of other astrometers are:

Σ.   Pos. 219° 12'   Dist. 1″·61   Ep. 1828·60
H.   224° 19'   1″·99   1830·05

CLXXXI.   295 P. IV. TAURI.

Æ   4h 58m 21s   Prec. + 3s·64
Dec. N 24° 02'·9   —— N 5″·33

Position 195°·5 (w 2)   Distance 28″·0 (w 2)   Epoch 1831·94

A double star, between the horns of Taurus; where a line from Sirius passed close to the west of Betelgeuze, and led nearly as far again, will find it, lying between a and β Tauri. A 6, pearly white; B 13, pale blue. This is No. 114 ℍ. v., and though a widish object in a bare field, it is fine and delicate. ℍ., who calls A 203 Tauri, has not mentioned the quadrant in which B is placed, but assuming his angle 72° 24' to be *sp*, his measures will be thus:

Pos. 197° 36'   Dist. 30″·03   Ep. 1782·94

CLXXXII. 66 ERIDANI.

Æ. 4ʰ 58ᵐ 51ˢ   Prec. + 2ˢ·96
Dec. S 4° 52′·5   —— N 5″·29

Position 13°·8 (w 1)   Distance 47″·0 (w 1)   Epoch 1832·01

A coarse double star, close to the shin of Orion, where it will be seen closely preceding and north of β Eridani. A 6, white; and B 11, lilac; these two nearly pointing upon a third small star, near the south vertical, and there are other stars in the field. It appears on the Dorpat Catalogue, No. 642, but without measures or description: and in the edition of 1837, has the "*rej.*" against it.

CLXXXIII. β ERIDANI.

Æ. 4ʰ 59ᵐ 59ˢ   Prec. + 2ˢ·95
Dec. S 5° 17′·9   —— N 5″·19

Position 147°·5 (w 2)   Distance 120″·0 (w 1)   Epoch 1830·98

A bright star with a distant telescopic companion, on the shin-bone of Orion. A 3, topaz yellow; B 12, pale blue. I examined this object with anxious care, because in Σ.'s first Catalogue, No. 647, A is marked "vicinæ;" and he moreover considered it to be formed of two close stars of the 7th magnitude. All my endeavours, however, could only raise a round disc, and Σ. has since declared it "simplex." This star is readily found from its vicinity to Rigel, being just above it, and in the direction of the Hyades; the poetaster of these matters tells us:

> Where Rigel shows the Hero's foot,   north-westerly—not far—
> Against his leg in glory shines   the River's second star.

Many writers think this River, which, according to Sherburne, flows over the meridian at midnight in November, was originally intended to typify the Nile, and that the vanity of the Greeks led them to call it Eridanus. By other sage authorities we gather that the river represents either the Spartan *Fluviorum rex*, or the Po, or the Granicus, or Orion's river, or some other stream; while Ptolemy merely terms it the Ποταμοῦ ἀστερισμὸς, or *asterism of the river*, which is followed in the *Fluvius* of the Latins. In the early wood-cut figures of illustrations to Hyginus, Eridanus is represented as a reclining female; while in the MS. of Cicero's *Aratus*, it is delineated as a river-god, with his urn and aquatic plant. At all events it is one of the old 48 constellations, and its members have been thus successively enumerated:

Ptolemy	. . . 34 stars	Bayer	. . . . . 42 stars
Copernicus	. . 34	Bullialdus	. . . 39
Tycho Brahé	. . 19	Flamsteed	. . . 84
Kepler	 39	Bode	 343

This star is called Cursa, from the Arabic *al-kursá*, a chair or throne, and is the principal individual of the asterism seen in this hemisphere; *a* being far down in the south, though not quite at the end of the River, *ultima Fluvii*, as its name *Achernar* implies, it being from *ákher-nahr*, the latter part. It is also called *Dhalim*, the ostrich, a name given by the Bedawí Arabs, very probably before the Greek constellations were known to their countrymen; and while some called λ, *β*, ψ Eridani and τ Orionis the Giant's throne, others termed it *udh-hí*, the little nest, or place in the sand where the ostrich's egg is laid, which, by an error of transcription, became *az-ha*.

CLXXXIV. 4 ♓. VII. TAURI.

Æ   5ʰ   2ᵐ 50ˢ   Prec. + 3ˢ·45
Dec. N 16° 30'·1   —— N 5"·03

Position 60°·7 *(w 2)*   Distance 25"·0 *(w 2)*   Epoch 1837·73

A very delicate double star preceding a tolerably condensed cluster, over the right arm of Orion. A 8, yellow; B 11, bluish. This object is an outlier of a rich gathering of small stars, which more than fills the field; it was registered by ♓. in February, 1784, under an estimation of 20' or 25' of diameter, but he did not notice the pair here measured. However, Sir John Herschel thus describes it, No. 349: " Large rich cluster; stars 12 to 15 m.; fills field. Place that of a D *. The most compressed part is 42ˢ·5 foll. the D *, and 3' south of it." The whole may be fished up by carrying a line from the foremost star in Orion's belt, Mintaka, through Bellatrix, and there intersecting it by another from Aldebaran, due east towards γ Geminorum.

CLXXXV. ι LEPORIS.

Æ   5ʰ 04ᵐ 50ˢ   Prec. + 2ˢ·79
Dec. S 12° 03'·9   —— N 4"·78

Position 336°·9 *(w 4)*   Distance 15"·0 *(w 1)*   Epoch 1836·93

A fine and delicate double star, in the Hare's left ear; where a line from Betelgeuze through ε—the middle star of the belt—and extended rather more than as far again into the south-west, will pick it up. A 4½, white; B 12, pale violet, with a reddish distant star nearly north, which is the one mentioned by Piazzi, Nota 11, Hora V. This is 67 ♓. III., and No. 655 of Struve's great Catalogue, by whom these measures are given:

♓. Pos. 359° 31'   Dist. 12"·34   Ep. 1782·69
Σ.   337° 39'   12"·81   1832·25

Little of a decided character, however, can as yet be deduced from the observations of so difficult a star.

CLXXXVI. α AURIGÆ.

Æ 5ʰ 04ᵐ 53ˢ   Prec. + 4ˢ·40

Dec. N 45° 49'·7   ——— N 4"·77

Position A B 150°·9 (w 3) Distance 165"·0 (w 1)⎫
——— A C 346°·9 (w 3)   ——— 450"·0 (w 1)⎭ Epoch 1831·77

A standard Greenwich star, with two distant companions, on the right shoulder-blade of Auriga. A 1, bright white; B 12, pale blue; C 9, grey; AB being 30 Ḥ. vi., and AC No. 51 of H. and S., under these measures:

Ḥ.   Pos. 151° 23' Dist. 169"·0 Ep. 1780·69
H. and S. 348° 02'   454"·2   1821·22

Here the principal star is Capella, a name considered to allude to the goat and kids, which Auriga, the waggoner, has charge of; but it is sometimes called *el-'áyyúk*, a word of doubtful origin and signification. The Arabs distinctly termed it the Guardian of the Pleiades; and many astronomers treated it as a single constellation, under the name of Hircus, or Capra, the goat. Capella is a brilliant object, and as it never sets at Bedford, and my view was unobstructed, it described a noble circle, of which both the upper and lower transits used to be taken. This is one of those stars which Piazzi attacked with the intention of detecting parallax, as detailed in the *Memoirs of the Italian Society*. Sir William Herschel measured its diameter, and concluded it to be 2"·5. Sir John Herschel says, " I have a strong impression that Capella, within my recollection, has increased in brightness. M. Struve is of the same opinion." Its proper motions in space have been valued as follows:

P.... Æ + 0"·12   Dec. − 0"·44
B....   + 0"·20   − 0"·41
A....   + 0"·14   − 0"·43

Auriga is one of the original forty-eight asterisms, though it has gone by divers other denominations, as Heniochus, Myrtillus, Elasippus, and Erichthonius. It is thought to have been the Horus of the Egyptians; but there is a want of apparent connection between the goat, kids, and carter, and the potent son of Isis. The Arabians drew a mule, instead of the human form; but they knew the latter figure also, and called it *Mumsiki-l a' inan*, or holder of the reins. It has been thus tabulated:

Ptolemy . . . 14 stars   Bullialdus. . . . 27 stars
Ulugh Beigh . . 13   Hevelius 40
Tycho Brahé . . 27   Flamsteed . . . 66
Bayer . . . 32   Bode 239

The goat in this constellation has been recognised as Amalthæa, the nurse of Jupiter, and mother of the Ερίφοι, Hædi, or two stars ζ and η in the arm of Auriga, emphatically termed " horrida et insana sydera:" with a third star they form an isosceles triangle. ζ and η were termed *al-'anz*, the goat, by the Arabians, and the former was *Dhát-al-'inán*, corrupted to Sadatoni in the Alphonsine Tables. The Hædi were regarded by mariners of yore, as affording presages of the weather: and they were so much dreaded, that they are said to have closed navigation at their rising.   Hence, in an Epigram of the *Anthologia*, Callimachus says:

> Tempt not the winds, forewarned of dangers nigh,
> When the Kids glitter in the western sky.

When the day of their peculiar influence was passed, a festival with sports and games was celebrated, under the denomination of *Natalis Navigationis*.   Even Germanicus calls them unfriendly stars to seamen, *nautis inimicum sydus in undis*.   In the early and well-known Venetian illustrations to Hyginus, Heniochus is delineated in a car drawn by two oxen and two horses, with a goat on his right shoulder, and two kids on his right hand: "In manu duas quæ Hædi appellantur stellis prope occidentibus formati."

Capella, the shepherd's star, is a brilliant insulated object, and therefore of easy alignment.   A line drawn from Polaris perpendicular to the line of the Pointers, and on the opposite side to Ursa Major, passes, at 44° distance, through it.   It will also be found by a ray projected through a and δ, the two most northern stars of the Great Bear's body, into the irregular pentagon formed by Auriga.   If looking from the southward for it, take the rhymester's advice:

> From Rigel rise, and lead a line,   through Bellatrix's light,
> Pass Nath, upon the Bull's north horn, and gain Capella's height—
> Where a large triangle is form'd   (isosceles it seems),
> When *beta* is with *delta* join'd   to lustrous *alpha's* beams.

CLXXXVII. ϱ^1 ORIONIS.

Æ 5h 04m 55s	Prec. + 3s·13
Dec. N 2° 39'·9	—— N 4"·77

Position 61°·8 (w 7)   Distance 6"·8 (w 5)   Epoch 1835·89

A pretty double star, between the right arm and thigh of Orion; in a line with the stars of the belt, preceding it by exactly double its length.   A 5, orange; B 8½, smalt blue—the tints are so decided as to bear out Σ.'s remark, " colores insignes."   This object was discovered by H. in December, 1784, but was first micrometrically attacked by S., whose No. 469 it is.   The principal measures of other astrometers may be thus stated:

	Pos.	Dist.	Ep.
S.	61° 59'	7"·05	1825·12
D.	61° 05'	7"·31	1831·14
Σ.	63° 28'	7"·05	1832·05

So that there has been no appreciable change in ten years.   There are several other small stars in the field, of which two bright ones in the *sp* quadrant form a coarse pair, at an $\angle$ from A = 240°, with $\triangle$ Æ = 29s.   A natural index for the future detection of proper motions in the star ρ, is offered us in its just preceding and being nearly equidistant between two small stars, the one north and the other south of it.

CLXXXVIII.   14 AURIGÆ.

Æ   5ʰ 04ᵐ 59ˢ   Prec. + 3ˢ·90
Dec. N 32° 29'·8   —— N 4"·76

Position A B 224°·5 (w 8)   Distance 13"·5 (w 6) ⎫
—————— A C 340°·0 (w 1)   —————— 15"·1 (w 1) ⎬ Epoch 1832·81
                                                            ⎭

A fine triple star, over Auriga's right knee; about 15° down on the line which runs from Capella to Rigel. A 5, pale yellow; B 7½, orange; C 16, purple; A and C pointing to a distant fourth star in the *np* quadrant. The two principal individuals of this object form 19 Ḥ. � ɪᴠ., and B was thus noticed by Piazzi: "1" temporis 20" ad austrum præcedit telescopica." The former measures of A and B are as follows:

Ḥ.   Pos. 232° 22'   Dist. 16"·09 ±   Ep. 1780·73
H. and S.   224° 23'   14"·61   1822·10

From these determinations, a change in the angle of position was inferred, but my measures, after an interval of ten years, do not confirm it. Σ. also examined this object, and discovered the delicate companion C, which had escaped the gaze of all other observers, and requires the most careful attention even to be perceived by occasional glimpses, but when seen, has a peculiar deep purple tint, which strikes singularly on the eye from so excessively minute an object. Σ.'s measures are:

Pos. A B 225°·48   Dist. 14"·653 ⎫ Ep. 1830·55
A C 342°·37   12"·577 ⎭

CLXXXIX.   ϰ LEPORIS.

Æ   5ʰ 5ᵐ 51ˢ   Prec. + 2ˢ·77
Dec. S 13° 08'·0   —— N 4"·69

Position 359°·5 (w 3)   Distance 3"·7 (w 2)   Epoch 1835·02

A close double star, at the root of the animal's left ear, and may be readily fished up about 5° south of Rigel, on a line run from Bellatrix through the latter. A 5, pale white; B 9, clear grey, pointing towards a distant star on the northern verge of the field. This exquisite object was on Σ.'s list of 1827, No. 661; and on the arrival of the grand Catalogue of 1837, the mean of his measures was found to be:

Pos. 358°·68   Dist. 3"·053   Ep. 1832·23

From these results, the general fixity of the components might be inferred; but, as the weights show, I do not place great confidence in my measures, which were troubled with variable refractions.

CXC.   β ORIONIS.

Æ   5ʰ 6ᵐ 51ˢ   Prec. + 2ˢ·88
Dec. S 8° 23′·5   —— N 4″·61

Position 199°·4 (w 9)   Distance 9″·5 (w 6)   Epoch 1832·07

A standard Greenwich star, double, in the Hero's right foot, at the commencement of the flexuous Eridanus; it is familiarly termed Rigel, from the Arabic *Rijl-al-jauzá*, the giant's leg; and Recorde assures us, it was called "*Algebar* by the Arabitians." A 1, pale yellow; B 9, sapphire blue. This splendid object, which is 33 Ḥ. II., is somewhat difficult to measure on account of the component's disparity in magnitude, and the brilliance of the large star. Still the results are in gratifying accordance, being:

	Pos.	Dist.	Ep.
Ḥ.	201° 48′	6″·48	1781·76
H. and S.	200° 41′	8″·87	1822·10
D.*	199° 48′	9″·86	1831·15
Σ.	199° 46′	9″·14	1831·53

Here Ḥ.'s positions are from those printed in the *Philosophical Transactions* for 1785; but H. has made extracts from the original MS. observations, by a mean of which he obtains:

Pos. 200° 45′   Ep. 1791·60   Dist. 9″·53   Ep. 1781·81

an agreement hardly to be expected under the difficulty of estimating exactly the position of the occult line passing through the centres of two stars so close, and so very unequal. Ḥ. remarks, " The small star not wanting apparent magnitude, is better seen with my power of 227 than with 460." The proper motion of A has been thus registered:

P.... Æ − 0″·05   Dec. − 0″·02
B.... + 1″·07   − 0″·01
A.... + 0″·02   − 0″·01

β Orionis has been designated *Rá'i al-Jauza* in Arabian astrognosy, as shepherd of the Jauza, whose herds, or thirst-allaying camels, are represented by *a*, γ, δ, and κ. Zahn tells us, in his *Oculus Artificialis*, 1702, that Francis Grindel observed through his telescope, that two stars in the right foot of Orion were surrounded with great splendour, as though emulous of the Sun; and that a phenomenon resembling them in splendour, cannot be found in the whole firmament. Now, as I cannot conceive either λ or τ to have been thus shining in the field with β, I can only impute the remark to a spurious image in a bad instrument, coloured by the same enthusiasm which showed Padre de Rheita the seamless coat of our Lord and a chalice, in this same asterism.

Independent of the "nautis infestus Orion" character of the constellation, Rigel had one of his own; for it was to the astronomical rising of this "marinus aster" in March, that *St. Marinus* and *St. Aster* owe

* The Rev. W. R. Dawes has shown me a diagram which he made of this delicate object, with a two-foot telescope, of 1⁶⁄₁₀ inches aperture, made by Dollond, having a pancratic eye-piece charged with a magnifying power of seventy times. This same little instrument showed the companion to Polaris distinctly.

their births in the Romish calendar. It is easy to find. A line run from the head of Leo through Procyon, arrives at Rigel; as does one from Castor, by Betelgeuze; and the locale of the star is thus expressed:

> With glittering gems Orion's belt,   his sword, his shoulders, blaze;
> While radiant Rigel on his foot   pours forth its silver rays.

It will be recollected, that this was one of the stars selected by Count d'Assas de Montardier, a captain in the French navy, for his investigations of parallax; and that he concluded he had detected an amount of from one to two seconds. But as he merely observed its appulse and disappearance behind an iron frame fixed on a mountain at different periods of the year, it would be difficult to prove such a quantity, right or wrong, even if the frame were absolutely immovable during the intervals, and insensible to the variations of temperature.

CXCI.   20 P. V. TAURI.

$\mathbb{R}$ 5^h 7^m 23^s Prec. + 3^s·50
Dec. N 18° 15'·3   —— N 4''·56

Position 168°·5 (w 8)   Distance 2''·1 (w 4)   Epoch 1834·89

A neat double star, on the Bull's southern horn; where a line run from the cluster in Orion's sword, and extended as far again to the north, passes upon it. A 8, and B 8½; both bluish, and lying between two stars in the *sp* and one in the *nf* quadrant, and nearest to the latter. It was discovered by Σ., and is No. 670 of the great Dorpat Catalogue, where the registered measures are:

> Pos. 171°·13 Dist. 2''·327 Ep. 1830·53

M. Struve styles A, *alba;* but in noticing so slight a difference of shade, even on so small an object, it is requisite to know to what degree his field of view was illuminated, and in what manner. It is possible that colour may interfere with our exact perception of size, which points out the necessity of obtaining greater accuracy of expression in the language of sidereal astronomy.

CXCII.   λ AURIGÆ.

$\mathbb{R}$ 5^h 7^m 53^s Prec. + 4^s·16
Dec. N 39° 57'·0   —— N 4''·52

Position 30°·2 (w 8)   Distance 102''·8 (w 5)   Epoch 1835·88

A star with a distant companion, on the Waggoner's loins; and rather more than 6° down a line drawn from Capella to Bellatrix. A 5, pale yellow; B 9½, plum colour. This object is 22 H. v., but as he described it merely " — multiple, — 2 within 30," it is impossible to identify them in the group of small stars of the galaxy wherein they are placed. I therefore measured to B, although there were two or three minute stars nearer, because it is the second of Sir James South's No. 472; and a little coarsely-double star about 3ᵐ in the *nf* quadrant, not far from the parallel, is his C; which are thus registered:

$$\text{Pos. A B } 34° \ 36' \quad \text{Dist. } 102'''\cdot14 \atop \text{A C } 81° \ 31' \qquad\quad 193''\cdot94 \Big\} \ \text{Ep. } 1825\cdot10$$

λ Aurigæ has a very sensible movement in space, which, though it escaped Piazzi, has had the following values assigned to it:

$$B.... \ \text{Æ} + 0'''\cdot71 \qquad \text{Dec.} - 0''\cdot66$$
$$A.... \quad\ + 0''\cdot68 \qquad\qquad - 0''\cdot67$$

CXCIII. 25 P. V. TAURI.

$$\text{Æ} \quad 5^h \ 8^m \ 03^s \qquad \text{Prec.} + 3^s\cdot54$$
$$\text{Dec. N } 19° \ 57'\cdot2 \qquad\text{------} \ \text{N } 4''\cdot51$$

Position 148°·4 (*w* 3) Distance 10''·0 (*w* 3) Epoch 1839·76

A neat double star, in the middle of the Bull's southern horn; and about 11° along a line projected from Aldebaran towards Pollux. A 8, bright white; B 11, bluish; and there are other companions, as mentioned at 37 P. v. This delicate object is one of Σ.'s Third Class, being No. 674 of the Dorpat Catalogue, where it is thus registered:

Pos. 147°·33 Dist. 10'''·547 Ep. 1828·19

CXCIV. 33 ♄. VII. AURIGÆ

$$\text{Æ} \quad 5^h \ 9^m \ 02^s \qquad \text{Prec.} + 4^s\cdot15$$
$$\text{Dec. N } 39° \ 10'\cdot2 \qquad\text{-- --} \ \text{N } 4''\cdot42$$

Position 42°·2 (*w* 1) Distance 25''·0 (*w* 1) Epoch 1836·71

A very delicate double star in a group, on the Waggoner's loins. A 7½, pale white; B 13, dusky. A fine field of small stars in ·a rich neighbourhood, with but little disposition to form. The most prominent member is a bright orange-coloured star of the 7·8 magnitude, forming a scalene triangle, with two others to the *sf;* near it, in the *np* quadrant, is the delicate pair above estimated, while on the northern verge of the field is a triplet of 10th-magnitude stars.

This object, which is H.'s No. 350, was first pointed out by Sir William Herschel, in 1785, who describes it as a pretty compact cluster, " with one large star, the rest nearly of a size;" but he makes no mention of the strong colour seen both by his son and myself. It is about 7° on the line from Capella towards Bellatrix, or nearly one-sixth of the distance between those stars.

CXCV.   37 P. V. TAURI.

Æ   5ʰ 9ᵐ 47ˢ   Prec. + 3ˢ·56
Dec. N 19° 57′·7   —— N 4″·37

Position 204°·1 (w 2)   Distance 9″·0 (w 2)   Epoch 1830·81

A very delicate double star, in the middle of the Bull's southern horn; at nearly one third of the distance between ζ and Aldebaran. A 7, deep yellow; B 11, bluish. It is the following of a curious series of six stars nearly in the same declination; the one immediately preceding it, being 25 P. v., before described. No. 37 is one of Σ.'s Third Class, and No. 680 of the Dorpat Catalogue, under the following measures:

Pos. 201°·77   Dist. 8″·720   Ep. 1827·85

CXCVI.   τ ORIONIS.

Æ   5ʰ 9ᵐ 50ˢ   Prec. + 2ˢ·91
Dec. S 7° 01′·3   —— N 4″·35

Position A B 255°·0 (w 1)   Distance 15″·0 (w 1) ⎫
—— A C 65°·0 (w 1)   —— 20″·0 (w 1) ⎭   Epoch 1835·97

An elegant and extremely delicate triple star, on Orion's right instep; where it is the vertex of an obtuse-angled triangle, formed with Rigel and β Eridani. A 4, pale orange; B 15, blue; and C 12, lilac; the three lie nearly in a line sp and nf, between two brightish stars at either end of the parallel. This is 25 ℍ. v.; registered October, 1780, but without measures; and it is No. 2259 of H.'s 20-foot Sweeps, who thus records it:

Pos. A B 250°·4   Dist. A B 18″·0
A C 63°·8   A C 18″·0

CXCVII.   23 ORIONIS.

Æ   5ʰ 14ᵐ 25ˢ   Prec. + 3ˢ·15
Dec. N 3° 23′·1   —— N 3″·96

Position 27°·9 (w 9)   Distance 32″·3 (w 9)   Epoch 1835·17

A neat double star, in Orion's right arm-pit; where a line carried from the Pleiades through the Hyades, will find it about 3° south of Bellatrix. A 5, white; B 7, pale grey. This is a fine object for telescopes of moderate power, and is 84 ℍ. iv. It has been thus measured:

	Pos.	Dist.	Ep.
ℍ.	30° 27′	26″·16	1782·75
H. and S.	27° 20′	33″·04	1822·05
Σ.	28° 15′	31″·71	1831·44

Hence, taking all the probable errors of observation into consideration, there is perhaps no appreciable change in position; nor indeed in distance, since H. has shown from his father's MSS. that, at the above date, the stars were 32".80 apart.

CXCVIII.   111 TAURI.

$\mathcal{R}$ 5^h 15^m 05^s Prec. + 3^s·47
Dec. N 17° 13'·8   —— N 3"·90

Position 271°·2 (w 4)   Distance 63"·0 (w 1)   Epoch 1832·95

A star with a distant *comes*, below the middle of the Bull's southern horn, in a poor field; and in the mid-distance between γ Orionis and β Tauri. A 6, white; and B 8½, lilac. Something seems to be the matter with the distance of 110 Ḥ. v., or it must have had an annual increase of nearly 0"·37 per annum; but there is no reason to suppose any change in the relative position angle of these stars in half a century, the former measures being:

Ḥ.   Pos. 273° 48'   Dist. 46"·70   Ep. 1782·87
S.   271° 17'   61"·76   1825·06

CXCIX.   β TAURI.

$\mathcal{R}$ 5^h 16^m 11^s Prec. + 3^s·78
Dec. N 28° 28'·0   —— N 3"·81

Position 225°·0 (w 1)   Difference $\mathcal{R}$ = 14^s·5 (w 1)   Epoch 1836·65

A standard Greenwich star, with a distant companion, and three other small stars in the field, forming a regular figure with the two preceding and two following β. A 2, brilliant white; B 10, pale grey. This object, β, is on the very tip of the horn of Taurus, and therefore at the greatest distance from the hoof: can this have given rise to the otherwise pointless sarcasm, of not knowing B from a bull's foot? This position gained it the name of Nath, from *Al-nátih*, the butting; and as it is also in the Waggoner's left ancle, it was called *Kab'dhí-l-'inán*, i. e., heel of the rein-holder, and entered on several Catalogues as γ Aurigæ. The proper motion assigned to this star—small as it is—may be stated:

P.... $\mathcal{R}$ − 0"·03   Dec. − 0"·17
B....   + 0"·12   − 0"·19
A....   + 0"·08   − 0"·20

In finding Nath by alignment, it must be sought about half-way between the Pleiades and Pollux; or, following the poet's dogma, a line sent

From centre of Orion's belt   to where Capella's seen,
Will point to the observant eye   Nath in mid-way between.

CC. γ ORIONIS.

℞ 5ʰ 16ᵐ 33ˢ   Prec. + 3ˢ·21
Dec. N 6° 12'·0   —— N 3"·78

Position 150°·0 (w 3)   Distance 178"·0 (w 2)   Epoch 1838·85

A bright star, with a minute distant companion, on Orion's right shoulder; and it is one of ⩩.'s insulated objects. A 2, pale but clear yellow; B 15, grey; a third star precedes by about 34ˢ, in the *sp* quadrant. γ Orionis was rejected from the Greenwich List in 1830, there being no fewer than four others of this constellation retained as standards: it is called Bellatrix, or the female warrior, and is the smaller of the two upper stars in Orion. The gender of this star puzzles Hood, who knoweth not why it should be female, "excepte it be this, that women born under this constellation shall have mighty tongues." Bellatrix is the *Al-nájid*, or subduer, of the Alphonsine Tables; but Ulugh Beigh calls it *Al-mirzam al-nájid*, though other Asiatic astronomers give the first epithet, *Al-mirzam*, to α Orionis. Hyde, in his notes on Ulugh Beigh (Syntagma I. 59), explains the Arabic words as signifying " the conquering lion;" but his interpretation is doubtful. γ Orionis is the north-west star of the four at the corners of this asterism, so to speak; and an ideal line, carried from Sirius over Aldebaran to the Bull's ear, passes over it in the mid-distance; the rhymester then directs,

From Bellatrix now pass a line,   to Betelgeuze the red,
And, to the north, three little stars   will mark Orion's head.

A friend considered my distance as "much too large," and the colour of A to be "reddish." On referring to Mr. Challis, that gentleman examined the object with the great Northumberland equatoreal, pronounced A to be "yellowish," and made the following measures:

Pos. 148° 04' Dist. 180"·46 Ep. 1841·19

Bellatrix has a small though sensible movement in space; but the doctors differ respecting its amount and direction. Thus the values of those most worthy of attention are the following:

P.... ℞ − 0"·17   Dec. − 0"·03
B....   + 0"·09   + 0"·01

CCI. 84 P. V. ORIONIS.

℞ 5ʰ 16ᵐ 53ˢ   Prec. + 3ˢ·11
Dec. N 1° 46'·4   —— N 3"·75

Position 322°·5 (w 9)   Distance 2"·6 (w 6)   Epoch 1835·11

A close double star on Orion's right side; where a line from Orion's belt towards Aldebaran passes it at about 4½° below Bellatrix. A 8, silvery white; B 10, grey. This delicate object is 52 ⩩. I., and was placed by Σ. in his First Class, "plurium maxima." There are few close double stars whose fixity for upwards of half a century has been more satisfactorily proved than this, the other measures being as follows:

	Pos.		Dist.		Ep.	
Hͪ.	322°	48'	2″·0±		1782·76	
S.	320°	48'	2″·98		1825·06	
Σ.	323°	13'	2″·61		1831·81	

CCII.   39 Hͪ. VII. AURIGÆ.

Æ	5ʰ 17ᵐ 18ˢ	Prec. + 3ˢ·99
Dec.	N 35° 10'·3	—— N 3″·71

Position 235°·0 (w 1)   Distance 5″·0 (w 1)   Epoch 1836·79

A minute double star announces this cluster, on the robe under the left thigh of Auriga. A 9½ and B 11, both grey. The object is a compressed oval cluster of 10th to 14th-magnitude stars, about 3' in diameter, trending *sf* and *np*, with a pair of 10th-magnitude to the north; in a splendid district of the heavens. It was discovered by Hͪ. in January, 1787; but the neat double star here estimated, is No. 699 of H.'s Third Series of Sweep Observations. It is about 12° down on the line which the eye projects from Capella towards Betelgeuze, and is there intercepted by another line drawn from Bellatrix through β Tauri, and extended 6½° beyond.

CCIII.   79 M. LEPORIS.

Æ	5ʰ 17ᵐ 50ˢ	Prec. + 2ˢ·47
Dec.	S 24° 39'·9	—— N 3″·67

Mean Epoch of the Observation . . . . 1835·98

A bright stellar nebula, of a milky white tinge, under the Hare's feet, the following edge of whose disc just precedes a line formed by two stars lying across the vertical, and it is followed nearly on the parallel by a 9th-magnitude star. It is a fine object, blazing towards the centre, and was discovered by Méchain, in 1780. It was resolved by Hͪ. into a mottled nebulosity, in 1783, with a seven-foot telescope; but on applying the twenty-foot in the following year, he fairly made it a "beautiful cluster of stars nearly 3 minutes in diameter, of a globular construction, and certainly extremely rich." The mean apparent place is obtained by differentiation from ξ Leporis, which is a fine white star, with a red companion of the 7th magnitude in the *np* quadrant. An imaginary line run from Betelgeuze before *a* Leporis and over β, will hit this object about 4° south-west of the latter.

CCIV. 38 M. AURIGÆ.

Æ 5ʰ 18ᵐ 41ˢ   Prec. + 4ˢ·01

Dec. N 35° 44'·9   —— N 3"·59

Position 251°·0 (w 1)   Difference Æ 14"·5 (w 1)   Epoch 1835·80

A rich cluster of minute stars, on the Waggoner's left thigh, of which a remarkable pair in the following part are here estimated. A 7, yellow; and B 9, pale yellow; having a little companion about 25" off in the *sf* quarter. Messier discovered this in 1764, and described it "a mass of stars of a square form without any nebulosity, extending to about 15' of a degree;" but it is singular that the palpable cruciform shape of the most clustering part did not attract his notice. It is an oblique cross, with a pair of large stars in each arm, and a conspicuous single one in the centre; the whole followed by a bright individual of the 7th magnitude.

The very unusual shape of this cluster, recalls the sagacity of Sir William Herschel's speculations upon the subject, and very much favours the idea of an attractive power lodged in the brightest part. For although the form be not globular, it is plainly to be seen that there is a tendency towards sphericity, by the swell of the dimensions as they draw near the most luminous place, denoting, as it were, a stream, or tide of stars, setting towards a centre. As the stars in the same nebula must be very nearly all at the same relative distances from us, and they appear to be about the same size, Sir William infers that their real magnitudes must be nearly equal. Granting, therefore, that these nebulæ and clusters of stars are formed by their mutual attraction, he concludes that we may judge of their relative age, by the disposition of their component parts, those being the oldest which are the most compressed.

To fish up this object, a line from Rigel must be carried northwards through β Tauri, on the tip of the Bull's left horn, and about 7° beyond, where it will be intersected by the ray from Capella to Betelgeuze.

CCV. 118 TAURI.

Æ 5ʰ 19ᵐ 25ˢ   Prec. + 3ˢ·68

Dec. N 25° 00'·8   —— N 3"·53

Position 195°·5 (w 5)   Distance 5"·3 (w 3)   Epoch 1833·78

——— 195°·9 (w 9)   ——— 5"·0 (w 9)   ——— 1838·91

A very neat double star, between the tips of the Bull's horns; and mid-way between the Pleiades and δ Geminorum. A 7, white; B 7½, pale blue. This elegant object is 75 ɪʜ. ɪɪ., and was noticed thus by Piazzi:

"Duplex; minor ad austrum; medium observatum,"—which method perhaps prompted him to assign the slight proper motion he has registered to A, but which has not been confirmed. A consideration of all the measures of this star, and allowing for probable small errors of observation, identifies its fixity. These are the data:

Ḥl.	Pos. 192° 45′	Dist. 5″·03	Ep. 1782·94
H. and S.	194° 01′	5″·66	1821·97
Σ.	196° 46′	4″·89	1829·63
D.	196° 20′	5″·15	1832·87

A proper motion is ascribed to A, in which it is probable that both stars partake. The best valuations give:

$$P.... \text{Æ} + 0''·11 \quad \text{Dec.} - 0''·06$$
$$B.... \quad + 0''·08 \quad\quad\quad - 0''·07$$

CCVI.   261 Ḥl. I. AURIGÆ.

Æ 5ʰ 20ᵐ 51ˢ	Prec. + 3ˢ·97
Dec. N 34° 06′·9	—— N 3″·41

Mean Epoch of the Observation . . . . 1835·80

A resolvable nebula, on the lower garment of Auriga, about 2ᵐ·5 sf 38 Messier. This very curious object was discovered by Ḥl. in February, 1793, who remarked that it "seems to have one or two stars in the middle, or an irregular nucleus." This object was next examined by H., No. 355, who described it as a nebula including a triple star, surrounding them like an atmosphere. With these premonitions, I attacked it under most favourable circumstances. The nebula is situated in a rich field of minute stars, with five of the 10th-magnitude, disposed in an equatoreal line above, or to the south of it, and preceded by a bright yellow 7½ magnitude star in the same direction. After intently gazing, under moderate power, the triangle rises distinctly from the star-dust, and presents a singular subject for speculation.

CCVII.   109 P. ORIONIS.

Æ 5ʰ 21ᵐ 02ˢ	Prec. + 2ˢ·87
Dec. S 8° 30′·7	—— N 3″·39

Position 295°·0 (w 1)   Distance 20″·0 (w 1)   Epoch 1834·71

A delicate double star, in the space between Orion's right heel and left knee; where it may be found by drawing a line from the third star in Orion's belt, over the sword cluster, and carrying it nearly as far again beyond. A 7½, pale white; and B 10, blue. This pretty object was Σ.'s Fourth Class, No. 722, but is not placed among his measured stars;

being branded in the great Catalogue with "*rej.*" It is in a barren but brightish field, in which an occasional glow seems to verify the "diffused nebulosity" which H.'s powerful light-grasping 20-foot reflector saw, No. 2268 of his Sweeps.

CCVIII. β LEPORIS.

$\mathcal{R}$ 5^h 21^m 23^s PREC. + 2^s·57

DEC. S 20° 53'·5   —— N 3"·36

POSITION 67°·5 (*w* 1)   DISTANCE 210"·0 (*w* 1)   EPOCH 1832·00

A star with a distant telescopic companion, between the legs of Lepus. A 4, deep yellow; and B 11, blue, and in the centre of three small stars in the following part of the field. This star is often called Nihal, but the name is more properly applied to *a*, *β*, *γ*, and *δ*, the Arabian *Al-nihál*, or thirst-slaking camels; it will be identified by drawing a line from the middle star of Orion's belt, through the sword, and extending it 3° below *a* Leporis.

A difference from the general laws of precession has been exhibited by this star, which, though of no great amount, is deserving of being well watched, and the little *comes* here noted may be a direction. The values at present given are:

$$P.... \mathcal{R} - 0"·03 \quad \text{Dec.} - 0"·07$$
$$Br... \quad - 0"·04 \qquad\qquad - 0"·10$$
$$B.... \quad + 0"·06 \qquad\qquad - 0"·08$$

CCIX. 32 ORIONIS.

$\mathcal{R}$ 5^h 22^m 13^s PREC. + 3^s·20

DEC. N 5° 49'·3   —— N 3"·29

POSITION 205°·4 (*w* 3)   DISTANCE 1"·0 (*w* 2)   EPOCH 1831·13

—— 206°·2 (*w* 7)   —— 1"·0 (*w* 5)   —— 1839·20

A close double star, on Orion's right shoulder. A 5, bright white; B 7, pale white. This elegant object was discovered by H̶., and his observations compared with late results seemed to show a retrograde motion in the angle of position; but this has not been confirmed by the latest measures. It was considered too difficult for the five-foot equatoreal by H. and S., in 1822; and their measures were therefore cautiously advanced. Nor did H. place reliance on his angle, 214° 33', taken afterwards with the seven-foot telescope. I therefore re-attacked it under every favouring circumstance in 1839, and place pretty good confidence in the results obtained. The other observations are:

$$\text{H̶.} \quad \text{Pos. } 217° \; 50' \quad \text{Dist. } 1"·0 \pm \quad \text{Ep. } 1780·06$$
$$\Sigma. \qquad\quad 203° \; 45' \qquad\quad 1"·04 \qquad\quad 1830·96$$

A line from the leading star of Orion's belt carried towards β Tauri, passes 32 Orionis at rather more than 6°, where it will be seen just to the eastward of Bellatrix.

CCX. 33 ORIONIS.

Æ 5ʰ 22ᵐ 51ˢ	Prec. + 3ˢ·14
Dec. N 3° 09′·9	—— N 3″·24

Position 26°·7 (w 9)   Distance 1″·9 (w 9)   Epoch 1830·16

——— 25°·8 (w 9)   ——— 2″·0 (w 9)   ——— 1838·21

A close double star, on Orion's right shoulder, where it is a little more than one-third of the distance from Bellatrix to the last, or following star of Orion's belt. A 6, white; B 8, pale blue; with a distant 8th-magnitude star in the *np* quadrant, which must be 121 P. Hora V. This superb object is 22 ♓. I., and not of very difficult measurement, though rated as one of Σ.'s "vicinæ." The results of the former astrometers are:

♓.	Pos. 28° 37′	Dist. 0″·70	Ep. 1781·81
H. and S.	26° 11′	2″·02	1822·02
Σ.	25° 35′	1″·87	1831·22

These results, compared with my own, indicate no change in the angle; but as ♓. says they were only half the diameter of the small star apart, the distance may possibly have increased.

CCXI. δ ORIONIS.

Æ 5ʰ 23ᵐ 50ˢ	Prec. + 3ˢ·06
Dec. S 0° 25′·4	—— N 3″·15

Position 359°·9 (w 9)   Distance 53″·2 (w 9)   Epoch 1835·11

A standard Greenwich star, coarsely double; it is the leader of the three "bullions" in Orion's girdle or belt, and nearly on the equator. A 2, brilliant white; B 7, pale violet. This object is 10 ♓. v., and has been thus registered:

♓.	Pos. 358° 10′	Dist. 52″·96	Ep. 1779·77
H. and S.	0° 03′	54″·87	1822·97

Weighing the circumstances, these positions agree well enough with Piazzi's description: "alia 7, 8ᵅ magnitudinis in eodem verticali, 51″ ad boream;" and the same may be said of the distances, although ♓. has recorded that he gave "full measure." The coincidence of these results proves the fixity of the large star, and militates against the large amount of proper motion which has been imputed to it; the later assigned values are:

$$P.... \, R - 0''·12 \quad Dec. + 0''·05$$
$$B.... \quad + 0''·08 \quad\quad - 0''·04$$

This star being the preceder of Orion's beautiful belt, has been popularly distinguished under various names. Among astronomers it is usually known as Mintaka, from the Arabian *Mintakah-al-jauza*, the giant's belt; which some people also designated *al-lekat*, the gold grains or spangles. It was also called, with its associates, *Jacob's staff*, perhaps from the traditional idea mentioned by Eusebius, that Israel was an astrologer. It was also the *Golden Yard* of seamen, the *Three Kings* of soothsayers, the *Ell-and-yard* of tradesmen, the *Rake* of husbandmen, and *Our Lady's Wand* of the Catholics. The belt points on one side to Sirius, the brightest of all the stars; and on the other to the Hyades and Pleiades; and the rhymester points out the individual before us:

In the blue vast, Orion's Belt   shines with its bullions three,
And of those bright conspicuous gems   the first as *delta* see.

CCXII. 1 M. TAURI.

$$R \quad 5^h \, 24^m \, 51^s \quad\quad Prec. + 3''·60$$
$$Dec. \, N \, 21° \, 54'·2 \quad\quad —— N \, 3''·06$$

MEAN EPOCH OF THE OBSERVATION . . . . 1836·99

A large nebula, pearly white, about a degree north-west of the star ζ on the tip of the Bull's southern horn, and on the outskirts of the galaxy. It is of an oval form, with its axis-major trending *np* and *sf*, and the brightest portion towards the south. Sir John Herschel registers this in his Catalogue of 1833, as a "barely resolvable cluster;" and figures it with a fair elliptical boundary. He applied his 7, 10, and 20-foot reflectors, and endeavoured to ascertain its relative distance by a modification of their space-penetrating capacity. "As all the observations," he concludes, "of the large telescopes agree to call this object resolvable, it is probably a cluster of stars at no very great distance beyond their gauging powers; its profundity may therefore be of about the 980th order." All this shows the difficulty of what, to my view, is rather a milky nebulosity than a cluster. The powerful telescope constructed by Lord Rosse, however, not only displays the component stars distinctly, but also shows several fringy appendages around, and a deep bifurcation to the south. So do sidereal wonders increase with our means of optical practice!

This fine nebula is remarkable as having been discovered by M. Messier—the comet-ferret of Louis XV.—while observing ζ Tauri and a comet in 1758, when he caught up a "whitish light, elongated like the flame of a taper." This accident induced him to form his well-known and useful Catalogue of nebulæ and clusters, from the observations of himself, La Caille, and Méchain, in order to prevent astronomers from mistaking any of those objects for comets; and the List of 103 which he furnished to the public, was considered to have scraped them all together, as far as climate permitted.   Whence D'Alembert, speaking of Messier, observed, "on ne peut s'empêcher de regretter qu'un Obser-vateur si exact et si plein de zèle, n'ait pas été placé dans un climat plus heureux." But the progress of astronomy has not depended upon climate, as the names of Tycho, Römer, Flamsteed, Bradley, Hevelius, Huygens, Schroeter, Olbers, and others of the ἱερὰ φάλανξ, abundantly testify. Indeed, in the department before us, within twenty years of Messier's publication, the illustrious Sir William Herschel increased the 103 by

2500 new members, in the decried climate of England, thus affording a strong instance how moral causes can control the physical.   Piazzi, whose observatory in the Conca d'Oro was to the eye most charmingly situated, was so troubled with a peculiar flickery hot aerial refraction, that one night he exclaimed to me, " Ah, Greenwich is the paradise for an observer!"

It is rather curious, on recollecting that this nebula was first caught up in seeking the comet of 1759, that it was also a mare's nest to more than one astronomical tyro in August, 1835, when on the look-out for the return of Halley's comet, in the very month in which it had first been seen seventy-seven years before: and ζ Tauri was also the star which served as a " pointer," on that interesting advent.

CCXIII.   α LEPORIS.

℞   5ʰ 25ᵐ 40ˢ   Prec. + 2ˢ·64
Dec. S 17° 56'·5   —— N 2"·99

Position 261°·0 (w 1)   Difference ℞ 17ˢ·4 (w 1)   Epoch 1834·01

A Greenwich star of 1830, with a distant companion, on the body of Lepus. A 3½, pale yellow; B 9½, grey; a bright 6th magnitude in the *np* quadrant.   This object is easily found by alignment; for a ray carried from ε, the central star of Orion's belt, through θ and its nebulous patch on the sword, as low down as Sirius, falls upon α Leporis; it is thus recorded in galley-rhymes:

Orion's image, on the south,   has four stars—small but fair;
Their figure quadrilateral   points out the timid Hare.

This asterism is one of the old 48, and is said to have been placed immediately below Orion, as emblematic of caution and celerity. The Arabians called α, Arneb, from *al-arneb*, the hare; it was also, in con-junction with β, γ, and δ, named Kursa, from *Kursá-l-jaúzá*, or '*Arsh*-

al-jaúzá, the giant's chair or throne,—for *al-jaúzá,* the belted-sheep, seems to be here used as the proper name of the giant. 'Abdr rahmán Súfí designates the throne—one of the many which the Arabs had in their heavens, although a squatting rather than a sitting people—*al-muakhkherah,* the succeeding, as following that formed by λ, β, ψ Eridani, and τ Orionis. Ideler mentions its having this name, and angrily adds, "und Gott weiss wie sonst noch." It is a poorish constellation—if such a term may be applied to those wondrous assemblages—and has been thus registered:

Ptolemy	. . . 12 stars	Hevelius	. . . 16 stars
Tycho Brahé	. . 13	Flamsteed	. . . 19
Bayer	 14	Bode	 66

CCXIV. 36 M. AURIGÆ.

Æ 5ʰ 25ᵐ 44ˢ   Prec. + 3ˢ·96
Dec. N 34° 01'·9   —— N 2''·99

Position 308°·7 (*w 5*) Distance 12''·0 (*w 1*) Epoch 1836·71

A neat double star in a splendid cluster, on the robe below the Waggoner's left thigh, and near the centre of the Galaxy stream. A 8, and B 9, both white; in a rich though open splash of stars from the 8th to the 14th magnitudes, with numerous outliers, like the device of a star whose rays are formed of small stars. This object was registered by M. in 1764; and the double star, as H. remarks, is admirably placed, for future astronomers to ascertain whether there be internal motion in clusters. A line carried from the central star in Orion's belt, through ζ Tauri, and continued about 13° beyond, will reach the cluster, following φ Aurigæ by about two degrees.

CCXV. λ ORIONIS.

Æ 5ʰ 26ᵐ 19ˢ   Prec. + 3ˢ·30
Dec. N 9° 49'·3   —— N 2''·94

Position 42°·5 (*w 8*) Distance 4''·6 (*w 6*) Epoch 1833·17
——— 43°·0 (*w 9*)   ——— 4''·5 (*w 9*)   ——— 1843·19

A neat double star, in Orion's ear; where it will be seen at about 5° on a line shot from Betelgeuze to Aldebaran, being the northern of the three small stars forming Orion's head. A 4, pale white; and B 6, violet. This fine object is 9 Ḥ. ɪɪ., and appears to be, from the following measures and my own, without any appreciable motion:

	Pos.	Dist.	Ep.
Ḥ.	44° 46'	5''·83	1779·88
H. and S.	40° 46'	5''·57	1822·19
Σ.	40° 32'	4''·24	1830·81
D.	43° 02'	4''·65	1832·95

This double star, and the two ϕ's in Orion's head, forms, says Kazwíní, an *atháfi* constituting the Vth Lunar Mansion; the peculiar aspect of which gained λ the name of Heka, from *al-hek'ah*, a white spot. On the early application of the telescope to this spot, Galileo found it to consist of twenty-one stars; but this definition of it does not seem to have obtained generally. "It is evident," says ʜ., "the whole appeared nebulous to Flamsteed for no other reason than because his telescope had not sufficient power to distinguish them." Hence the term, *in capite nebulosa*, of the Catalogues. It forms the apex of a triangle, the base of which extends between a and γ Orionis.

CCXVI. θ^1 ORIONIS.

Æ	5ʰ 27ᵐ 25ˢ	Prec.	$+ 2^s{\cdot}94$
Dec. S	5° 30′·0	——	N 2″·84

Position AB 311°·1 *(w 8)*	Distance 13″·0 *(w 6)*	
—— AC 60°·2 *(w 8)*	—— 13″·5 *(w 6)*	Epoch 1834·07
—— AD 344°·7 *(w 8)*	—— 16″·7 *(w 8)*	
—— BE 350°·0 *(w 1)*	—— 5″·0 *(w 1)*	

A multiple star, the beautiful trapezium in the "Fish's mouth" of the vast nebula in the middle of Orion's sword-scabbard. A 6, pale white; B 7, faint lilac; C 7½, garnet; D 8, reddish; and E 15, blue. This was entered 1 ʜ. iii., in November, 1776, and had the honour of being the object to which the grand forty-foot reflector was first directed, in February, 1787, under the designation of "quadruple." As a trapezium it was gazed at, measured, and delineated, for upwards of fifty years, when Σ. announced it "quintuplex," by the addition of the little star E. Now when we consider the eye of ʜ., the measures of S., and the rigorous examination of H., this little companion must be looked upon as variable; indeed nothing can exceed the confidence with which H. assured me, of its not being visible when he made the beautiful drawing of 1824, confirmed by himself and Mr. Ramage on the 3rd of March, 1826; and yet in 1828 it was not to be overlooked but by wilful inattention. Mr. Dawes afterwards saw it well with his five-foot telescope. The best measures for comparison with my epoch, are those of Σ. and S.; and by adjusting the latter's uncials and quadrants, they will stand thus:

S. 1824·50			Σ. 1836·15		
AB Pos. 310° 48′	Dist. 13″·453		Pos. 311° 14′	Dist. 12″·983	
AC	60° 04′	13″·582	60° 07′	13″·467	
AD	345° 03′	16″·685	342° 10′	16″·780	
BE	*(not seen)*		353° 42′	3″·860 (1832·53)]	

Ptolemy, Tycho Brahé, and Hevelius, ranked θ of the 3rd magnitude, as did Bayer in his *Uranometria*, all evidently supposing the two contiguous stars and the bright spot constituted a single star. The effulgent nebula in which it is placed, familiarly called the Fish's head, with its streaming appendages, certainly has an irregular resemblance to the head of some monster of the polyneme genus. Its brilliancy is not equal throughout, but the glare of the brighter parts gives intensity to the darkness which they bound, and excites a sensation of looking through it into the luminous regions of illimitable space, a sensation not entirely owing to any optical illusion of contrast. This supposition must have forced itself upon Huygens, independently of any recollection of the empyrean heaven of the ancients; and had Voltaire seen the object under powerful means, he would hardly have lashed.Dr. Derham for asking, whether nebulæ be not this shining region, seen through a chasm of the *primum mobile.* Another wonderful singularity is, that the nebulous and apparently attenuated matter seems to recede from the stars of the trapezium, so as to leave a black space around each, between them and the glow, as though they were either repelling or absorbing it.

This is a most splendid object under any telescope, but the greater the optical power applied, the more inexplicable does it become. My own telescope showed it to very great advantage, but it is here where the light-grasping quality of reflectors is brought advantageously to bear. Thus in the twenty-foot telescope at Slough, Sir John Herschel gained perceptions of its modification which were not decided to my view: " I know not," he says, " how to describe it better than by comparing it to a curdling liquid, or a surface strewed over with flocks of wool, or to the breaking up of a mackerel sky, when the clouds of which it consists begin to assume a cirrous appearance. It is not very unlike the mottling of the sun's disc, only, if I may so express myself, the grain is much coarser, and the intervals darker; and the flocculi, instead of being generally round, are drawn into little wisps. They present, however, no appearance of being composed of stars, and their aspect is altogether different from that of resolvable nebulæ." Such, at present, are the only ascertained peculiarities of the wondrous mass. It is pronounced to be of the singular nature termed milky nebulosity by Sir William Herschel: " to attempt," he remarks, " even at a guess at what this light may be, would be presumptuous. If it should be surmised, for instance, that this nebulosity is of the nature of the zodiacal light, we should then be obliged to admit the existence of an effect without its cause. An idea of a phosphorical condition is not more philosophical, unless we could show from what source of phosphorical matter such immeasurable tracts of luminous phenomena could draw their existence, and permanency: for though minute changes have been observed, yet a general resemblance, allowing for the difference of telescopes, is still to be perceived in the great nebulosity of Orion, ever since the time of its first discovery." This illustrious astronomer was, at first, inclined to consider all the nebulæ as resolvable, but this *milky* instance, with that in Andromeda, contradicted the notion, and led him to inferences respecting nebulous matter, and its possible gradation to stars by condensation, so as to form a distinct and plausible theory of cosmogony; with the originality of which neither the *Á'kásah,* or *fifth element* of the Brahmans, of which the heavens are formed, nor the vague notions of Tycho Brahé and Kepler, can properly be said to interfere. From these bold and almost overwhelming ideas we may yet become conscious, as well of the operations of the powerful agents by which whole systems are formed, as of those tremendous forces by which others are destroyed.

We are told that this nebula was one of the first fruits of Galileo's telescope; but it is certain that Huygens discovered it by accident in 1656, as stated in his *Systema Saturnium*, where he notes, "Portentum, cui certe simile aliud nusquàm apud reliquas fixas potuit animadverti." From a comparison of the descriptions and drawings of this object, since his time, great alterations might be inferred; but astronomical delineation was not then sufficiently advanced to render the diagrams at all satisfactory, nor were the instruments sufficiently powerful. Thus, while one man thinks his 3½ foot telescope indicated "myriads upon myriads" of stars in its composition, Lord Rosse, with the most powerful and perfect instrument extant, gained no appearance of re-solution. It may therefore be concluded, that the first rigidly accurate representation of it, is that by Sir John Herschel; and he who wishes to acquire all the actual knowledge we at present possess on the subject, cannot refer to a better description than that contained in his paper, published in the second volume of the *Memoirs of the Astronomical Society*. "Several astronomers," says Sir John, "on comparing this nebula with the figures of it handed down to us by its discoverer, Huygens, have concluded that its form has undergone a perceptible change; but when it is considered

how difficult it is to represent such an object duly, and how entirely its appearance will differ even in the same telescope, according to the clearness of the air, or other temporary causes, we shall readily admit that we have no evidence of change that can be relied on." To the drawing which illustrates that account, posterity will refer with confidence, in order to "catch Nature in the fact:" meantime, it seems clear, that if the parallax of this nebula be no greater than that of the stars, as one hypothesis assumes, its breadth cannot be less than a hundred times that of the diameter of the Earth's orbit: but if, as is still more probable, at a vast distance beyond, its magnitude must be utterly inconceivable.

This luminous spot is so well known to all star-gazers, that it is hardly necessary to add, that a line projected from α Orionis, through ζ, the third of the belt, will pass upon θ and the nebula, in the sword-scabbard. The portion called the Fish's mouth, with the well-known trapezium, may be rudely sketched as in the preceding figure *.

θ^2 Orionis, which is 133″ from θ^1, on an angle $= 135°$, is coarsely double, of the 6th and 7th magnitudes. At the epoch above named, viz. 1834·07, the components measured 91°·5 as the angle of position, and 52″ for the distance.

CCXVII.   362 H. ORIONIS.

Æ 5^h 27^m 36^s Prec. $+ 2^s$·97
Dec. S $4°$ 27′·6   —— N 2″·83

Position 58°·6 (w 2)   Distance 5″·0 (w 2)   Epoch 1835·79

A delicate double star in the wide-spread cluster on Orion's sword. A 6, lucid white; B 9, pale blue. The principal members of this group of stars, are of the 6th and 7th magnitudes, with some smaller; and from their brightness and disposition form a capital test for the light of a telescope. It was examined by H., and entered on his Catalogue of 1830; whence it may lay claim to being an aggregated and connected assemblage, and, comparatively speaking, not very remote from us.

CCXVIII.   ι ORIONIS.

Æ 5^h 27^m 36^s Prec. $+ 2^s$·93
Dec. S $6°$ 01′·2   —— N 2″·83

Position AB 141°·7 (w 8)   Distance 11″·5 (w 6)⎱ Epoch 1832·13
——— AC 102°·8 (w 3)   ——— 48″·9 (w 3)⎰

A fine triple star, in a good field on Orion's sword-scabbard; and 5° south of the middle star in the belt. A 3½, white; B 8½, pale blue; and C 11, grape red. Piazzi says of ι, in his *Notæ*, " Duplex: comes 0″·4 temporis sequitur, et vix distingui potest,"—but his instrument being fully equal to distinguish such a magnitude as that of B, his remark excites a suspicion that it may be variable. There is a glow about this

---

* Since my observations were made, a sixth star has been seen just outside A, nearly in the line with A and B. But it is a very *intensiva* of vision, and therefore quite escaped me. The Rev. W. R. Dawes managed to measure it, and he kindly communicated to me the following results:

Mean of 6 Observations of Pos. $= 127°$ 48′⎱ Ep. 1842·16
——— 4 ——————— Dis. $=$ 2″·79⎰

"My attention," he writes, "was directed to this star by Mr. W. Lassell, Jun., of Star-field. He lately saw the small companion with a newly-figured 9-inch Newtonian metal, of 112 inches focus, made by himself."

object, when viewed under favouring circumstances; yet I cannot assert that the nebulosity in which it is enveloped, is clearly seen. But under proper means it is well worth scrutiny; for nebulous stars are certainly among the most remarkable objects in the heavens, and perhaps should be distinguished from stellar nebulæ in being of a less doubtful character, as to the state of condensation, the central matter in such being suddenly vivid, and sharply defined. ι Orionis is 12 Ⴙ. III., and was thus measured when first classed:

$$\begin{array}{ll} \text{A B Pos. } 133°\ 51' & \text{Dis. } 12''\cdot50 \\ \text{A C } \quad\ 101°\ 19' & \phantom{\text{Dis. }}48''\cdot31 \end{array}\Bigg\} \text{ Ep. } 1779\cdot77$$

When Sir James South re-examined this star, in 1824, A and B were considered, from the apparent change of angle in fifty-five years, to have a direct orbital motion $= + 0''\cdot202$ per annum; but more recent observations do not support the inference. The measures with which I compared mine, are:

	Pos.	Dist.	Ep.
S.	141° 58′	12″·08	1824·74
D.	141° 21′	11″·89	1831·16
Σ.	142° 10′	11″·32	1831·86

CCXIX.   ε ORIONIS.

Æ	5ʰ 28ᵐ 06ˢ	Prec.	+ 3ˢ·04
Dec. S	1° 18′·6	——	N 2″·78

Position 67°·9 (w 1)   Distance 160″·0 (w 1)   Epoch 1835·02

A standard Greenwich star, in the centre of Orion's belt, with a distant companion. A 2½, bright white, and nebulous; B 10, pale blue. This fine star, rated a full second magnitude by Flamsteed, is in a neat trapezium of the 8th magnitude, in a rich vicinity. It is often called Alnilam, from the Arabic *Al-Niḍhám*, or *Niẓám*, the string of pearls, in allusion to its situation between ζ and δ, forming, as Robert Recorde says, the bullions set in Orion's girdle. It may assist the alignment of the vicinity to state, that the belt extends exactly 3°, or 1½° on each side of this star.

As neither ζ nor δ could have offered much peculiarity to Padre de Rheita's binocular telescope in 1643, the treble-bodied star which he saw in or near Orion's belt, may have been ε or σ,—" in aut propè cingulum Orionis vidisse se tricorpoream stellam." The worthy Bohemian's visions and views sadly interfere with the exactness of his real discoveries in *cœlo stellifero*. The galley-poet tells us:

Our Lady's wand is bless'd by all   who watch those gems on high,
And centre of that brilliant zone   *epsilon* meets the eye.

The attractions of this beautiful constellation have thus afforded five objects in close succession; and numerous others deck this comparatively compact region. It is a wonderful spot; and there is food for the theorist in the brilliant oblique zone exhibited by Taurus and Orion, coming to a full stop at Sirius.

CCXX. 26 AURIGÆ.

Æ. 5ʰ 28ᵐ 21ˢ   Prec. + 3ˢ·84
Dec. N 30° 23′·4   —— N 2″·76

Position 267°·8 (w 9)   Distance 12″·3 (w 9)   Epoch 1833·09

A neat double star, on the Waggoner's left shin; where a line from the cluster in Orion's sword, led through the middle star of the belt, through ζ Tauri, will hit it at less than 10° beyond the latter. A 5, pale white; and B 8, violet. This fine object is 64 Ħ. III., and from an error in the original entry, was erroneously supposed to have changed its quadrant, in an interval of forty-three years, in an orbital motion *npsf* or retrograde. These are the measures with which mine are compared:

Ħ.   Pos. 272° 36′   Dist. 13″·41   Ep. 1782·68
S.   268° 22′   12″·32   1825·03
Σ.   268° 02′   12″·34   1828·61

Σ.'s epoch is a mean of four years, and should a retrograde motion be hereafter established, it will prove the delicacy of his measures, and the sterling talent of Ħ.; for when he established those admirable epochs, there were no spider-line micrometers, &c.

CCXXI. 124 TAURI.

Æ. 5ʰ 29ᵐ 32ˢ   Prec. + 3ˢ·64
Dec. N 23° 13′·5   —— N 2″·66

Position AB 240°·8 (w 1)   Distance 98″·0 (w 1)
——— BC 315°·0 (w 7)   ——— 5″·0 (w 3)   Epoch 1835·65
——— AD 170°·0 (w 1)   ——— 82″·0 (w 1)

A coarse quadruple star, in the space over the Bull's southern horn. A 8½, garnet; B 8, and C 9, both pale white, and forming a very delicate object; D 10, bluish. This star does not appear upon the British Catalogue, but was well observed by Piazzi, who remarked—" Præcedit telescopica ad Austrum, nec alia inventa." This group was examined because it happened to be near the spot where I was on the look-out for Halley's comet, on its most welcome return to our neighbourhood, in August, 1835. Several stars in this vicinity, which I used as *comet-pointers*, were meridionally observed for me by my friend Mr. Henderson, in the Royal Observatory at Edinburgh.

124 Tauri is rather more than one-third of the distance from Alde-baran to Castor; and about two degrees north, very slightly following ζ on the tip of the right horn.

CCXXII. σ ORIONIS.

R.   5ʰ 30ᵐ 43ˢ   Prec. + 3ˢ·01
Dec. S 2° 41′·8   —— N 2″·55

Position Aa 235°·9 (w 2)   Distance 12″·0 (w 1)
—————— A B   84°·2 (w 7)   ——————   12″·5 (w 5)
—————— A C   60°·8 (w 8)   ——————   41″·8 (w 8)
—————— A D 321°·6 (w 3)   —————— 211″·5 (w 2)
—————— D E 266°·8 (w 8)   ——————   8″·5 (w 6)
—————— D F   21°·8 (w 6)   ——————   67″·8 (w 5)

Epoch 1832·20

A multiple star, just below the belt of Orion, forming a scalene triangle with ζ and ε. A 4, bright white; a 11, ash-coloured; B 8, bluish; C 7, grape red; D 8½, dusky; E 9, white; F 8, pale grey. This is a fine group of 10 members, forming 10 and 11 ♂. ɪɪ., where it is denominated "a double-treble star, or two sets of treble stars almost similarly situated;" H. and S. call it "a very pretty double-triple star;" but Professor Barlow, with greater precision, says it is "double-qua-druple, with two very fine stars between the sets." As some of these lie at a great distance, I measured to the uncials of H. and S., with only the addition a, to the bright, or following set. By reducing Σ.'s obser-vations to one epoch, and arranging his letters and quadrants to quadrate with ours, the scale of comparison will be :

	♂. 1779·77				H. and S. 1823·13		
A a	(not seen)			A a	(not seen)		
A B	Pos. 84° 55′	Dist.	13″·44	A B	Pos. 83° 19′	Dist.	12″·91
A C	60° 55′		43″·20	A C	61° 03′		42″·77
A D	(not measured)			A D	322° 57′		210″·26
D E	267° 00′ ±			D E	266° 21′		11″·14
D F	23° 25′			D F	21° 49′		68″·26

	Σ. 1831·06		
A a	Pos. 236° 52′	Dist.	11″·01
A B	84° 30′		12″·86
A C	59° 55′		41″·60
A D	(not measured)		
D E	267° 51′		8″·35
D F	21° 35′		68″·08

As this is a good object for trying the light and definition of a telescope, and the following of its groups is both delicate and pretty as a quadruple set, the explorer is recommended to examine it when in apparition. Nor need he be very much annoyed with his instrument, should he be unable to distinguish the minute *comes* a; since it is so small a point of light, that it escaped even the searching eye of ♂. This group may be readily fished up, as it forms the southern vertex of a triangle with the two last stars in the belt, as above stated; and it is rather less than a degree from ζ, in the direction of Rigel.

CCXXIII. ζ ORIONIS.

Æ 5ʰ 32ᵐ 41ˢ	Prec. + 3ˢ·02
Dec. S 2° 02′·0	— N 2″·38

Position AB 148°·8 (w 9)	Distance 2″·5 (w 9)	Epoch 1839·19
—— AC 7°·8 (w 4)	—— 56″·0 (w 2)	

A fine triple star, the last or lowest on Orion's belt, formerly one of the Greenwich List, but rejected in 1830. A 3, topaz yellow, and very bright for its magnitude; B 6½, light purple; and C 10, grey. The principal star is designated on Piazzi's and other Catalogues, Alnitak, the Arabian *al-niták*, the girdle; otherwise in conjunction with δ and ε, *minṭakah al-jaúzá*, the giant's belt. A slight difference from the general laws of precession has been exhibited to the following values:

$$P.... \text{Æ} - 0''·12 \quad \text{Dec.} - 0''·08$$
$$B.... + 0''·09 \quad - 0''·01$$

As this was classed 21 Ḥ. iv., in 1780, Sir William could not have seen the large star double; and yet it seemed difficult to account for his overlooking so remarkable and elegant a pair, wherefore it has been surmised, that the *comes* was under occultation at the time. Later observations do not countenance this singular evolution; and I took such pains to establish an epoch for future comparison, that I have every confidence in the results. The other measures which I have carefully consulted, are:

H. and S.	Pos. 150° 03′	Dist. 2″·62	Ep. 1822·61
H.	145° 52′	2″·60	1830·18
Σ.	151° 18′	2″·35	1831·22
D.	148° 23′	3″·00	1832·56

CCXXIV. 34 Ḥ. IV. ORIONIS.

Æ 5ʰ 33ᵐ 21ˢ	Prec. + 3ˢ·28
Dec. N 9° 00′·2	— N 2″·33

Mean Epoch of the Observation .... 1837·65

A planetary nebula, of a bluish white tint, on the nape of Orion's neck; and about 3½° on the line from Betelgeuze towards the three small stars forming Orion's head. This is a small and pale, but very distinct object, with a faint disc, discovered by Ḥ. in December, 1785, and is No. 365 of his son's Catalogue; wherein it is described as "rather oval, and perhaps of a mottled light:" a power of vision beyond what my means afforded. It was differentiated with *a* Orionis, and is preceded by several small stars, the foremost of which is coarsely double, and of the 8th and 10th magnitudes.

CCXXV.   γ LEPORIS.

Æ   5ʰ 37ᵐ 48ˢ   Prec. + 2ˢ·52
Dec. S 22° 30'·2   —— N 1"·94

Position AB 349°·0 (*w* 7)   Distance 92"·9 (*w* 5)⎫
——— BC 345°·0 (*w* 2)   ——— 45"·0 (*w* 1)⎬ Epoch 1832·06
                                        ⎭

A wide triple star, in a barren field, in the Hare's left hind foot; where a line passed from δ Orionis, the preceding star of the belt, through the sword cluster, and carried 16° beyond, hits upon it.   A 4, light yellow; B 6½, pale green; C 13, dusky; and a fourth star, of the 12th magnitude, following at △ Æ = 21ˢ. This poor object was only examined because, under 50 ♓. v., we are told there is a companion within 40' of A, of course meaning forty seconds.   This escaped my search, and also that of the Astronomer Royal, who obligingly examined it at my request, and forwarded me Mr. Main's diagram, which is identical with my own. B is mentioned by Piazzi in the notes to Hora V., No. 219; and by reducing his Æs and Decs. the results agree very fairly, epoch considered, with the micrometrical measures above registered:

Pos. 348° 0'   Dist. 95"·8 Ep. 1800

On the whole, though γ Leporis is of a fine lustre, I have little doubt of B and C being the stars which ♓. classed.   The proper motion of A has been thus stated:

P.... Æ — 0"·42   Dec. — 0"·40
B....   — 0"·29   — 0"·35
A....   — 0"·34   — 0"·39

CCXXVI.   78 M. ORIONIS.

Æ   5ʰ 38ᵐ 33ˢ   Prec. + 3ˢ·07
Dec. N 0° 00'·7   —— N 1"·87

Position 32°·0 (*w* 2)   Distance 45"·0 (*w* 1)   Epoch 1836·79

Two stars in a "wispy" nebula, just above Orion's left hip; where a ray from Rigel carried between the centre and last stars of the belt, and extended 2° farther, picks it up.   A 8½, and B 9, both white.   This object was first fixed by Messier in 1780; and described as "two bright nuclei surrounded by nebulosity." It is a singular mass of matter trending from a well defined northern disc into the *sf* quadrant, where it melts away.   The nebula lies equatoreally between two small stars, which are nearly equidistant from it, in a blankish part of the heavens; and in its most compressed portion is the wide double star.   This was beautifully drawn by H., and is figure 36 of his Catalogue of 1830.

CCXXVII. 52 ORIONIS.

Æ 5ʰ 39ᵐ 24ˢ   Prec. + 3ˢ·22
Dec. N 6° 23'·6   —— N 1'''·80

Position 199°·9 (w 7)   Distance 1''·8 (w 5)   Epoch 1838·27

A close double star, on Orion's left shoulder; about 2° south-west of Betelgeuze. A 6, pale white; B 6½, yellowish. From a comparison of all the measures of this very elegant object, it appears to have remained unaltered, both in position and distance, for upwards of half a century. The other registered results are:

	Pos.	Dist.	Ep.
Ḧ.	200° 19'	1''·00 ±	1781·76
S.	200° 41'	1''·65	1824·18
Σ.	200° 01'	1''·75	1831·23

CCXXVIII. 225 P. V. AURIGÆ.

Æ 5ʰ 39ᵐ 31ˢ   Prec. + 3ˢ·89
Dec. N 31° 43'·7   —— N 1'''·79

Position 61°·5 (w 8)   Distance 3''·8 (w 5)   Epoch 1832·00

A very neat double star, on the Waggoner's left shin; lying in the line formed between β Aurigæ and δ Orionis, the preceding star in Orion's belt; nearly in mid-distance between β and the three small stars forming Orion's head. A 8, creamy white; B 8½, pale grey. This fine object was discovered by Σ., and is No. 796 of the great Dorpat Catalogue, where its position and distance are thus given:

Pos. 61°·16   Dist. 3''·596   Ep. 1830·79

CCXXIX. ν AURIGÆ.

Æ 5ʰ 40ᵐ 24ˢ   Prec. + 4ˢ·15
Dec. N 39° 05'·6   —— N 1'''·71

Position 201°·9 (w 1)   Distance 85''·0 (w 1)   Epoch 1833·75

A coarse double star, on Auriga's left arm. A 5, rich yellow; B 12, dusky red; herein agreeing more than usual with Sir William Herschel's colour. This object is 90 Ḧ. v., thus tabulated:

Pos. 208° 12'   Dist. 53''·71   Ep. 1782·68

The position of B, and even the colour, identify it with Sir William's star; but the discordance in the distance is very great, for mere error of estimation. Sir William says, that the small star is not visible till after some minutes attention; but I found no difficulty in seeing the one here registered. It is readily distinguished by glancing about 10° along a line passed from Capella to Procyon, over ε in the knee of Castor.

CCXXX.   37 M. AURIGÆ.

Æ 5ʰ 41ᵐ 46ˢ   Prec. + 3ˢ·92
Dec. N 32° 30′·1   —— N 1″·59

Position 357°·0 (w 1)   Distance 25″·0 (w 1)   Epoch 1836·79

A double star in a cluster in front of Auriga's left shin. A and B, both 10th-magnitude, and pale yellow. A magnificent object, the whole field being strewed as it were with sparkling gold-dust; and the group is resolvable into about 500 stars, from the 10th to the 14th magnitudes, besides the outliers. It was found and fixed by Messier in 1764, who described it as "a mass of small stars, much enveloped in nebulous matter." This nebulous matter, however, yields to my telescope, and resolves into infinitely minute points of lucid light, among the distinct little individuals. It is immediately preceded on the parallel by another small double star: and is about half a degree north east of 225 P. v., whose alineation is already described.

CCXXXI.   α ORIONIS.

Æ 5ʰ 46ᵐ 30ˢ   Prec. + 3ˢ·24
Dec. N 7° 22′·3   —— N 1″·18

Position 155°·0 (w 1)   Distance 160″·0 (w 1)   Epoch 1832·75

A standard Greenwich star, with a distant *comes*, on Orion's left shoulder. A 1, orange tinge; B 11, bluish, and the two point nearly upon a pale small star in the *np* quadrant, at Δ Æ 15ˢ·7. The object forms 39 H. vi., and was thus registered:

Pos. 152° 18′   Dist. 161″·72   Ep. 1780·78

It is called Betelgeuze, from *ibṭ-al-jauzá*, the giant's *axilla*, or shoulder, whence it is also *menkib-al-jauzá;* and it has likewise been dèsignated *al-mirzam*, the roarer. It is the northernmost of the four bright stars forming the *corners* of this constellation, and cannot be mistaken by the most casual observer: moreover, with Sirius and Procyon, it forms a conspicuous triangle, which is nearly equilateral; while Procyon makes a right-angled one with Betelgeuze and Pollux. It is hardly necessary to diagram this well-known and splendid group; but possibly there may be a beginner who would wish for the following figure, as a guide.

H. has recently pointed out this fine star as being variable and periodic, and he thinks the most obvious conclusion is, an annual, or nearly annual period; but further observations are necessary for the confirmation: on his star-list the maximum was stated as above Rigel, the minimum below Aldebaran. It was suspected of a wide proper motion, but the ordeal of the best observations reduces it so greatly, that it is now barely entitled to registry.

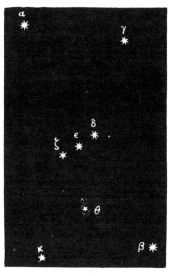

Orion may be considered the most beautiful and brilliant of all the constellations, without disparaging the Great Bear; and when just over our meridian, is so well accompanied, as to present the finest view of the heavens in this hemisphere. The principal stars of Orion, when joined by imaginary lines, form two inverted cones, and resemble a clepsydra, or hour-glass. He is usually represented as a classic warrior; but Paulus Venetus, *De côpositione Mūdi*, equips him in knightly armour, with a huge club in one hand, a formidable human-faced shield in the other, and a long Toledo sword by his side: and this is also the style in which he figures among the illustrations to Julius Firmicus, in 1497. It is a paranatellon of Taurus, and as the ecliptic passes nearly through its middle, it is visible to all the world; while its figure, belt, and pendant sword, so well described by Manilius, render it of easy recognition: hence it is written:

> Orion's beams! Orion's beams!
> His star-gemmed belt, and shining blade;
> His isles of light, his silvery streams,
> And gloomy gulfs of mystic shade.

No constellation was more noted among the ancients than Orion. As it occupies an extensive space in the heavens, this circumstance may have probably given Pindar his notion that Orion was of a monstrously large size; and hence the *jugula* of Plautus, the *magni pars maxima cœli* of Manilius, and the *jebbér* of the Arabians. Hood tells that "the reason why this fellow was placed in heaven," was to teach men not to be too confident in their own strength. But though his name was long ago bettered from Oarion or Arion, and he has been notorious as the Candaen of the Bœotians, the Hyreides of the old astrologers, and what not, the world will not yet agree in the nomenclature; as may be seen in the astronomical glossary of that redoubtable anti-Newtonian, *highte* Sir Richard Phillips, late sheriff of the good city of London. Following Naṣiru-l dín, the name is *El-Jebbár*, the hero; but, says Ideler, not *Algebra*, as is sometimes written in the astrognostic books. A disciple of the unhappy Lieutenant Brothers proposed to designate the whole asterism Nelson: and in 1807, the

University of Leipsic resolved, that the stars belonging to the belt and sword of Orion, as well as the intermediate ones, "shall in future be called the constellation NAPOLEON." Was that learned body in possession of a copy of Thomas Hood's treatise?

The present appellation, however, is of too long a standing, and has too firm a hold on men's minds, to be easily shaken; and, despite of his dirty origin, it seems "this fellow" must stand. Both the Septuagint and the Vulgate call it Orion, according to the Greeks and Romans. It is mentioned in Job, Ezekiel, and Amos; and the Mosaicists persist that it represented Nimrod, as mighty a hunter as Orion, and the author of the post-diluvian heresy*. From his terrible and threatening gesture, as much as from his time of rising, he was held to portend tempests and misfortune, and was therefore so much dreaded by the mariners of yore, as to give rise to the ancient proverb "Fallit sæpissime nautas Orion." Polybius attributes the loss of the Roman fleet in the first Punic war, to the obstinacy of the consuls, who, despite of the pilots, would sail between the risings of Orion and Sirius, always a squally time. The Latin writers are full of invective against *pluviosus et tristis* Orion; while the *nimbosus* of Virgil, the *nautis infestus* of Horace, the *aquosus* of Propertius, the *horridus sideribus* of Pliny, and the like sage allusions, fill the imagination with storms, hail, and deluges of rain. Added to this, we are reminded by Hood that this asterism was "the verie cutthrote of cattle:" and Hood was not addicted to astrology.

This constellation is a rich mine for the practical astronomer, as containing a wondrous universe of bright stars, double stars, clusters, and nebulæ, within itself. The Capuchin de Rheita asserted that, with his binocular instrument, he found more than 2000 stars in it; and where he is not dwelling upon Teutonic crosses and seamless tunics, he will be found worthy of credit. What may be telescopically obtained will not be decided, perhaps, until some amateur astronomer undertakes to map and tabulate it; for such work is out of the line of duty of the regular meridian observatories. The number of stars whose mean apparent places have been noted, are:

Ptolemy	. . . 38 stars	Bullialdus	. . . 61 stars
Ulugh Beigh	. . 38	Hevelius	. . . 62
Tycho Brahé	. . 62	Flamsteed	. . . 78
Bayer	 49	Bode	 304

The reader of course will remember that the equinoctial circle cuts the middle of Orion; which is also about 8° west of the solstitial colure, or *solis statio*. Nor will he forget the trimming which Halley gave Père Souciet, about the Dodecatemorion of Aries, Newton's chronology, and the equinoctial colure.

---

* Orion was designated *Khesil*, or *Kesil*, by the Hebrews, which the learned say comes from *chasel*, to be inconstant, to stir up, in allusion to the unsettled weather supposed to attend this constellation. Hence Rabelais has pleasantly called the grand Council of Trent, the Council of Chesil, to denote that it was a stormy, fickle, and troublesome meeting. Has the Australian term of being *chiselled*, any affinity with this?

CCXXXII. β AURIGÆ.

Æ 5ʰ 47ᵐ 48ˢ   PREC. + 4ˢ·40
DEC. N 44° 55'·3   —— N 1"·06

POSITION 38°·2 (w 2)   DISTANCE 185"·0 (w 2)   EPOCH 1837·70

A bright star with a distant companion, on the Waggoner's left shoulder. A 2, lucid yellow; B 10½, bluish. This object forms 88 ♓. vɪ., and was thus measured:

Pos. 35° 48' Dist. 169"·10 Ep. 1782·23

As a discussion arose upon this difference in the distance, I requested Mr. Challis to try it with the great Northumberland equatoreal, with which he kindly complied; and the following results constitute the last epoch:

Pos. 37° 56' Dist. 183"·45 Ep. 1841·96

This fine star—familiarly known as Menkalinan, from the Arabic *Menkib-dhi-l'inán*, the rein-holder's shoulder—lost much of its importance on being rejected from the Greenwich Catalogue in 1830. It may be picked up by projecting an imaginary line from the Præsepe, in Cancer, through Castor, which is nearly half-way: or a ray from Rigel through Bellatrix, led rather more than three times as far to the north, hits it. Our friend the galley-rhymester submits a third alineation, thus:

From the Pole-star direct a glance,   with Betelgeuze to mix,
About mid-distance, near the Goat,   Menkalinan you'll fix:
And there behold how neat it forms   with Capra bright a base,
While *delta* as a vertex stands,   the triangle to grace.

CCXXXIII. θ AURIGÆ.

Æ 5ʰ 48ᵐ 48ˢ   PREC. + 4ˢ·08
DEC. N 37° 11'·7   —— N 0"·98

POSITION 289°·0 (w 2)   DISTANCE 30"·0 (w 1)   EPOCH 1832·64

A neat double star, in the Waggoner's left wrist; where a line from Procyon through ε in Castor's knee, and 14° beyond, will find it, in the direction of the brilliant Capella. A 4, brilliant lilac; B 10, pale yellow; and lower down in the same quadrant, near the vertical, is a yellowish star of the 9th magnitude, which is that observed with A in the Catalogue of H. and S., No. 68, when its distance was found to be = 125". This proves it to be 34 ♓. vɪ., classed in September, 1780, but not measured; and A B are 89 ♓. v., thus registered:

Pos. 286°·00 Dist. 35"·30 Ep. 1782·68

A comparison of the best meridional results detects a small proper motion in space:

P.... Æ + 0"·11   Dec. − 0"·04
B....   + 0"·12   − 0"·11

CCXXXIV. 59 ORIONIS.

Æ   5ʰ 50ᵐ 06ˢ   Prec. + 3ˢ·11
Dec. N 1° 48'·9   —— N 0"·87

Position 201°·0 (w 1)   Distance 42"·0 (w 1)   Epoch 1833·69

A small star, with a very minute companion, between Orion's left side, and the Galaxy. A 6, white; B 13, blue. This delicate though wide object is 100 ♆. v., who remarked that the small individual is " a point requiring some attention to be seen." His measures were:

Pos. 205° 0'   Dist. 37"·25   Ep. 1782·76

A is preceded in the *np* quadrant by an 8th-magnitude star, whose angle is about 290°, with a distance of 178": this is 282 P. v. 59 Orionis may be picked up by a line shot from Rigel through ζ, the third star of the belt, and carried nearly 6° beyond.

CCXXXV. 35 CAMELOPARDI.

Æ   5ʰ 51ᵐ 48ˢ   Prec. + 4ˢ·75
Dec. N 51° 34'·2   —— N 0"·72

Position 14°·4 (w 2)   Distance 30"·0 (w 1)   Epoch 1833·66

A small double star, which, though absurdly chronicled in the Camelopard, is in the Waggoner's eye; and it is nearly in the line between β and δ Aurigæ. A 7, white; B 10, lilac. It was picked up by H., and is No. 2292 of his Sweeps; and our results, weighing the conditions, are not harshly discordant, his magnitudes being registered 6½ and 11, with Pos. 13°·8 and distance 35"·0.

CCXXXVI. 35 M. GEMINORUM.

Æ   5ʰ 59ᵐ 01ˢ   Prec. + 3ˢ·67
Dec. N 24° 21'·3   —— N 0"·09

Mean Epoch of the Observation . . . . 1836·80

A cluster, near Castor's right foot, in the Galaxy, discovered and registered by Messier in 1764. It presents a gorgeous field of stars from the 9th to the 16th magnitudes, but with the centre of the mass less rich than the rest. From the small stars being inclined to form curves of three or four, and often with a large one at the root of the curve, it somewhat reminds one of the bursting of a sky-rocket.

Under favourable circumstances this cluster can be distinguished by the naked eye; it therefore may be comparatively near us. It must be sought on the line between Castor and ζ on the tip of the Bull's southern horn, at exactly one-quarter of the distance from the latter: or a line led from a Leporis through Betelgeuze, and extended 18° beyond, will strike upon it.

This object being so handy to the point assumed by Hipparchus, as the north extreme of the ecliptic, I swept for anything which might be on the actual spot, under the necessary corrections, in Æ 6^h, and Dec. N 23° 27'. After some search, I found a star of the 12th magnitude, but too small for having its place differentiated for any permanent purpose.

CCXXXVII. 41 AURIGÆ.

Æ 5^h 59^m 21^s PREC. + 4^s·59

DEC. N 48° 44'·1 —— N 0"·06

POSITION 351°·4 (w 6) DISTANCE 8"·5 (w 4) EPOCH 1831·13

—— 352°·8 (w 9) —— 8"·2 (w 9) —— 1837·97

A neat double star, in front of the Waggoner's chin; where a line from β Tauri, on the tip of the Bull's northern horn, led through β Aurigæ, and carried 4° beyond, strikes upon it. A 7, silvery white; and B 7½, pale violet, and it probably partakes of the proper motion assigned to A:

P.... Æ − 0"·15	Dec. − 0"·17
B.... + 0"·18	− 0"·12
T... + 0"·23	− 0"·24

This pair is Piazzi's Nos. 333 and 334, Hora V., as well as 82 Ⱶ. III.; and there appears no appreciable motion in the lapse of 55 years.

	Pos.	Dist.	Ep.
Ⱶ.	350° 00'	8"·53	1782·85
P.	350° 00'	9"·00	1800·00
H. and S.	353° 16'	8"·81	1822·53
Σ.	353° 07'	7"·99	1830·31

CCXXXVIII. 24 Ⱶ. VIII. ORIONIS.

Æ 5^h 59^m 25^s PREC. + 3^s·40

DEC. N 13° 58'·6 —— N 0"·05

POSITION 108°·8 (w 8) DISTANCE 2"·4 (w 9) EPOCH 1837·02

A close double star in a small cluster, on Orion's left hand. A 7½, and B 8½, both lucid white. This elegant little triangular group, has many glimpse stars thronging about the two lower angles. The cluster was discovered by Ⱶ. in 1784; but the very neat pair here measured, was first enrolled by Σ., No. 848, as *lucida acervi*, and the following measures have been obtained:

S.	Pos. 109° 33'	Dist. 2″·75	Ep. 1825·10
H.	108° 15'	2″·86	1830·92
D.	107° 20'	2″·59	1831·13
Σ.	108° 50'	2″·35	1833·19

This, therefore, must be merely an optical object. A line from the central star of Orion's belt passed close over Betelgeuze, and prolonged 7° —or rather more than as far again—beyond, picks it up between the Pleiades and Procyon.

These gatherings occurring indifferently upon the *Via Lactea* and off it, awaken still more our admiration of the stupendous richness of the Universe, in every department of which there appears such a profusion of creation, if we may so express ourselves of the works of the ALMIGHTY, in which our utmost ken has yet never detected any redundancy, much less anything made in vain.

CCXXXIX. 25 ♅. VII. ORIONIS.

Æ 6ʰ 03ᵐ 35ˢ	PREC. + 3ˢ·20	
DEC. N 5° 28'·9	—— S 0″·31	

POSITION 355°·0 (*w* 1)   DISTANCE 5″·0 (*w* 1)   EPOCH 1833·00

A neat but minute double star, in a cluster, under Orion's left shoulder and in an outcropping of the Galaxy. A 9½ and B 10, both pale yellow. This is a tolerably rich and compressed mass of stars, from the 9th to the 16th magnitudes, with numerous stragglers. It was discovered by Herschel in 1786, and is No. 384 of his son's great Catalogue; but the reference in the latter to No. 2288 of the Double-star Sweeps, ought to have been 2301.

To fish up this object, pass a line from Rigel through the lower star of Orion's belt, and carry it a little more than as far again to the northeast, where it will strike the cluster at about 4° south-east of Betelgeuze.

CCXL. 4 LYNCIS.

Æ 6ʰ 07ᵐ 51ˢ	PREC. + 5ˢ·33	
DEC. N 59° 25'·8	—— S 0″·68	

POSITION 90°·2 (*w* 6)   DISTANCE 1″·0 (*w* 3)   EPOCH 1837·89

A close double star, in the animal's snout; at about 30° from Polaris, on a line through Sirius, and closely north-east of 2 Lyncis. A 6 and B 7½, both white. This elegant but difficult object was discovered by Σ., and is one of his *vicinæ;* being No. 881 of the great Dorpat Catalogue, where it is thus registered:

Pos. 88°·97   Dist. 0″·815   Ep. 1830·28

This would imply a direct movement in angle; but the pair is too impracticable to merit reliance on epochs of short periods. It is only from accurate and continued observations, that an orbit worthy of confidence will emerge.

The Lynx, *seu Tigris,* is one of the new asterisms formed by Hevelius, from the *stellæ informes* of the neighbourhood, and added to the old 48 constellations. It is pretty extensive, occupying the vast space between Auriga and the Greater Bear, above the Twins; but though it contains many beauties for assisted vision, there are few remarkable objects to the naked eye. Hevelius started it with 19 stars, Flamsteed gave it 44, and Bode 149.

Hevelius defends the location he has assigned to this animal, and in a set paragraph, *De Loco Lyncis,* tells us that he cast it between the Great Bear and Auriga, where an empty space was found on the globes, which was wont to be filled up with title and dedication. He acknowledges that the 19 components he assigned it are small and insignificant, but thinks that those who would examine the Lynx ought to be lynx-eyed. He formed a symbol for this, as well as for the other asterisms, giving permission to those carpers who dislike them, to make new ones if they choose: "Si cuidam Momo fortè displicent, liberum ipsi per me esto, alios characteres effingere."

CCXLI.   58 P. VI. MONOCEROTIS.

Æ   6ʰ 10ᵐ 43ˢ   Prec. + 3ˢ·36
Dec. N 12° 21 ·2   —— S 0″·94

Position 295°·0 *(w* 1)   Distance 20″·0 *(w* 1)   Epoch 1839·10

A most delicate double star, close to Orion's left hand, and in the Galaxy. A 8 and B 13, both dull yellow; followed at about 10ˢ by a coarse pair, of the 9th and 10th magnitudes, which constitute the No. 892 rejected from Σ.'s list. The object here estimated is No. 891 of the same Catalogue, where it is thus registered by its discoverer:

Pos. 292°·23   Dist. 21″·903   Ep. 1830·33

Here is another of those cases where illumination is out of the question; but the rock-crystal micrometer enabled me to catch up a tolerably fair angle. The instrument is easily managed on practice.

This star, though placed on the Unicorn's horn by various map-makers, is filched from Orion, and in Sir John Lubbock's Map is placed on that gentleman's club; so that in any reform of the heavens, the matter must be gravely looked to. It will be found by carrying a line from Rigel through ε, in the middle of Orion's belt, which, passed under Betelgeuze on his left shoulder, and extended 7½° beyond, will strike upon the little star in question.

L 2

CCXLII.   5 LYNCIS.

Æ   6ʰ 12ᵐ 50ˢ   Prec. + 5ˢ·25
Dec. N 58° 29'·7   —— S 1"·12

Position A B 130°·0 (w 1)   Distance 25"·0 (w 1))
———— A C 271°·9 (w 7)   ———— 96"·0 (w 2)}   Epoch 1833·77

A coarse triple star, on the animal's lower jaw; it is a little to the north of a line running from Capella towards Dubhe, and about 18° from the former star, where it precedes 6 Lyncis. A 6, orange tinge; B 13, blue; and C 9, pale garnet, being No. 61 of the Palermo Catalogue. The larger individuals of this object form 102 ℍ. vi., whose searching scrutiny overlooked B. The following are the registered measures of A and C: and to these may be added the testimony obtained by reducing Piazzi's meridian observations.

ℍ.   Pos. 272° 00'   Dist. 88"·33   Ep. 1782·87
P.   271° 30'   98"·6   1800·00
S.   272° 07'   95"·45   1825·05

Sir John Herschel, in his notes, alludes to the difference in distance between his father's measures and more recent ones, saying, " it may be remarked once for all, that there is great reason to suspect a considerable instrumental error in all the measures of that early period, exceeding 40", the result being constantly (or most commonly) in defect, and that not unfrequently to a very large amount. The cause probably lies in the construction of the micrometer used; and its effect is to throw a great uncertainty on the earlier distances of all stars of the Fifth and Sixth Classes. Fortunately these are the least replete with interest." Here, however, my friend's opinion,—and with the utmost deference be it said,— is, like one of Homer's prayers, only to be in part received.

CCXLIII.   μ GEMINORUM.

Æ   6ʰ 13ᵐ 17ˢ   Prec. + 3ˢ·62
Dec. N 22° 35'·5   —— S 1"·16

Position 89°·0 (w 1)   Distance 80ˢ·0 (w 1)   Epoch 1831·98

A Greenwich star of 1830, with a distant *comes*, on Castor's right instep; a glance from Orion's sword cluster through ζ, the lowest of the belt, carried closely to the east of Betelgeuze, and 16° beyond, will rest upon μ. A 3, crocus yellow; B 11, bluish; there are two other companions in the *sf*, and a group of small stars follow at Δ Æ = 25ˢ. This object is known as Tejat *post*, from *Tahyáh*, a word used by the Arabs, as the name of a constellation formed by the two stars η and μ, in the anterior feet of Gemini, also called *Al-hen'ah*, and *Al-nuhhátaï*. The latter word is the dual of *Nuhát*, side or inclination; which affording but little clue, *Nahátaï* (from *nahát*) two strangers, has been suggested. The proper motion of the præses has been thus valued:

$$P.... \text{Æ} + 0''\cdot13 \qquad \text{Dec.} - 0''\cdot16$$
$$B.... \quad + 0''\cdot15 \qquad\qquad - 0''\cdot13$$

CCXLIV.   ζ CANIS MAJORIS.

Æ 6ʰ 14ᵐ 10ˢ   Prec. + 2ˢ·30
Dec. S 29° 59′·9   ―― S 1″·24

Position 338°·0 (*w* 4)   Distance 167″·0 (*w* 1)   Epoch 1833·81

A star with a distant companion, on the Greater Dog's left hind paw; where it will be found in a manner insulated, by running a line from Pollux into the south-west, closely shaving Sirius, and carrying it about 14° beyond that lustrous star. A 3, light orange; B 7, pale grey; these components being Piazzi's Nos. 81 and 80, Hora VI. This object is called Phurud, and is supposed to be from *Al-furúd,* the single ones; but it is probably an error of transcription, easily made in Arabic, for *Al-kurúd,* the monkeys, *i. e.* from 2 to 5 of the smaller stars of Canis Major, with θ, κ, and λ Columbæ.

Comparisons of the best observations have indicated that A has a movement in space independent of the general laws of precession, and the amount has been thus conflictingly stated:

$$P.... \text{Æ} - 0''\cdot20 \qquad \text{Dec.} - 0''\cdot25$$
$$B.... \quad + 0''\cdot06 \qquad\qquad + 0''\cdot03$$

CCXLV.   8 MONOCEROTIS.

Æ 6ʰ 15ᵐ 17ˢ   Prec. + 3ˢ·18
Dec. N 4° 40′·1   ―― S 1″·34

Position 23°·8 (*w* 8)   Distance 12ˢ·9 (*w* 8)   Epoch 1834·19

A neat double star, in the Unicorn's nostril; where a glance from Aldebaran, passed closely over the head and shoulders of Orion, will find it at about 7½° east of Betelgeuze. A 5½, golden yellow; B 8, lilac. This fine object is composed of Piazzi's Nos. 84 and 85 of Hora VI.; and it was classed 29 ℌ. iii., in 1781, but no measures were then taken. Subsequent observations afford reasonable presumption of its retrograding, in the approximate ratio of − 0·75:

	Pos.	Dist.	Ep.
H. and S.	25° 21′	14″·379	1823·04
Σ.	25° 52′	13″·865	1831·74

CCXLVI.   β CANIS MAJORIS.

Æ   6ʰ 15ᵐ 40ˢ   Prec. +   2ˢ·64
Dec. S 17° 52′·9   —— S 1″·37

Position 339°·0 (ʷ 1)   Distance 104″·0 (ʷ 1)   Epoch 1833·76

A bright star with a distant companion, on the Dog's fore-paw.
A 2½, fine white; B 9, dusky grey, and another of the same magnitude
in the *sp* quadrant. An almost imperceptible movement in space is
attributed to the leader, of which the following are the most accurately
investigated quantities:

$$P.... \text{Æ} - 0''·04 \qquad Dec. - 0''·06$$
$$B.... \quad + 0''·05 \qquad\qquad 0''·00$$

β Canis Majoris is called Mirzam, the roarer, a term applied to the
camel as well as the lion. *Al-mirzam* is the name of this star, of β in
Canis Minor, and of γ and a in Orion; the two former being called
al-mirzamáni, the two roarers. A line dropped from Nath, on the
northern horn of the Bull, to Betelgeuze, and from thence nearly the
same distance southwards, will fall upon the star under discussion; the
rhymer remarks:

Where Sirius blazes in the south,   and leaves the ship behind,
Look west-south-west, just four degrees,   and *beta* there you'll find.

CCXLVII.   15 GEMINORUM.

Æ   6ʰ 18ᵐ 14ˢ   Prec. +   3ˢ·58
Dec. N 20 52′·9   —— S 1″·59

Position 205°·4 (ʷ 5)   Distance 33″·2 (ʷ 3)   Epoch 1832·04

A fine double star, on Castor's right heel; very nearly in mid-distance
of an imaginary line between Castor and Bellatrix, where it is the
northern member of a trapezium of small stars. A 6, flushed white;
B 8, bluish, the latter being Piazzi's No. 99, Hora VI. This object is
classed twice over by Sir William Herschel, being 52 and 56 Ḥ. v.;
Flamsteed erred in its entry, as shown by Mr. Baily, and it was mistaken
by Mr. Taylor, at Madras, for 17 Geminorum, which is among the *non
inventa* of Piazzi. Both stars were well determined at Palermo, by
repeated observations; and as they had been the cause of such repeated
inadvertencies, I gave A a more than usual attention with the meridional
instruments. A careful comparison of the results impresses a belief, that
the proper motion in Æ is inappreciable; but that a slight annual
movement in declination actually exists. The micrometric measurements
afford presumptive proof that they are relatively unaltered:

Ḥ.   Pos. 210° 00′   Dist. 32″·65   Ep. 1782·08
H. and S.   204° 39′   32″·69   1822·10

CCXLVIII. 104 P. VI. MONOCEROTIS.

$\mathcal{R}$ 6^h 18^m 30^s Prec. + 3^s·08

Dec. N 0° 32'·6   —— S 1"·62

Position AB 151°·5 (*w* 6)   Distance 67"·8 (*w* 2)⎫

——— BC 170°·0 (*w* 2)   ——— 0"·6 (*w* 1)⎬ Epoch 1833·14

A coarsely triple star, between the boundary line of Orion and the Unicorn's chest; it is about 17° from Procyon, on a ray carried to the west-south-west through Orion's sword cluster and Rigel. A 7½, topaz yellow; B and C 8½, both of a plum tinge. B is most exquisitely double, Piazzi's No. 105, Hora VI., and 910 of the Dorpat Catalogue, where it was classed among the *vicinissimæ*. When Struve first subjected it to measurement, under a power of 540, it was:

Pos. 168° 48'   Dist. 0"·78   Ep. 1825·12

On considering the great difficulty of an object which my best powers only elongated, these observations may be considered to correspond. Σ. afterwards added A to B C, under the following determinations, with magnifying power 320:

Pos. 150°·57   Dist. 66"·15   Ep. 1831·68

CCXLIX. 10 MONOCEROTIS.

$\mathcal{R}$ 6^h 20^m 03^s Prec. + 2^s·96

Dec. S 4° 40'·2   —— S 1"·75

Position 225°·0 (*w* 1)   Distance 72"·0 (*w* 1)   Epoch 1832·99

A wide double star in an elegant group, on the Unicorn's right fore-knee: it is about 12¼° in an occult line carried from Sirius a little to the west of Capella, and directly between β in the Lesser Dog and α Leporis. A 6, pale yellow; B 9, orange, with a *comes* to the south. Though this object is a capital one for testing the performance of a telescope, it has not been classed among the clusters. Piazzi, in his note upon 116 vi., says, " plures telescopicæ simul conspicuæ."

Monoceros was introduced into the firmament by Bartschius, among the delineations on his four-foot globe; it was, perhaps, out of regard to the husband of Kepler's daughter, retained by Hevelius, being now considered as one of his constellations. It is concocted of the *stellæ informes* scattered about in the large space between Orion, Hydra, and the two Dogs, over a portion of the Milky Way. But though extensive it is not conspicuous, few of its gems rising to the 4th magnitude. It has been pretty well ransacked since its first appearance in the *Prodromus Astronomiæ*, and many capital pairs, nebulæ, and clusters have been reaped. The stars have been thus successively tabulated:

Hevelius	. . . . 19 stars	Piazzi	. . . . 95 stars
Flamsteed	 31	Bode	 220

CCL. 11 MONOCEROTIS.

Æ   6^h 21^m 04^s Prec. + 2^s·91

Dec. S 6° 56'·1   —— S 1"·84

Position AB 130°·3 $_{(w\ 7)}$ Distance 7"·2 $_{(w\ 5)}$

—— AC 121°·6 $_{(w\ 8)}$ —— 9"·6 $_{(w\ 3)}$ } Epoch 1834·02

—— BC 102°·3 $_{(w\ 7)}$ —— 2"·8 $_{(w\ 4)}$

A fine triple star, in the Unicorn's right fore-leg: a ray shot from the Bull's eye through Bellatrix, and extended rather more than as far again into the south-east, will pick it up in the out-cropping of the Milky Way. A 6½, white; B 7, and C 8, both pale white. Two constituents of this object appear bracketted in Piazzi's Catalogue as double—Nos. 121 and 122, Hora VI.—the stars he saw and determined being A and B; and about 250" away in the *np* quadrant, at an angle of 340°, is the little star alluded to in the Palermo Catalogue—"alia 8^æ magnit. præcedit ad boream." Sir William Herschel, who discovered it in 1781, classed it a "curious treble star," pronouncing it to be "one of the most beautiful sights in the heavens;" but the next observers, his son and Sir James South, registered it quadruple. This is 10 Ⱨ. ɪ. and 17 Ⱨ. ɪɪ.; 71 of H. and S.; and 919 of Σ.; and the several measures are so coincident, on comparison—notwithstanding the nearness of magnitudes creates an anomaly of quadrants—as to prove the general fixity of the individuals. But a slight degree of proper motion is imputed to A, of the following varying values:

P.... Æ − 0"·06   Dec. − 0"·12

B....   + 0"·06   + 0"·05

CCLI. 2 Ⱨ. VII. MONOCEROTIS.

Æ   6^h 22^m 45^s Prec. + 3^s·19

Dec. N 5° 03'·1   —— S 1"·99

Mean Epoch of the Observation . . . . 1836·17

A tolerably compressed cluster, between the Unicorn's fore legs. This is a brilliant gathering of large and small stars, from the 7th to the 14th magnitude; the latter running in rays. It was first registered by Sir William Herschel in the summer of 1784, and is No. 392 of his son's Catalogue. Its place is differentiated from *a* Orionis; and it may be found nearly in mid-distance between Pollux and *a* Leporis, where it is crossed by a line led from Procyon to the west, and passed between Orion's belt and his right shoulder, about 2° below Bellatrix.

CCLII.   20 GEMINORUM.

ℛ   6ʰ 22ᵐ 57ˢ   Prec. + 3ˢ·50
Dec. N 17° 53'·2   — S 2"·00

Position 209°·2 (w 8)   Distance 20"·4 (w 8)   Epoch 1833·99

A neat double star, on Castor's left heel, about 1½° to the north-north-west of the bright star γ Geminorum; where a line carried from Rigel over the middle component of Orion's belt, will point it out at about 14° beyond Betelgeuze. A 8, topaz yellow; B 8½, cerulean blue.

A little explanation may be requisite on the identity of the individuals constituting this pair, for Piazzi asserts that Bradley, with but one exception, always observed 21 Geminorum for 20. Mr. Baily says that 21 does not exist, at least in the position given by the British Catalogue, observing: "The observation, from which it has been deduced by Flamsteed, was made on February 19, 1696, at 7ʰ 26ᵐ 50ˢ; as may be seen MSS. vol. xxiii. page 43. Most of the modern astronomers have supposed it to be Piazzi vi. 135; or the second of the two stars forming the double star 20 Geminorum. But this is on the assumption that Flamsteed has made an error of 1ᵐ in recording the time; and that 7ʰ 26ᵐ 50ˢ ought to be read 7ʰ 25ᵐ 50ˢ. Although this is very probable, (a similar mistake having *certainly* been twice committed on the same day with two previous stars,) yet there is nothing in the original MS. entry to warrant the conclusion. It is the only solution, however, of the difficulty."

This is a very fine object, and was classed 46 ♓. iv., in 1781, but no measures of position were given, and only an estimated distance of 25". H. and S., therefore, afford the earliest micrometrical point of departure; and their measures, compared with those of Σ. and my own, show that no appreciable change has occurred in twelve years:

H. and S. Pos. 208° 57'   Dist. 19"·454   Ep. 1822·05
Σ.   209° 48'   20"·012   1830·00

CCLIII.   14 MONOCEROTIS.

ℛ   6ʰ 26ᵐ 06ˢ   Prec. + 3ˢ·25
Dec. N 7° 41'·5   — S 2"·28

Position 210°·0 (w 1)   Distance 10"·0 (w 1)   Epoch 1833·89

A most delicate double star, in the Unicorn's eye. A 6, yellowish white; B 16, dusky. This is indeed a difficult object, B being the *minimum visibile* of my instrument, and with it only seen by such transient glimpses, that but for a distant pale 10th-magnitude star in the *sp* quadrant, nearly in the same line of bearing, my estimations must have been much wider. There is also a dusky 12th star in the *sf*, at about 100" from A; so that the whole forms a wide quadruple object. It is nearly mid-way between Procyon and the three little stars which form Orion's head—and also of a line passed from Castor to a Leporis—in the middle of the Milky Way.

About 2′ preceding, and 40′ south of this, is a sprinkle of brilliant stars, registered 2 Ⅱ. vii. in 1784, and described as tracing " winding lines around 12 Monocerotis."

CCLIV. γ GEMINORUM.

Æ 6ʰ 28ᵐ 28ˢ   Prec. + 3ˢ·46
Dec. N 16° 31′·8   —— S 2″·49

Position AB 335°·0 (w 1)   Distance 75ˢ·0 (w 1) ⎱
———— AC 290°·0 (w 1)   ———— 110ˢ·0 (w 1) ⎰ Epoch 1830·80

A coarse triple star, on the right foot of Pollux, in a rich field. A 3, brilliant white; B 13, and C 12, both pale plum colour; followed nearly on the parallel, Δ Æ = 40ˢ, by a neat 9th-magnitude star. This object, with ξ on the other foot, is called Alhena, from *al-hen'ah*, a ring or brand on a horse's neck, and the two form the VIth Mansion of the Moon. These stars are also ξ *Al-zerr*, the button; and γ *Al-meïsán*, the proud marcher; and they are described by 'Kazwíní as two whitish glimmering stars in the Milky Way; but Tízíní makes the Alhena include also η, μ, and v, in Castor's right foot; and some of his countrymen termed η and v, *Al-nuhhátaï*, the expressed dual of *al-nuhháh*, a large camel's hump. η is called Προπους by Ptolemy, as being in the forefoot, but the name is best applied to H Geminorum, as the ante-foot, or star preceding the feet.

The alignment of this star is easy. A ray from Rigel through the middle star of Orion's belt, will pass clear of Betelgeuze to Alhena. The same imaginary line continued, passes close under Castor: it is also about two-thirds of the distance between Pollux and Betelgeuze.

Proper motions are detected in γ Geminorum, but not to the amount which was suspected, the most authentic values being:

P.... Æ + 0″·05   Dec. - 0″·08
B....   + 0″·08   - 0″·02

CCLV. v^1 CANIS MAJORIS.

Æ 6ʰ 29ᵐ 23ˢ   Prec. + 2ˢ·62
Dec. S 18° 32′·0   —— S 2″·56

Position 260°·2 (w 5)   Distance 17″·2 (w 5)   Epoch 1830·83

A neat double star, in the Greater Dog's left fore knee, and about 3° to the south-west of Sirius. A 6½, pale garnet; B 8, grey. The last is

Piazzi's No. 178 of Hora VI., and the pair is followed in the *sf* quadrant by ν^2. This is 81 Ḥ. IV., recorded as having a distance of 18″·32, but with a position "nearly preceding" in 1782. From Ḥ.'s remark this object was sedulously attacked at Palermo, and the place of both components in Æ and Dec. were established; though the angle of position and distance derived from such data, is rather too vague for relying upon. But H. and S. then micrometrically measured it, and the results stand thus:

	Pos.	Dist.	Ep.
P.	255° 30′	19″·01	1800·00
H. and S.	259° 52′	17″·24	1821·22

Sir William's expression of "nearly preceding," being irreconcileable with a deviation of 10° from the parallel, it was concluded, that an obvious and considerable change had occurred in the angle of position: but this was not confirmed by my observations. Yet, as a friend considered the angle to be = 256° in 1839, I requested the Astronomer Royal to re-examine it, and he kindly handed me the following measures by the Rev. Robert Main:

Pos. 261° 36′   Dist. 17″·34   Ep. 1842·82

CCLVI. 174 P. VI. LYNCIS.

Æ   6ʰ 30ᵐ 42ˢ   PREC. + 5ˢ·33
Dec. N 59° 35′·6   —— S 2″·68

POSITION 134°·2 (*w* 8)   DISTANCE 4″·0 (*w* 6)   EPOCH 1835·11

A neat double star, under the animal's eye, nearly in mid-distance between Dubhe and Capella, where it is crossed by a line passing from Polaris a little to the westward of Procyon. A 7½, bright white; B 10, blue: the magnitude of the latter star was carefully estimated by my usual method of assuming Piazzi's brightness of A as the standard; and also by referring directly to the companion of Polaris. This delicate object was discovered by Σ., and is No. 946 of the great Dorpat Catalogue. S. measured it in 1825, and from the difficulty he experienced, I expected to find the companion much smaller than I did. It appeared to him of the 12th magnitude, and Σ.—from finding it 8·5 in 1827, 10 in 1831, and 8·5 again two years afterwards—asks, *Num minor variabilis?* Their measures are:

	Pos.	Dist.	Ep.
S.	136° 48′	4″·063	1825·07
Σ.	133° 28′	4″·197	1830·58

Now these results, in the brief period of ten years, afford an indication of evolution, the nature of which must be shown by a train of future observations. But the possibility of the *comes* being variable, awakens considerations of peculiar interest; it having been surmised, that certain small acolyte stars shine by reflected light, a point which is still to be ascertained. But sidereal science is yet in its infancy.

CCLVII.　12 LYNCIS.

Æ　6^h 32^m 5^s　　PREC. + 5^s·32
DEC. N 59° 35'·6　　—— S 2"·80

POSITION A B 154°·3 (*w* 9)　DISTANCE 1"·6 (*w* 6)⎫
—— A C 305°·1 (*w* 9)　　—— 8"·6 (*w* 9)⎬　EPOCH 1832·96

—— A B 149°·5 (*w* 9)　　—— 1"·6 (*w* 9)⎫
—— A C 305°·6 (*w* 9)　　—— 8"·9 (*w* 9)⎬　EPOCH 1839·27

A neat triple star, on the animal's cheek, so exactly following 174
P. vi., above described, that the alineation there given will answer for
both. A 6, white; B 6½, ruddy; C 7½, bluish. This curious object, of
which A and C are Piazzi's 185 and 184 of Hora VI., was discovered
to be triple in 1780, and registered 6 Ħ. i., and 22 Ħ. iii. By a
comparison of the measures then made, with the subsequent ones of H.
and S., Σ. D. and myself, it will be seen that the two close stars have
undergone so great a retrograde orbital change as to promise to bring the
three stars into a straight line in about half a century, C having remained
relatively unaltered with the primary. The other measures are:

Ħ. 1780·68
Pos. A B 181° 23'　Dist. 1"·5±
　　A C 302° 33'　　9"·38

H. and S. 1822·93
Pos. A B 158° 39'　Dist. 2"·59
　　A C 306° 50'　　9"·85

Σ. 1831·10
Pos. A B 153° 42'　Dist. 1"·52
　　A C 304° 12'　　8"·67

D. 1833·13
Pos. A B 153° 19'　Dist. 1"·64
　　A C 304° 06'　　8"·88

From a rough-cast geometrical treatment of these, there results an *annus
magnus* of nearly seven of our centuries.

CCLVIII.　15 MONOCEROTIS.

Æ　6^h 32^m 10^s　　PREC. + 3^s·30
DEC. N 10° 02'·2　　—— S 2"·81

POSITION A B 206°·2 (*w* 6)　DISTANCE 2"·5 (*w* 3)⎫
—— A C　15°·0 (*w* 1)　　—— 15"·0 (*w* 1)⎬　EPOCH 1835·13

A delicate triple star, in a magnificent stellar field, between the
Unicorn's ears, at one-third of the distance from Procyon towards Aldebaran. A 6, greenish; B 9¼, pale grey; C 15,'blue. This very fine
object is one of Σ.'s First Class, and No. 950 of the Dorpat Catalogue,
where these measures are registered:

Pos. A B 208°·66　Dist. 2"·76⎫
　　A C　12°·90　　　16"·58⎬　Ep. 1831·67

The bright star is crowned by three pairs, of which the *sf* is the nearest.
Piazzi had noted—"Duplex videtur. Multæ simul conspiciuntur."

CCLIX. ε GEMINORUM.

ℛ 6ʰ 34ᵐ 05ˢ   Prec. + 3ˢ·69
Dec. N 25° 16′·9   —— S 2″·97

Position 94°·1 (w 5)   Distance 110″·6 (w 3)   Epoch 1831·98

A star with a distant companion, on Castor's right knee; it is about 26°, or rather more than one-third of the distance, from Procyon towards Capella, where a line led from Rigel through Betelgeuze also reaches it. A 3, brilliant white; B 9½, cerulean blue.   This wide object is 73 ℍ. vi.; registered by Sir William in 1782, with a distance of 110″·48, but no angle of position given.   It was first measured by S.:

Pos. 93° 42′ Dist. 111″·57 Ep. 1825·04

This star is called Mebsuta, from *al-dhirá' al mebsútah*, the outstretched arm; *i.e.* Castor and Pollux, the bright stars of whose heads form the VIIth Lunar Mansion.

CCLX. 56 AURIGÆ.

ℛ 6ʰ 35ᵐ 12ˢ   Prec. + 4ˢ·33
Dec. N 43° 43′·7   —— S 3″·07

Position 17°·1 (w 9)   Distance 56″·8 (w 6)   Epoch 1831·92

A wide double star; it is just to the north of an imaginary line carried from Capella eastward through β Aurigæ, and extended as far again as the distance between those two stars.   A 6, silvery white; B 8½, lilac.   This is an object which, though belonging to Auriga, is on Telescopium Herschelii, an asterism proposed by the Abbé Hell to commemorate the discovery of the planet Uranus, in this spot, 13th March, 1781.   It is No. 107 ℍ. v., and appears, from a comparison of determinations, to have remained stationary for fifty years.   The other measures are:

ℍ. Pos. 17° 24′ Dist. 52″·95 Ep. 1782·80
S.   17° 08′   55″·38   1823·20

CCLXI. 215 P. VI. LYNCIS.

ℛ 6ʰ 36ᵐ 46ˢ   Prec. + 4ˢ·83
Dec. N 53° 12′·1   —— S 3″·20

Position 69°·1 (w 4)   Distance 23″·0 (w 2)   Epoch 1831·78

A delicate double star, on the neck of the Lynx; it is about 11° on a line shot from β Aurigæ towards Dubhe, or nearly one-third of that distance. A 8, bright yellow; B 11, dusky green. This is rather a troublesome object to treat, but on the whole the measures are satisfactory from coincidence *inter se.* It was first detected and registered by Σ., and these results appear in the Dorpat Catalogue:

Pos. 66°·40   Dist. 21″·93   Ep. 1829·21

CCLXII. α CANIS MAJORIS.

Æ 6ʰ 38ᵐ 06ˢ   PREC. + 2ˢ·68

DEC. S 16° 30′·1   —— S 3″·32

POSITION 45°·0 (*w* 1)   DISTANCE 150″·0 (*w* 1)   EPOCH 1835·80

A standard Greenwich star, with a distant companion, in the Greater Dog's mouth. A 1, brilliant white; B 10, deep yellow, other distant small stars in the field; and a line through the two here cited passes nearly upon that mentioned by Piazzi, "alia 8ᵃᵉ magnit. præcedit 3″ temporis, 3′ ad Boream." A, or Sirius, is subject to a large proper motion, the values of which have been stated as follows:

P....	Æ − 0″·51	Dec.	− 1″·14
B....	− 0″·48		− 1″·23
A....	− 0″·53		− 1″·23

Sirius, the dog-star, and one of Orion's hounds, is the brightest of all the stars in the firmament, and therefore regarded as their chief; for I have frequently compared it with Canopus, the next in brilliance, when both were nearly on the meridian together, and the latter yielded the palm to Κύων. From this brilliance there is little probability of its being mistaken for any of its stellar neighbours; but it may be noted, that a line from the Pleiades through Orion's belt passes, at about 20° beyond the latter, through Sirius. The geometrical diagram here presented to the gaze, was not lost to the rhymester:

Let Procyon join with Betelgeuze,	and pass a line afar,
To reach the point where Sirius glows—	the most conspicuous star;
Then will the eye delighted view	a figure fine and vast,
Its span is equilateral,	triangular its cast.

This star derived its Greek name from Σείριος, in allusion to the brightness, heat, and dryness assigned to it; though Dr. Hutton gravely informs us that the term is from *Siris*, which he says is the most ancient appellation of the Nile, for when this star rose heliacally, and became visible to the Egyptians and Ethiopians, their year commenced, and with it the inundation of their fecundating river. As that beneficial flood was attributed to the influence of the beautiful star, it was therefore worshipped as Sothis, Osiris, and Latrator Anubis*; and was viewed as the abode of the soul of Isis. Jacob Bryant insists, that the word Sirius was borrowed by the Greeks from the Egyptian *Cahen*

---

* Bainbridge, who was well versed in Arabian astronomy, wrote a treatise, *Canicularia*, together with a demonstration of the heliacal rising of Sirius for the

Sehor; and others recognise in it the *Mazzaroth* of Job; while Novidius, who gave a scriptural meaning to each constellation, says it alludes to Tobit's dog: "and so it may," ejaculates Moxon, "because he hath a tayle." It is first mentioned as a star by Hesiod, though Wyllyam Salysbury, 1550, and Heyschius, contend that the name applies equally to the Sun and the dog-star; and Homer, albeit he does not cite Sirius by name, compares the brightness of Achilles' armour to the pernicious blaze of the dog-star;

> Whose burning breath
> Taints the red air, with fevers, plagues, and death.

Some of the ancients asserted that a star in the head of the Dog, perfectly distinguishable from Sirius, perhaps meaning γ, was designated Isis, in former ages; but they were assuredly in error, as may be inferred from Diodorus and Plutarch, and all the honours of the constellation were vested in the dog-star. *Lælaps,* one of Actæon's kennel, was, however, slipped in, and moreover the Latins called it *Canis Candens,* and *Canicula;* which last should seem to apply to the Lesser Dog, but that, among the many opinions on this serious topic, the shew of hands is for Sirius. Yet Horace, inviting Mæcenas to quit the "Fumum et opes strepitumque" of Rome, (one would think London was meant,) for the country, during hot weather, thus describes the aspect of the heavens:

> Jam clarus occultum Andromedæ pater (*Cepheus*)
> Ostendit ignem; jam PROCYON furit,
> Et stella vesani Leonis (*Regulus*),
> Sole dies referente siccos.

There is no end to the evil influences which the ancients attributed to this star, though Geminus considered the bulk of them as rather resulting from the Sun; yet he was borne down by those who held Sirius to be an object equally terrible and splendid. While Virgil and others considered the unhealthy and oppressive period, which followed the summer solstice in Italy, was owing to the presence of the dog-star, Manilius thought it was a distant sun to illumine remote bodies; and thus he speaks, through the means of Sherburne:

> 'Tis strongly credited this owns a light
> And runs a course not than the Sun's less bright,
> But that remov'd from sight so great a way
> It seems to cast a dim and weaker ray.

From its heliacal rising the ancients reckoned their *dies caniculares,* or dog-days, which, however, in our climate, often commenced a fortnight after the veritable dog-days were ended; they have been frequently

---

parallel of Lower Egypt. This was published at Oxford in 1648, five years after his death. The sonorous *tetrandryan monogynian* bard commemorates the dog-star's advent:

> Sailing in air, when dark monsoon inshrouds
> His tropic mountains in a night of clouds,
> High o'er his head the beams of Sirius glow,
> And dog of Nile, Anubis, barks below.
>   *   *   *   *
> Her long canals the sacred waters fill,
> And edge with silver every peopled hill;
> O'er furrow'd glebes and green savannahs sweep,
> And towns and temples laugh amid the deep.

shifted and adjusted, and now seem to be established among the Almanacks, from the 3rd of July to the 11th of August; *i. e.* before Sirius rises! An extraordinary influence in engendering diseases among men, and madness among dogs, was assigned to the canicular days; hence their advent was of paramount importance, and Theon Alexandrinus has left a full formula, to find the exact time of the dog-star's rising; twenty days before which, and twenty days after, included the period of perspiration, hydrophobia, and other evils.

Canis Major is situated in the Southern Hemisphere, below Orion's feet; and the appellation of the principal star was frequently applied to the whole asterism, as an emblem of watchfulness and fidelity; hence its name Alshira, from the Arabic *Ash-shi'rá-l-Yemeníyah,* the bright shining star of Yemen, or Arabia Felix. This *Shi'ra,* it will be remembered, is largely complimented by Mahomet, in the fifty-third Sura of the Koran. The Greater Dog is one of the old 48 constellations, and has been thus tabulated:

Ptolemy . . . 29 stars		Hevelius . . . . 22 stars
Tycho Brahé . . 13		Flamsteed . . . 31
Bayer 19		Bode 161

Mr. Barker, of Lyndon, in the fifty-first volume of the *Philosophical Transactions,* considered that Sirius has changed colour, from red to white, in the lapse of ages; and quotes Aratus, Cicero, Virgil, Ovid, Seneca, Horace, and Ptolemy, in proof. The ancients, however, used the names of colours with the utmost latitude. *Splendescere, purpurascere,* signified to shine brightly; ποικίλος of Aratus expresses a glittering object; and the *rubra Canicula* of Horace may allude to heat. Mr. Barker's evidence for the mutation has therein more learning than point; but Seneca has an admission that the redness of Sirius was so strong as to exceed that of Mars; and Ptolemy says it was of the same colour as Cor Scorpii. These witnesses, both men of character and trust, are directly opposed to Hyginus, who asserts that the star was white, *flammæ candorum.* This Barker gets over, by considering that *candor* may be used for brightness, without regard to colour; and he might have called in Eratosthenes, a witness of high credit, to prove that Sirius at first signified bright, glittering, sparkling, and was afterwards given exclusively as the name of the most brilliant of the fixed stars. At all events, such a variation would be the more remarkable, since the other principal stars are unchanged in colour. Ptolemy calls Arcturus, Aldebaran, Pollux, Betelgeuze, and Antares, ὑπόκιῤῥος, or reddish, as they now actually are.

Sirius holds a leading place among the insulated stars, and is considered to be free from disturbance, although it seems to be obvious, that no two stars in the universe can be altogether out of the sphere of each other's attraction: but upon the supposition that the masses of Sirius and our Sun are equal, and that the former has a parallax of 1″, it would take about forty millions of years for them to fall to one another by their mutual action.

The brilliance of Sirius has long attracted the attention of philosophers, and every practical astronomer must be conversant with its superiority over its compeers. Sir William Herschel says, that when this star was about to enter his large telescope, the announcing light was equal to that on the approach of sunrise, and upon gaining the field of view, the star appeared in all the splendour of the rising sun, so that it was impossible to behold it without pain to the eye. By Sir John Herschel's photometric experiments on the apparent brightness of stars, the light of Sirius was found to be about 324 times that of an average star of the 6th magnitude. Consequently, if both bodies be assumed as of similar proportions, light diminishing as the square of the distance of the luminous body increases, their respective distances from us must be in the ratio of 57·3 to 1.

Another word upon such astonishing luminosity. My regretted friend Dr. Wollaston, in his skilfully ingenious researches in this branch of photometry, says, " From a comparison which I made in the year 1799 of the light of the Sun with that of the Moon, I should estimate the direct light of the Sun as being nearly one million times greater than that of the Moon; and consequently the direct light of the Sun as very many millions times greater than that afforded us by all the fixed stars taken collectively. Such then being, to our visual organs, the vast disproportion in radiance between the Sun and the whole starry firmament, it is not to be expected that we should assign very accurately how much greater the light of the Sun is, than that exceedingly minute quantity of it which shines upon us from any one, even the most brilliant of the fixed stars." We must refer the reader to the 119th volume of the *Philosophical Transactions*, for the details of his method of obtaining results by approximate ratios, and leap to his conclusion, that " we are not warranted by these experiments in supposing that the light of Sirius exceeds a 20,000,000,000th part of the Sun's light." Dr. W. therefore assuming the low limit of possible parallax of half a second, and consequently its distance from the Earth to be 525,481 times the radius of our orbit, concluded its intrinsic splendour to be nearly equal to that of fourteen suns: and that, if the star were placed where the Sun is, it would appear nearly four times as large as that luminary! Well might Voltaire make Micromegas, one of its inhabitants, to be eight leagues in stature.

The wit of Ferney drew his vast ideas of the magnitude of the dog-star from the several computations of modern philosophers. Maginus of Padua, considered the magnitude of Sirius to be equal to 10′, Kepler supposed 4′, and Tycho thought it was 2′; Ricciolus, however, brought it down to 18″: on which assumption its true magnitude was thus tabulated, according to the distance in the Copernican Hypothesis, maintaining the parallax of the fixed stars made by the Earth's motion, not to exceed 10″, and imagining the diameter of the annual orbit to be such as upon those principles it is stated to be:

AUTHORITIES.	Distances in semi-diameter of the Earth.	TRUE MAGNITUDE OF SIRIUS.	
		Diameter of Sirius contains diameters of the Earth :	The body of Sirius contains the Earth's body :
Copernicus . .	47,439,800	4170	71,677,171,300
Galilæus . . .	49,832,416	4380	88,427,672,000
Bullialdus . .	60,227,920	5300	148,877,000,000
Keplerus . . .	142,746,428	12550	1,967,656,371,000

Exorbitant as this appears, Vendelinus made his distance vastly greater, namely 605 millions of semi-diameters of the Earth. Yet Schickard says, "The speculations that represent the starry heavens the farthest removed from us, and consequently most amplify the stars, are more favourable to truth, for more confined ones would by no means admit of the annual parallax of our globe."

Astronomy is indebted to Sirius upon many counts, but perhaps in none of higher scientific interest, than' that of investigating the knotty question of Parallax. The dazzling splendour of this star, had long created a notion of its being nearer to us than any other of the stellar host, and therefore the fittest for determining the annual parallax of the Orbis Magnus. Huygens, assuming the Sun and Sirius to be of equal magnitude, made some ingenious but rather unsound optical experiments, from whence he concluded the light and diameter of the former to be 27,664 times greater than those of the latter; and that, consequently, the star's distance must be 27,664 times beyond the distance of the Sun from the Earth. From the varieties of the zenith distances observed at Paris, 130 years ago, Cassini II. inferred a parallax in declination amounting to 6″ in space, an inference which, though it gave the star still a diameter of 380 millions of leagues, excited the approbation of astronomers; and, from similar variations in the observations of La Caille, at the Cape of Good Hope, a parallax of 4″ was deduced. In 1760 Dr. Maskelyne made a proposition to the Royal Society for discovering this desideratum, the finding out of which, would be "the fullest and directest proof of the Copernican System;" the most striking objection to which was, that the enormous displacement of the spectator's place which that system supposed, was not supported by a corresponding change in the positions of the fixed stars. This proposal seems to have had little effect, and the matter slumbered till Piazzi revived it in a confirmation, by the Palermo observations, of La Caille's amount of the parallax of Sirius. This was announced formally to the Italian Society of Sciences; and in the notes to Hora VI. of the Catalogue, Piazzi says, "Juxta meas observationes, quæ cum iis conveniunt, quas ad Caput Bonæ Spei tentavit La Caille, *Sirii paralaxis* statui probabiliter potest quatuor secundorum circiter*." The question then rested till the recent admirable operations of Messrs. Henderson and Maclear, whose zeal and ability have been so applied as to produce a result, which must ever keep their names on the *Fasti* of Science. For the observations of these gentlemen, we must direct the reader to the *Memoirs of the Royal Astronomical Society*, and shall merely give the important result. On resolving, by the method of minimum squares, the two sets of equations, and combining the results according to their relative weights, the greatest effect of parallax in declination is found, from the whole of the 231 observations, $= + 0''{\cdot}15$; and the greatest effect of aberration in decli-

---

* Piazzi's discussion was communicated to the Italian Society in 1805, under the title—Ricerche sulla paralasse annua di alcune delle principale fisse; cioè la Capra, Aldebaram, Procione, Sirio, Arturo, ed Atair: and he concludes Sirius with—"In ogni maniera, da tutto ciò sembrami che si possa ben conchiudere, che se la paralasse di 4″ non è pienamente sicura, non lascia però di essere molto probabile." He states, moreover, his confidence in the well-known circle which he used on the occasion—"opera dell' immortale Artefice, Ramsden."

nation, $= 13''\cdot07$. These quantities are to the total effect of parallax and aberration in the proportion of $13''\cdot13$ to $20''\cdot50$, whence the final results are:

Parallax of Sirius, (or the angle subtended by the radius of the
Earth's orbit, at a distance equal to that of the star,) . . $= 0''\cdot23$
Constant of Aberration . . . . $= 20''\cdot41$

The possible error of this determination of the parallax may be estimated not to exceed a quarter of a second, as it is almost certain that the constant of aberration is not in error to a greater amount. On the whole, it may be concluded that the parallax of Sirius is not greater than half a second of space, and that it is probably much less. See 61 Cygni. The rigorous investigations by the same astronomers on a Centauri, are equally successful, and are still closely attended to.

CCLXIII.   14 LYNCIS.

Æ   6^h 38^m 57^s PREC. $+$ 5^s·32
DEC. N 59° 37'·6   —— S 3''·39

POSITION 50°·0 (w 2)   DISTANCE 1''·0 (w 1)   EPOCH 1833·31

A close double star, under the Lynx's eye; between Dubhe and Capella. A 5½, golden yellow; B 7, purple. This is one of Σ's *per-vicinæ*, and No. 963 of the great Dorpat Catalogue, where its measures are thus registered:

Pos. 50°·51   Dist. 0''·897   Ep. 1830·88

It is a very delicate and pretty object, and only seen with dark notches at intervals, being in contact in general, yet with the colours distinct. Piazzi pronounced this a double star, but the term was meant to include the *preceding* small star as B. His note states: "Duplex. Comes telescopica 1'' temporis præcedit."

CCLXIV.   31 ⨗. VIII. MONOCEROTIS.

Æ   6^h 39^m 42^s PREC. $+$ 3^s·14
DEC. S 3° 00'·1   —— S 3''·45

MEAN EPOCH OF THE OBSERVATION . . . . 1836·17

A loose cluster in the Galaxy, on the Unicorn's breast; 15° on a line from β Canis Majoris towards Pollux. It was discovered and registered by ⨗. in January, 1785, and is No. 408 of his son's Catalogue of 1830. It is a region of stars extending far beyond the field, with the principal members from the 8½ to the 11th magnitudes, curiously studded in pairs and triplets. Between these a certain glow indicates numbers of others still smaller.

M 2

CCLXV. 41 M. CANIS MAJORIS.

Æ 6ʰ 39ᵐ 55ˢ   Prec. + 2ˢ·57

Dec. S 20° 34′·8   —— S 3″·48

Position 85°·0 (ᵂ¹)   Distance 45″·0 (ᵂ¹)   Epoch 1836·17

A double star, in a scattered cluster, on the Greater Dog's chest. A 9, lucid white; B 10, pale white. This was registered by Messier in 1764, as a "mass of small stars;" but it is divided into five groups, of which the central one is the richest, and marked by three bright stars forming a crescent. In the *np* is the open double star which is here estimated; and 41 Messier may be struck upon by running a ray from Aldebaran, through ε in the centre of Orion's belt, and from thence between Sirius and Mirzam to about 4° in the south-east space beyond them. But as a beacon is rather acceptable in so low a declination, the tyro may hit his object by first directing his telescope—charged with a low power—upon Sirius, and then depressing it 4° 5′, when in about a minute a pair of 8th magnitudes will appear, constituting 233 and 236 P. Hora VI., and in about another minute, the cluster under discussion will follow.

CCLXVI. 59 AURIGÆ.

Æ 6ʰ 42ᵐ 00ˢ   Prec. + 4ˢ·13

Dec. N 39° 03′·2   —— S 3″·64

Position 222°·9 (ᵂ⁵)   Distance 22″·0 (ᵂ²)   Epoch 1833·10

A delicate double star, between the Waggoner's left arm and the Lynx. A 6, pale yellow; B 11, livid. This is the object described by ⩜. as the apex of an isosceles triangle, and classed 102 ɪᴠ. The following are its registered measures:

⩜.	Pos. 216° 57′	Dist. 23″·50	Ep. 1782·85
S.	221° 41′	21″·60	1825·02
Σ.	222° 38′	22″·26	1831·11

This star is certainly one of no easy measurement, but our results are sufficiently strong to warrant the inference of a slow *spnf*, or direct angular motion. A glance from the Hyades through Nath, at the tip of the Bull's left horn, carried about 22° into the north-east, will strike upon three small stars, of which the most northern is the one under discussion. This place will also be intersected by a line from Procyon through δ Geminorum; and by another from Orion's sword cluster, through the lowest star of the belt and Betelgeuze, and extended three times further north-east-ward. The alignment is therefore of very ready accomplishment.

CCLXVII. 27 ♅. VI. MONOCEROTIS.

Æ 6ʰ 43ᵐ 33ˢ   Prec. + 3ˢ·08
Dec. N 0° 38'·6   —— S 3"·79

Position 10°·5 (w 2)   Distance 15"·0 (w 1)   Epoch 1835·22

A compressed cluster in the Via Lactea, on the Unicorn's neck. A 8½, pale straw-colour; B 9½, light grey. This object was first classed by ♅. in 1786, and is broken into three several rich groups, occupying a very considerable space. Near the centre is the double star here observed, but, from having a small *comes* in the *np* quadrant, it ought rather to be registered triple. A trapezium of brighter stars follows; and it is to be fished up about one-third of the distance between Procyon and Rigel, where it is intersected by a transverse line from Pollux to about 1° west of β Canis Majoris.

CCLXVIII. 38 GEMINORUM.

Æ 6ʰ 45ᵐ 37ˢ   Prec. + 3ˢ·38
Dec. N 13° 22'·6   —— S 3"·97

Position 171°·8 (w 8)   Distance 6"·0 (w 7)   Epoch 1836·10
——— 170°·7 (w 9)   ——— 5"·8 (w 7)   ——— 1839·17

A neat double star, on the left instep of Pollux. A 5½, light yellow; B 8, purple. This is a very fine object, and the colours so marked, that they cannot be entirely imputed to the illusory effect of contrast. It is 47 ♅. III., and from a comparison of all the measures, a slight but constant diminution in the angle may be inferred. These are the astrometric results:

	Pos.	Dist.	Ep.
♅.	179° 54'	7"·95	1781·99
H. and S.	174° 24'	5"·53	1822·67
Σ.	174° 52'	5"·74	1829·24
D.	172° 25'	5"·95	1832·93

which suggest a retrograde slow motion of − 0°·16 per annum; and the distance appearing stationary, hints a period of upwards of 2000 years. A glance from Rigel carried below ζ—the southern star of Orion's belt—and prolonged rather more than twice as far again, till it meets a line cast between Procyon and Nath, will have just passed over it. 38 Geminorum exhibits a sensible aberration from the common laws of precession, which has been thus valued:

P.... Æ + 0"·06   Dec. − 0"·08
B....   + 0"·08   − 0"·06

CCLXIX. 2 ♊. VI. GEMINORUM.

Æ   6ʰ 45ᵐ 56ˢ   Prec. + 3ˢ·50
Dec. N 18° 10'·5   —— S 3"·99

Mean Epoch of the Observation . . . . 1837·91

A compressed cluster, on the calf of Pollux's right leg, one-third of the distance from Pollux to Rigel, on a line carried from the former

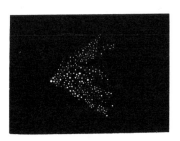

star between the second and third " bullions" of Orion's belt to the latter: discovered by ♊. in 1783, and forming No. 415 of his son's Catalogue. It is a faint angular-shaped group of extremely small stars—say 12 to 16 magnitudes—which only under the most favourable circumstances can I discern with satisfaction. The region around is immensely rich, and not at all wanting in double stars. Differen-tiated with γ Geminorum for a mean place; and when best seen, it is something like the hasty sketch herewith given.

CCLXX. π² CANIS MAJORIS.

Æ   6ʰ 48ᵐ 08ˢ   Prec. + 2ˢ·51
Dec. S 20° 12'·4   —— S 4"·18

Position AB 149°·0 (w 5)   Distance 45"·0 (w 5)
——— AC 182°·5 (w 3)   ——— 52"·5 (w 3)   Epoch 1834·14
——— AD 185°·0 (w 3)   ——— 125"·0 (w 1)

A coarse quadruple star, on the chest of Canis Major; where it is the middle one of three small stars, about 4½° to the south-south-east of Sirius. A 6, flushed white; B 9½, ruddy; C 10, ruddy; D 11, dusky. A and B were classed as a double star, 65 ♊. v.; and Piazzi, note 222 Hora VI., says, "Binæ sequuntur 10ᵃ magn. 1" circiter ad austrum." Herschel's measures were:

Pos. 154° 12' Dist. 44"·93 Ep. 1782·17

When Sir James South examined this object, he included the two companions in the sp quadrant, and registered it quadruple, thus:

AB Pos. 147° 57' Dist. 45"·03 Ep. 1825·04
AC   184° 18'   52"·96   1825·07
AD   185° 16'   128"·36   1825·10

On weighing all these results, there seems to have been some error in ♊.'s angle, at the first epoch.

CCLXXI.   μ CANIS MAJORIS.

Æ   6ʰ 48ᵐ 46ˢ   Prec. + 2ˢ·75
Dec.   S 13° 50'·5   ——   S 4"·23

Position 342°·9 (w 8)   Distance 3"·5 (w 5)   Epoch 1834·15

A neat double star, on the Dog's right ear; where a line through Orion's belt will meet it, at nearly 4° north-east of Sirius.   A 5½, topaz yellow; B 9½, grey. This elegant object was discovered by Σ., and is No. 997 of the Dorpat Catalogue; but the earliest measures I met with are those of H., with his 7-foot telescope, which formed the only comparison until the arrival of Σ.'s great work, when the results were found to run thus:

H.   Pos. 341° 17'   Dist. 3"·63   Ep. 1830·18
Σ.   343° 31'   3"·22   1831·20

CCLXXII.   14 ♄. VII. CANIS MAJORIS.

Æ   6ʰ 52ᵐ 10ˢ   Prec. + 2ˢ·76
Dec.   S 13° 29'·2   ——   S 4"·52

Mean Epoch of the Observation . . . . 1838·32

A tolerably compressed cluster at the back of the Greater Dog's head, principally composed of stars from the 8th to the 11th magnitudes, of which the four principal form the letter Y; there are also some glimpse stars, but to no great extent. Yet to ♄.'s powerful "ken," it appeared to be 20' in diameter, when he observed it in February, 1785. It may be fished up by first finding μ, the object above registered; when it will appear in the nf quadrant, well within a degree's distance.

CCLXXIII.   ε CANIS MAJORIS.

Æ   6ʰ 52ᵐ 20ˢ   Prec. + 2ˢ·35
Dec.   S 28° 45'·5   ——   S 4"·54

Position 84°·5 (w 3)   Difference Æ = 24ˢ·1 (w 2)   Epoch 1834·83

A Greenwich star with a distant companion, on the Greater Dog's belly: it will be readily found by running a line from the middle of Orion's belt through β, the bright star to the west of Sirius, and extending the same 14° further into the south-east quarter.   A 2½, pale orange; B 7, violet, with no appearance of the suspected nearer comes. My meridian observations afford no confirmation of the proper motion assigned by Piazzi, on comparing his results with those of Bradley and La Caille: but Mr. Baily's recent investigation shows a discrepancy with precessional law, although the direction is shaken; the assigned values being:

$$P.... \; \mathcal{R} \; - \; 0''\text{·}05 \qquad \text{Dec.} + 0''\text{·}07$$
$$B.... \qquad + \; 0''\text{·}06 \qquad \qquad + \; 0''\text{·}01$$

This star is called Adara, from *al 'adhára*, the virgins; *o*, *η*, *δ*, and *ε*, on the shoulder, tail, and between the tail and legs. Adjacent to these Royer cut away a portion of Canis Major, and constructed Columba Noachi therewith in 1679. The part thus usurped was called Muliphein, from *al-muhlefeïn*, the two stars sworn by, because they were often mistaken for *Soheïl*, or Canopus, before which they rise: these two stars are now *a* and *β* Columbæ. Muliphein is recognised as comprehending the two stars called *Hadár*, ground, and *al-wezn*, weight, astonishing sidereal names, says Ideler, of which the Arabians were ignorant of the proper location, for while some placed them to the two bright stars in the Dove, others stuck them in the fore-foot of the Centaur, and a third party assigned them to *ζ*, *λ*, and *γ*, in Argo. Four of the Greater Dog's *informes* are termed *El-Kurûd*, the apes, by Ḳazwíní, which term applies principally to those which Ptolemy described as standing in a line; but 'Abd-u-rahmán Ṣúfí calls them *el-furûd*, bright and insulated. See *δ* Canis Majoris, No. CCLXXVIII. The galley rhymes allude to Royer's robbery—

Where Canis Major, from the south,   th' horizon moves above—
The stars that deck'd his hinder feet   now form the Patriarch's Dove.

CCLXXIV.   301 P. VI. LYNCIS.

$\mathcal{R}$	6ʰ 52ᵐ 56ˢ	Prec.	+ 4ˢ·79
Dec.	N 52° 59′·4	——	S 4″·59

Position 158°·9 (*w* 8)   Distance 3″·2 (*w* 6)   Epoch 1833·21
——— 159°·4 (*w* 9)   ——— 3″·0 (*w* 9)   ——— 1843·19

A neat double star, on the animal's neck; where a ray conducted from Polaris to the westward of Castor, passes over it at 35° from the pole, or rather more than half-way, on the line between *β* Aurigæ and *β* Ursæ Majoris. A 6, and B 6½, both white. This pretty object is 69 H. I.; and from the first measures taken, as compared with Sir James South's observations, a change in direction *npsf*, or retrograde motion, was to be expected, to the annual amount of $-0°\text{·}252$; but the more recent results do not countenance such a change. The measures are more discordant than might have been expected:

	Pos.	Dist.	Ep.
H.	167° 24′	3″·50±	1782·87
S.	156° 54′	3″·89	1824·59
Σ.	159° 11′	2″·94	1830·34
D.	160° 37′	3″·32	1831·15

CCLXXV. ζ GEMINORUM.

Æ 6^h 54^m 37^s Prec. + 3^s·56

Dec. N 20° 47'·9   —— S 4"·73

Position AB 355°·0 (*w* 4)   Distance 90"·0 (*w* 2)⎫
——— AC 85°·0 (*w* 1)   ——— 65"·0 (*w* 1)⎬ Epoch 1831·81

A coarse triple star, on the right knee of Pollux. A 4, pale topaz; B 8, violet; and C 13, grey. This was registered as a double star 9 Ḥ. vi., and re-examined as a pair by H. and S.; but the third star lying too handy to be omitted, when once seen, I entered it. Including a deduction drawn from Piazzi's mean places of the components, the previous measures are thus tabulated:

	Pos.	Dist.	Ep.
Ḥ.	351° 14'	91"·86	1779·77
P.	354° 30'	88"·30	1800·00
H. and S.	355° 27'	91"·03	1821·23

This star is called Mekbuda, from *al-maḳbúdah*, contracted, or rather *mut-a-kabbidah*, a culminating star; it comprehends the two bright stars of Castor and Pollux, and forms the VIIth Lunar Mansion, called by the Arabs *al-dhirá'*, the arm or paw of the lion. It is easily seen on running a line between the cluster in Orion's sword and Pollux, for it passes over ζ at 9° from the latter star; and it is near the mid-distance between ζ Tauri, the tip of the southern horn, and the Præsepe in Cancer.

CCLXXVI. 50 M. MONOCEROTIS.

Æ 6^h 55^m 11^s Prec. + 2^s·88

Dec. S 8° 06'·7   —— S 4"·78

Position 170°·0 (*w* 1)   Distance 5"·0 (*w* 1)   Epoch 1833·25

A delicate and close double star in a cluster of the Via Lactea, on the Unicorn's right shoulder. A 8 and B 13, both pale white. This is an irregularly round and very rich mass, occupying with its numerous outliers more than the field, and composed of stars from the 8th to the 16th magnitudes; and there are certain spots of splendour which indicate minute masses beyond the power of my telescope. The most decided points are, a red star towards the southern verge, and a pretty little equilateral triangle of 10th sizers, just below, or north of it. The double star here noted was carefully estimated under a full knowledge of the vertical and parallel lines of the field of view: this was made triple by H., whose 2357 of the Fifth Series it is; but he must be mistaken in calling it Σ. 748, which is θ Orionis. It is sufficiently conspicuous as a double star, and though I perceive an infinitesimal point exactly on the vertical of A, I cannot ascertain whether it is H.'s C.

This superb object was discovered by Messier in 1771, and registered "a mass of small stars more or less brilliant." It is 9° north-north-east of Sirius, or rather more than one-third of the distance between that star and Procyon.

CCLXXVII. 33 ♅. VIII. MONOCEROTIS.

Æ 7ʰ 00ᵐ 44ˢ   Prec. + 2ˢ·83
Dec. S 10° 22'·6   —— S 5"·25

Position 300°·0 (w 1)   Distance 15"·0 (w 1)   Epoch 1833·12

A double star in a loose cluster, under the Unicorn's chest, and about 8° north-east of Sirius, in a direction pointed out by leading a line from Aldebaran over Bellatrix, and nearly twice as far again. A 9, yellow; B 12, dusky. This is a scattered group of brightish stars, in an irregular lozenge form, and consists chiefly of three vertical rows, having four individuals in each; several are of the 9th magnitude, and reddish. It was registered by ♅. in 1785.

CCLXXVIII. δ CANIS MAJORIS.

Æ 7ʰ 01ᵐ 53ˢ   Prec. + 2ˢ·44
Dec. S 26° 08'·6   —— S 5"·35

Position 224°·0 (w 1)   Distance 165"·0 (w 1)   Epoch 1832·90

A star with a distant companion on the loins of Canis Major; where a line from Betelgeuze to the south-south-east, through Sirius, intercepts it at 12° below that star. A 3½, light yellow; B 7½, very pale; other small stars in the field, and np is the one mentioned by Piazzi, "Alia 9ᵃ magnitud. præcedit 45"·5 temporis, 1' 48" ad boream." My observations are, of course, not sufficiently nice for an authority, but still they countenance the slight proper motion attributed to this star, both in Æ and declination. It is considered variable; having been registered 2nd magnitude by Hevelius, La Caille, and Brisbane; 2½ by Halley; 3 by Ptolemy, Tycho, and Flamsteed; and 3½ by Piazzi and Johnson. From comparisons made at the above epoch, the comparative brightness was similar to that recorded in the Palermo Catalogue.

δ is called Wezen, al-wezn, weight, from appearing to rise with difficulty above the horizon, as if chained to the ground. The same sluggishness was applied to α and β Centauri, which 1000 years ago, under the 30th parallel of latitude, only obtained a meridian altitude of 4°. The most general application, however, of the name, will be found under ε Canis Majoris. The group of which δ may be considered as the centre, and which consists of ε, η, δ, o, and ι, were called El-Zára, the virgins, by the early Arabians.

CCLXXIX. 34 ♓. VIII. MONOCEROTIS.

Æ 7ʰ 06ᵐ 58ˢ   Prec. + 2ˢ·84

Dec. S 10° 00′·8   —— S 5‴·77

Position 22°·0 (w 6)   Distance 21″·0 (w 3)   Epoch 1837·91

A neat double star, on the following boundary of a loose cluster in the Galaxy, between the Unicorn and the Greater Dog's head. A 8, and B 8½, both silvery white. This is a very rich field of stars, in the which is a brilliant oval mass, bounded by a sapphire-tinted 6th-magnitude star, in the *sf* quadrant, and the pair here measured a little north of it. ♓., who discovered this group in 1785, makes no mention of the latter; nor of two other pairs which are in the field, one above and the other below the object measured. A line from Pollux, passed by Gomeisa (β Canis Majoris) to nearly as far again, will find this object posited 9° east-north-east of Sirius.

CCLXXX. λ GEMINORUM.

Æ 7ʰ 08ᵐ 54ˢ   Prec. + 3ˢ·45

Dec. N 16° 49′·5   —— S 5″·93

Position 29°·2 (w 8)   Distance 10″·3 (w 2)   Epoch 1838·79

A delicate double star on the left thigh of Pollux, about 12° on a line from Procyon towards β Aurigæ, in the north-north-west, and rather less than a third of the distance between Castor and Sirius. A 4½, brilliant white; B 12, yellowish: the pair observed under the most favourable circumstances of weather and instruments, but the companion was seen best under an averted eye. This fine object was discovered by Σ., and is No. 1061 of the Dorpat Catalogue, where it is thus registered:

Pos. 30°·93′   Dist. 9″·56   Ep. 1829·86

λ Geminorum has been placed on the variable class, but I could detect no difference in its brightness as compared with 22 Monocerotis, ν Orionis, and 2 Lyncis: it was also considered to have a large spacial movement, but recent investigations have diminished it to a question of mere instrumental error; the best values are:

P.... Æ − 0″·10   Dec. − 0″·04
B....   + 0″·03   0″·00

This being one of those objects impracticable to artificial light, rendered it necessary to apply a non-illuminating micrometer; and the mean of angles carefully taken with Dollond's spherical rock-crystal, enabled me to form a position of high weight.

CCLXXXI. 19 LYNCIS.

Æ 7ʰ 09ᵐ 46ˢ   Prec. + 4ˢ·93

Dec. N 55° 34'·6   —— S 6"·01

Position A B 312°·4 (w 9)   Distance 14"·6 (w 9) ⎫
—————— A C 358°·2 (w 3)   ———— 215"·2 (w 2) ⎬ Epoch 1833·77
—————— A B 313°·8 (w 9)   ———— 14"·8 (w 9) ⎭ —— 1839·65

A coarse triple star, in the nape of the Lynx's neck, nearly in mid-distance of an imaginary line thrown from Polaris to Pollux. A 7, white; B and C, both 8, and plum coloured. This is 83 ℍ. III., and Nos. 47, 48, and 49 P. VII. It is also No. 78 of H. and S.; but it is curious that these astronomers have each assigned what I deem, on repeatedly comparing the light of the components, a wrong quadrant to B. By shifting it from the *sp* to the *np*, their measures will compare with those of Σ. and myself thus:

	Pos.	Dist.	Ep.
ℍ.	316° 54'	14"·19	1782·86
H. and S.	313° 05'	14"·54	1821·22
Σ.	313° 50'	14"·72	1829·51

This star was made micrometrically triple by H. and S., by including C, which lies near the north vertical, exactly in the same Æ with B. The whole appear to be unchanged, though A is suspected of proper motions to the following amount:

P....Æ − 0"·16   Dec. − 0"·08
B....   − 0"·03   − 0"·05

CCLXXXII. 20 LYNCIS.

Æ 7ʰ 09ᵐ 59ˢ   Prec. + 4ˢ·61

Dec. N 50° 26'·4   —— S 6"·03

Position 253°·3 (w 9)   Distance 15"·2 (w 9)   Epoch 1835·39

A neat double star, on the animal's chest; and 16° east-north-east of Capella. A and B, both 7½, and silvery white. This object is No. 61 of ℍ.'s List of 145, but no measures were given; and Piazzi noted it double, in these words, "Duplex. Comes ejusdem magnit. 2" circiter temporis sequitur, 15" circiter ad boream," thus assuming the opposite quadrant to that which has been adopted. The other measures are:

H. and S. Pos. 252° 39'   Dist. 15"·96   Ep. 1823·33
Σ.   253° 25'   15"·04   1830·55

From a comparison of these results the relative fixity of these stars may be deduced; but B has a slight proper motion in declination, amounting perhaps to − 0"·07 per annum, which is a mean of several authorities.

CCLXXXIII. δ GEMINORUM.

Æ. 7ʰ 10ᵐ 34ˢ   PREC. + 3ˢ·59
DEC. N 22° 16′·3   — S 6″·07

POSITION 198°·5 (w 8)   DISTANCE 7″·1 (w 5)   EPOCH 1833·15
——— 196°·8 (w 9)   ——— 7″·2 (w 6)   ——— 1838·92

A Greenwich star of the second rank, double, on the right hip of Pollux; it is exactly half-way between the Præsepe and ζ Tauri, on the tip of the Bull's southern horn, and nearly on the line from Castor towards Sirius. A 3½, pale white; B 9, purple. This delicate object is rather troublesome to measure in distance, from disparity, but certainly with my instrument is not "one of the most difficult stars in the heavens." It is 27 Ⴙ. II., and was thus registered at its discovery:

Pos. 184° 09′   Dist. 6″·50±   Ep. 1781·20

The same astronomer re-examining this star in 1802, and finding the angle to be 195° 17′, concluded that a large change had taken place in twenty-one years; subsequent observations, however, indicate some error in the original entry; the later results being all remarkably coincident. The measures of other astronomers are:

H. and S.   Pos. 195° 24′   Dist. 7″·25   Ep. 1822·14
Σ.   196° 54′   7″·15   1829·72
D.   196° 55′   7″·13   1831·02

This star is known as Wasat, from the Arabic al-wasat, the middle or centre, and it has a small spacial movement, to the following values:

P.... Æ. − 0″·06   Dec. − 0″·05
B....   + 0″·08   − 0″·05

CCLXXXIV. 12 Ⴙ. VII. CANIS MAJORIS.

Æ. 7ʰ 10ᵐ 35ˢ   PREC. + 2ˢ·72
DEC. S 15° 21′·4   — S 6″·07

MEAN EPOCH OF THE OBSERVATION .... 1837·02

A tolerably compressed but extensive cluster, on the boundary between the Unicorn, and the Greater Dog. It was discovered by the indefatigable Miss Herschel, in 1785; and consists of a singular group of very lucid specks, formed of stars nearly all of 10th-magnitude. The most compressed portion occupies a third of the field with power 66; and it is followed by a solitary yellowish star, of the 8th magnitude. It can be fished up, under a moderately magnifying eye-piece, at 7°½ west-north-west of Sirius; where an imaginary line from Aldebaran passed over Bellatrix, will intersect it.

CCLXXXV. 30 CANIS MAJORIS.

Æ 7ʰ 12ᵐ 04ˢ   Prec. + 2ˢ·48

Dec. S 24° 40'·0   —— S 6"·20

Position 73°·0 (w 1)   Distance 85"·0 (w 1)   Epoch 1834·83

A star with a companion, in cluster 17 ʜ. vɪɪ., on the Greater Dog's back; where a line from Bellatrix through Sirius, and 12° beyond, will find it. A 6½, white; B 9, pale grey. The whole has a beautiful appearance, the bright white star A being surrounded by a rich gathering of minute companions, in a slightly elongated form, and nearly vertical position. The latest investigations into the proper motions of this star, destroy the amount formerly attributed to the declination, and reduce that in Æ to about + 0"·06 per annum.

CCLXXXVI. 61 GEMINORUM.

Æ 7ʰ 17ᵐ 31ˢ   Prec. + 3ˢ·54

Dec. N 20° 34'·3   —— S 6"·65

Position AB 110°·0 (w 1)   Distance 60"·0 (w 1) }
—— DC 42°·4 (w 8)   —— 6"·5 (w 7) } Epoch 1835·85

A course double star pointing to a neat pair in the *np* quadrant, on the loins of Pollux, and about 2° to the south-east of δ Geminorum, the alignment of which has been given. A 7½, deep yellow; B 9, yellowish; C 8, blue; D 9, bluish; and besides these, the field is very rich in small stars. A is the individual selected by ʜ. as the director to C D, or 48, ɪɪɪ., which was thus registered when first discovered:

Pos. 46° 06'   Dist. 6"·25   Ep. 1781·99

It was next examined by Sir James South, who made it:

Pos. 39° 16'   Dist. 6"·52   Ep. 1824·21

whence Sir John Herschel inferred, that a notable retrograde change of position had occurred = − 0°·166 per annum; a surmise not confirmed. He also states that the above "very exact" results of ʜ. were taken in 1783; but the printed account in the *Philosophical Transactions*, gives the same angle at a date two years earlier. Indeed, to save anomalous comparisons of dates, I may here state, that all the epochs of ʜ. quoted by me, are from the lists inserted in that work.

61 Geminorum exhibits evidence of proper motions, the values of which have been thus registered:

P.... Æ + 0"·13   Dec. − 0"·02
B....　　+ 0"·06   　　　− 0"·02
T....　　+ 0"·06   　　　0"·00

CCLXXXVII.   η CANIS MAJORIS.

Æ   7ʰ 17ᵐ 46ˢ   Prec. + 2ˢ·37
Dec. S 28° 59'·7   —— S 6"·67

Position 285°·0 (w 2)   Distance 169"·0 (w 1)   Epoch 1833·82

A star with a distant companion, at the root of the Greater Dog's tail; where an imaginary line from the three small stars forming Orion's head, passed through Sirius, will strike it at 17° beyond. A 3, pale red; B 7½, dull grey; two small stars following. A is called Aludra, from the Arabian al-'adhrá, which is the singular of al-'adhára: see ε Canis Majoris. B proved to be No. 103 of Piazzi's Hora VII., whose reduced places give 286° for the angle of position, on a distance of 174", for the year 1800.

According to the *Megale Syntaxis*, Hipparchus found that the solstitial colure passed through the caudine star of the Greater Dog, which appears to have occasionally served, by its arrival at the meridian, to indicate the zero for reckoning the hours; its Æ, which was then exactly 90°, rendering it convenient for that purpose. If the longitude assigned by Hipparchus to this star be compared with its present place, the annual precession will be 50"·7; and both the theory of gravitation and the deductions of modern operations coincide in indicating 50"·1, as the mean annual value. This will very nearly agree with the Platonic year, or complete revolution of the equinoxes in 25,920 years, as given by Ricciolus and approved by Flamsteed, at the rate of a degree in about seventy-two years. Well may Hipparchus be dubbed the Præses of ancient astronomers! See α Leonis.

CCLXXXVIII.   63 GEMINORUM.

Æ   7ʰ 18ᵐ 14ˢ   Prec. + 3ˢ·57
Dec. N 21° 46'·1   —— S 6"·71

Position 325°·0 (w 2)   Distance 50"·0 (w 1)   Epoch 1831·95

A wide double star on the back of Pollux; following Wasat, δ Geminorum, within 2°, about east by south. A 6, yellow; B 10, reddish, and with two telescopic stars in the *sp* quadrant, they form a regular curve. This is 53 H. v., registered in 1781, but without giving an angle of position. It was then examined by H. and S., who measured the angle, but not the distance. The two results, however, lead to the inference that they are only optical neighbours. The large star probably has the proper motion assigned to it:

$$P....\text{Æ} - 0''·06 \qquad \text{Dec.} - 0''·12$$
$$B.... \quad - 0''·02 \qquad \qquad - 0''·07$$

CCLXXXIX. β CANIS MINORIS.

ÆR 7ʰ 18ᵐ 28ˢ   PREC. + 3ˢ·26
DEC. N 8° 36'·4   —— S 6"·73

POSITION AB 80°·0 (w 2) DISTANCE 35"·0 (w 1)⎫
———— AC 312°·0 (w 2) ————— 105"·0 (w 1)⎭ EPOCH 1831·84

A wide triple star, on Procyon's neck; rather better than 4° to the
north-north-west of its lucida, where its magnitude readily points it out.
A 3, white; B 12, orange; C 10, flushed; the last is coarsely double
with one of the same magnitude, and there are other stars in the field,
of which the white one preceding is that alluded to by Piazzi, "Alia
8ᵃ magnitudin. præcedit 43" temporis, 2½' circiter ad boream." The
large individual is named Gomeisa, from *al-ghomeïsá*, watery-eyed; and
it is *Al-Mirzam*, one of the "roarers," mentioned under β Canis Majoris.

CCXC. 45 ⩑. IV. GEMINORUM.

ÆR 7ʰ 19ᵐ 43ˢ   PREC. + 3ˢ·56
DEC. N 21° 13'·9   —— S 6"·83

POSITION 355°·0 (w 1) DISTANCE 95"·0 (w 1) EPOCH 1836·22

A star enveloped in an atmosphere, with a distant companion on the
loins of Pollux. A 7½, greyish white; B 8, dusky blue; other stars
following. This was observed by ⩑. in 1787, as a "star of the 9th-
magnitude, with a pretty bright nebulosity, equally dispersed all around.
A very remarkable phenomenon." H., whose No. 450 it is, describes it,
"a star of the 8th magnitude, in the centre of an exactly round
bright hemisphere 25" in diameter." The beauty of this is, in great
measure, lost to my instrument, for I could only bring it to bear as a
burred star: it lies about 2° to the east-south-east of Wasat, δ Geminorum, whose alignment is already treated of.

CCXCI. 116 P. VII. MONOCEROTIS.

ÆR 7ʰ 20ᵐ 21ˢ   PREC. + 2ˢ·82
DEC. S 11° 14'·2   —— S 6"·89

POSITION 315°·0 (w 3) DISTANCE 20"·0 (w 2) EPOCH 1834·11

A delicate double star, under the Unicorn's belly, where a line from
β Canis Majoris, led through Sirius about 11° to the east-north-east,

will meet it. A 7, yellow; B 9½, violet, a third star close to the south vertical, of the 14th magnitude, and clear blue. This was discovered by Σ., and is No. 1097 of the Dorpat Catalogue, with these data:

Pos. 312° 12′ Dist. 20″·20 Ep. 1832·15

This object is close to the gap in the fanciful boundary which marks out Argo's northern limb on our maps; which gap cuts a narrow slice of about 13° long by 1° broad, right through the body of Monoceros, in order to catch up a star pertaining to Canis Minor, which Flamsteed, by some mistake, registered as 13 Navis. A and B point upon a distant telescopic group in the *np*.

CCXCII. α GEMINORUM.

Æ 7ʰ 24ᵐ 23ˢ   Prec. + 3ˢ·85
Dec. N 32° 14′·0   — S 7″·21

Position	Distance	Epoch
AB 258°·8 (*w* 5)	4″·7 (*w* 5)	Epoch 1830·95
AC 162°·0 (*w* 4)	72″·9 (*w* 3)	
AB 257°·9 (*w* 8)	5″·1 (*w* 3)	— 1832·25
AB 256°·3 (*w* 5)	4″·7 (*w* 5)	— 1834·24
AB 255°·2 (*w* 6)	4″·8 (*w* 8)	— 1836·31
AB 255°·1 (*w* 5)	4″·7 (*w* 2)	— 1837·35
AC 162°·8 (*w* 5)	73″·1 (*w* 5)	
AB 254°·9 (*w* 7)	4″·8 (*w* 4)	— 1838·33
AC 162°·2 (*w* 7)	72″·4 (*w* 5)	
AB 252°·3 (*w* 9)	4″·9 (*w* 9)	— 1843·13
AC 162°·6 (*w* 7)	73″·0 (*w* 5)	

A standard Greenwich star, neatly double, in the head of Castor, and about half way between Regulus and Aldebaran. A 3, bright white; B 3½, pale white; C 11, dusky, and there is another very small acolyte at a distance, in the *sp* quadrant of the field, with which a physical connection has been also suspected. This very interesting object, when classed 1 H. ɪɪ., was thus:

Pos. 302° 47′ Dist. 5″·16 Ep. 1778·27

When H. and S. examined it, the secondary had changed its quadrant from *np* to *sp*, and stood as follows:

Pos. 267° 07′ Dist. 5″·35 Ep. 1821·21

showing a rapid retrograde motion of 0°·97 per annum; and the more recent observations so fully confirm the binary system, that Sir J. Herschel concludes the small star will pass its periastre at the close of 1855, at a distance of 0″·66; which brings its *annus magnus* to about 250 years.

At the moment it was made, this was a bold prediction; and although the state of mutual approach and accelerated angular motion do not circumstantially corroborate it, there is good evidence that the theory is substantially right.

To arrive at his deductions, Sir John Herschel gathered together all the observations he could rely upon, it being a question he was deeply interested in, because, he says, Castor is "the largest and finest of all the double stars in our hemisphere, and that whose unequivocal angular motion first impressed on my father's mind a full conviction of the reality of his long cherished views on the subject of the binary stars." By the alignments of Pound and Bradley, he was able to carry the angle back for upwards of 100 years; and by computations which approximate as near as the present state of the subject will allow, he has deduced the following elements of the elliptic orbit of the secondary round the primary; major semi-axis $8''·086$; eccentricity $0·7582$; position of perihelion $169° 10'$; inclination of the real orbit to the apparent orbit on the sphere of the heavens, $70° 03'$; mean motion, $1°·425$; period of revolution 253 years. "This star," he adds, "seems on the point of undergoing, within the ensuing twenty-four years, a remarkable change similar to that of which γ Virginis has already furnished a striking instance during the last century, and passing from a distant double star of the second class to a close one of the first, and ultimately to one of extreme closeness and difficulty, such as only the very finest telescopes, with all the improvements we may expect in them, will be capable of showing otherwise than single." But there are some orbital anomalies still in the way.

Using Herschel's bow, albeit with hardly vigour to bend it, I attempted an orbit of this revolver, notwithstanding I soon found that the values of its annual changes are violently discordant. The projection brought out an ex-centricity of $0·7781$, an inclination $= 70° 36'$, and a period of 240 years; the last condition being obtained by H.'s novel and ingenious process of cutting out the graphic orbit from card-board, and weighing both it and its requisite sectors in a balance. These are the previous angles used:

Bradley and Pound	. . .	Pos. 355° 53′	Ep. 1719·84
Bradley and Maskelyne	. .	323° 47′	1759·80
Herschel the Elder	. . .	293° 03′	1783·64
Herschel the Elder	. . .	284° 19′	1800·27
Struve	. . . . .	272° 52′	1813·83
Herschel Junior	. .	270° 00′	1816·97
Herschel Junior and South	.	264° 59′	1823·11

Several years previous to the combined operations in the Blackman-street observatory, Sir John Herschel had measured various double stars at Slough, with a 7-foot reflector; and as some of the synoptic results only were printed in the register published by him and Sir James South, in 1824, I requested of him, and obtained, the full details from his journal. With the accustomed diffidence of real merit, Sir John places "no vast confidence" in his observations of 1816, except in the above-cited measurement of Castor, "which," he tells me, "from the circumstances described, *must* be correct, and is valuable. And I well remember comparing (with my father) this particular result with his former measures, much to his and my own satisfaction, as a verification of its orbital movement." The circumstances alluded to are thus given in the Journal: "Dec. 20, 1816, α Geminorum. Double. Unequal. White—both stars. Position of the small star, exactly preceding—I made both stars run along the wire repeatedly from one side to the other, and both were covered.   I then made them run above and below it, and could perceive no deviation from exact parallelism.   The evening is perfectly still, clear, and frosty."

Bradley appears to have made his estimations upon the parallelism of the line of direction of the pair, to that joining Castor and Pollux, in 1759, "at all times of the year," evidently intending to notice whether any annual oscillation might be observed.   This induced Roger Long, Lowndes's Professor of Astronomy at Cambridge, to attack Castor with telescopes of fourteen and seventeen feet in length, with a view to Galileo's suggestion on parallax; but the prospect of success soon became so hopeless, that he was "persuaded the stars would always be found to appear the same."   This gentleman was more happy in the construction of an enormous astronomical machine—the very A^1 of orreries—at Pembroke College.   It is a hollow sphere, about eighteen feet in diameter, with its polar axis parallel to the mundane axis, upon which it is readily turned by a winch and rack-work; thus it can be made to revolve, while about thirty persons conveniently attend a scientific lecture in the interior, and contemplate the orderly march of the constellations painted on the moving concavity above them, the stars being pierced through the metal according to the several magnitudes, so that the light penetrates and each assumes a curious radiated, or rather stellated form.   This sphere was completed, with considerable expense as well as ingenuity, in 1758; but although six pounds per annum is allowed to a keeper, who is generally an under-graduate, it was suffered to fall so much out of order as to mar the projector's intention of popularizing astronomy; and many a good man and true has lived and learned in Cambridge, without even being aware of its existence.   Of this I could tell a story or two, but shall only add, that it was lately brushed up a bit; and I had the satisfaction of being on its floor with a party of Cambridge *savans* of the first magnitude, in whom the shade of Long must have delighted.

Δίδυμοι, Gemini, Tindaridæ, or Gemelli, is the third constellation of the zodiac, and one of the ancient 48; lying nearly mid-way between Orion and the Great Bear, in a region long viewed as the centre of the heavens.   Among the Orientals it was represented as a pair of kids, denoting that part of spring when these animals appear; but the Greeks changed them to two children with their feet on the Galaxy; and the Arabians, whose tenets prohibited the human form in delineations, afterwards altered them to a couple of peacocks.   Paulus Venetus, and the early Venetian illustrators of Hyginus, represent them as two winged angels.   Among the ancients every sign had its tutelary deity, and Phœbus had charge of Gemini, which gave rise to the astrological jargon about the connexion between the sun and this asterism; to the disparagement of the latter, for many inuendos are on record, and we are told, in the manuscript Almanac of 1386, that whoever happens to be born under the aërial triplicity of the Twins, shall be "ryght pore and wayke, and lyf in mykul tribulacion."

Astronomers, however, view it in a different light; for though it is not splendidly conspicuous nor thickly studded, it is fine, and contains bright individuals, which, with its numerous double stars, clusters, and

N 2

nebulæ, render it interesting and important; and, from its being the sign of St. Paul's ship, we see that it was esteemed propitious by ancient mariners.   It has been thus tabulated:

Ptolemy . . . 25 stars		Bayer . . . 33 stars
Copernicus . . 25		Hevelius . . . 38
Tycho Brahé . . 29		Flamsteed . . 85
Kepler . . . 30		Bode 190

Castor was called by the later Arabians, *Rás-al-tawum al-mokaddem*, the head of the foremost twin; and with Pollux it constitutes *al-dhirá' al-mebsútah*, the outstretched arm, forming the VIIth Lunar Mansion. This *dhirá'* is intended to mean the drawn-in paw of the large lion alluded to by Kazwíní; an allusion which Ideler ascribes rather to the disciples of confusion and ignorance, than to astrognosts.   This huge monstrosity may be thus figured.   The two stars in the heads of the Twins, and in the Lesser Dog, form its paws, the *Præsepe*, its nose— ζ, γ, η, and *a* in the Greek Lion, its forehead—Arcturus and Spica, its shin-bones—β, η, γ, δ, and ε Virginis, the hips—and the stars in Corvus, its hind quarters.   No wonder that Ideler indignantly asks, who could have made such a mistake as placing the nose on the forehead, the legs on the hip?   " Welches Missverhältniss!   Der Nase zur Stirn, der Schienbeine zur Hüfte!"

To know this star by alignment is easy, as a ray from Rigel, led through ε, the middle star of Orion's belt, and under Betelgeuze, will, at about twice that distance further on, rest upon Castor: or, if taking the poetaster's advice:

From *gamma* on the Great Bear's flank	let a long ray be cast,
Conduct it under Merak's blaze	to south-west regions vast;
Across the Lynx to Gemini	this line will thus be led,
And carried further on will reach	bright Betelgeuze the red.

A proper motion in space has been assigned to the principal star, to the following amount:

P.... Æ	— 0″·16	Dec. - 0″·10
B....	— 0″·12	- 0″·07
A....	— 0″·21	- 0″·08

CCXCIII.   52 ℍ. VIII. MONOCEROTIS.

Æ 7ʰ 25ᵐ 53ˢ	Prec. + 2ˢ·79
Dec. S 12° 42′·2	—— S 7″·30

Position BC 140°·0 (w 5)   Distance 12″·0 (w 2)   Epoch 1836·83

The principal star of a loose Galaxy cluster, under the Unicorn's belly, and pointing to a double star in the *np*, on an angle 288° and △ Æ 14ˢ·5.   A 7, faint yellow; B and C, both 10, and both dusky; the two latter point to a 9th-magnitude star preceding them by about 6ˢ, and the field contains the cluster and its outliers, with several 8th magnitudes grouped near the centre.   It was first registered at Slough, in 1786, and may be fished up by dropping a line from Pollux close by the west side of Procyon, and extending it 19° to the south, where it stands 12° east-north-east of Sirius.

CCXCIV. 149 P. VII. ARGO NAVIS.

$\mathbb{R}$ 7ʰ 27ᵐ 33ˢ   PREC. + 2ˢ·54

DEC. S 23° 07′·7   —— S 7″·47

POSITION 284°·1 (w 8)   DISTANCE 10″·2 (w 3)   EPOCH 1831·92

A neat double star in the Galaxy, over the aplustre of the Argo's poop; where a line sent from Pollux through Procyon, and 28° beyond, will hit it; as will a cross line from Orion's sword-cluster carried through Sirius, and 14° into the south-east quarter. A and B, both 6, and both topaz-tinted. This appears to be No. 19 of Ḥ.'s 145, registered in 1784; and though the results of Piazzi's circle observations do not quite coincide with my micrometer measures, the star here noted as B, adjusting the quadrant, must be his No. 147 of Hora VII. Sir James South, No. 552, gives these data:

Pos. 284° 53′   Dist. 9″·01   Ep. 1825·01

CCXCV. 1 Ḥ. VI. GEMINORUM.

$\mathbb{R}$ 7ʰ 28ᵐ 57ˢ   PREC. + 3ˢ·57

DEC. N 21° 55′·7   —— S 7″·59

MEAN EPOCH OF THE OBSERVATION . . . . 1835·90

A compressed cluster under the left shoulder of Pollux; and rather more than one-third of the distance from β Geminorum to β Canis Minoris, following Wasat nearly on the parallel, at about 4°. This was first registered by Ḥ. in 1783; and was described as a "beautiful cluster of many large and compressed small stars, about 12′ in diameter." My telescope only shows a faint mass of very small stars, inclining from *sp* to *nf*, but of indistinct figure, the objects being from the 10th to the 16th magnitude. It is No. 458 of H.'s Catalogue; and was carefully differentiated with Wasat (δ Geminorum).

CCXCVI. 38 Ḥ. VIII. ARGO NAVIS.

$\mathbb{R}$ 7ʰ 29ᵐ 12ˢ   PREC. + 2ˢ·73

DEC. S 14° 08′·3   —— S 7″·60

POSITION 303°·8 (w 8)   DISTANCE 8″·0 (w 4)   EPOCH 1834·21

A double star in a loose cluster of the Milky Way, over the Argo's stern; and one of those seized by Bode to make his *Officina Typographica*. A 7½, and B 8, both bright bluish white. It inhabits a very splendid field of large and small stars, disposed somewhat in a lozenge-shape, and preceded by a 7th magnitude with a companion about $20''$ *nf* it. The cluster was not registered till 1785, but the double star is 63 Ħ. II., the former measures of which are:

Ħ.	Pos.	300° 12′	Dist.	6″·50 ±	Ep. 1782·78
S.		303° 20′		7″·44	1825·02
Σ.		304° 44′		7″·46	1831·44

To fish this object up, run a line about 12° east-by-north from Sirius, and intersect it by another from Pollux through Procyon, and continued 20° lower down. It is in a very rich vicinity.

CCXCVII.　159 P. VII. CAMELOPARDI.

Æ	$7^h 30^m 37^s$	Prec.	+ 5ˢ·78
Dec.	N 65° 31′·7	——	S 7″·72

Position 5°·1 (w 5)	Distance 15″·4 (w 5)	Epoch 1831·01
—— 4°·7 (w 9)	—— 15″·6 (w 9)	—— 1839·27

A neat double star, in front of the Greater Bear's head; where a line run from Capella through δ in Auriga's head, and extended 15° to the north-east, will meet it. A and B, both 8, and both white, in a rich neighbourhood. This fine object is formed by Piazzi's Nos. 159 and 160, Hora VII.; and the fixity shown by my observations, is also indicated by a comparison of the other measures and deductions from Piazzi, which are:

P.	Pos.	6° 00′	Dist.	13″· 8	Ep. 1800·00
S.		4° 17′		16″·17	1825·05
Σ.		4° 52′		15″·46	1830·59

CCXCVIII.　α CANIS MINORIS.

Æ	$7^h 30^m 55^s$	Prec.	+ 3ˢ·19
Dec.	N 5° 37′·8	——	S 7″·74

Position 85°·0 (w 1)	Distance 145″·0 (w 1)	Epoch 1833·81

A standard Greenwich star with a distant companion, on the loins of the Lesser Dog. A 1½, yellowish white; B 8, orange tinge; several small stars in the field. A is a splendid star, though very considerably less bright than Sirius, which accounts for the latter being called the Greater Dog, quite as well as the assigned reason, as to rising time. Authorities have differed as to Procyon's magnitude; Ptolemy and Hevelius designating it 1, Tycho Brahé 2, and most of the others 1½. Hunters after parallax will recollect, that this is one of those stars upon which Piazzi bestowed such labour to detect the angle which the mean diameter of the Earth's orbit subtends from them, as related in the xⅢth volume of the *Italian Society's Memoirs;* and that an infinity of observations induced him to assign 3″± as the value of Procyon. It has also a large proper motion in space; it is variously assigned; but the following values are the most coincident:

P....	℞ − 0″·71	Dec.	− 0″·98
B....	− 0″·63		− 1″·05
A....	− 0″·69		− 1″·05

Cânis Minor, though a small asterism, is one of the old 48, and, as well as its *lucida*, was called Προκύων, the precursor-dog, because it appeared in the morning dawn before Sirius: though Jacob Bryant persists that the Greek filched the word from the Egyptian *Pur Cahen.* Hence also its name of *Ante-Canis;* and it was popularly considered as Orion's second hand, or *Canicula,* which title Horace, Pliny, Hyginus, and Galen, support against Germanicus, Julius Firmicus, and Appian, who are all for Sirius. The Arabians recognised its quality of forerunner to the Dog-star in *al-kelb-al-mutekaddem,* the antecedent dog; but they also called it *Al-shï ra-l-shámiyah,* the bright star of Syria; *ghomaïsá,* watery-eyed; and *Al-kelb-al-asghar,* the lesser dog. All this shows that the constellation was one of much interest, and the regard of the ancients descended to the astrologers of later ages: " What meteoroscoper," demands old Leonard Digges, " what meteoroscoper, yea, who learned in matters astronomical, noteth not the great effects at the rising of the starre called the *Litel Dogge?*"

This constellation stands to the north-east of the Greater Dog, so that the Milky Way passes between them; and under Gemini. They were anciently in closer connection, but the intrusion of Monoceros between them, by Hevelius, has parted them. The number of stars given to this asterism, in successive Catalogues, has been as follows:

Ptolemy	. . . . 2 stars	Bayer	8 stars
Hyginus	 3	Hevelius	13
Tycho Brahé	. . . 5	Flamsteed . . .	14
Kepler	5	Bode	55

Procyon is a member of the magnificent equilateral triangle formed in conjunction with Sirius and Betelgeuze, as well as a right-angled one with Betelgeuze and Pollux. A perpendicular raised at Sirius to a line drawn from that star to Orion's belt, will also pass through Procyon to the northward; or, as the alignment is expressed:

Orion's belt from Taurus' eye,   leads down to Sirius bright,
His spreading shoulders guide you east,   'bove Procyon's pleasing light.

CCXCIX. 170 P. VII. CANIS MINORIS.

℞ 7ʰ 31ᵐ 37ˢ	Prec. + 3ˢ·19
Dec. N 5° 35′·7	—— S 7″·80

Position 132°·9 (*w* 8)   Distance 1″·4 (*w* 4)   Epoch 1833·22

A close double star, in a fine vicinity on the loins of the Lesser Dog, closely *sf* of Procyon. A 7, white; B 8, ash-coloured, with a minute

blue star preceding it about 2', and another of the 11th magnitude in the
sp quadrant. This very pretty star resembles η Coronæ, but is smaller;
and to see it well, we are directed by ♅. to observe it when Procyon is
near its meridian altitude. The components of this object are both close
and oblique, which may account for the results of the several astrometers
being too discordant to admit of a decision upon the apparent variations;
though a direct and increasing angular motion must be inferred, now
amounting to about 1° per annum, according to some observers. The
registered details of comparison are:

♅.	Pos. 117° 21'	Dist. *(caret)*	Ep. 1781·91
H. and S.	127° 08'	*(caret)*	1823·13
Σ.	132° 06'	1"·46	1829·43
D.	134° 13'	1"·33	1831·01

An almost imperceptible movement in space is attributed to the
leader, apparently in the same direction with Procyon, from which it
stands at an angle of about 105°, and a distance of 10'; of this movement,
the most rigidly investigated values are:

P....	Æ − 0"·42	Dec.	0"·00
B....	− 0"·02		− 0"·04
T....	+ 0"·11		− 0"·06

From the above-cited extract from Σ.'s observations, I do not clearly
understand my friend Amici's remark in the twelfth volume of Baron
De Zach's *Correspondance Astronomique:* " si potrebbe con ragione
sospettare che il gran cannocchiale achromatico del Signor Struve non
abbia tanta luce distinta, quanto un mio telescopio di undici pollici di
diametro; imperocchè egli annunzia l'osservazione sua della doppia
stella 3 Canis Minoris, ossía Herschel i. 23, e non fa parola di un
altra stelletta vicina, che l'accompagna."

CCC.   46 ♅. VIII. ARGO NAVIS.

Æ 7ʰ 32ᵐ 09ˢ	Prec. + 2ˢ·72
Dec. S 16° 00'·3	—— S 7"·84

Mean Epoch of the Observation . . . . 1836·24

A loose cluster outlying the Galaxy, over the Argo's stern, where a
ray led from Arneb (a Leporis) between Mirzam and Sirius, and as far
again beyond, will strike upon it; and a line from Castor passed half a
degree to the west of Procyon, and extended 22° beyond, also picks it
up. It comprises a rich field of scattered stars, with occasional glows of
star-dust, so that the magnitudes may range from 9 to 16, and smaller
still; and the stragglers run into the south-south-east quarter, where is the
crowded group 47 ♅. viii. It was discovered by ♅. on the last evening
of the year 1785.

About a minute preceding this object, and 1½° to the north, is a
small faint cluster, which is probably 87 ♅. viii., described by Sir
William as consisting of small stars, and not rich.

CCCI. 175 P. VII. ARGO NAVIS.

Æ 7ʰ 32ᵐ 17ˢ   Prec. + 2ˢ·46
Dec. S 26° 26'·5   —— S 7"·85

Position 326°·8 (w 4)   Distance 9"·8 (w 2)   Epoch 1831·90

A neat double star, in the corymbus of the Argo's ουρά, or poop; where it may be picked up by a line from the lowest star in Orion's belt, through Sirius, and 19°, or nearly as far again, beyond. A and B, both 6½, and both topaz-yellow; but the tinge which it exhibited under observation, may be owing to its low altitude. This is a tolerably fair object, and is composed of Piazzi's Nos. 175 and 177 of Hora VII., a reduction from whose mean apparent places affords another gratifying proof of the excellence of his meridian observations:

Pos. 325° 0'   Dist. 10"·5   Ep. 1800

CCCII. 46 M. ARGO NAVIS.

Æ 7ʰ 34ᵐ 30ˢ   Prec. + 2ˢ·75
Dec. S 14° 27'·3   —— S 8"·03

Position 90°·0 (w 1)   Distance 15"·0 (w 1)   Epoch 1836·24

A very delicate double star in a fine cluster, outlying the Galaxy, over the Argo's poop. A 8½, and B 11, both pale white. A noble though rather loose assemblage of stars from the 8th to the 13th magnitude, more than filling the field, especially in length, with power 93; the most compressed part trending *sf* and *np*. Among the larger stars on the northern verge is an extremely faint planetary nebula, which is 39 Ḥ. IV., and 464 of his son's Catalogue. This was discovered by Messier in 1769, who considered it as being "rather enveloped in nebulous matter;" this opinion, however, must have arisen from the splendid glow of the mass, for judging from his own remark, it is not likely that he perceived the planetary nebula on the north. Ḥ., who observed it in 1786, expressly says, "no connexion with the cluster, which is free from nebulosity." Such is my own view on attentively gazing; but the impression left on the *senses*, is that of awful vastness and bewildering distance,—yet inducing the opinion, that those bodies bespangling the vastness of space, may differ in magnitude and other attributes.

To fish up this object, an occult line must be carried from *a* Leporis through Sirius, and extended 13½°, or nearly as far again, to the eastward; where a glance from Castor over Procyon passes through it. In the following field there is a coarse pair of 7th-magnitude stars, lying *sf* and *np* of each other.

CCCIII. 64 ♅. IV. ARGO NAVIS.

Æ 7ʰ 34ᵐ 46ˢ   Prec. + 2ˢ·67
Dec. S 17° 50′·2   —— S 8″·05

Mean Epoch of the Observation . . . . 1838·21

A bright planetary nebula, pale bluish-white, over the Argo's poop, and on an outlying wave of the Milky Way. This fine object exactly

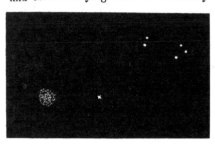

precedes a 7th magnitude, and is followed by some small stars, as in the annexed diagram; by which it is very readily identified when fished up, and this may be done by throwing a line from Castor through Procyon, and extending it 24° to the south, where it follows Sirius about 14° east by south.

This was registered by ♅. in March, 1790, and was only estimated at about 12″ or 15″ in diameter. But the inference from such a supposition is vast! "Granting," says H., "these objects to be equally distant from us with the stars, their real dimensions must be such as would fill, on the lowest computation, the whole orbit of Uranus." The mean apparent place of this nebula was obtained by differentiation with 4 Argo Navis.

CCCIV. ϰ GEMINORUM.

Æ 7ʰ 34ᵐ 47ˢ   Prec. + 3ˢ·73
Dec. N 24° 46′·5   —— S 8″·05

Position 231°·9 (w 6)   Distance 6″·0 (w 4)   Epoch 1838·98

A very delicate double star, on the left shoulder of Pollux, and about 3½° to the south of its lucida. A 4, orange; B 10, pale blue. This elegant object was discovered by H., No. 427 of his Sweeps, with the 20-foot reflector, and estimated at about an angle of 240°, with a distance of 5″; it is one of the remarkable points to which he expressly calls the attention of astronomers, as forming a case where possibly the *comes* is shining by a reflected light. Difficult as the pair seemed to be, they were attacked by the Rev. W. R. Dawes, who with only a 5-foot telescope, obtained these results:

Pos. 225° 10′   Dist. 6″·25 Ep. 1832·16

The large star is suspected of having a small movement in space, but the values assigned are not coincident; the best are:

P.... Æ − 0″·16   Dec. − 0″·02
B....   0″·00   − 0″·05

CCCV. β GEMINORUM.

Æ 7ʰ 35ᵐ 31ˢ   Prec. + 3ˢ·68

Dec. N 28° 24′·4   — S 8″·11

Position AB 66°·9 (w 8)   Distance 130″·0 (w 3) ⎫

—— AC 73°·6 (w 6)   —— 202″·7 (w 3) ⎬   Epoch 1832·31

A standard Greenwich star coarsely triple, or rather quadruple, in the eye of Pollux. A 2, orange tinge; B 12½, ash-coloured; C 11½, pale violet, and it has a minute *comes* to the *sp*, which, though unnoticed in former registers, is certainly now (1832) as bright as C: these companions form a neat triangle*. This wide object is 42 Ⱨ. vɪ., but at its first registry the two nearest only were measured:

Pos. 65° 32′   Dist. 116″·80   Ep. 1783·20

But Sir James South included the distant star, and made it triple; and by altering his uncial letters so that B becomes the nearest to the principal, according to the rule which I have adopted, the measures he obtained will stand thus:

Pos. AB 66° 23′   Dist. 132″·31 ⎫
AC 72° 40′   198″·47 ⎬   Ep. 1825·10

This star has been suspected of varying in lustre, since it is recorded as having at times been brighter than Castor, whence Bradley rated it of the 1st magnitude; others have classed it in the 3rd rank; but Ptolemy, Tycho, La Caille, Zach,and all the best authorities, classify it 2. Nor is this the only anomaly of Pollux, for the ancients represented it in colour ὑπόκιῤῥος, *subrufa*, reddish; Lichtenstein says, *Quæ trahit ad æram, et est cerea;* and certainly, in 1832, its tint was as I have mentioned, under an eyepiece magnifying 240 times. It has a considerable proper motion, the amount of which has been thus assigned:

P....Æ − 0″·72   Dec. − 0″·11

B....   − 0″·66   − 0″·06

A....   − 0″·71   − 0″·06

This star is well known as Pollux, the brother of Castor; but in the Alphonsine Tables, and in other old astronomical works, it is called *Ras-al-geuze*, the twin's head, from the doubtful word *jauzá* or *júzá*. It was, however, better known among the later Arabians as *Rás-al-tawum al muäkhkhar*, the head of the hindmost twin; and the two heads form the VIIth Lunar Mansion, in the Arabian constellation *al-dhirá' al-mebsúṭah*. See δ Geminorum. Pollux may generally be known by his connection with Castor; but for further identity, a line drawn from the Pleiades through Nath on the Bull's horn, passes to it; as will also

---

* While this is in the press, the Rev. W. R. Dawes has shown me an exact diagram which he made of this object, January 1, 1829, with a 3½-foot achromatic, charged with a Huygenian eye-piece magnifying 200 times. With this instrument he saw the *three* companions very distinctly, although two only were visible, and that but on remarkably fine nights, in Sir James South's 7-foot equatoreal, with an aperture of 5 inches.

a line from Rigel between ϵ and ζ Orionis, the two last stars of the belt.
Having found it, the brackish rhymes tell us:

If Betelgeuze and Procyon	with Pollux bright be cast,
Amid the glories of the sky,	shines a triangle vast;
To gauge with practised studious eye	the form that shines afar
The angle of twice forty-five,	shows 'tis rectangular.

CCCVI. π GEMINORUM.

Æ   7^h 37^m 11^s Prec. + 3^s·88

Dec. N 33° 48'·2   —— S 8"·24

Position AB 212°·5 (w 2)   Distance 25"·0 (w 1) ⎫
——— AC 340°·0 (w 2)   ——— 95"·0 (w 2) ⎬ Epoch 1839·12

A most delicate triple star, just above the heads of Gemini; where
it is reached by a line from Procyon through Pollux, and extended 5½°
to the north. A 5½, topaz yellow; B 13, bluish; C 12, dusky. The
two principal members form 53 ♄. iv., but no *set* measures appear till
Σ. entered it on the Dorpat Catalogue, with the following results:

Pos. 211° 72'   Dist. 22"·60   Ep. 1831·25

The individual C, in the *np* quadrant, is the star measured by H.
and S., No. 83, which was reckoned the *minimum visibile* of the 7-foot
equatoreal: the Rev. W. R. Dawes, however, has shown me a very neat
diagram which he made of it in 1828, with a 3½-foot telescope under
an erect eyepiece.

CCCVII.   93 M. ARGO NAVIS

Æ   7^h 37^m 44^s Prec. + 2^s·54

Dec. S 23° 29'·1   —— S 8"·29

Mean Epoch of the Observation . . . . 1836·13

A small galaxy cluster, in the aplustre of the Argo's poop; a line
from Orion's sword-cluster, led through Sirius, strikes upon it 16° beyond,
where it will be intersected by a ray from Castor through Procyon.
This neat group is of a star-fish shape, the *sp* portion being the brightest,
with individuals of 7 to 12 magnitudes: it was first registered by Messier,
in 1781, as a mass of small stars.

The unlucky Chevalier d'Angos, of the Grand-Master's observatory
at the summit of the palace at Malta, mistook this cluster for a comet:
from which, and some still more suspicious assertions, my excellent
friend, Baron de Zach, was induced to term any egregious astronomical
blunders—*Angosiades.*

CCCVIII. 2 ARGO NAVIS.

Æ 7ʰ 38ᵐ 08ˢ   Prec. + 2ˢ·76
Dec. S 14° 18'·3   —— S 8"·32

Position 338°·8 (ᵂ 9)   Distance 16"·8 (ᵂ 9)   Epoch 1836·20

A neat double star over the Argo's stern, where a line from Arneb through Sirius, and as far again to the east, will meet it intersected by a ray from Castor over Procyon. A 7, silvery white; B 7½, pale white; and another star in the *nf* quadrant. This fine object is 91 ℍ. iv., and Nos. 204 and 205 of Piazzi's Hora VII. It was measured by ℍ. thus:

Pos. 339° 12'   Dist. 17"·38   Ep. 1782·78

whence its fixity seems to be established, the increase of distance supposed by H. and S. not being confirmed by my observations, their result being 19"·66. From Piazzi's mean apparent place for 1800, the space between these stars was 17"·50.

CCCIX. 5 ARGO NAVIS.

Æ 7ʰ 40ᵐ 27ˢ   Prec. + 2ˢ·82
Dec. S 11° 48'·3   —— S 8"·51

Position 19°·0 (ᵂ 7)   Distance 3"·5 (ᵂ 5)   Epoch 1834·11

A close double star, over the Argo's stern, rather more than 2° north by east of the last object. A 7½, pale yellow; B 9, light blue. This fine object is No. 55 *Officina Typographica*, an asterism which Bode proposed to make by gathering 100 of the *informes* between Sirius and the hind legs of Monoceros, in commemoration of the art of printing. It must have been seen by Piazzi, who made this note to the principal star: "alia 9ᵉ magnit. præcedit ad austrum;" for though this implies an error of quadrant, such error is readily made. It has been measured by Σ., with these results:

Pos. 17°·8   Dist. 3"·30   Ep. 1835·24

CCCX. 14 CANIS MINORIS.

Æ 7ʰ 50ᵐ 03ˢ   Prec. + 3ˢ·12
Dec. N 2° 38'·8   —— S 9"·26

Position AB 64°·9 (ᵂ 8)   Distance 75"·0 (ᵂ 4) ⎫
—————— AC 153°·4 (ᵂ 5)   —————— 115"·0 (ᵂ 2) ⎬ Epoch 1831·14
                                              ⎭

A wide triple star on the boundary of the Lesser Dog, and 6° to the south-east of the star Procyon, near where the beforementioned section or slice of Argo penetrates through the Unicorn. A 6, pale white; B 8, bluish; C 9, blue. This object is 84 Ḥ. VI., and the two nearest were thus registered:

Pos. 63° 36'   Dist. 65"·46   Ep. 1782·11

The third star was stated to form an angle with the other two, in the *sf*, a little longer than a rectangle. The three were measured by H. and S.; and from a comparison of the whole, I am not inclined to impute any appreciable motion; their results were:

Pos. A B 65° 42'   Dist. 76"·021
A C 152° 50'   112"·168 } Ep. 1822·15

CCCXI. 37 Ḥ. VI. ARGO NAVIS.

Æ 7ʰ 52ᵐ 23ˢ   Prec. + 2ˢ·85
Dec. S 10° 20'·5   —— S 9"·44

Mean Epoch of the Observation . . .   1837·02

A compressed cluster of stars, from the 10th to the 16th magnitude, and even smaller, over the Argo's poop, in a rich vicinity of two or three fields; a line drawn from δ Geminorum over Procyon, and carried exactly as far again into the south-east, will strike upon it. It was registered by Ḥ. in February, 1791, and described as having some of the stars "next to invisible;" and H. considers the most compressed part to be 4' or 5' in diameter. In the preceding portion is a very minute double star.

This cluster is more susceptible to optical power than many of its class, and is apparently congregated by peculiar principles of attraction, independently of the innumerable outliers scattered around. It therefore offers a province for meditation as well as calculation, and suggests a most sublime conception of the boundless extent of the material universe, in the mysterious vastness which those suns beyond suns, and glorious systems of suns, probably with attendant planetary bodies, unfold! Hervey, *meditating* upon the immensity of the universe, has eloquently observed, " Could we wing our way to the highest apparent star, we should then see other skies expanded, other suns that distribute their inexhaustible beams of day, other stars that gild the alternate night, and other, perhaps nobler, systems established; established in unknown profusion through the boundless regions of space. Nor do the dominions of the GREAT SOVEREIGN end there; even at the end of this vast tour, we find ourselves advanced no farther than the frontiers of Creation, arrived only at the suburbs of the GREAT JEHOVAH's kingdom." This is inserted rather for the intended than the expressed sentiment; for the *alternate nights*, and *frontiers*, and *suburbs*, can only be viewed as the worthy rector's *maniéra di parlare*.

CCCXII. 11 CANCRI.

Æ 7ʰ 59ᵐ 02ˢ   Prec. + 3ˢ·68
Dec. N 27° 56′·4   —— S 9″·95

Position 213°·5 (w 5)   Distance 3″·2 (w 2)   Epoch 1839·70

A close double star, between the head of Pollux and the preceding claw of Cancer; following the former nearly on the parallel, at about 5°½ distance. A 7, pale yellow; B 12, lilac. This delicate object is not 11 ⩗. I., as supposed by H. and S., who mistook for it á very neat double star about 3′ preceding, and a little north of the parallel. On the arrival of the Dorpat Catalogue, it was found that Σ. had observed both; 11 ⩗. I. being Σ.'s 1177; and the one before us, his No. 1186, thus measured:

Pos. 218° 52′ Dist. 3″·18 Ep. 1828·26

This occasioned some confusion, as I had formerly measured 88 H. and S. with some degree of exactness at the equatoreal, and also, they being of the 7th and 8th magnitudes, with much facility; but on fixing the mean apparent place at the transit-circle, discovered the error. My measures of 1832 were:

Position 354°·2 (w 8)   Distance 3″·8 (w 6)   Epoch 1832·76

CCCXIII. 29 MONOCEROTIS.

Æ 8ʰ 00ᵐ 33ˢ   Prec. + 3ˢ·02
Dec. S 2° 31′·3   —— S 10″·06

Position AB 104°·7 (w 3)   Distance 30″·0 (w 2) ⎫
—— AC 243°·8 (w 8)   —— 66″·9 (w 9) ⎭ Epoch 1832·14

A delicate triple star, on the Unicorn's flank; it is about 11° to the south-east of Procyon, which is the last of the splendid host that adorns the three preceding hours. A 5½, light yellow; B 13, grey; C 9, pale blue, apparently the *comes* mentioned by Piazzi, Nota 316, Hora VII., "5ˢ temporis præcedit ad austrum." A and C point to a distant star of the 8th magnitude; and there are several companions in the field, of which one in the *nf* is coarsely double. The two nearest of this group constitute 97 ⩗. IV., and were thus registered:

Pos. 105° 12′ Dist. 29″·90 Ep. 1782·80

The small star B escaped detection with the instruments of H. and S.; but their measure of C., together with the observations of Σ. on the three, compared with my own, indicate that no appreciable motion has occurred in half a century.

CCCXIV.   ARGO NAVIS.

Æ 8ʰ 00ᵐ 44ˢ   Prec. +   2ˢ·56
Dec. S 23° 50′·8   —— S 10″·08

Position 191°·0 (ω 1)   Difference Æ = 3ˢ·8 (ω 1)   Epoch 1833·82

A Greenwich star, with a distant companion, in the aplustre of the ship's poop; where an imaginary line from Rigel, passed close under Sirius, will reach it in the south-east region, by doubling the distance between those stars. A 3½, pale yellow; B 10, greyish; other small stars in the field. This star is called *Tureïs*, the *scutulum* or *little shield*, corresponding to the ἀσπιδίσκη of Ptolemy. A spacial movement is assigned it, to the following effect:

P.... Æ − 0″·21   Dec. + 0″·09
B....   − 0″·03   + 0″·03

Argo is one of the old 48 constellations, occupying a very large space in the Southern Hemisphere, but its lucida, Canopus, as well as most of its more important stars, is always hidden from us. There are sound reasons for assigning the highest antiquity to this grand asterism, though the etymologists are crowding on when they derive the English word canopy, or covering, from Canopus, as hath lately been imprinted: such sages would readily see our " son of a gun," in the Greek παῖς Γυνῆς (Dor.) " This constellation," says Sherburne, " *sails* by our meridian at midnight, in January, she being deified for saving deities;" but he takes no notice of a strange peculiarity in the good ship's sailing properties. In the apparent motion of the sidereal system, this constellation actually dips stern foremost, as Aratus remarked, but which the old scholiast, whoever he is, assures us, does not *really* set before the prow. *Averso astro* is very properly applied to it, as those who do not distinguish between a stem and a stern may satisfy themselves, on looking for the *gubernaculum*. There is, moreover, no stem to *nobilis Argo*, as Manilius designates it, the ship sharing the sectional system so remarkable in Pegasus and Taurus, which section was termed ἡμίτομος by the Greeks; and Aratus (§ 54, Morel's ed. 1559) expressly says the ship was halved to the mast, ἴσον διχόωσα κατ᾽ αὐτόν. In the *Theatrum Cometicum*, 1667, by Lubienietzki, there is a large *tabula Uraniscopica*, on which Argo is represented as a goodly Argosie with three masts and a tier of ports; her courses and topsails are clean full with the wind aft, and yet she must be pulled back, and set stern foremost.

Owing to the great extent of this constellation, it is subdivided into four departments,—the hull, the keel, the stern, and the sail,—Argo navis, carinâ, puppi, velis. Ptolemy assigned 45 stars to Ἀργὼ; but as a large portion could never be observed in our hemisphere, the number continued small till Bode made it amount to 540, by gathering all those observed by Halley, La Caille, and other southern observers. My friend Sir Thomas Brisbane, however, has registered no fewer than 1330 stars in Argo; and as the Via Lactea sweeps directly across it, there is yet a rich harvest for future astronomers.

CCCXV. ζ CANCRI.

Æ 8ʰ 03ᵐ 02ˢ   Prec. + 3ˢ·44
Dec. N 18° 07′·5   —— S 10″·25

Position		Distance		Epoch
AB 28°·3 (*w* 4)		1″·3 (*w* 2)		
—— AC 149°·4 (*w* 6)		—— 5″·4 (*w* 4)		1832·23
—— AB 23°·8 (*w* 5)		—— 1″·0 (*w* 2)		
—— AC 147°·1 (*w* 5)		—— 4″·8 (*w* 4)		1834·36
—— AB 19°·6 (*w* 4)		—— 1″·3 (*w* 4)		
—— AC 148°·3 (*w* 6)		—— 5″·2 (*w* 5)		1835·28
—— AB 12°·8 (*w* 3)		—— 1″·2 (*w* 3)		
—— AC 146°·9 (*w* 3)		—— 5″·4 (*w* 3)		1837·11
—— AB 5°·2 (*w* 8)		—— 1″·3 (*w* 3)		
—— AC 148°·2 (*w* 9)		—— 5″·1 (*w* 6)		1839·32
—— AB 355°·1 (*w* 8)		—— 1″·2 (*w* 5)		
—— AC 147°·2 (*w* 9)		—— 5″·0 (*w* 6)		1843·11

A fine triple star, just below the after claws of the Crab, where a shot from Castor through Pollux, carried twice that distance to the south-south-east, will strike it; or it may be found by a ray from Sirius through Procyon, extended to half their distance beyond. But there is much shade on the authority which designates it *Tegmine*. A 6, yellow; B 7, orange tinge; C 7½, yellowish, No. 6 of Piazzi's Hora VIII. This very interesting physical object forms 24 ӈ. I., and 19 ӈ. III., and by modifying the dates and diameter of A for distance, we obtain a point of departure from whence an extraordinary retrograde motion is exhibited. The best interval measures for comparing with my own are as follow,— those of Piazzi being deduced from his Æs and Decs., which of course cannot demand implicit reliance:

			Dist.	Ep.
ӈ.	{	Pos. AB 3° 28′	1″·00 }	1781·90
	{	AC 181° 44′	8″·05 }	
P.		AC 159° 00′	6″·47	1800·00
H. and S.		AC 158° 17′	6″·24	1822·14
S.		AB 57° 50′	1″·09	1825·27
H.	{	AB 35° 35′	1″·01 }	1830·44
	{	AC 155° 52′	5″·35 }	
D.	{	AB 30° 48′	1″·09 }	1831·30
	{	AC 150° 17′	5″·59 }	
Σ.	{	AB 21° 30′	1″·13 }	1833·27
	{	AC 148° 18′	5″·48 }	

This curious and very interesting object has occasioned no little discussion, since ӈ. had pronounced it to be a "most minute treble star," and more than forty years afterwards H. and S. had declared A and B to be one—"beautifully defined and round." But Sir James South on a second trial saw A "unquestionably elongated," whence a supposition arose that the star B had come from behind A. The distance, however, appears to be very stationary, so that these remarks are extraordinary, and must be imputed to some anomaly. The Rev. W. R. Dawes, of Ormskirk, took it in hand with a 5-foot telescope only. Up to 1830, a *direct* motion of + 1°·25 per annum was assigned to the star B; whereas from the accurate measures of Sir John Herschel and Mr. Dawes, it was afterwards considered to have a *retrograde* one to the surprising mean amount of — 6° 51′. As both these astronomers corresponded with me on the subject, I determined to devote my best powers to it, and though the near object is difficult from convergence, I was able to assign considerable weights to the results: and from them I drew a scheme of the interpolated positions, angular velocities, and projected distances. Hence it follows, that the two close stars perform a binary revolution in about sixty years, while the outer one accomplishes a grand orbital ellipse in 500 or 600 years. It would seem that ℍ.'s angle of A and B in 1781, must have been quadranted wrong, for the retrocession from 183°·3, is more explicable than the *nf* position above given. The mean of my own observations, embracing a period of nearly eleven years, gives a retrograde march of only 2°·77 per annum: or from ℍ.'s epoch = 3°·78. Sir John Herschel thought, that an entire revolution would have occurred from the epoch of his father's observation to the end of March, 1837, in a periodic time of 55·34 years; which, though not precisely confirmed by me, will be seen to meet the remarkable phenomenon as nearly as the present sidereal knowledge has advanced. "If this be really a TERNARY system," said Sir John, "connected by the mutual attraction of its parts, its perturbations will present one of the most intricate problems in physical astronomy."

Mr. Dawes informs me, that A and B were quite vertical in 1841·32, for the angle of position was then exactly on the zero point; thereby affording a gratifying confirmation of his previous deductions. And it must be noted, that a movement in space has been detected in ζ, in which the *comites* doubtlessly partake, of the following value:

$$P....\text{Æ} + 0''·04 \quad \text{Dec.} - 0''·10$$
$$B.... \quad + 0''·14 \quad\quad - 0''·11$$

CCCXVI. 11 ℍ. VII. ARGO NAVIS.

Æ 8ʰ 03ᵐ 10ˢ   PREC. + 2ˢ·82
DEC. S 12° 23′·4   —— S 10″·26

MEAN EPOCH OF THE OBSERVATION . . . . 1832·97

A compressed cluster, in the space under the haunches of Monoceros, where a line from Arneb drawn through Sirius, and extended rather more than as far again to the east, will find it as the eastern point of a triangle, equilateral with Sirius and Procyon. It consists of a large and loose, but rich, group of small stars pretty equally strewed over the field, with a close double star in the middle, and a bright yellow one of the 6th magnitude to the *sf*. It was registered by ℍ. in January, 1785.

CCCXVII. 13 P. VIII. CANCRI.

ÆR 8ʰ 04ᵐ 47ˢ   PREC. + 3ˢ·30
DEC. N 11° 19′·7   —— S 10″·38

POSITION 338°·0 (w 3)   DISTANCE 2″·5 (w 2)   EPOCH 1832·27

A close double star, over the Crab's southern leg; where a line from the upper star in Orion's belt passed under β Canis Minoris, and extended 12° beyond, will find it. Five skips like that from Bellatrix to Betelgeuze will also pick it up. A 7½, lucid white; B 12, pale grey,— and there are several distant stars in the *sf* quadrant, with one in the *np* nearly pointed upon by a line through A and B. This object was discovered by Σ., and thus measured:

Pos. 335°·93 Dist. 2″·357 Ep. 1829·55

CCCXVIII. 22 ♅. VI. MONOCEROTIS.

ÆR 8ʰ 05ᵐ 40ˢ   PREC. + 2ˢ·96
DEC. S 5° 19′·2   —— S 10″·45

POSITION 205°·0 (w 1)   DISTANCE 4″·0 (w 1)   EPOCH 1834·29

A neat but minute double star, in a tolerably compressed cluster on the Unicorn's flank, and lying 14° south-east of Procyon. A 9½, and B 10, both white. This object is in the midst of a splendid group, in a rich splashy region of stragglers, which fills the field of view, and has several small pairs, chiefly of the 9th magnitude. It was discovered by Miss Herschel in 1783, and was classed by ♅. in February, 1786.

CCCXIX. φ¹ CANCRI.

ÆR 8ʰ 16ᵐ 43ˢ   PREC. + 3ˢ·67
DEC. N 28° 25′·0   —— S 11″·26

POSITION 22°·1 (w 4)   DISTANCE 120″·0 (w 1)   EPOCH 1830·99

A star with a distant companion, in the space above the Crab's northern legs; they lie on the parallel with β Tauri and Pollux, at about one-third of the distance eastward of the latter. A 6½, pale white; B 11, lilac. This object was registered 109 ♅. VI., in 1782; but no measures were given. It was, however, assiduously attacked by Sir James South, who obtained these results:

Pos. 21° 48′ Dist. 120″·945 Ep. 1825·13

A small quantity of proper motion has been detected in A, the amount of which is thus stated:

$$P.... \mathbb{R} - 0''{\cdot}07 \quad \text{Dec.} - 0''{\cdot}10$$
$$B.... \quad - 0''{\cdot}02 \quad\quad - 0''{\cdot}10$$

CCCXX. φ^2 CANCRI.

$$\mathbb{R} \quad 8^h 17^m 06^s \quad \text{Prec.} + 3^s{\cdot}64$$
$$\text{Dec. N } 27° 27'{\cdot}2 \quad\quad —— \text{S } 11''{\cdot}29$$

Position 212°·5 (w 9)   Distance 4″·8 (w 9)   Epoch 1833·25

———— 213°·9 (w 9)   ———— 4″·8 (w 9)   ——— 1843·19

A close double star, above the Crab's northern legs, where it may be fished up by the alignment of ϕ'. A 6, and B $6\frac{1}{2}$, both silvery white. This very pretty object is 40 ♓. II.; and the substantial agreement of all the measures indicate its fixity. Piazzi briefly says, "videtur duplex." The following are the registered results obtained by the astrometers, who preceded me:

	Pos.	Dist.	Ep.
♓.	213° 18′	5″·500	1782·09
H. and S.	211° 13′	5″·514	1822·48
Σ.	212° 01′	4″·563	1829·45

This star seems to have a spacial movement of similar value and direction with ϕ'. They may be in connexion.

CCCXXI. υ^1 CANCRI.

$$\mathbb{R} \quad 8^h 17^m 08^s \quad \text{Prec.} + 3^s{\cdot}58$$
$$\text{Dec. N } 25° 03'{\cdot}3 \quad\quad —— \text{S } 11''{\cdot}29$$

Position 37°·9 (w 4)   Distance 6″·0 (w 4)   Epoch 1831·17

———— 38°·6 (w 8)   ———— 5″·7 (w 6)   ——— 1837·26

———— 40°·1 (w 9)   ———— 5″·8 (w 8)   ——— 1843·18

A neat double star, on the Crab's northern middle leg; where a line carried from Sirius through Procyon, and extended rather more than as far again into the north-east, will reach it. A 7, pale white; B $7\frac{1}{2}$, greyish. This fine object is composed of Nos. 65 and 66 of Piazzi's Hora VIII., and it is 41 ♓. II. By a comparison of these measures:

	Pos.	Dist.	Ep.
♓.	57° 51′	4″·00	1782·09
H. and S.	37° 47′	6″·04	1822·12

it was inferred that a retrograde annual angular motion of $- 0°{\cdot}514$ had taken place. The subsequent observations, however, of Professor Struve, Sir John Herschel, and myself, afford no support to this supposed rotation; consequently some error must exist in ♓.'s register. And it should be remarked—albeit the process cannot demand implicit reliance—that a sifting of Piazzi's mean places for the mid-epoch, 1800, yields an angle $= 45° 30'$, and a distance of 5″·2.

CCCXXII. 67 P. VIII. CANCRI.

Æ 8ʰ 17ᵐ 20ˢ   Prec. + 3ˢ·23

Dec. N 8° 04'·9   —— S 11"·31

Position 328°·0 (w 2)   Distance 35"·0 (w 1)   Epoch 1837·92

A wide double star, on the Crab's hindmost right leg; it may be found by running a line from the centre of Orion's belt through Procyon, and 14°, or half as far again, beyond. A 6, pearl white; B 13, violet, with a glimpse star preceding it. This very delicate object is 109 Ḥ. v., and was thus first registered:

Pos. 325° 00′   Dist. 35"·40   Ep. 1782·85

CCCXXIII. 72 P. VIII. ARGO NAVIS.

Æ 8ʰ 18ᵐ 09ˢ   Prec. + 2ˢ·59

Dec. S 23° 31'·8   —— S 11"·37

Position 85°·4 (w 6)   Distance 45"·0 (w 3)   Epoch 1830·83

A coarse double star, close upon the compass with which the moderns have furnished the Argo. A 6, red; B 9½, green, which is Piazzi's No. 74, Hora VIII. A ray from Rigel passed below Sirius, and extended 25° to the east-south-east (rather more than as far again), will pick it up. This appears to be S. No. 568, whose micrometrical measures may be thus compared with reductions from the mean apparent places in the Palermo Catalogue, and the whole data yield evidence of fixity:

P.   Pos. 85° 36′   Dist. 48"·00   Ep. 1800·00

S.   85° 00′   40"·64   1825·16

CCCXXIV. 81 P. VIII. MONOCEROTIS.

Æ 8ʰ 20ᵐ 25ˢ   Prec. + 2ˢ·02

Dec. S 1° 59'·5   —— S 11"·53

Position 325°·0 (w 2)   Distance 15"·0 (w 1)   Epoch 1838·69

A very delicate double star, at the root of the Unicorn's tail; lying about 15° to the south-east of Procyon, on the line formed by that star and ζ Tauri, at the tip of the Bull's southern horn. A 7, pale topaz tint; B 11, violet; other stars in the field, of which the brightest and nearest is in the *sp* quadrant, near the vertical. This is one of Σ.'s discoveries, and was thus registered:

Pos. 331° 32′   Dist. 18"·20   Ep. 1828·71

CCCXXV. θ CANCRI.

Æ 8ʰ 22ᵐ 28ˢ   Prec. + 3ˢ·43
Dec. N 18° 37′·9   —— S 11″·67

Position 62°·4 (w 3)   Distance 65″·0 (w 1)   Epoch 1833·31

A star with a distant companion, in the middle of the Crab's body. A 5½, yellow; B 9, grey, and there are four other stars in the preceding part of the field, the nearest of which is of the 8th magnitude, and about 75″ distant. This object must be 59 ḤI. v., enrolled in February, 1782, without any measure of position; but a distance of 44″·88, casts a doubt on the identity of B. An imagined line from Sirius passed 3° east of Procyon, and extended nearly as far again to the north-east, will readily fish the object up.

A comparison of the best meridional observations of θ Cancri, show a slight movement in space, of which the several values are near each other, and the direction also coincident; they are:

P.... Æ	− 0″·15	Dec.	− 0″·05
Br...	− 0″·08		− 0″·05
B....	− 0″·02		− 0″·05

CCCXXVI. 108 P. VIII. HYDRÆ.

Æ 8ʰ 27ᵐ 20ˢ   Prec. + 3ˢ·20
Dec. N 7° 10′·5   —— S 12″·02

Position 24°·7 (w 4)   Distance 10″·5 (w 6)   Epoch 1831·19
——— 24°·9 (w 9)   ——— 10″·5 (w 9)   ——— 1839·06

A neat double star, between Hydra's head and Cancer, close to δ Hydræ, which is the preceding of three nearly equi-distant stars in that monster's head, and following Procyon by 16°. A 6, pale yellow; B 7, rose-tint; and there are several glimpse stars in the field, preceded by a 9th-magnitude at some distance in the *sp*. And Piazzi no doubt alludes to B in his note No. 108, Hora VIII. where he says: "Duplex. Comes 0″·5 temporis sequitur ad boream."

This is 49 ḤI. iii., in which there appeared to be a slight movement in the *npsf* direction; but such motion is not confirmed; and as my second epoch was taken under very advantageous circumstances, I have the utmost confidence in the results. The other registered measures are:

	Pos.	Dist.	Ep.
ḤI.	27° 12′	12″·50	1782·06
H and S.	24° 03′	10″·84	1822·64
Σ.	25° 45′	10″·33	1832·95

CCCXXVII. δ HYDRÆ.

Æ 8ʰ 29ᵐ 14ˢ   PREC. + 3ˢ·19
DEC. N 6° 15'·5   —— S 12"·15

POSITION 312°·0 (w 1) DISTANCE 210"·0 (w 1) EPOCH 1832·29

A star with a distant companion, in the head of Hydra. A 4, light topaz; B 9, livid—several other stars in the field. The alignment for δ Hydræ is described in the detail of the preceding object. I was led to a particular scrutiny of this star from an impression which obtained, that a gentleman had detected a *comes* close to A, that is, within 2". Now, as the discovery was stated to have been made with a 5-foot telescope, I may, with my means, safely pronounce it to be an error.

CCCXXVIII. 118 P. VIII. CANCRI.

Æ 8ʰ 29ᵐ 55ˢ   PREC. + 3ˢ·46
DEC. N 20° 14'·0   —— S 12"·20

POSITION 82°·7 (w 4) DISTANCE 55"·8 (w 2) EPOCH 1831·88

A wide double star, on the Crab's body. A 8, and B 8½, both pale white; a third star in the *np* quadrant, at about 3' distance. In general alignment, it will be seen about one-third of the distance from Pollux towards Regulus. This object was No. 1249 of Σ.'s first Catalogue, but was rejected from his grand work. A, small as it is, has been pretty well observed; and both Baily and Argelander assign it a slight proper motion. This, however, is not countenanced by my reductions.

CCCXXIX. 63 ♅. VII. PYXIS NAUTICA.

Æ 8ʰ 30ᵐ 31ˢ   PREC. + 2ˢ·47
DEC. S 29° 23'·4   —— S 12"·24

MEAN EPOCH OF THE OBSERVATION . . . . 1836·10

A compressed cluster, on the Argo's compass-card. The most gathering portion consists of stars from the 10th to the 15th magnitudes, with a glow of star-dust. It was discovered in March, 1793, and is 516 of H.'s Catalogue of 1830, where the shape is aptly likened to a flattened X. This object lying in a region devoid of large stars, is only to be fished up by running a line from the cluster in Orion's sword over Sirius, and extending it twice as far again into the south-east region.

The Mariner's Compass is an introduction of La Caille's; and as if the needle and card were not a sufficient anachronism, the classic old Argo was supplied by Bode with a log and line: so sixty-eight stars were constellated from the *informes*, and assigned to *Pyxis Nautica* and *Lochium Funis.*

CCCXXX.   124 P. VIII. CANCRI.

$\cancel{R}$ 8^h 30^m 39^s Prec. +   3^s·46

Dec. N 20° 06′·3   —— S 12″·27

Position AB 155°·8 (w 3)   Distance 45″·0 (w 2) ⎫
——— AC 242°·0 (w 5)   ——— 90″·5 (w 2) ⎬ Epoch 1830·92
⎭

A coarse triple star, on the Crab's body; with an alignment similar to that of 118 P. viii., before described.   A 7, pale yellow; B 7½, dusky; C 6½, lucid white.   This, though a wide object, forms a very fair scalene triangle, and is No. 571 of Sir James South's Catalogue of 1826, where it is thus registered:

Pos. A B 157° 01′   Dist. 45″·037⎫ Ep. 1825·13
      A C 240° 58′   92″·257⎭

A proper motion is detected in A, and it is not improbable that the three are physically connected, so as to partake in the same movement in space.   These are the latest deductions:

P.... $\cancel{R}$ − 0″·16   Dec. − 0″·01
B.... − 0″·08   + 0″·20

CCCXXXI.   44 M. CANCRI.

$\cancel{R}$ 8^h 31^m 02^s Prec. +   3^s·46

Dec. N 20° 29′·7   —— S 12″·27

Position 331°·0 (w 4)   Distance 150″·0 (w 2)   Epoch 1831·19

A very wide double star in the well known cluster called Præsepe, on the Crab's body, enrolled by Messier on his celebrated List of 103, in 1769.   A 6½, and B 7½, both white, being the *sf* extreme of a wavy line represented by nine small stars.

The Præsepe, metaphorically rendered Bee-hive, is an aggregation of small stars which has long borne the name of a nebula, its components not being separately distinguishable by the naked eye; indeed, before the invention of the telescope, it was the only recognised one, for though that in Andromeda must have been seen, it attracted but little notice till the days of Simon Marius, in 1612.   Whereas the Præsepe in Cancer engaged very early speculation; insomuch that both Aratus and Theophrastus tell us, that its dimness and disappearance during the progressive condensation of the atmosphere, were regarded as the first sign of approaching rain. The group is rather scanty in numbers, but splendid from the comparative magnitude of its constituents, which renders it a capital object for trying the light of a telescope. Yet Galileo discovered this cluster to consist of 36 small stars, when it was supposed that there were only three *nebulous* stars, which emitted the peculiar light. The Præsepe was called by the Arabians *al-ma'laf*, a stall or den; and also *al-nathrah*, the fissure between the Lion's whiskers,—a district which formed the VIIIth Lunar Mansion. See ε Cancri.

An occult line projected from Spica under Regulus, and about 22° beyond the latter, runs through the Præsepe; or it may be found by a ray from the Pleiades being brought mid-way between Procyon and Castor, which will pass over ε, on Castor's knee. A line from Castor through Pollux, and continued about three times the distance between them, also reaches this remarkable cluster.

CCCXXXII.   129 P. VIII. CANCRI.

Æ 8^h 31^m 13^s Prec. + 3^s·46
Dec. N 20° 13'·8   —— S 12''·28

Position 53°·9 (w 8)   Distance 20''·3 (w 8)   Epoch 1833·22
—— 53°·4 (w 9)   —— 20''·5 (w 9)   —— 1839·16

A neat double star close to the Præsepe on the Crab's body; and the "cloudy Præsepe" group is visible to the inquiring eye, at one-third of the distance between Pollux and Regulus. A 7, golden yellow; B 10, blue. This object is the southern member of a triangle, and is preceded by three pairs of stars, all of which were measured by Sir James South. The former registers of this are:

S.   Pos. 52° 58'   Dist. 20''·69   Ep. 1825·14
Σ.   53° 51'   20''·52   1831·31

The reductions of A show some slight aberrations from the common laws of precession, which are thus valued:

P.... Æ − 0''·10   Dec. − 0''·07
B....   + 0''·06   − 0''·07
T....   + 0''·02   − 0''·07

CCCXXXIII.   ε CANCRI.

Æ 8^h 31^m 16^s Prec. + 3^s·45
Dec. N 20° 06'·2   —— S 12''·29

Position 248°·9 (w 4)   Distance 133''·6 (w 4)   Epoch 1830·98

A star with a distant companion, on the Crab's body. A 6½, and B 7, both pale white; and there is a third star in the field, of nearly the same magnitude. There has been some little difficulty in identifying this object, since the letter ε was intended by Bayer to denote the whole nebulous appearance of the Præsepe, where 44 Messier may be termed the leader, and ε the whipper-in.   It follows the triple star S 571, a little south of the parallel, with △ Æ 30ˢ·8; and the principal, A, is Piazzi's No. 130, while B is No. 128, of Hora VIII.   Sir W. Herschel considers this cluster as belonging to a certain nebulous stratum, so placed as to lie nearest to us.   This stratum runs from ε Cancri over 67 Messier; and is under the same alignment as the preceding object, 129 P. VIII.; it is moreover crossed by an imagined ray or line from Procyon to the tail of the Great Bear.

CCCXXXIV.   131 P. VIII. LYNCIS.

Æ   8ʰ 32ᵐ 04ˢ   PREC. +   4ˢ·30
DEC. N 49° 25′·9   —— S 12″·34

POSITION 331°·5 (*7)   DISTANCE 9″·8 (*5)   EPOCH 1833·31

A neat double star, close to the forepaw of Ursa Major; being about 2° west-north-west of ι, a bright star in that paw.   A and B 8½, and both white.   This object was pointed out by Piazzi in these words: "Duplex. Comes ejusdem magnitud. sequitur 0″·2 temporis, 7″ ad austrum;" but I have altered the quadrant, the stars being of the same brightness, to meet the following measures:

S.   Pos. 330° 58′   Dist. 10″·31   Ep. 1825·05
Σ.   331° 22′   9″·62   1830·75

CCCXXXV.   δ CANCRI

Æ   8ʰ 35ᵐ 35ˢ   PREC. +   3ˢ·42
DEC. N 18° 44′·4   —— S 12″·58

POSITION 163°·0 (*1)   DISTANCE 25″·0 (*1)   EPOCH 1838·26

A very delicate double star, under the Crab's mouth.   A 4½, straw colour; B 15, blue, only seen by glimpses.   This was discovered by H., sweep No. 457, and is situated nearly equatoreally between two distant stars: pursuant to my plan, I assumed it, from Piazzi, as 4·5 in brightness; but H., in his table of the comparative lustre of the individuals of Cancer, in the *Philosophical Transactions*, classes both it and γ as of the 4th magnitude, a degree in which I should rate them myself.   A ray from Rigel glanced to the north-east through β Canis Minoris, and carried nearly as far again, will find it at about 2½° south-south-east of the Præsepe.   A has certainly a proper motion in declination, which I am inclined to rate at − 0″·22 per annum; but that in Æ is inappreciable by my results.   Other investigators, however, value them thus:

P.... Æ − 0″·10   Dec. − 0″·21
B ...   + 0″·05   − 0″·24

δ Cancri is the southernmost of the stars called *Aselli* by the Romans, and ὄνοι by the Greeks; γ Cancri being the northern one; and they may very readily be found by their connexion with the Præsepe, which they closely follow in a line, one to the north and the other to the south. See 44 M, above. These stars form a part of the VIIIth compartment of the Lunar Zodiac. Manilius calls them *Jugulæ;* but the Arabians, borrowing Ptolemy's term, designated them *Al-himáraïn*, the two asses, whence Dupuis, despising the story about the insane Bacchus, conjures up the emblem of the tribe of Issachar: " Le Cancer," dit il, " où sont les étoiles appelées les ânes, forme l'empreinte du pavillon d'Issachar, que Jacob assimile à l'âne." As with the Præsepe, their dimness was anciently held to be an infallible prediction of rain.

Cancer is one of the ancient 48 constellations; but as its lucida is only of the 4th magnitude, it is neither conspicuous nor brilliant, whence it was of old represented of a black colour and without eyes; but Bartschius, in his *Planisphærium Stellarum*, 1661, and some others of still later date, converted it to a lobster. Indeed, mythology even seems to apologize for placing so poor an asterism on the solar rail-road, by stating that ox-eyed Juno exalted the creature, for the inconsiderable service of pinching the toes of Hercules in the Lernæan marsh: whence Columella designates it *Lernæus.* Yet, on the whole, there is scarcely one of the signs of the zodiac that has been the subject of more attention than Cancer, nor scarcely any one better determined. For the reason we have given under Leo, the Lion and the Crab were assigned as mansions of the sun and moon; and Cancer being also famous, according to Chaldaic and Platonic philosophy, as the supposed gate by which souls descended from heaven into human bodies, it, of course, obtained favour among mythologists. But the astrologers saw nothing but its " watery triplicity," and pronounced that all men born under it, shall be short, effeminate, and sickly. The successive enumerations of its component members, as optical means have progressed, are:

Ptolemy	. . . 13 stars	Kepler	. . . . . 17 stars
Copernicus	. . 13	Hevelius	 29
Tycho Brahé	. . 15	Flamsteed	 83
Chr. Clavius	. . 16	Bode	 179

Cancer, as the summer solstice, introduces the longest day in our hemisphere, and names the North Tropic; for as that " aisword beste," the Crab, walks obliquely, it is figurative of the sun's retrogression on arriving at its greatest northern declination in this sign. See *a*' Capricorni. It forms the fourth of the zodiacal signs, and designates one of the quadrants of the ecliptic : its name in Arabian records is simply *Al-sertán*, the crab, from the more ancient καρκῖνος and ὀκτάπους. In the fine copy of Albumazar's *Introductio in Astronomiam*, 1489, in the Bibliotheca Lambethana, Cancer is represented as a large crayfish; and in Lubienietzki's *Theatrum Cometicum*, 1667, it is figured as a huge lobster, between the tail of which and Gemini is a small shrimp-like companion, designated Cancer Minor.

An ardent antiquary, in his late inquiry into the meaning of *Choir Gaur*, said to be the British name of Stone-henge, is anxious to prove that those vestiges are the relics of a vast astronomical machine, or sort of Orrery. Overlooking the word χορὸς, he tells us that Calasius, in his Hebrew lexicon, translates "the radical *chor*, or *cor*, concha marina; which may be called Cancer, the crab-shell, from resembling more the quire of a church than any other." But why more than if it were rendered *ostrea?* And the nonsense is formally printed!

CCCXXXVI. ι CANCRI.

$\mathbb{R}$ 8^h 37^m 00^s PREC. + 3^s·65

DEC. N 29° 20′·4   —— S 12‴·68

POSITION 307°·8 (*w* 9)   DISTANCE 30″·1 (*w* 6)   EPOCH 1836·21

A double star, at the end of the Crab's northern claw; Piazzi's No. 158, Hora VIII., erroneously marked *v* in the Palermo Catalogue. A 5½, pale orange; B 8, clear blue, the colours finely contrasted. A line from ε Geminorum, carried through Pollux, and extended something more than as far again to the east-north-east, will readily find it; and its identity will be instantly made out by its forming another line to the south-west with Procyon and Sirius. It is 52 ⊞. IV., and the several observations made since its registry shew that no material change has occurred in a lapse of 54 years. The following are the other results:

⊞.	Pos. 309° 54′	Dist. 29″·90	Ep. 1782·11
H. and S.	307° 42′	29″·39	1822·26
Σ.	307° 06′	30″·46	1828·04

Though the values are small, there is a decided spacial movement in ι Cancri, of which the most authentic statements are:

P....$\mathbb{R}$ − 0″·12	Dec. − 0″·06
B.... + 0″·05	− 0″·07

CCCXXXVII. 160 P. VIII. HYDRÆ.

$\mathbb{R}$ 8^h 37^m 16^s PREC. + 3^s·03

DEC. S 2° 01′·4   —— S 12‴·70

POSITION 258°·9 (*w* 8)   DISTANCE 4″·9 (*w* 9)   EPOCH 1833·08

A neat double star, in the space interposed between Hydra's head, and the Unicorn's tail. A 7, silvery white; B 8, smalt blue. This object is No. 160 and 159 of Piazzi's Hora VIII., the last of which was only observed twice by him. It was registered No. 1270, Class II. of the Dorpat Catalogue. The several measures of this star have been remarkably coincident and satisfactory; being

S.	Pos. 258° 26′	Dist. 4″·95	Ep. 1824·53
Σ.	259° 06′	4″·70	1830·98

In $\mathbb{R}$ it slightly precedes ε Hydræ, the middle star in that creature's head being 10° to the southward of it, and nearly due south of the Præsepe.

CCCXXXVIII. ε HYDRÆ.

| Æ | 8ʰ 38ᵐ 18ˢ | Prec. + | 3ˢ·19 |
| Dec. | N 7° 00′·2 | —— S | 12″·77 |

Position 198°·4 (w 7)	Distance 3″·4 (w 4)	Epoch 1837·11
—— 199°·1 (w 9)	—— 3″·5 (w 6)	—— 1839·23
—— 203°·2 (w 9)	—— 3″·6 (w 8)	—— 1843·14

A secondary Greenwich star, double, and the middle one in the head of Hydra. A 4, pale yellow; B 8½, purple. A long ray from Betelgeuze brought over Procyon, and carried nearly as far again, finds it about 14° south of the Præsepe; and nearly mid-way between Pollux and Cor Hydræ. This beautiful object was discovered by Σ., and is No. 1273 of the great Dorpat Catalogue, under these measures:

Pos. 195° 34′ Dist. 3″·21 Ep. 1830·60

But it must be remembered that Σ.'s measures did not arrive in this country till the autumn of 1837, though a catalogue of places, without angles or distance, had been nearly ten years in circulation. Mr. Dawes was therefore unacquainted with any other measures of this star, when his own observations indicated a change both in angle and distance, and this detection was creditable both to himself and his instrument. His whole series was as follows:

Pos. 195° 16′	Dist. 4″·34	Ep. 1831·13
197° 36′	4″·26	1832·20
199° 10′	3″·65	1834·00

This accurate observer wished me to bestow some pains on the star, because he suspected it of rotation. " Indeed," he remarked, " were the small star visible fifty years ago, as it is now, it never could have escaped the scrutinizing eye of Sir W. Herschel." On this appeal, my observations, especially at the last epoch, were rigidly attended to, and the results corroborate the orbital motion. The distance seems decreasing, but this conclusion is not yet so evident as that of the angle, which, on weighing all the data, may have an annual progress of about + 0°·8 per annum, or a circuit of 4½ centuries.

CCCXXXIX. 67 M. CANCRI.

| Æ | 8ʰ 42ᵐ 26ˢ | Prec. + | 3″·29 |
| Dec. | N 12° 23′·6 | —— S | 13″·05 |

Mean Epoch of the Observation . . . . 1836·19

A rich but loose cluster, at the root of the Crab's southern claw; where a line from Rigel through Procyon, into the east-north-east, will find it about 5° north of ε Hydræ.   It consists principally of a mass of stars of the 9th and 10th magnitudes, gathered somewhat in the form of a Phrygian cap,

followed by a crescent of stragglers. It was first registered by Messier in 1780, and resolved by Sir W. Herschel in 1783. The place here given was obtained by differentiation from δ Cancri: and though differing so largely in Æ from Messier, it is evidently his object. With a power of 157 on his large telescope, Sir William Herschel saw above 200 stars at once, in the field of view; and on gazing at it with my refractor, of far inferior light, but excellent definition, charged also with a power magnifying 157 times, as shown by a dynameter, the object appears as herewith represented.

CCCXL.   200 ℍ. I. LEONIS MINORIS.

Æ   8ʰ 42ᵐ 44ˢ   Prec. +   3ˢ·75
Dec. N 34° 00′·6   —— S 13‴·06

Mean Epoch of the Observation . . . . 1836·29

A bright oval nebula, between Lynx and Cancer, but in a confusing gap given to Leo Minor.   It trends *nf* and *sp*, with a splendid centre, and is closely followed by a 9th-magnitude star, which is in a line with a coarse telescopic pair in the *np* quadrant, and the preceding of a trio in the *sf*.   It was registered by ℍ., in February, 1787, and to his powerful eye and instrument appeared " very beautiful, 8′ long, and 3′ broad."

The out-door observer may find this object, and it is worth fishing for, by directing his telescope, under a very moderate power, to about 16° north by east of the Præsepe, which forms the imaginary centre of three radii described by Aldebaran and Castor, Betelgeuze and Pollux, and Sirius and Procyon.

CCCXLI.   15 HYDRÆ.

Æ   8ʰ 43ᵐ 43ˢ   Prec. +   2ˢ·95
Dec. S   6° 34′·9   —— S 13‴·13

Position A B 345°·0 (*w* 1)   Distance 40″·0 (*w* 1) ⎫
—————— A C   43°·0 (*w* 1)   —————— 55″·0 (*w* 1) ⎬ Epoch 1835·89
                                                    ⎭

A most delicate triple star, between the Unicorn's tail, and the first bend of Hydra.   A 6½, pearl white; B 12, and C 13, both purplish;

other stars in the field. It is located in a region utterly destitute of large stars, to the south of Hydra's head, being about 12° east of Cor Hydræ; its place is therefore nearly pointed by a ray passed from β Canis Minoris through Procyon, and carried six times as far into the south-east void. The two nearest individuals of this object form 120 ♅. v., and were thus registered:

Pos. 340° 00′   Dist. 43″·03   Ep. 1782·99

CCCXLII.   σ² CANCRI.

Æ 8ʰ 44ᵐ 28ˢ   Prec. +   3ˢ·68
Dec. N 31° 10′·9   ——   S 13″·28

Position 335°·0 (w 6)   Distance 1ˢ·4 (w 3)   Epoch 1838·26

A close double star, over the Crab's northern claw. A 5½, white; B 7, yellow. This star is designated ι² by Flamsteed, and he is followed by ♅. H. S. and Σ.; but there is only one star distinguished by that letter in Bayer's Map, which is 48 Cancri, and Piazzi's No. 158, Hora VIII. To preserve identity it may be mentioned that the object before us is No. 192 of the Palermo Catalogue; and that Mr. Baily has restored the σ in his edition of Flamsteed: a most valuable contri-bution to pure astronomy.

For want of convenient naked-eye stars in this vicinity, the searcher for the point in dispute, must remember that it is about 18° east of and on the parallel of Castor; where its place is sufficiently indicated by a long line drawn up from Sirius in the south-west, and passed over Procyon.

The other measures of this exquisite object, which is 30 ♅. i., are indicative of fixity, though not co-incident, being as follows:

	Pos.	Dist.	Ep.
♅.	338° 12′	0″·85	1782·27
H. and S.	340° 11′	1″·89	1822·15
Σ.	333° 18′	1″·51	1829·71

CCCXLIII.   17 HYDRÆ.

Æ 8ʰ 47ᵐ 39ˢ   Prec. +   2ˢ·94
Dec. S 7° 21′·8   ——   S 13″·39

Position 358°·9 (w 9)   Distance 4″·8 (w 9)   Epoch 1830·29
——— 357°·8 (w 9)   ——— 4″·5 (w 9)   ——— 1838·12

A close double star, between the Unicorn's tail and Hydra's heart. A and B, both 7½, and both white. This beautiful object is 77 ♅. ii., and Nos. 214 and 215 of Piazzi's Hora VIII. A reduction of these and the other measures are as follows, and which, compared with my own, are sufficient proof of fixity:

	Pos.	Dist.	Ep.
Ḥ.	356° 30′	5″·00 +	1782·99
P.	360° 00′	4″·50	1800·00
H. and S.	356° 08′	5″·72	1822·14
Σ.	358° 50′	4″·33	1831·59

To align this object, draw a right angle at Procyon to a line brought from Sirius, and it will meet 17 Hydræ at 11° west of Cor Hydræ, or rather a greater distance than that between the two first-named stars.

CCCXLIV.   ι URSÆ MAJORIS.

Æ   8ʰ 48ᵐ 14ˢ   Prec. +   4ˢ·13

Dec. N 48° 39′·9   —— S 13″·43

Position 348°·0 (w 2)   Distance 12″·0 (w 1)   Epoch 1839·12

A Greenwich star of 1830, double, and in the Great Bear's right fore-paw; where it may be readily identified by shooting a west-south-west ray from β, the southern-most of the two pointers, which will pass through θ at nearly 12°, then at half that distance further on will strike ι. A 3½, topaz yellow; B 13, purple. This fine object was discovered by H., being No. 2477 of his 20-foot Sweeps; and as I was desirous of recent micrometric measures for comparison, Mr. Challis was kind enough to send me the following, taken expressly with the great Northumberland equatoreal:

Pos. 350° 02′   Dist. 10″·68   Ep. 1841·19

There is a star of the 8th magnitude preceding at an angle = 268°·7, with Δ Æ = 52ˢ·9, which is that mentioned by Piazzi in Note 212 of Hora VIII. A is charged with very sensible proper motions, to the following amount:

		Dec.
P.... Æ − 1″·05		− 0″·32
B.... − 0″·63		− 0″·28
A.... − 0″·66		− 0″·29

From strong impressions on his mind, confirmed by what he saw in the Southern Hemisphere, Sir John Herschel supposes that certain very minute companions to stars may possibly shine by reflected light; and ι Ursæ is one of those instanced as having a dull satellite.

This star has obtained the name of Talita, the third vertebra, the meaning of which is not quite clear. Ulugh Beigh has it *Al Phikra Al Thálitha*, perhaps for *Al Kafzah al-thálithah*, the third spring, or leap, of the ghazal. This was applied by the Arab star-gazers to ι and κ; the first leap, *al-úla*, being ν and ξ; and the second, *al-thániyah*, was δ and μ. As Ideler found this as well in Tízíní's Catalogue as on the Cufic globes at Dresden, he adopted the antelope, though the purport of it is not very obvious. Hyde contrives to render it either the vertebræ of the Great Bear, or the cavity of his heel: but, as Ideler remarks, the Arabic names of the stars indicate an ancient sphere, different from that which their later astronomers modified from Ptolemy.

CCCXLV. σ⁴ CANCRI.

Æ 8ʰ 51ᵐ 35ˢ   Prec. + 3ˢ·70
Dec. N 32° 52′·4   —— S 13″·64

Position 136°·4 (w 9)   Distance 4″·8 (w 7)   Epoch 1837·13

A close double star, *nf* the Crab's northern claw. A 6, lucid white; B 9, sky blue. This beautiful object is No. 1298 of the great Dorpat Catalogue, and, from the following measures, and my own, appears to have been pretty constant in the lapse of twelve years:

S.	Pos. 135° 25′	Dist. 4″·85	Ep. 1825·04
Σ.	137° 47′	4″·60	1831·16

There being some confusion in identifying Bayer's *sigmas*, it is as well to state that this is Piazzi's No. 226, Hora VIII.; and a ray from Sirius through Procyon, carried nearly double that distance into the north-north-east, will find it.

CCCXLVI. σ² URSÆ MAJORIS.

Æ 8ʰ 56ᵐ 13ˢ   Prec. + 5ˢ·42
Dec. N 67° 46′·7   —— S 13″·94

Position 262°·4 (w 7)   Distance 5″·0 (w 5)   Epoch 1835·27

A neat double star, in the Bear's forehead. A 5½, flushed white; B 9½, sapphire blue. This delicate object is 54 Ⱨ. III.; and from the measures previous to my own, viz.:

Ⱨ.	Pos. 283° 00′	Dist. 7″·95	Ep. 1782·42
Σ.	263° 33′	4″·59	1832·14

a retrograde orbital motion was indicated, as the secondary had changed its quadrant from *np* to *sp*. My measures are of use to the epoch, for though the feebleness of the small star rendered them difficult, they were so well under the power of the instrument, that they may be depended upon. Professor Argelander has given the large star a proper motion of − 0″·099 in Æ, and − 0″·119 in declination; and further meridian observations will very probably bear him out.

There is only one star designated by the letter σ, in Bayer's Map: "but," says Mr. Baily, in his notes to the British Catalogue, "as there may be a doubt whether such letter belongs to 11 Ursæ Majoris, or to the present star, Flamsteed has annexed it to each, which I have therefore retained." The Arabian astronomers classed these stars, together with ρ, π, δ, ο, and others in the eyes, ear, and nose of the Bear, as *Al-dhibá*, or *zibá*, the antelopes. It will be easily fished up nearly mid-way of a line from Polaris to Talita, or ι Ursæ Majoris, where it is the north-east vertex of a small triangle formed by the aforesaid two *sigmas* and ρ.

CCCXLVII. θ HYDRÆ.

Æ 9ʰ 06ᵐ 02ˢ   PREC. +   3ˢ·12
DEC. N   2° 59'·2   —— S 14"·54

POSITION 168°·6 (*w 3*)   DISTANCE 45"·0 (*w 1*)   EPOCH 1835·27

A wide double star in the fore part of Hydra's neck. A 4½, pale
yellow; B 12, ash-coloured. This difficult object is 54 ♓. v., who gave
no measure, but remarked that the stars were excessively unequal; and
that the small one, "a point," was about 75° *sf*, and nearly a minute
distant. The assigned proper motion to A has been:

$$P....\, Æ + 0''·10 \qquad Dec. - 0''·37$$
$$B....\qquad + 0''·20 \qquad\qquad - 0''·32$$
$$A....\qquad + 0''·15 \qquad\qquad - 0''·33$$

This star, with δ, ε, η, ξ, ω, and σ, the first seven of the Water-serpent,
are termed by the Arabians *min el-a'zal,* of the unarmed. It will be
met from the west by a line drawn from Betelgeuze through Procyon,
till it meets another from the north-east, running from δ Leonis
through Regulus. The point of intersection is visible enough.

CCCXLVIII. 38 LYNCIS.

Æ 9ʰ 08ᵐ 52ˢ   PREC. +   3ˢ·76
DEC. N 37° 28'·6   —— S 14"·71

POSITION 241°·6 (*w 9*)   DISTANCE 2"·8 (*w 9*)   EPOCH 1832·35

A close double star, in the animal's tail. A 4, silvery white; B 7½,
lilac. This beautiful and delicate object is 9 ♓. ɪ., and has thus been
severally measured:

		Pos.	Dist.	Ep.
♓.		244° 09'	2"·00±	1780·90
H. and S.		242° 40'	2"·89	1822·46
Σ.		240° 14'	2"·70	1829·17

These results are not so coincident as a star bearing illumination so
well as this does ought to have yielded; and I cannot but deem that
the proper motion assigned by Piazzi is somewhat countenanced by the
small change which has occurred in the angle of position, although it is
not entered in the Lists of Brioschi or Argelander. The amount given
in the Palermo Catalogue, and by Mr. Baily, may be thus stated:

$$P....\, Æ + 0''·28 \qquad Dec. - 0''·20$$
$$B....\qquad - 0''·02 \qquad\qquad - 0''·04$$

I was induced to be the more particular with this star, firstly,
because H. and S. relate that a doubt had arisen whether 9 ♓. ɪ., were
the same with 38 or 39 of the British Catalogue, both of which are
neatly double: and secondly, because Baron de Zach, in his discussions
of Σ.'s early observations (*Correspondance Astronomique,* vol. viii.,
p. 413), inferred that the angle had considerably augmented since the
year 1780. This star I distinctly saw double—as a choice test object—

in the telescope of Piazzi's circle, in March, 1814, though he has not entered it among his "binæ." It is to be found where a line from Regulus, carried over ε Leonis, and 23° further to the north-north-west, meets another from ζ and γ Ursæ Majoris.

CCCXLIX. 205 ♅. I. URSÆ MAJORIS.

$\mathcal{R}$ 9^h 10^m 54^s PREC. +   4^s·21
DEC. N 51° 40'·5   —— S 14"·83

MEAN EPOCH OF THE OBSERVATION . . . . 1836·29

A bright nebula in the animal's right fore-leg, of a pale creamy whiteness, with several bright stars in the northern part of the field. It is large, nucleated, and elliptical, with its major axis lying *np* and *sf*, about 4'. It was first classed by ♅. in March, 1788, and is No. 584 of his son's Catalogue. Differentiated with θ Ursæ Majoris to obtain its mean apparent place: it lies 1°½ west-south-west of θ Ursæ Majoris, and nearly on the line described by ι, θ, β, and δ of that constellation.

CCCL. 39 LYNCIS.

$\mathcal{R}$ 9^h 11^m 35^s PREC. +   4^s·14
DEC. N 50° 13'·2   —— S 14"·87

POSITION 319°·2 (w 5)   DISTANCE 5"·9 (w 4)   EPOCH 1832·91
——— 319°·5 (w 9)   ——— 6"·2 (w 9)   —— 1839·18

A neat double star, on the Great Bear's right leg, and about 2°½ to the south-south-west of θ in that constellation; whence, but for the map-makers, it must have pertained to Ursa Major. A 6½, lucid white; B 9, sapphire blue. It has required some trouble to trace unequivocally the identity of this object. Mr. Baily diminished its $\mathcal{R}$ above 3½', in order to correspond with modern observations, as there appeared to be some error in the British Catalogue reduction of Feb. 16, 1704. Piazzi, Note 47 to Hora IX., says that, by Flamsteed and La Lande, the proper annual motion in $\mathcal{R}$ of this star would be − 0"·8, which he vainly tried to confirm from Bradley, who observed it on March 14, 1757; and he therefore concluded that some error of $\mathcal{R}$ had crept in here, and at 40 Lyncis.

I gave great attention to the subject; and as 40 Lyncis is decidedly a single star, I make no doubt whatever that this is 84 ♅. III.; nor of H.'s having made a wrong entry, in marking it No. 1334 of the Dorpat Catalogue. It is No. 596 of S.; but indeed the original measures undeniably confirm this, being:

Pos. 318° 12' Dist. 7"·18 Ep. 1782·87

P 2

CCCLI.   21 URSÆ MAJORIS.

Æ   9ʰ 14ᵐ 15ˢ   Prec. +   4ˢ·32
Dec. N 54° 41'·9   —— S 15"·03

Position 311°·2 (*w* 4)   Distance 5"·9 (*w* 5)   Epoch 1831·32
———— 310°·9 (*w* 9)   ———— 6"·3 (*w* 9)   —— 1839·17

A neat double star, on the Bear's left fore-knee, where an occult line from Polaris to the west of θ will find it nearly on the parallel of declination with γ Ursæ Majoris. A 8, silvery white; B 9, violet tint, with a third star at a distance in the *np* quadrant. This fine object is 73 Ḥ. ɪɪ.; and as that observant astronomer gained these results:

Pos. 306° 45'   Dist. 5"·00 ±   Ep. 1782·87
317° 38'   ...   ...   1802·39

he concluded that a change of 10° 52' had taken place, in 19 years and 184 days. This, however, is not confirmed, and it is probable that the 317°, of 1802, ought to have been registered 307°. The other recent observations have been:

H. and S.   Pos. 309° 02'   Dist. 6"·47   Ep. 1822·12
Σ.   310° 56'   5"·69   1830·29

A false alarm having been sounded about this star, I attacked it again in 1839, and hope I have settled the question of 1802, because all the circumstances, observatorial and atmospherical, were so truly favourable, that I can safely assign the highest weight to the results.

CCCLII.   137 Ḥ. I. LYNCIS.

Æ   9ʰ 14ᵐ 32ˢ   Prec. +   3ˢ·69
Dec. N 35° 11'·9   —— S 15"·04

Mean Epoch of the Observation . . . . 1836·29

A bright nebula, on the fore-paws of Leo Minor; a line from ϵ under γ and through λ Ursæ Majoris, carried 16° beyond, marks its site. It is round, pale white, and sparkling at the centre; nearly all the stars in the field precede it, especially a yellow 7th-magnitude, which lies on the parallel. This may be liable to error of identity, if Ḥ. mistook 41 Lyncis for 40; but the place here given will be found, I hope, tolerably accurate. At all events, it is No. 593 of H.'s Catalogue of 1830. It was discovered by Ḥ. on the 28th March, 1786, and he remarked that the *chevelure*, or additional faint circular nebulosity surrounding the nucleus, was 3' in diameter: by my equatoreal, of course, such a magnitude could not be inferred. Still it was well seen—and pretty fairly observed—in the meridian instruments.

CCCLIII.   65 P. IX. HYDRÆ.

ÆR 9^h 14^m 56^s Prec. + 3^s·13
Dec. N   4° 10'·8   —— S 15"·07

Position 309°·8 (w 9)   Distance 21"·6 (w 9)   Epoch 1836·29

A neat double star, at the back of Hydra's neck; it is 12° north by west of Cor Hydræ, and 16° west-south-west of Regulus. A 8 and B 9, both white; the last being Piazzi's 64 Hora IX., and there are two other small stars in the *sf* quadrant. This is so easy an object under the lamp, that the several measures are remarkably coincident, and afford evidence that the components are merely optical. The best for comparison are:

S.	Pos. 310° 25'	Dist. 22"·16	Ep. 1825·01
Σ.	310° 28'	21"·29	1832·23

CCCLIV.   41 LYNCIS.

ÆR 9^h 18^m 09^s Prec. + 3^s·97
Dec. N   46° 17'·9   —— S 15"·25

Position 160°·8 (w 9)   Distance 86"·8 (w 5)   Epoch 1832·26

A wide pair of stars under the Great Bear's foot. A 6½ and B 8½, both bluish, being Nos. 78 and 81, P. Hora IX., and therefore classed after his magnitudes, though A seems to merit a higher rate. This object is No. 598 of Sir James South's Catalogue, who has registered it for 55 ꟸ. iv.; but that star is upwards of 3^m to the *nf*, with its acolyte at an angle of about 320°, and 15" from its primary. 41 Lyncis has shown a slight aberration from the common laws of precession, and the following registered amount will show that it has been narrowly watched:

P....	ÆR − 0"·06	Dec. − 0"·19
B....	+ 0"·05	− 0"·13
A....	− 0"·03	− 0"·16
T....	+ 0"·16	− 0"·02

The measures of 41 Lyncis are remarkably accordant between Sir James South and myself; and a reduction of Piazzi's mean apparent places affords a fresh proof of the excellence of his practice in observing, and his ability in obtaining conclusions.

P.	Pos. 161° 00'	Dist. 87"·00	Ep. 1800·00
S.	161° 31'	86"·C5	1824·72

Such a result from an altitude and azimuth circle, so little expected to be placed in juxta-position with rigorous metrical observations, is a gratifying warrant for the value of the Palermo Catalogue.

CCCLV. 23 URSÆ MAJORIS.

Æ 9ʰ 18ᵐ 50ˢ   Prec. + 4ˢ·82

Dec. N 63° 45'·4   —— S 15"·29

Position 271°·8 (w 9)   Distance 23"·0 (w 6)   Epoch 1833·26

A neat double star, in the Great Bear's neck; at rather better than one-third the distance between θ and Polaris. A 4, pale white; B 9½, grey. This is 29 Ⱨ. iv., and from the earlier measures seemed to be constant in angle, but to have had a slight increase of distance: yet from bearing illumination badly, the latter also is very questionable. The results of my predecessors are:

Ⱨ.	Pos. 273° 14'	Dist. 19"·22	Ep. 1781·32
H. and S.	270° 33'	27"·33	1822·13
Σ.	272° 45'	22"·81	1830·61

Piazzi attributed proper motions with minus signs to this star; but Mr. Baily, from most careful investigation, has altered both the values and directions; they are:

P.... Æ − 0"·09   Dec. − 0"·04

B....   + 0"·29   + 0"·04

CCCLVI. α HYDRÆ.

Æ 9ʰ 19ᵐ 44ˢ   Prec. + 2ˢ·95

Dec. S 7° 58'·1   —— S 15"·34

Position 155°·0 (w 6)   Distance 285"·0 (w 2)   Epoch 1833·16

A standard Greenwich star, with a distant companion, in Hydra's heart. A 2, but suspected of variability, orange tint; B 10, pale green, the latter perhaps being one of the two stars noted in the remarks to 111 Ⱨ. vi., but of which there are no measures given. Mr. Challis kindly re-examined this at my request, in 1841, with the Northumberland equatoreal; yet the difference was hardly greater from my results, than what might almost be imputed to the proper motion of the large star, the amount of which may be more than what is thus registered:

P.... Æ − 0"·15   Dec. − 0"·05

B....   + 0"·02   + 0"·03

A....   − 0"·02   + 0" 05

This star has been known as Cor Hydræ, and Lucida Hydræ, among the Latins; and also as Alphard, i. e. the Arabic Al-fard, the solitary, so termed perhaps, because there is no competitor in brightness near it. In Ulugh Beigh we find it designated Unuk al Shugja, which perhaps were better written Unk-esh-shujá', the serpent's neck. But it has been most familiarly known as Hydra's heart. It is readily found by drawing a line from γ and δ Ursæ Majoris, the two last in the square—

southwards by γ Leonis and through Regulus; or, as expressed in galley-stave heroics:

Thro' Cancer's sign, whence no bright stars	distinguish'd light impart,
Pollux from Castor leads you down	to hideous Hydra's heart.

Ύδρος, Hydra, *seu Serpens Aquaticus*, the water-snake, with *corvus* and the *crater* upon it, is figured after the same manner in most of the delineations, from the grand Farnese Globe and the MS. of Cicero's Aratus, down to Mr. Carey's maps and Miss Whitwell's drawings. While some term it Hydra, others use the designation Hydros, and a third party Anguis; but Riccioli names it *Al-havich* in his Almagest, with the concurrence of his brother Jesuit, Athanasius Kircher. In addition to these, it has been called Asina, Coluber, Anguis, Sublimatus, Furiosus,—in a word, all manner of names. In the Alphonsine Tables it is *Asuia*, whence the *Asina* cited by Bayer*. It is, indeed, a strange asterism, trailing to such a length, that but for the subdivisions afforded by the cup and the crow, it would seem interminable: the parts so treated become Hydra, Hydra et Crater, Hydra et Corvus, and Hydræ continuatio. The head of the reptile is to the south of Cancer, from whence its body winds eastward with many curves, under Leo and Virgo as far as Libra. Mythology calls it the Lernæan serpent; but the Mosaicists, taking the name literally, see in Hydra the flood, in Corvus Noah's raven, and in Crater the cup out of which the patriarch sinned with the juice of the grape. In the rare MS. of Cicero's Aratus, of the second or third century, Hydrus, Crater, and Corvus, are lumped together, with S̄. OMN̄. STELLÆ. XXXVI. It was one of the old 48 constellations, and has thus been gradually increased in constituents:

Ptolemy	. . . 27 stars	Bullialdus .	. . 33 stars
Tycho Brahé	. . 24	Hevelius .	. . 31
Clavius	. . . 34	Flamsteed	. . 60
Bayer	 29	Bode .	. . . 370

---

* These Tables are so intimately connected with the science of the Middle Ages, and are so often quoted, as to merit especial mention. Alphonso the Wise, prince of Castile, assembled the best astronomers of the age at Toledo, for the purpose of examining and correcting Ptolemy's Μεγάλη Σύνταξις. After four years of incessant application this undertaking was completed at an expense, we are told, of 400,000 ducats, though other more probable accounts reduce it to 40,000: either sum spent in the pursuit of knowledge, at that day, deserves immortality. But the chronicles leave us greatly in the dark; for they also say that the chief adviser was a Christian, a Jew, or a Moor; and that Alphonso died at 58, 63, or 81 years of age.

The epoch of the Alphonsine Tables was fixed to the 30th of May, 1252, the day of the prince's accession to the throne; and he himself wrote the preface to them. They are usually regarded as a mere compilation, but the following fac-simile of the head-piece to the black-letter edition of 1483, would imply something of practice: for there is King Alphonso,—and we may suppose that Isaac Hazen, Aben Said, Alcabitius, and Abn Ragel are admiring his majesty's grand armillary sphere.

CCCLVII. ω LEONIS.

| Æ | 9h 19m 53s | PREC. | + | 3s·22 |
| DEC. N | 9° 45'·0 | —— | S | 15"·35 |

POSITION	160°·0 (w 1)	DISTANCE	0"·5	(w 1)	EPOCH	1832·11
——	round (w 8)	——	round	(w 8)	——	1834·25
——	355°·0 (w 2)	——	elongated	(w 1)	——	1839·33
——	193°·0 (w 3)	——	0"·3	(w 1)	——	1843·14

An exquisite close double star, before the Lion's left fore-foot; being 26 ♄. ɪ., and one of the "pervicinæ" of Σ. A 6½, pale yellow; B 7½, greenish; at times both stars looking yellow. I am the more particular in stating these colours, as I was drawn to the subject by ♄.'s remark that it would be curious " if a considerable difference in the colours could have led us to discover which of the two stars is before the other! But the far greatest part of their diameters being spurious, it is probable that a different coloured light of two stars would join together, where the rays of one extend into those of the other; and so, producing a third colour by the mixture of it, still leave the question undecided."

A movement in space has been assigned to ω, in which, of course, the companion partakes; but the amount and course are variously given. These are the best:

| P.... | Æ | - 0"·12 | Dec. | 0"·00 |
| B.... | | + 0"·09 | | - 0"·04 |

When this object was first discovered to be double, in February, 1782, ♄. saw the stars hanging together, with part of the smaller one as it were emerged from behind the larger, being then in the *sf* quadrant of its orbit. My angular measures, such as they are, imply a direct motion into the *np* quadrant; but as it was rather cuneated than double, the distance I have given is a mere comparative estimation. Indeed the image, while under observation, was magnified as highly as its brightness would possibly bear; and hence the uncertainty. From a diagram I have deduced that in 1834 it was imperfectly seen, and therefore deemed round, so much so to the *senses* as to bear a high weight: from the same process it also seems, that in 1839 the wrong end of the egg was measured, for by subtracting 180° we obtain a regular progress in the angular positions, viz., 110°, 160°, 175°, 193°. However this may be, by allowing for ♄.'s having actually seen a division between them, with his twenty-foot telescope, and assuming his position, we gain a fair point of departure, which proves important:

♄. Pos. 110° 54' Dist. 0"·4 Ep. 1783·26

To fish up this interesting object by alignment, carry a ray from Castor through the Præsepe, and extend it just as far again in the southeast, where it is the middle one of Flamsteed's Nos. 6, 2, and 3; it lies in an open space about 11° to the west-south-west of Regulus, and 17° due north of Alphard.

CCCLVIII.   3 LEONIS.

Æ   9ʰ 19ᵐ 58ˢ   PREC. + 3ˢ·20
DEC.   N 8° 53'·0   —— S 15"·35

POSITION 78°·0 (w 1)   DISTANCE 25"·0 (w 1)   EPOCH 1834·25

A delicate double star, close to the Lion's left fore-paw, where it will be found by the above alignment. A 6½, pale yellow; B 13, blue, two or three other stars in the field, of which the nearest is about 2' distant in the *sf*. This object is 47 Ḥ. IV., classed in 1782, but without measures; and it was looked at by me principally as a focus adjustment for attacking ω, which is within a degree to the due north of it.

CCCLIX.   7 LEONIS MINORIS.

Æ   9ʰ 21ᵐ 02ˢ   PREC. + 3ˢ·65
DEC.   N 34° 21'·3   —— S 15"·41

POSITION 132°·9 (w 2)   DISTANCE 55"·0 (w 1)   EPOCH 1833·16

A wide double star, immediately under the animal's right fore-paw. A 6, bluish white; B 11, livid. This object was registered 69 Ḥ. V., in 1782, with a distance of 58"·30, but no angle of position; and it is No. 1116 of H.'s twenty-foot Sweeps. The companion is one of those minute and dusky objects which are best seen by averting the eye to the verge of the field; but there are many of much smaller magnitude, which shine quite sharply, and emit a strong blue ray. It may be found by carrying a line from Regulus close to the eastward of ε Leonis, and passing it exactly as far again into the north-north-west region.

CCCLX.   τ¹ HYDRÆ.

Æ   9ʰ 21ᵐ 02ˢ   PREC. + 3ˢ·03
DEC.   S 2° 04'·3   —— S 15"·41

POSITION 2°·9 (w 8)   DISTANCE 64"·9 (w 5)   EPOCH 1831·97

A wide double star, in the Serpent's fore-body, and 6° north of Alphard. A 5½, flushed white; B 8½, lilac, with a small star preceding it near the *np* vertical. This object is 71 Ḥ. VI., and these were the first measures:

Pos. 358° 36'   Dist. 61"·67   Ep. 1782·06

Nearly forty years afterwards H. and S. examined it micrometrically;

and to the results so obtained, we may place the deductions which follow from a treatment of Piazzi's mean apparent places of Nos. 94 and 95, Hora IX.:

P.	Pos. 3° 00′	Dist. 64″·00	Ep. 1800·00
H. and S.	3° 11′	66″·68	1821·23

From a discussion of the epochs of 1782 and 1821, this object was inferred to have changed its angle by 4° 35′; and to have increased its distance. But the late measures are so coincident as to show no alteration in upwards of thirty years; and it must therefore be concluded that Ƕ. wrote *np* for *nf.*

According to Ptolemy, the curve of the neck, καμπή, is formed by τ¹, τ², ι, and Λ; but Kazwíní calls them the knot, *'okdah*, of the throat.

CCCLXI.   θ URSÆ MAJORIS.

Æ   9ʰ 22ᵐ 8ˢ   Prec. +   4ˢ·05
Dec. N 52° 24′·2   —— S 16″·07

Position 245°·0 (*w 2*)   Difference Æ = 51″·9 (*w 1*)   Epoch 1834·74

A secondary Greenwich star, with only a very distant companion; on the animal's right fore-thigh, where a line from δ through β, extended more than as far again into the west-south-west, will pick it up by its splendour, and 5° further it will strike ι. A 3, brilliant white; B 10, dusky. This object was examined and entered for watching the rapid proper motions of the large star, the values of which have been thus severally stated:

P....	Æ − 1″·80	Dec. − 0″·60
B....	− 1″·55	− 0″·57
A....	− 1″·59	− 0″·57

This star, with six others in the Bear's throat, breast, and fore-knees, viz., τ, h, υ, φ, e, and *f*, were called by the Arabian astronomers *Serir benat na'sch*, throne of Benat na'sch, which last is said to be the Hebrew *aish*, which Ebn Ezra says is a cart or tumbrel. The space has also been termed *Al-khud*, the pond.

CCCLXII.   57 Ƕ. I. LEONIS.

Æ   9ʰ 23ᵐ 07ˢ   Prec. +   3ˢ·39
Dec. N 22° 12′·1   —— S 16″·07

Mean Epoch of the Observation . . . . 1834·27

A bright-class white nebula, in the Lion's lower jaw; first discovered by Ƕ. in 1784, and No. 604 of his son's Catalogue of 1830; the former describing it as a double nebula, each having a seeming nucleus, with their apparent nebulosities running into each other, and this is confirmed by the latter. It is vertically between two groups, of three small stars each, and it is elongated with a major axis lying *sp* and *nf*. To fish it up, run a line from Regulus to γ, and there draw another, perpendicular to it, which, carried nearly twice the length of the base, will strike it 2° south of λ Leonis. The upper or south part, is better defined than the lower; it requires, however, the closest attention and most patient watching, to make it a bicentral object, with my means; but the annexed is something of its aspect under the best vision.

This nebula is in the IXth Lunar Mansion, which is the district between ξ Cancri and λ Leonis in the Lion's jaws. This space was named *Elṭerf*, or *Al-ṭarf*, *acies oculi*, the Lion's glance.

CCCLXIII. 6 LEONIS.

℟ 9ʰ 23ᵐ 23ˢ   Prec. + 3ˢ·22
Dec. N 10° 25′·1   —— S 15″·54

Position 73°·6 (w 8)   Distance 37″·6 (w 3)   Epoch 1832·23

A double star, in the Lion's left fore-paw; lying 9° west by south of Regulus, on the line projected from θ Leonis through that luminary, and at half the length of that line. A 6, pale rose-tint; B 9½, purple. This object is 26 Ḥ. iv., and it has been thus measured:

Ḥ.   Pos. 77° 05′   Dist. 36″·15   Ep. 1781·14
H. and S.   74° 33′   38″·12   1822·16

in discussing which, the proper motion attributed to A must be considered. It is thus stated:

P.... ℟ − 0″·21   Dec. − 0″·01
B....   + 0″·05   − 0″·05

CCCLXIV. 7 LEONIS.

℟ 9ʰ 27ᵐ 08ˢ   Prec. + 3ˢ·29
Dec. N 15° 05′·4   —— S 15″·75

Position 80°·1 (w 7)   Distance 42″·6 (w 9)   Epoch 1832·23

A wide but very delicate double star, in the space opposite the Lion's neck; it is 8° west-north-west of Regulus, in the line formed by that luminary and η Virginis, a bright star lying about twice the distance in the east-south-east. A 6½, flushed white; B 8½, violet tint. This object is 58 Ḫ. v., entered as " supra pedem borealem anteriorem," and the observations which preceded mine are:

| Ḫ. | Pos. 81° 24′ | Dist. 42″·41 | Ep. 1782·10 |
| H. and S. | 80° 35′ | 44″·19 | 1821·23 |

CCCLXV.   78 Ḫ. I. URSÆ MAJORIS.

Æ   9ʰ 34ᵐ 52ˢ   Prec. +   5ˢ·66
Dec. N 73° 01′·2   ——   S 16″·16

Mean Epoch of the Observation . . . . 1836·29

A bright-class round nebula, above the Great Bear's ear, with several stars in the field from the 9th to the 12th magnitudes; of which a vertical pair precedes the nebula, and it is closely followed by a very minute one, which is caught only by glimpses. This object was discovered by Ḫ. in April, 1785, and is No. 629 of his son's Catalogue; the last of whom registers its diameter as = 50″. The mean apparent place was obtained by differentiation with λ Draconis; and it may be fished up by running a line to the north-east-ward from Mizar under λ Draconis, in the tails of the Bear and the Dragon, and carrying till it is nearly due south of Polaris. Here the observer will find Flamsteed's 27 Ursæ Majoris, a star of the 5½ magnitude, and closely following it is the nebula sought.

CCCLXVI.   ψ LEONIS.

Æ   9ʰ 35ᵐ 01ˢ   Prec. +   3ˢ·28
Dec. N 14° 45′·0   ——   S 16″·17

Position 140°·0 (w 2)   Distance 260″·0 (w 1)   Epoch 1834·19

A variable star, with a distant companion, on the Lion's left fore-knee; and it is about 10° due south of ε Leonis, and 6°½ west-north-west of Regulus. A 6, bright orange; B 10, bluish white; a third star, of the 11th magnitude, nearly 6′ distant on an angle of about 280°. ψ has proper motions attributed to it, of the following values:

| P.... Æ + 0″·04 | Dec. − 0″·08 |
| B.... + 0″·05 | − 0″·04 |

The star A is here designated of the 6th magnitude, as assumed from Piazzi; but was certainly very bright for its rate, at the epoch of my observations. It is said to vary from 5 to 8, but has not been so closely followed as such variability seemed to demand. Ptolemy enrolled it ε in lustre, and was followed by Tycho Brahé and Hevelius. Ulugh Beigh designates it 6, in which he is copied by Flamsteed, La Caille, and Mayer. They say it once disappeared altogether, after having been noticed by Montanari, in 1667; and that Miraldi saw it again very small, in 1691. Mr. Pigott says that he always perceived it of the 5·6th size. Mr. Challis, who was kind enough to re-examine this object for me in 1841, with the Northumberland equatoreal, diagrammed ψ of the 5th magnitude.

This variable star precedes the one, Mayer 420, mentioned by Koch as fading from the 7th to the 10th brightness, which Pigott was unable to find. But though there is a kink in the Æ, it is probably Piazzi's No. 176, Hora IX.

CCCLXVII.   161 P. IX. SEXTANTIS.

Æ   9ʰ 35ᵐ 09ˢ   Prec. +   3ˢ·12
Dec. N   3° 21'·4   ——   S 16"·17

Position 145°·0 (w 3)   Distance 4"·0 (w 1)   Epoch 1834·26

A delicate double star, just inside the upper frame of the Sextant, but also on the more ancient Lion's leg; where it will be found—"alone in its glory"—nearly in mid-distance, and closely west, of a line between Regulus and Alphard. A 8, yellowish white; B 13, blue, with two or three other small stars in the field, one of which nearly precedes. This object was discoverd by Σ., being his No. 1377, and is thus registered in the great Dorpat Catalogue:

Pos. 142°·20   Dist. 3"·317   Ep. 1830·24

CCCLXVIII.   ε LEONIS.

Æ   9ʰ 36ᵐ 46ˢ   Prec. +   3ˢ·43
Dec. N   24° 30'·5   ——   S 16"·26

Position 9°·1 (w 2)   Difference Æ = 1"·9 (w 1)   Epoch 1833·75

A Greenwich star of 1830, with a distant companion, on the Lion's ear; where it forms the north-west vertex of a nearly right-angled triangle with γ Leonis and Regulus. A 3, yellow; B 10, pale grey. A is designated Rás-al-asad-al-jénúbi, or Australis, to denote the south or undermost of two stars in the Lion's head; asad being one of the numerous words for that animal in Arabia. See μ Leonis. An almost imperceptible movement in space is attributed to ε, which is thus valued:

P.... Æ − 0"·03   Dec. + 0"·02
B....   0"·00   − 0"·04

CCCLXIX. 81 AND 82 M. URSÆ MAJORIS.

ÆR 9ʰ 42ᵐ 10ˢ   Prec. + 5ˢ·13
Dec. N 69° 51·8   —— S 16″·53

MEAN EPOCH OF THE OBSERVATION . . . . 1837·19

No. 81 is a fine bright oval nebula, of a white colour, in the Great
Bear's ear, which was first registered by M. Messier in 1781, and exhibited
a mottled nebulosity to ⩁. Its major axis lies *np* and *sf;* and it certainly is brightest in the middle. There are several minute companions
in the field, of which a close double star in the *sp* quadrant is No. 1386
of Σ.'s grand Catalogue, and by him marked *vicinæ;* the members are
both of the 9th magnitude, and trend *np* and *sf*, about 2″ apart, forming
a fine though difficult object.

With a low power, No. 82 M. can be brought into the north part of
the same field of view, although they are half a degree apart. It is very
long, narrow, and bright, especially at its northern limb, but rather
paler than No. 81. A line drawn through three stars in the *sp* to a
fourth in the *nf* passes directly through the nebula. The two nebulæ
precede λ, in the end of Draco's tail, by 25°, but as the vicinity is
deficient of large stars, they are not readily fished up.

The apparent place here taken, is that of a small star between the
two nebulæ, which was differentiated with 29 Ursæ Majoris, and every
care taken in the reductions. The bright star on the animal's chest,
south of 29, viz. φ, is pronounced to be double, both components being
of the 5th magnitude, and only half a second asunder.

CCCLXX. μ LEONIS.

ÆR 9ʰ 43ᵐ 39ˢ   Prec. + 3ˢ·45
Dec. N 26° 45′·5   —— S 16″·60

Position 235°·0 (*w* 1)   Difference ÆR = 21ˢ·0 (*w* 1)   Epoch 1833·75

A star with a distant companion, on the *os frontis* of Leo, where it
has its fellow in brightness 2°½ to the south-south-west, and with it
forms a reduced portrait of γ Leonis and Regulus, which are similarly
situated relative to each other, about 10° south-east of their miniature.
A 3, orange; B 10, pale lilac; forming an equilateral triangle with two
stars preceding it, of which that to the south is of the 8th magnitude.
This star is known as Rasalas, from the Arabian *Rás-al-Asad*, and
further designated by *Al shemáli*, or *borealis*, to denote the northern
star in Asad's head. See ε Leonis.

This object is charged with a proper motion, which appeared very
slight to Piazzi, on comparison of his observations with those of Bradley;

but later investigations show a very sensible value in $\mathcal{R}$. The following are the registered amounts:

$$P.... \; \mathcal{R} \; - \; 0''{\cdot}03 \qquad \text{Dec.} \; + \; 0''{\cdot}02$$
$$B.... \qquad - \; 0''{\cdot}27 \qquad\qquad\; - \; 0''{\cdot}06$$
$$A.... \qquad - \; 0''{\cdot}31 \qquad\qquad\; - \; 0''{\cdot}08$$

Nasíra-l-dín mentions that the two stars on the Lion's forehead, meaning ϵ and μ, are a whip's length apart. As this is a favourite term of measure among the Arabian writers, it may be stated that, in this instance, it is rather more than 2°.

CCCLXXI. 9 SEXTANTIS.

$\mathcal{R}$ 9^h 45^m 45^s PREC. $+ \; 3^s{\cdot}14$

DEC. N 5° $41'{\cdot}8$ —— S $16''{\cdot}70$

POSITION $293^\circ{\cdot}5$ (w 9)   DISTANCE $49''{\cdot}7$ (w 5)   EPOCH $1832{\cdot}18$

A double star on the right fore-leg of Leo, though crimped into the Sextant; it lies at one-third of the way from Regulus to Alphard. A 7 and B 9, both blue, and well defined. A reduction from the mean apparent places of Nos. 204 and 205, Hora IX., of the Palermo Catalogue, with the measures of Sir James South, stand thus for comparison:

$$P. \; \; \text{Pos.} \; 296^\circ \; 00' \quad \text{Dist.} \; 49''{\cdot}7 \quad \text{Ep.} \; 1800{\cdot}00$$
$$S. \qquad\quad 292^\circ \; 43' \qquad\quad 51''{\cdot}02 \qquad\quad 1825{\cdot}01$$

Desirous of assigning an asterism to the perpetual remembrance of celestial affairs, and especially wishing to commemorate the instrument so successfully used by Tycho Brahé at Uranienburg, about the year 1590, Hevelius gathered some *informes* between the Lion's fore-legs and Hydra, and called them *Sextans Uraniœ*. But, with more zeal than taste, he fixed the machine upon the Serpent's back, under the plea that the said Sextant was not in the most convenient situation, but that he placed it between Leo and Hydra because these animals were of a fiery nature, to speak with astrologers, and formed a sort of commemoration of the destruction of his instruments when his house at Dantzic was burnt in September, 1679; or, as he expresses it, when Vulcan overcame Urania. He who thus placed it in the heavens only mustered 12 stars, but Flamsteed made out 41, and Bode has increased them to 112. This, and some other of Hevel's denominations, have occasioned an ill-natured and groundless sneer from the redoubtable La Lande, himself a wholesale apotheosizer; but it assuredly speaks more for his flippancy than for his scientific gratitude.

The above ebullition is, however, to be strictly confined to the case in point; for whatever singularities or failings he possessed—and, from the testimony of his own *camarades*, these were neither few nor trivial— there were not many of his day to whom the "million" were more indebted for scientific supplies, than to Jerome Le Francais La Lande.

CCCLXXII. 286 ♅. I. URSÆ MAJORIS.

Æ 9ʰ 49ᵐ 30ˢ   Prec. + 5ˢ·00
Dec. N 69° 30′·4   —— S 16″·88

Mean Epoch of the Observation . . . . 1836·21

A bright-class round nebula, at the back of Ursa Major's left ear, preceding λ, at the end of Draco's tail, by 22°; it is lucid white, and lights up in the centre. There are two lines of three stars each across the field, of which the one preceding the nebula is of the 7th magnitude, and that following of the 10th; between these the sky is intensely black, and shows the nebula as if floating in awful and illimitable space, at an inconceivable distance. Dr. Derham, whose judgment led him to consider nebulæ as vast areæ of light "infallibly beyond* the fixed stars," thought that some of them might be openings in an opacity surrounding the visible system, which chasms allow us a sight of the empyreal sphere beyond it. The present object, under the favourable conditions in which I viewed it, would have almost countenanced his supposition.

This nebula was discovered by ♅. in November, 1801; and he says, that "on the *nf* side there is a faint ray interrupting the roundness."

CCCLXXIII. 163 ♅. I. SEXTANTIS.

Æ 9ʰ 57ᵐ 16ˢ   Prec. + 2ˢ·99
Dec. S 6° 56′·9   —— S 17″·24

Mean Epoch of the Observation . . . . 1836·19

An elongated bright nebula, on the radius or graduated limb of the Sextant, followed by two stars of the 11th magnitude, which are the only other objects in the field of view. Its major axis trends towards the vertical of the *sp* and *nf* quadrants; and the extremes appear pointed. It was discovered by ♅. on the 22nd of February, 1787, and is No. 668 of his son's Catalogue.

It is remarkable that this object was very clearly distinguished in my telescope; for H. says it was scarcely perceptible in his 20-foot when he gave it only six inches of aperture. It follows Alphard by about 10°, a little north of the parallel; where it precedes a knot of small stars, which are a couple of degrees further to the west.

---

* Yet a recent and greatly patronized treatise on Astronomy, written expressly to accompany the large perforated constellation-cards called *Urania's Mirror*, published in 1825, intrepidly asserts, that the nebulosities "are now generally considered to be at no very considerable distance from the Earth."

CCCLXXIV. α LEONIS.

Æ 9ʰ 59ᵐ 51ˢ   Prec. + 3ˢ·22
Dec. N 12° 44'·8   —— S 17"·35

Position 306°·8 (w 8)   Distance 175"·8 (w 6)   Epoch 1837·33

A standard Greenwich star, with a distant companion, in the Lion's breast. A 1, flushed white; B 8½, pale purple. This object is 11 ℍ. vi., and the details were thus, at its first registry:

Pos. 305° 05'   Dist. 168"·33   Ep. 1781·84

Piazzi then entered it in the Palermo Catalogue: "Duplex. Comes præcedit;" but there can be little doubt of this *comes* being his No. 249 of Hora IX., as I found, by repeated trials, that the Δ Æ was = 9ˢ·7. By obtaining data from a reduction of his mean apparent places, the details of Piazzi, and the results of H. and S., stand thus:

P.   Pos. 307° 00'   Dist. 178"·00   Ep. 1800·00
H. and S.   307° 07'   174"·96   1821·21

A comparison of the measures of ℍ., and H. and S., induced a belief, that a considerable alteration had occurred in the relative places of the two stars, in a lapse of forty years, showing a physical connection between them; but I am inclined rather to attribute the differences to proper motion and instrumental errors, than to inconstancy of angle or increase of distance. Indeed, it is a wide object for this system of measuring, and a long run upon the micrometer spring. The proper motion has been thus stated:

P.... Æ − 0"·28   Dec. − 0"·01
B....   − 0"·23   0"·00
A....   − 0"·27   + 0"·02

This star is well known as Καρδία λέοντος, Cor Leonis, the lion's heart. It is pointed to by Aldebaran and γ Geminorum, as well as by running a line from Orion's belt through Procyon, and carrying it nearly twice as far again to the east-north-east. The prolongation of the same line, or rather great circle, will lead to Denebola, in Leo's tail. Regulus and Denebola form the longest side of an extensive quadrilateral figure, with two other stars to the north of them; there is a still more remarkable square adjoining this, γ being a corner-stone of each. Regulus is also readily found by drawing a line southwards from γ and δ Ursæ Majoris, the last stars in the square; or, with the poetaster, reversing it:

From Hydra's pass through Leo's heart,   (which marks th' Ecliptic Line,)
You'll rise to where, in Ursa Great,   the third and fourth stars shine.

Ptolemy calls this star Βασιλίσκος, from an opinion of its influencing the affairs of the heavens; whence comes its Latin name Regulus, a word which appears to have been first used by Copernicus as the diminutive of *rex*. It is the lucida of the extensive northern constellation Leo, whose stars are well disposed and conspicuous, forming the fifth asterism in zodiacal order. The classic star-gazers viewed this as the apotheosis of the Nemæan Lion, and the emblem of heat; but Stower's celebrated manuscript Almanac of 1386, recognises in it one of Daniel's lions, and therefore " whoso es born in yat syne he schal be hardy and lytherus." Schickhard insisted that it represented the Lion of the tribe of Judah, mentioned in the Apocalypse. It was under the tutelary protection of Jove himself, whence the astrologers chattered largely about an alliance between the planet Jupiter and the constellation Leo. Macrobius—*De Somnio Scipionis*—says that the Lion was assigned as the Sun's house, and Cancer as the Moon's, because they were in those signs "in ipsâ geniturâ mundi nascentes." The Arabs called this "fiery trigon" *Kalb-al-Asad*, or lion's heart, and *Meliki*, or kingly; for this impression of greatness was as rife among the Oriental astronomers and their successors, as among their classic predecessors. Thus Wyllyam Salysbury, treating of the sphere, or frame of the world, in 1552, tells us, " The Lyon's herte is called of some men, the Royall Starre, for they that are borne under it, are thought to have a royall nativitie;" and in the *Tabule Astronomice Alfonsi Regis*, 1492, it is written against Regulus, " Que est super cor: et dicit. Rex." Yet after all Horace only sings of it as

Stella vesani Leonis.

Λέων, Leo, Nemeas alumnus, Bacchi sidus, Stella regia, are also names by which the Lion has been designated; and it is visible to the gazer by the large trapezium which it displays. Even should Regulus not be personally known, this trapezium is readily found by the universally-known pointers of the Great Bear; for as they serve to show Polaris to the northward, so also doth the line produced by them, prolonged southward about 45°, point to the Lion. It is one of the old 48 constellated groups, and has been thus catalogued:

Ptolemy . . . 35 stars		Maraldi . . . 60 stars
Tycho Brahé . . 40		Flamsteed. . . 95
Bayer 43		Hodell 276
Hevelius . . . 50		Bode 337

The retrograde motion, owing to the recession of the equinoctial points—from a slow vibration of the Earth's axis, occasioned by planetary attractions—by which the stars appear to go in *antecedentia*, or backwards from west to east, contrary to the order of the signs of the zodiac—affords data from which the march of those heavenly bodies, in a course parallel to the ecliptic, is easily traced. This motion was first detected and reduced to rule by Hipparchus, in discussing his own observations with those of Aristyllus and Timocharis; and the longitude of Regulus has, through successive ages, been made a datum-step, by the best astronomers of all nations. From these a few may be selected, in order to show the changes of that point, since it has been under observation:

ASTRONOMER.	DATE.	LONGITUDE.
Timocharis . . . . . B.C.	295 . . .	♋ 27° 54′·5
Hipparchus	128 . . .	29° 50′
Ptolemy . . . . . A.D.	136 . . .	♌ 2° 30′
Abd-r-ahman Súfí . .	964 . . .	15° 12′
Chrysococcas Persa . .	1115 . . .	17° 30′
Ulugh Beigh	1437 . . .	19° 55′
Tycho Brahé	1587 . . .	24° 06′
Flamsteed	1689 . . .	25° 31′·3
Maskelyne	1770 . . .	26° 38′
Airy	1840 . . .	27° 36′·3

The Astronomer Royal, Mr. Airy, sent me his position of this star to rigid exactness. "Our catalogue place of Regulus," he obligingly wrote, "from the observations of 1840, is as follows:

Mean Æ 1 Jan. 1840 = 9^h 59^m $50^s\cdot71$
Mean N.P.D. $= 77^h$ 15^m $12^s\cdot43$

"With these, and the mean obliquity $= 23° 27' 36''\cdot52$, the latitude and longitude will be computed thus:

Longitude . . . . . $147° 36' 20''\cdot15$
Ecliptic N.P.D. . . . . $89° 32' 25''\cdot12$"

These data afford a striking instance of the sagacity of the early astronomers. They, however, considered the equinoxes to be immovable; and ascribed the change of distance of the stars from it, to a real motion of the orb of the fixed stars, which they supposed to have a slow revolution about the poles of the ecliptic in the Platonic period of 25,920 years, a space not remotely different from that produced by moderns, from other principles. Sir Isaac Newton very ably demonstrated, that the physical cause of the precession arises from the broad or spheroidal oblate figure of the Earth; which satisfactorily proves the operation, and accounts for the effect.

CCCLXXV. 4 ☿. I. SEXTANTIS.

Æ 10^h 05^m 58^s Prec. $+ 3^s\cdot12$
Dec. N $4° 15'\cdot1$ ——   S $17''\cdot61$

Mean Epoch of the Observation . . . . $1837\cdot02$

A bright-class round nebula, on the frame of the instrument, with another, rather larger but more faint, at about 29^s on the following parallel; the latter being attended by three stars, the middle one of which is the smallest, and is closely *nf.* This was discovered by ☿. in December, 1783, but it is very remarkable, that, though he made four observations of the object, he did not notice that there were two nebulæ in the field. H., however, saw them both, and has described them under No. 685. The place is not very difficult to find, being about 9° south by east of Regulus, and in the line with that luminary and μ in the head of the Lion.

This object is on or near the spot where the Capuchin, De Rheita, fancied he saw the napkin of S. Veronica, in 1643, with an improved telescope which he had just constructed. It would be much easier to ascribe this strange discovery to a heated imagination, than to deliberate falsehood; but it happens unfortunately that there is no staring cluster or nebula near. However, in case any one still chooses to search for it, we may state, that in a letter to his friend J. Caramuelis, dated Cologne, 24th April, 1643, he mentions having detected most clearly, by means of his binocular telescope, with the greatest surprise, admiration, and delight, the sacred "sudarium Veronicæ sive faciem Domini maximâ similitudine in astris expressum," in the sign of Leo, between the equinoctial and the zodiacal circles.   And this is an accurate reduction of the figure which Zahn gives of it in the *Oculus Artificialis*.

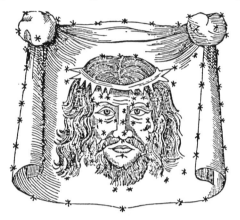

Padre de Rheita makes very respectable mention of this same apparition in his great work, *Oculus Enoch et Eliæ, sive Radius Sidereo-mysticus*, the very elaborate engraved title-page of which thus invites us, in the words of the Royal Psalmist, *Venite et videte opera Domini.* In craving permission to doubt his assertion, Sir John Herschel's words may be applied: "Many strange things were seen among the stars before the use of powerful telescopes became common."

CCCLXXVI.   γ LEONIS.

Æ 10ʰ 11ᵐ 08ˢ   Prec. +   3ˢ·30
Dec. N 20° 39'·0   —— S 17"·82

Position	Distance	Epoch
103°·2 (w 4)	2"·6 (w 4)	1831·36
102°·5 (w 3)	2"·8 (w 3)	1833·20
104°·9 (w 7)	2"·5 (w 9)	1836·42
106°·0 (w 9)	2"·6 (w 9)	1839·23
107°·2 (w 9)	2"·8 (w 9)	1843·18

A splendid double star, close to the Lion's mane, about $7°\frac{1}{2}$ to the north-north-east of Regulus, and nearly in the middle of the constellation. A 2, bright orange; B 4, greenish yellow, and there are two stars in a line with A in the *np* quadrant.   This most beautiful object is 28 Ⱨ. ɪ., and from a comparison of the several measures seems decidedly to have a slow progressive angular acceleration, which may perhaps give an *annus magnus* of about 1000 years; but the exact amount is involved in the question of proper motion, of which these are the assigned quantities:

P.... Æ + 0"·35   Dec. − 0"·20
B....   + 0"·35   − 0"·15
A....   + 0"·30   − 0"·14

The results which have been obtained by the best astrometers, since its registry in 1782, are:

	Pos.	Dist.	Ep.
♓.	83° 30′	3″·00 ±	1782·71
H. and S.	98° 24′	3″·24	1822·24
Σ.	103° 22′	2″·50	1831·51
D.	103° 41′	2″·64	1833·18

This star has been improperly called Algieba, from *Al jeb-bah*, the forehead; for no representation of the Lion, which I have examined, will justify that position. With *a*, ζ, and η, it forms the Xth Lunar Mansion.

CCCLXXVII. 58.P. X. URSÆ MAJORIS.

Æ 10ʰ 15ᵐ 51ˢ   Prec. + 3ˢ·86
Dec. N 53° 26′·0   —— S 18″·01

Position 85°·0 (w 6)   Distance 3″·6 (w 3)   Epoch 1832·49

A very neat double star, on the Great Bear's right shoulder. A 8, and B 8½, both white. This pretty but minute object was discovered by Σ., being No. 1428 of the Dorpat Catalogue; and the juxtaposition seems to be only optical. The measures which precede mine are:

	Pos.	Dist.	Ep.
H.	88° 43′	4″·09	1830·56
Σ.	84° 33′	3″·84	1831·69

To find this pair by alignment, run a line from the Lesser Bear's leading guard, β, through Dubhe, and the mid-distance on the northeast of that lucida will mark the place of κ Draconis, while a similar extent to the south-west of Dubhe will strike upon 58 P. x.

CCCLXXVIII. 27 ♓. IV. HYDRÆ.

Æ 10ʰ 17ᵐ 01ˢ   Prec. + 2ˢ·88
Dec. S 17° 50′·6   —— S 18″·05

Mean Epoch of the Observation . . . . 1837·18

A planetary nebula, pale greyish-white, nearly 2° south of μ, about 20° south-west by west of Regulus, and in the middle of Hydra's body. From its size, equable light, and colour, this fine object resembles Jupiter; and whatever be its nature, must be of awfully enormous magnitude. It was discovered by ♓. in February, 1785, and has four telescopic stellar companions, two of which are posited at nearly equal distances, *np* and *sf*, from the nebula. It was carefully differentiated with μ Hydræ; and as a line passing from a star in the *np* quadrant to another in the *sf*, just touched its disc, it was diagrammed as above.

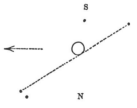

Though this remarkable nebula escaped H., his remarks on planetary nebulæ are so applicable to it, that they should be here transcribed. "Granting these objects," he observes, "to be equally distant from us with the stars, their real dimensions must be such as would fill, on the lowest computation, the whole orbit of Uranus. It is no less evident that, if they be solid bodies of a solar nature, the intrinsic splendour of their surfaces must be almost infinitely inferior to that of the Sun's. A circular portion of the Sun's disc, subtending an angle of 20″, would give a light equal to 100 *full moons;* while the objects in question are hardly, if at all, discernible to the naked eye. The uniformity of their discs, and their want of apparent central condensation, would certainly augur their light to be merely superficial, and in the nature of a hollow spherical shell; but whether filled with solid or gaseous matter, or altogether empty, it would be a waste of time to conjecture."

CCCLXXIX. 67 P. X. LEONIS.

Æ 10^h 17^m 09^s　　Prec. + 3^s·17

Dec. N 9° 35′·2　　—— S 18″·06

Position 64°·8 (w 5)　Distance 3″·0 (w 3)　Epoch 1831·18

—— 65°·3 (w 7)　　—— 3″·5 (w 5)　—— 1843·16

A very neat double star, on the Lion's right shoulder; about 5°½ to the south-west of Regulus, and exactly on the line described from that luminary to Algorab. A 8, white; B 9½, pale blue. This beautiful but delicate object is evidently only optical, and relatively fixed: it is 29 H. i., and has been thus measured:

	Pos.	Dist.	Ep.
H.	63° 28′	4″·00±	1782·13
S.	63° 59′	3″·63	1825·20
Σ.	65° 54′	3″·20	1832·56

The aberrations of A from the common laws of precession, may in time prove the independence thus indicated between the components of this object, the most authentic values of its proper motion being:

P.... Æ − 0″·13　Dec. − 0″·05

B....　　+ 0″·11　　　− 0″·20

CCCLXXX. 86 H. I. LEONIS MINORIS.

Æ 10^h 18^m 17^s　　Prec. + 3^s·39

Dec. + 29° 19′·1　　—— S 18″·10

Mean Epoch of the Observation .... 1836·19

A bright-class nebula, beneath the animal's belly, but pretty close to the old Lion's mane; where a north-north-east ray from Regulus carried closely before Algieba, and extending rather more than as far again, will find it in the centre of a trapezium of four stars, of which the two southern ones are the largest. This fine object is of an oval shape, with a palpable central nucleus; it was discovered by ♅. in April, 1785, and is No. 711 of his son's Catalogue.

CCCLXXXI. 49 LEONIS.

Æ 10ʰ 26ᵐ 38ˢ   Prec. + 3ˢ·16
Dec. N 9° 28'·5   —— S 18"·40

Position 158°·1 (w 5)   Distance 2"·5 (w 5)   Epoch 1838·37

A close double star, under Leo's right shoulder, close to ζ, and about 8° west-south-west of Regulus. A 6, silvery white; B 9, pale blue. This was discovered by Σ., and is thus registered in the great Dorpat Catalogue, No. 1450:

Pos. 161° 09'   Dist. 2"·39   Ep. 1830·76

These results would imply a retrograde angular motion, but that it is both too delicate and difficult an object to decide upon at once I am not aware of any other observations upon it, so that it must for the present remain questionable. The small spacial movement attributed to A by Piazzi, has disappeared under the test of recent examinations.

CCCLXXXII. 60 ♅. IV. URSÆ MAJORIS.

Æ 10ʰ 28ᵐ 45ˢ   Prec. + 3ˢ·79
Dec. N 54° 20'·4   —— S 18"·48

Mean Epoch of the Observation . . . . 1837·17

A planetary bluish-white nebula, in the Great Bear's right shoulder, having two stars of the 10th magnitude nearly between it and an orange-coloured companion in the *sf* quadrant. It is a small object but well defined, with a palpable un-attenuated round disc; this I note the more particularly as denoting the limit of my means upon such bodies, for I saw no symptom of the "very feeble atmosphere" with which H. says it is surrounded. It was discovered by ♅. in April, 1789, and is No. 731 of his son's Catalogue of 1830. The place is differentiated from β Ursæ Majoris; from which it bears south-west about 4°, and is nearly on the same parallel with γ.

Sir William Herschel considered the indistinctness on the edges sufficiently extensive, to make this a step between a planetary nebula and those bright in the middle.

CCCLXXXIII. 37 LEONIS MINORIS.

Æ 10ʰ 29ᵐ 42ˢ   Prec. + 3ˢ·40

Dec. N 32° 48'·3   —— S 18'''·51

Position A B 270°·0 (w 3)   Difference Æ = 21ˢ·9 (w 1) ⎫

—— A C 21°·4 (w 2)   ——————— 10ˢ·2 (w 1) ⎬ Epoch 1833·30

—— A D 291°·0 (w 2)   ——————— 18ˢ·2 (w 1) ⎭

A star with three very distant companions, on the Little Lion's right side; where it will be seen on a line produced from Regulus through Algieba, and extended to nearly double that distance into the north-north-east. A 4, yellowish white; B 7½, pale grey; C 13, reddish; D 12, violet tint. An almost insensible movement of A in space, has been detected by the perseverance of investigators, and this is the assigned value:

P.... Æ − 0''·03   Dec. + 0''·01
B.....   + 0''·06   0''·00

Piazzi noted the two principal members of this object, remarking, "Alia 7·8ᵃᵉ magnitud. præcedit 22'' temporis, 3' circiter ad boream;" and he has entered A by its old name Præcipua, as the *lucida*, or principal star of Leo Minor, registered of the 3rd magnitude by Hevelius, and continued so by Bode. Mr. Baily, however, in his recent edition of Flamsteed, has rated it 5¼ in lustre, under the following plea: "This star," he says, "is marked as of the 3rd magnitude, in the British Catalogue: but in the *original* entries it is designated three times as of the 6th, once of the 4th, and once as 4½, but nowhere greater. I have taken the mean of the whole." I have never seen it but as given above, from Piazzi.

This asterism was formed by Hevelius, from 18 *sporades* between Leo of the Zodiac, and the Great Bear; the constituents of which were increased by Flamsteed to 53 stars, and by Bode to 96. It was first announced as a constellation in the *Prodromus* of the former, 1691; and the author tells us, that he selected the place in order not to disturb the circles, notions, or rules of astrologers: "Since they esteem the Bear and the Lion as the hottest and fiercest animals, I wished to place there some quadruped of the same nature."

CCCLXXXIV. 35 SEXTANTIS.

Æ 10ʰ 35ᵐ 02ˢ   Prec. + 3ˢ·12

Dec. N 5° 35'·2   —— S 18'''·68

Position 240°·1 (w 9)   Distance 6''·9 (w 9)   Epoch 1834·27

—— 239°·6 (w 9)   —— 6''·8 (w 9)   —— 1839·19

A neat double star, on the north extreme of the graduated limb of the instrument, and three-fifths of the distance between Alphard and

Denebola. A 7, topaz yellow; B 8, smalt blue. This fine object is No. 36 of Ḥ.'s N. 145; and No. 141 of P. Hora X., where it is thus described: "Duplex. Comes telescopica præcedit 0″·4 temporis, 2″ ad austrum;" but the first micrometric measures, as far as I know, are those of H. and S.:

Pos. 237° 34′   Dist. 7″·87   Ep. 1822·33

The subsequent observations were:

Σ.   Pos. 240° 47′   Dist. 6″·75   Ep. 1825·20

D.   240° 26′   6″·93   1831·21

showing, on consideration, no angular motion, and the suspected decrease of distance to be in a very small ratio, if at all. Moreover, A has a slight proper movement in space, thus oppositely valued:

P.... Æ − 0″·05   Dec. + 0″·01

B....   + 0″·03   − 0″·06

CCCLXXXV. 95 M. LEONIS.

Æ 10ʰ 35ᵐ 31ˢ   Prec. + 3ˢ·18

Dec. N 12° 31′·9   —— S 18″·70

Mean Epoch of the Observation . . . . 1836·19

A lucid white nebula, on the Lion's ribs, with only two small stars, *np* and *nf*, in the field. Its place is almost due east of Regulus, with a distance of 9°, where it forms the southern vertex of a triangle nearly equilateral with γ and δ Leonis. This nebula is round and bright, and perhaps better defined on the southern than on the northern limb, a phenomenon worthy of remark, and observable in the great nebula of Andromeda, and other wonderful masses. It was discovered by Méchain in 1781, and registered by Messier as a "feeble nebula, without a star."

Nearly a degree to the eastward of this object, follows another round but not equally well defined nebula, large, and of a pale white colour. It is Messier's No. 96, and was also discovered by Méchain in 1781; it constitutes the intersecting point of a rectangle formed by five stars, of which the nearest is in the *sp* quadrant, and of the 11th magnitude.

CCCLXXXVI. 159 P. X. HYDRÆ.

Æ 10ʰ 39ᵐ 46ˢ   Prec. + 2ˢ·95

Dec. S 14° 47′·0   —— S 18″·83

Position 10°·0 (w 5)   Distance 31″·5 (w 9)   Epoch 1836·22

A double star, near the cup on the Hydra's back, where an east-south-east ray from Alphard towards the middle of the little square that constitutes Corvus, will meet it in the half-way. A 8, pale white; B 9, light blue. Piazzi first described this object's duplicity: "Duplex. Comes 9ᵃ magnit. sequitur 1″ temporis, ¼′ ad boream;" and the micrometrical results of it, previous to my own, are:

Σ.   Pos. 9° 45′   Dist. 31″·39   Ep. 1822·16
S.   11° 03′   31″·65   1825·17

from which it may be inferred to have undergone no appreciable change in a lapse of fourteen years, especially as the measures are rather difficult, from being teased with variable refraction.

CCCLXXXVII.   18 ♑. I. LEONIS.

Æ 10ʰ 39ᵐ 49ˢ   Prec. + 3ˢ·18
Dec. N 13° 28′·0   —— S 18″·83

Mean Epoch of the Observation . . . . 1837·22

A pair of bright-class nebulæ, *sp* and *nf* of each other, on the Lion's belly, discovered by ♑. in March, 1783, and No. 758 of his son's Catalogue; while at a small distance to the *nf* is a neat but minute double star. These are two of the three nebulæ described by both the Herschels; but the third I cannot distinguish, unless it be a glow in the *sf*, in a vertical line with two small stars. We now approach a region where these mysterious luminous masses are scattered over the vast concavity of the heavens, in truly boundless profusion; and in them, all true Herschelians must view mighty laboratories of the Universe, in which are contained the principles of future systems of suns, planets, and satellites!

The objects here treated of, are among the nebulæ included within a round patch of about 2° or 3° in diameter, in the apparently starless space of the Lion's loins. Now the observer unprovided with an equatoreal instrument—and unfortunately many of Urania's most zealous followers are in that predicament—may wish to fish it up. If his telescope be of capacity for grasping sufficient light, the field *may* be found, under a moderate power, south of the line which joins Regulus and θ Leonis about 10° east of, and nearly on the parallel with, the former.

CCCLXXXVIII.   41 SEXTANTIS.

Æ 10ʰ 42ᵐ 17ˢ   Prec. + 3ˢ·01
Dec. S 8° 03′·1   —— S 18″·90

Position AB 310°·0 (w 1)   Distance 20″·0 (w 1)⎫
——— AC 120°·4 (w 2)   ——— 290″·0 (w 1)⎭   Epoch 1838·31

A most delicate triple star below the Sextant, of which the third is not seen by me, though I have measured a distant companion in the same quadrant.   A 6, white; B 16, dusky; C 10, bluish.   It lies exactly on the parallel, and 20° to the east, of Alphard, nearly "alone in its glory," so that, like an oasis in a desert, it is tolerably visible to the inquiring eye.   A line from ϵ, in the Lion's head, through Regulus, prolonged more than as far again to the south-east, strikes upon it.   This object was forwarded to me by Sir J. Herchel, as an *experimentum crucis* of my optical power, on mounting the large telescope, there being a minute point in the *sf* of the 17th or 18th magnitude, which baffled all my endeavours to detect it.   Indeed the one in the *np* quadrant, B, is only caught by transient glimpses and keen gazing, so that the estimated angle and distance are next to mere guesses.

41 Sextantis has been held as having a very sensible proper movement, but recent comparisons have reduced the values of it; yet even the best are not agreed as to which course it is taking:

$$P.... \; \mathbb{R} - 0''·06 \qquad \text{Dec.} - 0''·03$$
$$B.... \qquad + 0''·03 \qquad\qquad + 0''·02$$

CCCLXXXIX.   362 ♓. II. LEONIS MINORIS.

$\mathbb{R}$ 10ʰ 42ᵐ 27ˢ	Prec. +	3ˢ·31
Dec. N 28° 49′·2	——	S 18″·91

Mean Epoch of the Observation . . . . 1836·26

A faint round nebula, pale white, on the ham of the Little Lion's hind-leg; it is preceded nearly on the parallel by an 8th-magnitude star, and there are several other small ones in the field, of which four following ones cross the parallel in a neat arc.   This was discovered and registered by ♓. in April, 1785; and is H.'s No. 773.   The mean apparent place is differentiated from ξ Ursæ Majoris, and it may be fished for nearly in mid-distance between γ Leonis and ξ Ursæ.

CCCXC.   179 P. X. LEONIS.

$\mathbb{R}$ 10ʰ 43ᵐ 50ˢ	Prec. +	3ˢ·13
Dec. N 8° 18′·7	——	S 18″·95

Position 305°·6 (w 6)   Distance 11″·8 (w 5)   Epoch 1836·26

A neat double star, under Leo's belly.   A 8½, and B 9, both bluish white.   Piazzi noticed the duplicity of this object: "Videtur duplex; præcedens ad austrum;" but the earliest measures I meet with are:

Σ.	Pos. 304° 42′	Dist.	Ep. 1822·43
S.	303° 21′	12″·51	1824·92

whence, on comparison with my results, a direct angular motion might be inferred; but the object is too difficult to expect a greater coincidence. It is to be picked up about 13° on a line conducted from Regulus to Spica Virginis.

CCCXCI. 54 LEONIS.

Æ 10ʰ 46ᵐ 56ˢ   Prec. + 3ˢ·27

Dec. N 25° 36'·1   —— S 19"·03

Position 102°·5 (*w* 8)   Distance 6"·5 (*w* 5)   Epoch 1832·26

———— 102°·7 (*w* 8)   ———— 6"·2 (*w* 8)   ———— 1839·33

A neat double star just over the Lion's back, where it is preserved from the Lesser Lion by one of the map-maker's nooks; it will be found about 15° north-east of Regulus, on the line produced towards Alkaid, at the end of the Great Bear's tail. A 4½, white; B 7, grey. This beautiful object is 30 Ḥ. III., whose notification of the colours is identical with mine, in this instance; and a comparison of the results of those astronomers who preceded me in measuring it, afford testimony that little change, if any, has taken place in fifty-eight years:

	Pos.	Dist.	Ep.
Ḥ.	99° 14'	7"·10	1781·14
H. and S.	98° 19'	7"·02	1821·68
Σ.	102° 48'	6"·18	1830·35

CCCXCII. 87 Ḥ. I. LEONIS MINORIS.

Æ 10ʰ 51ᵐ 38ˢ   Prec. + 3ˢ·29

Dec. N 29° 50'·0   —— S 19"·16

Mean Epoch of the Observation . . . . 1837·11

A large bright-class orbicular nebula, on the Little Lion's haunch, lying 4° on a line from ξ Ursæ Majoris into the south-west space towards Regulus; it was discovered by Ḥ. in April, 1785, and is No. 805 of his son's Catalogue of 1830. This remarkable object, with the exception of a 7th magnitude in the north, is in a field strewed with glimpse stars, from the most remote of which it may still be inconceivably remote, proceeding by analogy. II. observed it closely, and says, "no doubt a distant globular cluster;" in other words, not only suns beyond suns, but glorious systems of suns arranged in harmonious order. Where facts are still wanting, we can only form our opinions upon general principles. Now, when the dot which includes our system occupies a range of 3,600,000,000 of miles in diameter, besides a larger space which it controls,—should it be taken for an average among the millions of suns around, what imagination can grasp the immensity of creation! Indeed, where system thus stretches beyond system, the space must be infinite, or infinitely near it; and in such contemplation we become conscious of our own littleness. But no subject whatever, except revelation, can give a more exalted conception of the Eternal Fountain of all intelligence.

CCCXCIII. α CRATERIS.

$$\text{Æ.}\quad 10^h\ 52^m\ 00^s \qquad \text{Prec.} + \quad 2^s\text{·}95$$
$$\text{Dec. S}\ 17°\ 26'\text{·}9 \qquad \text{——}\ \text{S}\ 19''\text{·}17$$

Position AB 97°·0 (w 2) Difference Æ = 42ˢ·1 ⎫
—— BC 268°·2 (w̄ 3) ——————— 4ˢ·9 ⎬ Epoch 1835·38
⎭

A star with two very distant companions in the *sf*, on the base of the Cup. A 4, orange tint; B 8, intense blood colour; C 9, pale blue,— a fourth star away in the *sf* quadrant. This object may once have been brighter, since it acquired a name—Alkes—and was lettered *a*; but δ is now the lucida, and wears the Greenwich honours. It may be found by carrying an occult line from Arcturus, through δ Virginis, and rather more than the same distance to the south-west. The large star has a very considerable proper motion, the amount of which has been thus assigned:

$$\begin{array}{lll}
P\ldots & \text{Æ} - 0''\text{·}59 & \text{Dec.} + 0''\text{·}06 \\
B\ldots & \quad -\ 0''\text{·}44 & \quad +\ 0''\text{·}14 \\
A\ldots & \quad -\ 0''\text{·}48 & \quad +\ 0''\text{·}14
\end{array}$$

Κράτηρ, Crater, though a small and inconsiderable asterism, is one of the old 48; and is easily made out by six stars of the 4th magnitude in an annular form, on Hydra's back, forming Cicero's *fulgens Cratera*. The scholiast on *Germanicus* termed it *Urna*, and the Arabians *Bátiyah*, a large cup, and *al-Khas*, the shallow basin; which last was corrupted to *Alhas* by the framers of the Alphonsine Tables, but Scaliger properly suggested that the word should be *Alkes*, the name now used for the star *a*. While one party looked upon this goblet as Noah's, and others as the bowl of Bacchus, Schickhard, a reformer of the sphere, declared it to be the cup of Joseph. The number of its constituents have been thus stated:

Ptolemy	. . . 7 stars		Hevelius	. . . 10 stars
Tycho Brahé	. . 8		Flamsteed	. . . 31
Bayer	 11		Bode	 95

CCCXCIV. β URSÆ MAJORIS.

$$\text{Æ.}\quad 10^h\ 52^m\ 08^s \qquad \text{Prec.} + \quad 3^s\text{·}67$$
$$\text{Dec. N}\ 57°\ 14'\text{·}3 \qquad \text{——}\ \text{S}\ 19''\text{·}17$$

Position 172°·6 (w 2) Distance 75''·0 (w 1) Epoch 1831·37

A bright star with a distant companion, on the Greater Bear's body. A 2, greenish white; B 11, pale grey,—other stars in view. Though Piazzi has registered but a small quantity from his Bradleian deductions, this star appears to have a very perceptible proper motion according to later comparisons; they are thus:

$$\begin{array}{lll}
P\ldots & \text{Æ} + 0''\text{·}12 & \text{Dec.} + 0''\text{·}06 \\
Br\ldots & \quad +\ 0''\text{·}26 & \quad +\ 0''\text{·}05 \\
B\ldots & \quad +\ 0''\text{·}23 & \quad +\ 0''\text{·}03
\end{array}$$

Although the name Helice, a winding spiral figure in geometry, pertained to the whole asterism of the Greater Bear—as will presently be seen—it was also specially applied to β, the southernmost of the two pointers; this star has always been a favourite with ancient and modern seamen, because, by a line from it through α—both stars being the farthest from the tail—the Pole-star is always readily found.   It has since then obtained the name of Merak, from the Arabian *Meráḳ al-dubb-al-akbar*, the loins of the Greater Bear:

> Where Charles's Wain adorns the sky,   if Merak you would know,
> The Pole-star led through Dubhe's light   will mark it just below.

CCCXCV. α URSÆ MAJORIS.

Æ 10ʰ 53ᵐ 48ˢ   Prec. +   3ˢ·80
Dec. N 62° 36′·8   —— S 19″·21

Position 203°·8 (*w* 7)   Distance 380″·6 (*w* 3)   Epoch 1832·41

A standard Greenwich star, with a distant companion, on the Great Bear's back. A 1½, yellow; B 8, yellow, being No. 214 of Piazzi's Hora X.; from whose Catalogue these remarkably coincident results are obtained by reduction:

Pos. 205° 0′   Dist. 384″·0   Ep. 1800

A, the northern pointer, which was marked β or second magnitude by Ptolemy, is suspected by H. of being variable, and he asked me in October, 1838, to compare it with ε in the same asterism; but my slight examination was *res infecta*.   A proper motion has been given it, to the following value:

P.... Æ − 0″·24   Dec.   0″·00
B....   − 0″·20   − 0″·09
A....   − 0″·26   − 0″·09

Ἄρκτος μεγάλη, Arctos Major, the Great Bear, rivals Orion in beauty, and is the most splendid and conspicuous of those asterisms in the Northern Hemisphere which never set; and is, of course, one of the ancient constellated groups.   But the "doers into English" have certainly injured the purity of its descent to our times, for Job is made to talk about Arcturus, whereas Bochart assures us that the Hebrew word is derived from an Arabic one for bier; but Eben Ezra maintains it to be *agalah*, a waggon.   Both these renderings apply to the succeeding denominations of the Greeks, Romans, Italians, Germans, and English, in the Ἄμαξα, Plaustrum, Triones, Feretrum, Cataletto, Wagen, and David's Car, the Plough, and Charles's Wain.   In the latter, the two pointers are termed the hind wheels, the other two the fore wheels, and the three in the tail are the horses.   The Egyptians, we are assured, called this constellation the Hippopotamus, whence my intelligent friend, Professor Leemans, says, " Ursa Major, quæ secundum Champollionem dicebatur Canis Typhonis, in tabulis astronomicis indicatur figurâ hippopotami: Horus Apollo."   It was also sometimes styled Ἑλίκη in Greece, a name which, dropping the mythological fable, alludes to its circumvolution round the pole, whence Aratus, speaking through Germanicus, says:

Dat Graiis *Helice* cursus majoribus astris,
Phœnicas Cynosura regit.

Homer's description, however, of this revolving course, by which the asterism watches Orion from its arctic den, is but lamely rendered by Pope. In those early times the name of bier, or sarcophagus, was directly applied to the four bright stars disposed in the form of a quadrangle on the bear's body; and the three which we call the horses, or tail—for this bear actually has a tail of 20° projecting from his stern-frame—symbolized the children of the deceased in attendance. Kircher, to be sure, claims the four stars of "the square" as the bier of Lazarus, and the three of the tail as Maria, Martha, and Magdalen; while Schiller sees, in the same group, the ship of St. Peter. Our popular name of Charles's Wain (*ceorl* unde *churl*) is familiarized from the Gothic Karl-wagen, the *charl*, or peasant's cart; and it is applied to the seven well-known stars *a*, *β*, *γ*, *δ*, *ε*, *ζ*, and *η*, which are disposed in the form of a quadrangle joined by one of its corners to a triangle. Here the classic astronomer will recognise the Septentriones, of which Cicero says:

Quas nostri Septem soliti vocitare triones.

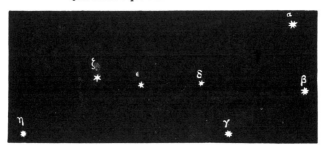

This constellation guided the nocturnal path of ships, whence it is introduced into the beautiful picture of night in Apollonius Rhodius; and Manilius tells us,

Seven equal stars adorn the Greater Bear,
And teach the Grecian sailors how to *steer*.

Modern navigators, of course, resorted to the same asterism, wherefore King James in his *Prentise*, describing the azure gown of Urania as decorated with fixed stars, says,

Heir shynes the Charlewain, there the Harp gives light,
And heir the Seaman's Starres, and there Twinnis bright.

Nor is it less an object of regard with our present seamen, by whom it is usually referred to in alignment, as a known figure; though as Neptune obligingly consented that it should never set within his domain, both he and Juno must have been dwelling in Europe, for it has been my fate to see it set often enough, as well as to lose it entirely. And there is little in southern celestial scenery to balance the loss, either in beauty or utility; look to the rhymes for its prime use in alignment:

Where yonder radiant hosts adorn   the northern evening sky,
Seven stars, a splendid glorious train,   first fix the wand'ring eye.
To deck great Ursa's shaggy form,   those brilliant orbs combine;
And where the first and second point,   there see Polaris shine.

But it must be admitted, that our poets have done little to foster or exalt the taste for astronomy. Perhaps the theme is too sublime to be shackled by metre; and as to the nonsensical

> Glittering stars in borrowed lustre

of Young*, and the

> Blue etherial vault of space

of a Parnassian contemporary, it were better to leave us to such effusions as Captain Sturmy enriched the old *Mariner's Magazine* with:

> The army of the starry sky
> Declares the glory of God most high;
> Seen and perceived among all nations
> In eight-and-forty constellations.

To return. The principal star in this constellation is called Dubhe, from *Dubb*, the Arabic for a bear, the name of the whole asterism, and erroneously entered in the Alphonsine Tables for α Ursæ Majoris only. It was also designated *Dhuhr dubb-al-akbar*, the back of the Great Bear. When Bayer facilitated the arrangement of the fixed stars, in 1603, he marked those in each constellation by the Greek alphabet, according to their degrees of brightness. But he made an exception in Ursa Major, so that the principal stars are lettered nearly in their order of Æ. The constituents of this grand asterism have been thus numbered, as progressive power has been applied:

Ptolemy	. . . 35 stars	Griemberger . .	57 stars
Copernicus	. . 35	Hevelius . . .	73
Tycho Brahé .	. 56	Flamsteed . .	87
Kepler	56	Bode	338

CCCXCVI. 88 ♓. I. LEONIS MINORIS.

Æ 10ʰ 54ᵐ 29ˢ   Prec. + 3ˢ·28
Dec. N 28° 49'·9   —— S 19"·23

Mean Epoch of the Observation . . . . 1836·26

A bright-class nebula, on the Little Lion's haunch, with some glimpse stars in the field, of which the principal are in the *sp* quadrant. It is pale white, elongated, and has the semblance of a nucleus. It was first registered in April, 1785, and is No. 810 of H.'s Catalogue, who says it is a resolvable distant cluster. Differentiated with ξ Ursæ Majoris, which star is also useful in its alignment, should such be attempted; a north-east ray from Regulus towards ξ will pass its site at four-fifths of the distance from the Lion's heart. ξ Ursæ will readily be made out, with ν 2° north of it, between Regulus and Alkaid.

---

* Did he of the *Night Thoughts* pick up his science from Blundvil? This sage philosopher, writing on the sphere, in 1594, has an apposite passage:
Q. "Why are not the stars seene as well in the day as in the night?"
A. "Because they are darkened by the excellent brightnesse of the Sunne, from whom they borrow their chiefest light."
We are, however, inclined to forgive Young's *borrowed lustre* when we recollect his
> One sun by day, by night *ten thousand* shine;
> And light us deep into the Deity.

CCCXCVII. 229 P. X. LEONIS.

Æ 10^h 55^m 44^s Prec. + 3^s·10
Dec. N 4° 30'·0 —— S 19"·26

Position 280°·0 (w 2)   Distance 1"·3 (w 2)   Epoch 1836·29

A very neat double star, preceding the Lion's hind-legs. A 8, and
B 8 , both white. It closely follows 58 Leonis, a star of the 5th mag-
nitude, which lies a little south of a line produced from Regulus to
Spica, at one-third of the distance. This pretty object was discovered
by Σ., who marked it one of his "pervicinæ" under No. 1504 of his
Catalogue of 1827; and he has since thus measured it:

Pos. 275°·68 Dist. 1"·076 Ep. 1829·13

A slight movement in space has been detected in A, of which we
must for the present suppose that B partakes; it has been thus valued:

P.... Æ - 0"·15   Dec. - 0"·02
B....   - 0"·08   - 0"·08
T....   + 0"·15   - 0"·18

CCCXCVIII. 13 ♓. I. LEONIS.

Æ 10^h 57^m 37^s Prec. + 3^s·07
Dec. N 0° 49'·6 —— S 19"·31

Mean Epoch of the Observation . . . . 1836·26

A bright-class nebula, preceding the Lion's hind-paws, with an
8th-magnitude star following in the sf quadrant, and four of the 10th
magnitude form a trapezium in the nf, between which and the nebula is
one of the 13th lustre. This object was discovered by ♓. in February,
1784, and is No. 818 of his son's Catalogue; being large, elongated in
direction np and sf, pale white, and well defined, with the brilliance
increasing in the sp region. It closely follows 62 Leonis, a star of the
6th magnitude, which is 20° south-east of Regulus, and about 11°¼
west by south of β Virginis, the nearest bright star to the east.

This enormous mass of luminous matter is an outlier of the vast
nebulous tract which appears to be posited nearly at right angles to the
Galaxy; but in irregular occurrence. This wonderful zone consists
mostly of groups of spherical nebulæ; and skilful inference shows, that
they are as much beyond our sidereal system, as the distance of the stars
exceeds that of our planet from the Sun! As to our own apparently
vast distance from the solar orb, it may be deemed pitifully minute and
almost infinitesimal in comparison.

Besides the more condensed masses, diffused nebulosity exists in an
abundance which exceeds all imagination; and the indefatigable ♓.
examined more than 150 square degrees of it. His conclusion is, that

the high degree of rarefaction of the nebulous matter, should not be
considered an obstacle to the theory of its finally being compressed
into a body of the density of our Sun: for, supposing the nebula to
be about 320 billions of miles distant, and its diameter subtending
an angle of 10', then must its magnitude exceed that of the Sun by
more than 2 trillions of times! This presents magnitude and mass vast
and inconceivable; and has staggered many a tyro. Now several important astronomical truths have been strongly conceived, and adopted
by vigorous understandings, long before their evidence became indubi-
table; and such will be received by the wise, without that "itching
morbus demonstrandi" of which Thomas Lydyat so bitterly complained,
when he determined to oppose, against Jesuit or Papist, the bringing in
of the Gregorian year. The developments which crown I.H. with imperishable fame, will for ages draw forth both practical and theoretical
talent, so that his reasonings and conclusions on the condensation of
nebulous matter into suns and planets, will be rigorously reviewed and
tested; and there is no doubt but that future exertions will create
progressive advances in means. Already has the Earl of Rosse, as
hath been mentioned, page 16, produced the most perfect telescope that
ever was constructed; and he has now undertaken another of 6 feet
aperture and 50 feet focus, with every prospect of attaining perfection.
May diligent observation and faithful records follow, in the true
Herschelian spirit of advancing sidereal astronomy.

CCCXCIX.   239 P. X. LEONIS.

$Æ$ $10^h 58^m 17^s$ Prec. $+$ $3^s·12$
Dec. N $7° 59'·9$ —— S $19''·32$

Position $164°·5$ (w 8)   Distance $8''·0$ (w 6)   Epoch $1831·25$
—— $164°·7$ (w 9)   —— $8''·2$ (w 9)   —— $1839·16$

A neat and delicate double star, close to the Lion's hind-legs. A 8,
topaz yellow; B 11½, cerulean blue. These places do not quite quadrate
with Piazzi's; and there is some shade of doubt as to identity, this being
placed before No. 238, Hora X., in the Palermo Catalogue; but the
object here observed is 78 I.H. II., and its relative fixity seems conclu-
sively established, by comparing the above results with the former
measures:

I.H.	Pos. 165° 21'	Dist. ...	Ep. 1783·00
S.	164° 01'	$8''·63$	1825·29
Σ.	164° 46'	$8''·03$	1833·23

This small pair closely follows χ Leonis, a star of the 4th magnitude,
nearly in mid-distance between Regulus and η Virginis: it will, therefore, be readily caught up by the practical gazer.

CCCC. 46 ℍ. V. URSÆ MAJORIS.

Æ 11ʰ 02ᵐ 02ˢ   Prec. + 3ˢ·57
Dec. N 56° 31'·8   —— S 19"·41

Mean Epoch of the Observation . . . . 1835·29

A large milky-white nebula, on the body of the Great Bear, with a small star at its *sp* apex, and an 8th-magnitude preceding it at double the distance; there is also a brightish group in the *np* quadrant. It is easily found, since it lies only about 1° south-east of β, Merak. This object was discovered by ℍ. in April, 1789; and is No. 831 of his son's Catalogue. It is faint but well defined, being much elongated with an axis-major trending *sp* and *nf* across the parallel, and a small star, like a nucleus, in the centre. As ℍ. considers this star to be unconnected with the nebula, it follows that it is between us and it, and therefore strengthens to confirmation our belief in the inconceivable remoteness of those mysterious bodies.

CCCCI. 9 P. XI. LEONIS.

Æ 11ʰ 05ᵐ 17ˢ   Prec. + 3ˢ·19
Dec. N 21° 00'·3   —— S 19"·48

Position 288°·6 (*w* 7)   Distance 1"·2 (*w* 3)   Epoch 1833·31

A neat double star, on the Lion's loins; and closely to the south-west of δ, Zosma, a star of the 3rd magnitude. Both 7½, and both faint yellow. This beautiful object, which resembles η Coronæ, was discovered by Σ., and is No. 1517 of the Dorpat Catalogue, where the results preceding mine are:

Pos. 287°·80   Dist. 1"·052   Ep. 1829·70

There appears to be a small though sensible proper motion of this star through space, of which these values have been estimated:

P.... Æ − 0"·37   Dec. − 0"·07
B....   − 0"·38   − 0"·11
T....   − 0"·20   − 0"·18

CCCCII. 97 M. URSÆ MAJORIS.

Æ 11ʰ 05ᵐ 24ˢ   Prec. + 3ˢ·53
Dec. N 55° 52'·9   —— S 19"·48

Mean Epoch of the Observation . . . . 1837·16

A large planetary nebula, or globular collection of nebulous matter, found by M. in 1781, on the Great Bear's flank, with several stars in the field, one of which is pretty close. It lies about 2° to the south-east of β, Merak, and just south of an imaginary line from β to γ. This very singular object is circular and uniform, and after a long inspection looks like a condensed mass of attenuated light, seemingly of the size of Jupiter. The mean apparent place was obtained by a differentiation with that of ψ Ursæ Majoris, and this diagram was sketched. Sir

William Herschel discovered this orb in 1789, and found it a globular body of equal light throughout: he also says, "From the observation of the 20-foot telescope, it appears that the profundity of this object is beyond the gauging power of that instrument; and as it must be suffi-ciently distant to be ambiguous, it cannot well be less than of the 980th order." The 980th order!

CCCCIII. δ LEONIS.

ℛ 11ʰ 05ᵐ 35ˢ   Prec. +   3ˢ·19

Dec. N 21° 24'·1   —— S 19"·48

Position A B   50°·0 (w 1)   Difference ℛ = 4ˢ·9 (w 1)

——— A C 345°·0 (w 1)   ——————— 2ˢ·9 (w 1)   } Epoch 1836·21

A standard Greenwich star, with companions forming a coarse triple object, in a black field, at the root of the Lion's tail. A 3, pale yellow; B 13, blue; C 9, violet; a fourth and most minute star is suspected nearly in the line of C, and about a third of the distance, but this is not confirmed by Mr. Dawes, who also examined the object for me.

δ Leonis and 6 Virginis are the stars which Flamsteed observed, in 1690, with the object which has since proved to be Uranus. δ has a proper motion, which has been thus registered:

P.... ℛ + 0"·18   Dec. − 0"·11
B....   + 0"·26   − 0"·14
A....   + 0"·22   − 0"·14

This star is called *Zosma*, from ζῶσμα or ζώμα, a tunic or girdle, but why so designated, deponent sayeth not: it is not mentioned by Ideler. With θ it forms the XIth Lunar Mansion, and is named *al-*

zubrah, the mane or hair on the lion's back. Zosma will be readily distinguished 18° north-east of Regulus, and 5° due north of θ, where it forms a fine scalene triangle with θ and β.

CCCCIV. 50 ♅. II. LEONIS.

Æ 11^h 08^m 28^s PREC. + 3^s·17
DEC. N 18° 55'·0   —— S 19"·54

MEAN EPOCH OF THE OBSERVATION . . . . 1836·35

A fine round white nebula, at the root of the Lion's tail, well-defined, and with a brightish centre. A little to the north of it is another rather smaller, which is 51 ♅. II.; and there are some telescopic stars between them. They are followed by a triangle of three stars, and the whole forms a field of high interest.

This object was discovered by ♅. in March, 1784, and described as a "triple" nebula, but I can only see the above-mentioned: they form Nos. 845 and 846 of H.'s Catalogue of 1830. The mean apparent place was carefully differentiated with δ Leonis, from which it bears about 2°½ to the south-south-east.

CCCCV. φ LEONIS.

Æ 11^h 08^m 32^s PREC. + 3^s·05
DEC. S 2° 46'·6   —— S 19"·54

POSITION 285°·5 (w 4)   DISTANCE 105"·0 (w 2)   EPOCH 1831·27

A star with a distant companion, under Leo's hind-paw; to pick it up, drop a ray from δ through θ, and prolong it south till it intersects a line produced from η Virginis to Alphard. A 5, pale yellow; B 8½, violet; two other distant stars in the field; which did not escape the steady gaze of Piazzi, who remarks, Nota 23, Hora XI., "Binæ telescopicæ præcedunt." This object was merely examined to ascertain whether an increase of distance between the two had actually taken place, to the extent denoted by a comparison of the following registered measures:

♅.	Pos. 281° 00'	Dist. 98"·58	Ep. 1782·11
H. and S.	286° 56'	106"·25	1821·24

φ Leonis has a very slight movement in space attributed to it, of the following contending values:

P.... Æ	− 0"·20	Dec. + 0"·09
B....	− 0"·05	− 0"·05

CCCCVI. ξ URSÆ MAJORIS.

Æ 11ʰ 09ᵐ 38ˢ   Prec. + 3ˢ·25
Dec. N 32° 25'·8   —— S 19"·56

Position	Distance	Epoch
207°·5 (w 4)	1"·8 (w 3)	1830·94
—— 196°·9 (w 7)	—— 1"·9 (w 3)	—— 1832·29
—— 190°·6 (w 9)	—— 2"·1 (w 3)	—— 1833·34
—— 182°·6 (w 7)	—— 1"·8 (w 4)	—— 1834·97
—— 180°·2 (w 5)	—— 1"·9 (w 5)	—— 1835·37
—— 170°·9 (w 8)	—— 1"·8 (w 3)	—— 1836·33
—— 165°·5 (w 7)	—— 1"·8 (w 6)	—— 1837·28
—— 160°·7 (w 9)	—— 2"·1 (w 4)	—— 1838·48
—— 156°·9 (w 8)	—— 2"·0 (w 5)	—— 1839·23
—— 143°·2 (w 9)	—— 2"·3 (w 6)	—— 1843·16

A binary star of the most interesting description, in the Bear's left hind-paw, directly under *ν*, or *Al úla Borealis;* the two forming a miniature of δ and θ Leonis, which are 10° due south of them. A 4, subdued white; B 5½, greyish white, and both very bright. It is usually designated Al Ula Australis, from the Arabian *Al-kafzah-al-úla,* the gazelle's first spring or leap; and has had this proper motion assigned it:

P.... Æ - 0"·52   Dec. - 0"·64
B....   - 0"·45   - 0"·57
A....   - 0"·51   - 0"·60

This extraordinary pair forms an object of the gravest importance, since its motion is so rapid as to admit of being demonstrated by measurements at short intervals. So far back as 1825, H. most strongly recommended it for *constant* and careful observation. "This done," said he, " there is no doubt of our arriving at a precise knowledge of the elements and position of the orbit described by each about their common centre of gravity; and the question of the extension or non-extension of the NEWTONIAN law of gravity to the sidereal heavens—the next great step which physical astronomy has yet to make—will be effectually decided." The effect of such a call, was to animate exertion; and its consequence has been a series of excellent measures by various astronomers. From the whole the following are selected, as my discussion points:

Ḥ.   Pos. 143° 47'   Dist. 3"·50   Ep. 1780·33
H. and S.   256° 27'   2"·81   1823·29
Σ.   229° 30'   1"·82   1827·26

There is not a binary star which goes further to prove that there is yet much to be accomplished in the art of measuring, than *ξ*; for the discordancies in the value of its annual movement in orbit are great. This, and the gap in the period of the starting points, make it difficult to investigate the elements by the process of gathering the radii vectores of the revolver from the angle of position—a method founded by Herschel on the condition, that they are equal to the square roots of the apparent angular velocities. Still I essayed it, and brought out a period of sixty-five years, which, perhaps, is not offensively out.

Sir John Herschel had also predicted that, between 1839 and 1841, this star would have completed a full revolution from the epoch of the first measurement of its position in 1780, in a periodic time of about 59 years. M. Savary elaborately computed its orbit in 1830, making its period to be $58\frac{1}{4}$ years, and in his acute discussion of the details, adduces an equation due to the finite velocity of light. Indeed, of all the stellar orbital periods yet discussed, this of ξ Ursæ Majoris is admitted to be the most rigorously determined; and did we but know its parallax, and thence its absolute distance from the Earth, we might readily decide upon the linear extent of its orbit. This long-sought and ardently-wished desideratum, however, now appears to be close at hand, and the admirable labours of Bessel and Henderson will enable us to set about weighing the masses of those stars whose relative movements become known: since the prevalence of universal gravitation is unequivocally established by the elliptic forms of the orbits of binary systems. But more of this at 61 Cygni.

From what is advanced respecting stars of the first lustre, it may be inferred that ξ Ursæ, which is of the 4th magnitude, may be about $7\frac{1}{2}$ millions of solar distances from us. With this deduction, which, though hypothetical, is not arbitrary, the geometer proceeds to view this binary system as united by attraction under masses equal to 117 and 42, which together embrace a space 159 times as large as our solar system; and if we can accord a density equal to that of the Sun, the diameter of the two bodies will be $4\frac{9}{10}$ and $3\frac{1}{4}$ times greater than his. A path slightly elliptic is indicated by the one moving round the other, in about sixty years, having an orbital inclination at an angle of $37°\ 45'$ with the plane of our visual ray, and a mean separation of $83\frac{2}{3}$ of our distance from the Sun. According to this hypothesis, the apparent diameters of these two stars are only $\frac{1}{400}''$ and $\frac{1}{585}''$; wherefore, if we see them in our best telescopes under an angle of half a second, the apparent enlargement must be attributed to the dispersion of light in the atmosphere, in the instrument, and in the eye.

CCCCVII. ν URSÆ MAJORIS.

Æ	11ʰ 09ᵐ 49ˢ	PREC. +	3ˢ·26
DEC. N	33° 58′·0	—— S	19″·57

POSITION 147°·2 (w 5)   DISTANCE 7″·8 (w 2)   EPOCH 1834·31

A delicate double star, on the Bear's left hind-foot, immediately above ξ, and therefore called *Al Ula Borealis*. The six stars in the three feet—ν and ξ, λ and μ, ι and κ—were designated *Kafzát al-dhibá*, springs of the gazelle; and the two in each foot one spring. These antelopes seem to have been the *informes* since gathered up as Leo Minor; and the springs were owing to the fear of the Greater

Lion's tail. Thus on the Borgian globe stands *Al-dhibá wa-auládu-há*, the gazelles and their young, where now the Lesser Lion figures. A 4, orange tint; B 12, cerulean blue, preceded exactly on the equatoreal line by a 7th-magnitude star, with $\triangle \mathbb{R} = 21^s \cdot 5$. This elegant object was discovered by Σ., No. 1524, and thus measured:

Pos. 146° 56′   Dist. 7″·096   Ep. 1830·69

and A has had a slight proper motion assigned, thus:

$$P.... \mathbb{R} - 0''\cdot03 \qquad Dec. + 0''\cdot05$$
$$B.... \quad + 0''\cdot06 \qquad\quad\; + 0''\cdot04$$
$$A.... \quad - 0''\cdot01 \qquad\quad\; + 0''\cdot05$$

It is mentioned under No. CCCCVI. that this star with ξ forms a miniature of δ and θ Leonis, 10° to the south of them; and for further identity it may be added that a west-south-west ray from Cor Caroli to ϵ Leonis passes them in mid-distance.

CCCCVIII. δ CRATERIS.

$\mathbb{R}$ 11ʰ 11ᵐ 21ˢ   PREC. + 3ˢ·00

DEC. S 13° 54′·8   —— S 19″·59

POSITION 94°·1 (w 1)   DIFFERENCE $\mathbb{R}$ = 19ˢ·8 (w 1)   EPOCH 1834·33

A secondary Greenwich star, with a very distant companion on the Goblet; midway between Alphard and Spica, but a little south of the line produced by them. A 3½, pale orange; B 11, pale blue,—other small stars in the field. The magnitude of A is here adopted from Piazzi, and he is followed by most of the recent catalogues; but on comparing it with others of similar grade, it hardly appears sufficiently bright to have been lifted out of the 4th magnitude, where Ptolemy, Tycho Brahé, Hevelius, and Flamsteed, placed it. It is now, however, the lucida of the asterism; and the value of its proper motion in space is thus registered:

$$P.... \mathbb{R} - 0''\cdot19 \qquad Dec. - 0''\cdot06$$
$$B.... \quad - 0''\cdot06 \qquad\quad\; + 0''\cdot17$$
$$A.... \quad - 0''\cdot10 \qquad\quad\; + 0''\cdot20$$

CCCCIX.   39 P. XI. CRATERIS.

$\mathbb{R}$ 11ʰ 11ᵐ 38ˢ   PREC. + 3ˢ·04

DEC. S 6° 01′·4   —— S 19″·60

POSITION 315°·0 (w 4)   DISTANCE 8″·0 (w 4)   EPOCH 1836·29

A neat but minute double star, between the Cup and the Lion's hind-feet; and exactly 8° due north of δ Crateris, the alignment for which has just been given: a ray projected from δ Crateris to θ Leonis passes over it at nearly one quarter of the distance. A 8½, and B 9, both bluish white,—other small stars in the field. This object was detected double by Σ., No. 1530 of the Dorpat Catalogue, under these measures:

Pos. 314° 60′ Dist. 7″·65 Ep. 1830·23

CCCCX. 66 M. LEONIS.

Æ 11ʰ 11ᵐ 48ˢ   Prec. + 3ˢ·14
Dec. N 13° 52′·4   —— S 19″·60

Mean Epoch of the Observation . . . . 1835·31

A large elongated nebula, with a bright nucleus, on the Lion's haunch, trending *np* and *sf;* this beautiful specimen of perspective lies just 3° south-east of θ Leonis.
It is preceded at about 73ˢ by another of a similar shape, which is Messier's No. 65, and both are in the field at the same time, under a moderate power, toge- ther with several stars. They were pointed out by Méchain to Messier

in 1780, and they appeared faint and hazy to him. The above is their appearance in my instrument.

These inconceivably vast creations are followed, exactly on the same parallel, at Δ Æ = 174ˢ, by another elliptical nebula of even a more stupendous character as to apparent dimensions. It was discovered by H., in sweeping, and is No. 875 of his Catalogue of 1830.

The two preceding of these singular objects were examined by Sir William Herschel, and his son also; and the latter says, "The general form of elongated nebulæ is elliptic, and their condensation towards the centre is almost invariably such as would arise from the superposition of luminous elliptic strata, increasing in density towards the centre. In many cases this increase of density is obviously attended with a diminu- tion of ellipticity, or a nearer approach to the globular form in the central than in the exterior strata." He then supposes the general constitution of those nebulæ to be that of oblate spheroidal masses of every degree of flatness from the sphere to the disc, and of every variety in respect of the law of their density, and ellipticity towards the centre. This must appear startling and paradoxical to those who imagine that the forms of these systems are maintained by forces identical with those which determine the form of a fluid mass in rotation; because, if the nebulæ be only clusters of discrete stars, as in the greater number of cases there is every reason to believe them to be, no pressure can be propagated through them. Consequently, since no general rotation of such a system as one mass can be supposed, Sir John suggests a scheme which he shows is not, under certain conditions, inconsistent with the law of gravitation. "It must rather be conceived," he tells us, "as a *quiescent form*, comprising within its limits an indefinite multitude of individual constituents, which, for aught we can tell, may be moving one among the other, each animated by its own inherent projectile force, and deflected into an orbit more or less complicated, by the influence of that law of internal gravitation which may result from the compounded attractions of all its parts."

CCCCXI. ι LEONIS.

$\mathbb{R}$　11^h 15^m 35^s　　Prec. + 3^s·12

Dec. N 11° 24′·8　　—— S 19″·67

Position 90°·5 (w 5)　Distance 2″·4 (w 2)　Epoch 1836·40

——— 87°·7 (w 6)　　——— 2″·4 (w 3)　——— 1839·32

——— 86°·0 (w 8)　　——— 2″·5 (w 4)　——— 1843·38

A binary star on the Lion's flank; 7° south-west of Denebola, with which star and θ it forms a neat scalene triangle, of which it is the southern vertex. A 4, pale yellow; B 7½, light blue. This beautiful object was discovered by Σ., whose earliest measures were:

Pos. 97° 0′　Dist. 2″·30　Ep. 1827·28

and afterwards, by treating the result of his observations by the method of least squares, he inferred a retrograde angular motion. My own measurements in 1836 and 1839 were so satisfactory, that Σ.'s views are confirmed, and I scruple not to designate it a binary system; but I am not aware that any other telescopes have yet been directed to its investigation; however, as great interest must attach to its movements, it may be proper to show ground for entertaining the suspicion, which is, that from 1827 to 1836, the observations gave a change of angle = − 0°·72 per annum, and from 1836 to 1839 one = − 0°·93.

Since this was written, the Rev. W. R. Dawes has shown me some forthcoming observations which he has taken of this object; and on the night of April 24th, 1843, we saw it very neatly from Mr. Bishop's observatory, in Regent's Park. My measures at Hartwell just afterwards were, to the senses, very satisfactory.

Piazzi's comparisons of ι Leonis with Bradley and Mayer, yielded no proper motion in $\mathbb{R}$; but more recent examinations show a positive movement; the values stand:

P.... $\mathbb{R}$　0″·00　Dec. − 0″·03

Br...　+ 0″·25　　　− 0″·06

B....　+ 0″·23　　　− 0″·06

CCCCXII. 219 ♅. I. URSÆ MAJORIS.

Æ 11ʰ 16ᵐ 01ˢ   Prec. +   3ˢ·28
Dec. N 39° 38'·2   —— S 19"·68

Mean Epoch of the Observation . . . . 1836·37

A bright-class nebula, before the animal's left hind-leg, nearly in a line with four telescopic stars to the south from the 9th to the 11th magnitudes—two of which precede, and two follow; the latter are the smallest and by far the nearest. It is small, round, and lucid white; and H. says it is resolvable. Assuredly it is most wonderful that this object—apparently about 40" or 50" in diameter—should present a remote universe; yet the resolvability implies the existence of an immense number of stars at a proximity apparently much greater than those in our own Via Lactea. Indeed it has been shown, that clustering collections of stars may easily contain upwards of 50,000 of them!

A ray from Polaris, through the tip of Draco's tail, and prolonged 30° due south, will arrive at this nebula's site; which is nearly in mid-distance between Alioth and Regulus, and exactly between Flamsteed's No. 55 and 57 Ursæ Majoris.

CCCCXIII. γ CRATERIS.

Æ 11ʰ 16ᵐ 54ˢ   Prec. +   2ˢ·99
Dec. S 16° 48'·3   —— S 19"·69

Position 102°·5 (w 2)   Distance 3"·0 (w 1)   Epoch 1838·26

A close double star, in the centre of the Goblet, and 3° south-south-east of δ, the present lucida. A 4, bright white; B 14, grey, a star of the 11th magnitude following nearly on the parallel, in the line of A and B, at about 25ˢ; and the 8th-magnitude star mentioned by Piazzi, Note 62, Hora XI., is at a distance in the np. This fine but delicate object, erroneously lettered χ in the Palermo Catalogue, was discovered by H. in his 20-foot Sweeps: the acolyte was sufficiently visible in my telescope for the rock-crystal, or for exact estimation, but was utterly obnubilated under the slightest artificial light. The proper motion of A has been asserted, and then doubted; but my observations countenance the following quantities at least, although they were not of a sufficiently exact nature to decide the question:

P....	Æ	− 0"·29	Dec.	+ 0"·02
B....		− 0"·09		+ 0"·04
A....		− 0"·19		+ 0"·07

CCCCXIV.   194 ⛢.   I. URSÆ MAJORIS.

Æ   11ʰ 17ᵐ 21ˢ   Prec. +   3ˢ·31
Dec. N 44° 27′·9   ——— S 19″·70

Mean Epoch of the Observation . . . . 1837·25

A large elongated nebula, between the Greater Bear's hind knees, with two minute stars about twice as far to the south of it, as they are from each other. This nebula was discovered by ⛢. in January, 1788, and is No. 887 of his son's Catalogue. It is pale white, brightish towards the centre, and its axis of extension is preceded by star-dust; but it presents an ill-defined surface, and has the appearance of a flat stratum seen obliquely.

A ray from Regulus into the north-east to γ Ursæ Majoris, will reach the site of this nebula at about two-thirds of the distance, where it is *nf* 56 Ursæ, about one degree.

CCCCXV.   83 LEONIS.

Æ   11ʰ 18ᵐ 39ˢ   Prec. +   3ˢ·09
Dec. N   3° 53′·1   ——— S 19″·72

Position 150°·5 (*w* 9)   Distance 29″·5 (*w* 9)   Epoch 1831·33
——— 150°·8 (*w* 9)   ——— 29″·8 (*w* 9)   ——— 1839·22

A neat double star, on Leo's right hind-leg, closely *np* τ Leonis, at 21° east-south-east of Regulus. A 8, silvery white; B 9, pale rose-tint. This object is 13 ⛢. iv.; and a comparison of the several measures, while it shows the distance to be stationary, does not confirm the suspected direct change in the orbital angle. The difference, therefore, between ⛢. and the recent observers, must be attributed to instrumental error. These are the results with which I compare my own, for drawing so decided a conclusion.

⛢.	Pos. 144° 55′	Dist. 29″·08	Ep. 1780·27
H. and S.	151° 07′	29″·54	1821·20
Σ.	150° 01′	29″·58	1832·71

The object is of very easy measurement, and therefore its fixity may be deemed to be established: indeed, under the present modes of observing, greater coincidence can hardly be looked for than that which appears from 1821 to 1839. It has a very sensible proper motion in space—presumptively common to both components—which is thus registered:

P....	Æ − 0″·78	Dec. + 0″·22	
B....	− 0″·77	+ 0″·16	
A....	− 0″·80	+ 0″·16	

CCCCXVI. 57 URSÆ MAJORIS.

Æ 11ʰ 20ᵐ 26ˢ   Prec. + 3ˢ·26

Dec. N 40° 13′·0   —— S 19″·75

Position 9°·9 (w 8)   Distance 5″·9 (w 5)   Epoch 1835·42

A neat double star, on the Bear's left hind-leg, nearly midway on a line produced between ε Ursæ Majoris and Regulus. A 6, lucid white; B 9, violet. This beautiful object is 86 Ḥ. iii., discovered in 1782, but as he gave no measures of distance, it was probably not rigidly observed. Still, however, as he mentions the angle of position to have been 75° 36′ *nf*, which was upwards of 4° less than Sir James South found it in 1825, a slight orbital change was inferred. This is not confirmed by my results, and the amount of proper movement in space is too insignificant for consideration; the micrometric conclusions for comparing with my epoch, stand thus:

S.   Pos. 10° 15′   Dist. 6″·294   Ep. 1825·25
Σ.   10° 42′   5″·373   1831·91

Another remark is called for: Ḥ. says that the small star is "a red point without sensible magnitude;" and S., upwards of half a century afterwards, rates it of the 10th lustre, as shown by his 7-foot telescope. In the summer of 1835 it was very distinct, being a bright 9th-size, bearing illumination admirably. Is it variable?

CCCCXVII. λ DRACONIS.

Æ 11ʰ 21ᵐ 50ˢ   Prec. + 3ˢ·68

Dec. N 70° 12′·8   —— S 19″·77

Position 345°·0 (w 1)   Difference Æ = 17ˢ·5 (w 1)   Epoch 1834·32

A bright star with a distant companion on the tip of Draco's tail. A 3½, orange tint; B 12, white; several other small stars in the field under power 157. Though neither Brioschi nor Argelander have placed λ on their lists, I cannot but suspect it of proper motions in space, even to greater values than those assigned, which are:

P.... Æ − 0″·15   Dec. + 0″·09
B....   − 0″·06   − 0″·08

This is the Giauzar of the Catalogues, from *al-jaúzá*, a word of doubtful origin and signification, but interpreted *al-júza*, the central, from being nearly midway between the Pole-star and the Pointers. Others have rendered it *Jauzahr*, as written by the Arabs for the Persian *Gau-zahr*, the poison-place. But this related to a notion that the nodes, or points where the Moon crosses the ecliptic, were poisonous, because those nodes happened to be called the head and tail of the

Dragon. λ, however, is merely registered here because it is of some importance in several of the polar alignments; it is readily known by its position between the Pointers and Polaris, as it follows the produced line at nearly 8° from Dubhe.

CCCCXVIII. 91 P. XI. LEONIS.

Æ 11ʰ 22ᵐ 43ˢ   Prec. + 3ˢ·05
Dec. S 5° 50′·2   —— S 19″·78

Position 330°·2 (*w* 5)   Distance 9″·5 (*w* 3)   Epoch 1834·30

A fine but very delicate double star, in a barren field, under the Lion's hind-paw; it lies south of a line from η Virginis to Alphard, and one-fourth of the way, where it is also two-thirds of the distance between Denebola and δ Crateris. A 8, creamy white; B 11, greenish, and rather more difficult under illumination than its magnitude quite warrants. This object was discovered by Σ., and the following are the measures assigned in the great Dorpat Catalogue:

Pos. 331°·85   Dist. 9″·38   Ep. 1831·65

CCCCXIX. 88 LEONIS.

Æ 11ʰ 23ᵐ 29ˢ   Prec. + 3ˢ·13
Dec. N 15° 15′·3   —— S 19″·79

Position 319°·8 (*w* 9)   Distance 14″·9 (*w* 9)   Epoch 1835·38

A neat double star, on Leo's flank, nearly midway between β and θ. A 7, topaz yellow; B 9, pale lilac; a third star of the 10th magnitude follows in the *sf* quadrant. This is a good object, although it rather weakens under illumination; it is 51 I.J. III., and was first enrolled with its distance scored as "a little inaccurate." Time, however, has proved it to be otherwise; for it has been thus measured:

I.J.	Pos. 317° 33′	Dist. 14″·63	Ep. 1782·11
II. and S.	320° 14′	14″·67	1823·28
Σ.	319° 50′	15″·31	1829·02

From these results, probable errors considered, compared with the very satisfactory observations I obtained, the object appears to be unchanged in a period of fifty-three years. A proper motion to the following amount has been assigned to A, in which B probably partakes:

P.... Æ	– 0″·38	Dec.	– 0″·25
B....	– 0″·27		– 0″·20
A....	– 0″·30		– 0″·17

CCCCXX. 17 CRATERIS.

$$\text{Æ } 11^h 24^m 21^s \qquad \text{Prec. } +\ 2^s \cdot 96$$
$$\text{Dec. S } 28^\circ 23' \cdot 0 \qquad ——\ \text{S } 19'' \cdot 80$$

Position 207°·8 (w 3)   Distance 10″·1 (w 2)   Epoch 1833·21

A neat double star, of which A is $5\frac{1}{2}$, lucid white; and B 7, violet tint. This object is situated in the far south, about 15° south by east of δ Crateris, its lucida; and there it is also pointed to by a ray from Spica through a in the Raven's beak. It is formed by Nos. 95 and 96 P. Hora IX., who designates it 17 Hydræ; but though on Hydra's back, it is in the Crater's boundary, and albeit Bayer's stars in that asterism do not exceed 11, Flamsteed numbered it as above, in the British Catalogue, having carried his numeration up to 31. It is 96 H̶. III., and was thus first registered:

Pos. 205° 33′ Dist. 9″·78 Ep. 1783·03

The observations are as coincident as can be expected, under the variable refractions with which the place of this star is troubled. A is also subject to a small amount of proper motion, thus:

$$P.\ldots\ \text{Æ } + 0''\cdot 24 \qquad \text{Dec. } + 0''\cdot 15$$
$$B.\ldots\ \quad\ \ + 0''\cdot 06 \qquad\qquad\ \ + 0''\cdot 15$$

Now had not B been physically connected with A, it seems clear that their situations regarding each other ought to have varied in forty years. Even Mr. Baily's reduced value, although it would let the position alone, would have increased the distance to $15''\cdot7$.

CCCCXXI. 90 LEONIS.

$$\text{Æ } 11^h 26^m 23^s \qquad \text{Prec. } +\ 3^s \cdot 13$$
$$\text{Dec. N } 17^\circ 40' \cdot 9 \qquad ——\ \text{S } 19'' \cdot 83$$

Position AB 209°·1 (w 9)   Distance 3″·5 (w 8) ⎫
——— AC 233°·9 (w 5)   ——— 58″·8 (w 3) ⎭ Epoch 1835·38

A triple star in the root of Leo's tail, *infra eductionem caudæ*, where it will be found 4° west-north-west of Denebola, nearly on the line shot from that star towards Zosma. A 6, silvery white; B $7\frac{1}{2}$, purplish; C $9\frac{1}{4}$, pale red. This fine object is 27 H̶. I., and must be classed as optical; for the excellent and coincident measures previous to mine, when compared, already indicated fixity. They were:

H̶. 1782·29	H. and S. 1822·27
Pos. AB 208° 51′ Dist. 3″·00±	Pos. 208° 52′ Dist. 4″·45
AC 234° 48′ 53″·72	233° 19′ 60″·75

CCCCXXII.   111 P. XI. URSÆ MAJORIS.

Æ   11ʰ 27ᵐ 52ˢ   Prec. +   3ˢ·17
Dec. N 28° 40'·0   —— S 19"·85

Position AB 340°·1 (w 8)   Distance   1"·4 (w 5)   ⎫
—— AC 145°·0 (w 2)   ——   17"·0 (w 1)   ⎬ Epoch 1834·31
                                                ⎭

A fine and delicate triple star, under the left hind-leg of Leo. A 6, and B 7, both pale blue; C 13, plum colour. This object forms Σ.'s No. 1555; and the two principals have undergone these micrometrical measurements:

Σ.	Pos. 339° 22'	Dist. 1"·25	Ep. 1829·12
H.	338° 00'	(omitted)	1830·26
D.	340° 18'	1"·45	1832·24

Here a typographical error has crept into H.'s first series of 7-foot measures, 111 Piazzi being designated 3. At a distance in the *sf* is the star mentioned in the Palermo Catalogue: " Alia 8ᵃ magnitud. sequitur 57" temporis, 20" circiter ad austrum." It is situated in a very vacant space to the eye, about 8° from δ Leonis, in a north-north-east direction towards Mizar; but, to the powerful reflectors now in use, is in a very ocean of nebulæ.

CCCCXXIII.   126 P. XI. VIRGINIS.

Æ   11ʰ 30ᵐ 14ˢ   Prec. +   3ˢ·06
Dec. S 1° 33'·1   —— S 19"·87

Position 280°·9 (w 5)   Distance 5"·0 (w 1)   Epoch 1833·27

A fine but very delicate double star, between Leo's hind-paw, and the Virgin's wing. A 7, pale orange; B 12, reddish,—with a distant dull star in the *sf*. This beautiful object, far too delicate for metrical observation with a small instrument, was discovered by Σ., and is No. 1560 of the Dorpat Catalogue, the last edition of which records these results:
Pos. 280°·57 Dist. 5"·093 Ep. 1831·58
This star may be fished up about 5° south-south-west of β Virginis.

CCCCXXIV.   94 ♌. I. URSÆ MAJORIS.

Æ   11ʰ 32ᵐ 48ˢ   Prec. +   3ˢ·19
Dec. N 37° 25'·9   —— S 19"·90

Mean Epoch of the Observation . . . . 1837·24

A first-class nebula, at the back of the Bear's hind-leg, of a pale white tint. It is elliptical, and though large, so faint as not to be readily made out, till the equatoreal clock fixes the telescope upon it, when it *rises* to view, lying slightly across the parallel, with a following star. It was discovered in April, 1785, and was re-examined by H., No. 945: the space around is apparently blank and starless; but the spot is to be sought by a line projected from Regulus through ξ Ursæ Majoris, and carried 5°¼ beyond, where it is intersected by a ray from θ Leonis to δ Ursæ, the north-east corner of the square.

CCCCXXV. β LEONIS.

Æ. 11ʰ 40ᵐ 54ˢ   Prec. + 3ˢ·10
Dec. N 15° 28′·0   —— S 19″·98

Position 114°·0 (*w* 2)   Distance 298″·0 (*w* 1)   Epoch 1833·47

A standard Greenwich star, with a distant companion, on the switch of Leo's tail. A 2½, bluish; B 8, dull red; preceded by a 7th-magnitude star in the *np*. A sensible proper motion is attributed to A, thus:

$$P.... \text{Æ} - 0''·53 \quad \text{Dec.} - 0''·08$$
$$B.... \quad - 0''·48 \quad\quad - 0''·10$$
$$A.... \quad - 0''·52 \quad\quad - 0''·09$$

This star is named Denebola, from the Arabian *dhanab-al-asad*, the lion's tail. It is likewise designated Serpha, from *al ṣarfah*, the changer (of the weather), being the XIIth Lunar Mansion; and has moreover been known as Daphira, from *al-dafírah*, the tuft of hair at the tail's extreme. A line from Procyon through Regulus passes θ, and over β Leonis, the latter being about 25° from Regulus; or, for eye measurement in aligning, half as far from Regulus, as the latter is from Procyon. If seeking it from the eastward, drop a line from Alkaid, lead it through Cor Caroli, and extend it about double the distance between those two stars into the south-west. The brackish rhymes point out a nearly equilateral figure, thus:

From Deneb, in the Lion's tail,   to Spica draw a line,
Then will these two with Arcturus   a bright triangle shine.

CCCCXXVI. β VIRGINIS.

Æ. 11ʰ 42ᵐ 22ˢ   Prec. + 3ˢ·10
Dec. N 2° 40′·0   —— S 19″·99

Position 285°·3 (*w* 2)   Difference Æ = 13ˢ·1 (*w* 1)   Epoch 1833·31

A discarded Greenwich star, with a small companion, in the upper part of the Virgin's right wing. A 3½, pale yellow; B 11, light blue; and these two are followed at a distance, near the parallel, by a 9th-magnitude star. A is charged with a large spacial motion, of which the following is the registered amount; besides its having been selected by Mayer, in his *Commentatio de motibus propriis:*

P.... Æ + 0″·76　　Dec. - 0″·30
B....　 + 0″·80　　　　 - 0″·28
A....　 + 0″·75　　　　 - 0″·28

This star is comparatively not bright for its magnitude; but may be found by the star-gazer's dropping a fancied line from γ Ursæ Majoris, the southern star of the square in the Greater Bear, through Deneb, and carrying it about 13° beyond. Piazzi calls it Zavijava, which is corrupted from *Záwiyat-al-'auwà*, the retreat of the barker. Ulugh Beigh has it *Min-al-'auwà*, *i. e.*, the stars of the barker, or barking bitch. These stars, β, γ, δ, and η, and according to Tízíní ε also, form the XIIIth Lunar Mansion; of which γ is termed by Kazwíní *Záwiyah-al-'auwà*, the barker's corner, being at the angle of those stars.

CCCCXXVII.　170 P. XI. LEONIS.

Æ 11ʰ 44ᵐ 31ˢ　　Prec. + 3ˢ·10
Dec. N 16° 19′·8　　—— S 20″·00

Position 13°·4 (w 3)　Distance 35″·0 (w 2)　Epoch 1832·99

A double star, in the brush of Leo's tail, and following Denebola at about 1° to the north-east. A 7½, pearl white; B 9½, livid. This object is 60 Ḥ. v.; and a comparison of my measures with the preceding ones, indicates that a slow angular change is in progress. The results alluded to are:

Ḥ.　　　Pos. 19° 12′　Dist. 37″·24　Ep. 1782·09
H. and S.　　14° 03′　　　37″·11　　　1823·28

which certainly show a retrocession of greater regularity and amount, than would be fairly owing to the anomalies included under the vague term "personal equation."

CCCCXXVIII.　173 Ḥ. I. URSÆ MAJORIS.

Æ 11ʰ 44ᵐ 36ˢ　　Prec. + 3ˢ·14
Dec. N 37° 52′·8　　—— S 20″·00

Mean Epoch of the Observation . . . . 1837·14

A bright-class nebula, of a pale white tint, with a central blaze, between the Bear's hind-legs, and the Hounds. It was discovered in March, 1786, and is No. 1005 of H.'s Catalogue. When seen by strong moonlight, it looks like a star in a burr; but in dark nights has a very large apparent diameter. The above mean apparent place hangs upon ξ Ursæ Majoris; its site is known by glancing from Alkaid, at the tip of the Greater Bear's tail, towards Regulus, and it is passed at rather less than half the distance. But the immediate vicinity is very poor to the unassisted eye.

CCCCXXIX. γ URSÆ MAJORIS.

Æ	11ʰ 45ᵐ 23ˢ	Prec. +	3ˢ·19
Dec.	N 54° 35′·1	——	S 20″·00

Position 34°·7 (w1)   Difference Æ = 15ˢ·3 (w1)   Epoch 1833·37

A standard Greenwich star, with a distant companion, on the Bear's right ham. A 2, topaz yellow; B 9, ashy paleness, with a fine group of stars in the field, of which one near the parallel of the *sp* quadrant is coarsely double. The reduction of my own meridional observations does not warrant the assumption of proper motion to A, but a small quantity has been assigned, of these values:

P....	Æ + 0″·06	Dec. −	0″·03
B....	+ 0″·24	−	0″·02
A....	+ 0″·19	−	0″·01

This star is called Phecda, from the Arabian *Fekhāh-al-dubb-al-akbar*, the thigh of the Great Bear; and being at the south-east angle of the conspicuous stellar square in that constellation, is useful in some of the alignments by which particular stars are noted. Thus we are told by the oft-quoted word-spinner:

> He who would scan the figured skies, its brightest gems to tell,
> Must first direct his mind's eye north, and learn the Bear's stars well.

CCCCXXX. 45 ⅏. V. URSÆ MAJORIS.

Æ	11ʰ 45ᵐ 25ˢ	Prec. +	3ˢ·18
Dec.	N 53° 13′·6	——	S 20″·00

Mean Epoch of the Observation . . . . 1838·24

A large pale-white nebula, on the Bear's right haunch, about 1°¼ south of γ; discovered in April, 1789. It has a peculiar appearance in the field, from there being a coarse small double star to the north of it, and from its being followed by a vertical line of five equidistant telescopic stellar attendants. This object is fine, but, in my instrument, faintish; it brightens towards the middle; and ⅏. says there is, in that part, an *unconnected* star, the which I cannot make out.

From every inference this nebula is a vast and remote globular cluster of worlds, for H. assures us it is actually resolvable. By its blazing towards the centre, proof is afforded that the stars are more condensed there than round its margin, an obvious indication of a clustering power directed from all parts towards the middle of the spherical group. In other words, the whole appearance affords presumptive evidence of a wonderful physical fact,—the actual existence of a central force.

CCCCXXXI. 65 URSÆ MAJORIS.

Æ 11ʰ 46ᵐ 45ˢ   Prec. + 3ˢ·15
Dec. N 47° 22′·0   —— S 20″·01

Position AB 35°·8 (w 7)   Distance 3″·8 (w 7) ⎫
——— AC 115°·0 (w 4)   ——— 63″·5 (w 4) ⎬ Epoch 1837·39

A triple star, on the Bear's left thigh, of which the sf member, or C, is Piazzi's No. 184, Hora XI. A 7, bright white; B 9½, pale purple; C 7, white. This is 72 Ḥ. ɪ., and the various measures hitherto taken indicate fixity. They are:

Ḥ. 1782·89		H. and S. 1821·30	
Pos. AB 36° 15′ Dist. 4″·00 ±		Pos. AB 36° 46′ Dist. 3″·71	
AC 112° 21′ 60″·08 —		AC 114° 17′ 62″·18	

The magnitude which I have assigned, on mature comparison, to B, does not altogether quadrate with Ḥ.'s description, of its being a mere point, which would hardly be suspected. It may be variable; and I have reason also to think C is. Probably all three are physically connected, in which case they will partake of the slow proper motions of A, which have been thus registered:

P.... Æ − 0″·05   Dec. + 0″·15
B.... + 0″·09   − 0″·02
T.... + 0″·02   0″·00

There is therefore reasonable ground for supposing that this object will increase in interest. It is easily fished up by carrying a ray from the Pole-star, between κ and λ Draconis, through γ Ursæ Majoris, and 7° south of it, where it will meet a cross line from ψ to η.

CCCCXXXII. 62 Ḥ. IV. URSÆ MAJORIS.

Æ 11ʰ 47ᵐ 03ˢ   Prec. + 3ˢ·18
Dec. N 56° 00′·7   —— S 20″·01

Mean Epoch of the Observation . . . . 1837·24

A planetary nebula, in a barren field, on the Bear's hind-quarter. It is small, and uniformly of a pale bluish-white colour, but exceedingly well-defined, without the haziness mentioned by H., No. 1017; but I certainly had a splendid night for the examination, with the instrument in capital working order. This neat object was discovered with a moderate reflector, at Slough, in April, 1789. There is a lilac-tinted 10th-magnitude star on its south vertical. The mean apparent place was obtained by differentiation from that of γ Ursæ Majoris, from which it is only about 1°½ to the north by east.

CCCCXXXIII.   2 COMÆ BERENICIS.

$\mathbb{R}$ 11ʰ 56ᵐ 05ˢ   Prec. + 3ˢ·08
Dec. N 22° 21′·1   —— S 20″·04

Position 240°·8 (*w* 6)   Distance 3″·3 (*w* 3)   Epoch 1832·31
—— 239°·9 (*w* 9)   —— 3″·6 (*w* 9)   —— 1839·37

A neat double star just over the Lion's tail; and nearly mid-way between its own lucida and Denebola, but preceding the imaginary line so produced. A 6, pearly white; B 7½, lilac tint. This beautiful object is 47 Ḥ. ɪɪ., and having been rigidly examined, is concluded to have no motion appreciable in fifty-seven years. The previous measures were:

	Pos.	Dist.	Ep.
Ḥ.	242° 18′	4″·00 ±	1782·30
H. and S.	238° 45′	3″·68	1823·14
H.	234° 47′	4″·88	1828·42
Σ.	240° 37′	3″·73	1829·54

My last measures of this star were so satisfactory, that I place the greatest reliance on them. The night was truly superb, and the definition of the objects so exquisite, that they resembled two jewels fixed in the field. Indeed, under the Claude Lorraine illumination, they were admirably sharp and tranquil; the vision therefore might almost be styled perfect. Under such circumstances, with the instrument in the finest working order, and the eye so turned, by inclining my head, as to have its principal section parallel to the wires, the results could hardly fail of being among the best I ever obtained.

CCCCXXXIV.   195 Ḥ. I. URSÆ MAJORIS.

$\mathbb{R}$ 11ʰ 58ᵐ 51ˢ   Prec. + 3ˢ·08
Dec. N 43° 57′·3   —— S 20″·04

Mean Epoch of the Observation . . . . 1837·24

A bright-class nebula, in a poor field, behind the Greater Bear's left hind-leg, at rather more than one-third of the distance from δ towards Denebola, where it is within a degree to the east by north of 67 Ursæ, a star of 5½ magnitude. It is of a lucid white colour, and narrow, being elongated in the direction of *np* and *sf*. Ḥ. discovered it in January, 1788; and it is No. 1088 of his son's Catalogue of 1830. In the *nf* quadrant is a fine wide double star, the individuals of which point exactly to the centre of the nebula, on a line forming an angle of about 230° with the meridian. The mean apparent place is obtained  by a careful differentiation with 2 Canes Venatici, and the annexed was its appearance in the field of view.

CCCCXXXV. 98 M. VIRGINIS.

Æ 12$^\text{h}$ 06$^\text{m}$ 01$^\text{s}$ Prec. + 3$^\text{s}$·06

Dec. N 15° 47'·2   —— S 20"·04

Mean Epoch of the Observation . . . . 1837·25

A fine and large, but rather pale nebula, between Virgo's left wing and Leo's tail; with the bright star, 6 Comæ Berenicis, following in the next field exactly on the parallel. M., who discovered it in 1781, merely registered it as " a nebula without a star, with an extremely faint light;" but on keeping a fixed gaze it brightens up towards the centre. It is elongated, in the direction of two stars, the one *np* and the other *sf* of the object; with another star in the *nf* quadrant pretty close. Differentiated with β Leonis, which star it follows by 6°$\frac{1}{2}$ in the direction of Arcturus; it lies on the outskirts of the vast region of nebulæ that adorns the Virgin's wing.

CCCCXXXVI. δ URSÆ MAJORIS.

Æ 12$^\text{h}$ 07$^\text{m}$ 28$^\text{s}$ Prec. + 3$^\text{s}$·00

Dec. N 57° 55'·3   —— S 20"·04

Position 127°·0 (*w* 2)   Difference Æ = 20$^\text{s}$·4 (*w* 1)   Epoch 1832·41

A fine star—suspected of variability—with a distant companion, on the Greater Bear's stern-frame. A 3, pale yellow; B 9, ash-coloured, other stars in the following part of the field. This was enrolled by Ptolemy, Ulugh Beigh, Hevelius, La Caille, Bradley, and Piazzi, of the 3rd magnitude; but Tycho Brahé and the Prince of Hesse designate it of the 2nd. Flamsteed records it of 2$\frac{1}{2}$, Pigott of the 4th, and I have, on careful comparison, sometimes thought it too bright for a 3rd rank. It may therefore prove to be variable from the 2nd to the 4th lustre; and that at long periods.

This star, the north-east one of the brilliant square, is Megrez of the Palermo and other Catalogues; a word abbreviated from the Arabian *Maghrez-al-dubb-al-akbar*, the root of the Great Bear's tail, since it is "à la naissance de la queue," rather than "in radice caudæ," a berth given sometimes to the neighbouring star Alioth, with its little companion, called the Fox, to the *nf*. ε Ursæ Majoris, was also called *al-hawar*, intensely bright, and *al-jaun*, the black horse; but its most usual name, Alioth, first appears in the Alphonsine Tables. This being also the reported site of Hevel's nebula of 1660, and Messier's No. 40, of 1764, I searched for them by *fishing*, but found only a couple of small stars lying *np* and *sf*, with gleams of others. This object, however, resolved by my telescope, may have been the one seen by those astronomers.

By Mr. Baily's recent investigation, Megrez appears to have a larger proper motion in Æ than was attributed to it by Piazzi, on a comparison with Bradley; these are the values:

P.... Æ − 0″·06   Dec. − 0″·08
B....   + 0″·29   − 0″·06

CCCCXXXVII.   95 ♓. I. CANUM VENATICORUM.

Æ 12ʰ 07ᵐ 35ˢ   Prec. + 3ˢ·04
Dec. N 37° 12′·8   —— S 20″·03

Mean Epoch of the Observation . . . . 1837·27

A fine white nebula, between the two Hounds, discovered by ♓. in April, 1785, and No. 1146 of his son's Catalogue. It is in a very poor field, with two small stars preceding, one on each side of the parallel, and a 10th-magnitude one pretty closely *sf*. On attentively gazing, especially when the equatoreal clock is applied, the nebula *comes up* very fairly defined, and is of a slightly oval shape, with its elongation from *np* to *sf*; but despite of all my coaxing, I was unable to see the two remarkable nuclei, so beautifully figured by H. This object is one of a nebulous group located between Cor Caroli and ξ Ursæ Majoris, and bearing west-south-west from the former, distant about 8°.

CCCCXXXVIII.   35 ♓. I. VIRGINIS.

Æ 12ʰ 07ᵐ 37ˢ   Prec. + 3ˢ·06
Dec. N 14° 02′·8   —— S 20″·03

Mean Epoch of the Observation . . . . 1837·24

A long pale-white nebula, among some telescopic stars, on the upper part of Virgo's left wing; announced in the preceding verge of the field by a 9th and a 10th-magnitude star, closely on each side of the parallel. It was discovered by ♓. in April, 1783, and is No. 1148 of his son's Catalogue, where it is erroneously synonymed as 109 ♓. i.—but it is delicately figured at No. 59 of the engraved illustrations. This is a very curious object, in shape resembling a weaver's shuttle, and lying across the parallel; the upper branch is the faintest, and the centre exhibits a palpable nucleus, which in my instrument brightens at intervals, as the eye rallies. It is an outlier of the vast and wonderful nebulous region passing through Virgo, and is one-third of the way from β Leonis to ε Virginis.

Nearly 3′ following this, and 1°¼ to the north, is 99 M., a large round nebula, which, though pale, is well defined in my instrument.

CCCCXXXIX.   2 CANUM VENATICORUM.

ℛ 12ʰ 08ᵐ 06ˢ   Prec. +   3ˢ·03
Dec. N 41° 33′·1   —— S 20″·03

Position 259°·8 (w 8)   Distance 11″·1 (w 5)   Epoch 1831·37
—— 259°·5 (w 9)   —— 11″·3 (w 7)   —— 1839·31

A neat double star, near Chara's mouth, and in a barren naked-eye spot, about 9° south-west of Cor Caroli, and one-third of the distance between that star and δ Leonis. A 6, golden yellow; B 9, smalt blue. This very fine object is 85 Ηℓ. iii., and notwithstanding the supposed connection between strong colours and motion, its fixity is fully established. Indeed, all the recorded observations are eminently coincident:

	Pos.	Dist.	Ep.
Ηℓ.	259° 00′	12″·20	1782·87
H. and S.	259° 31′	11″·53	1822·18
Σ.	259° 38′	11″·42	1832·16

CCCCXL.   32 P. XII. VIRGINIS.

ℛ 12ʰ 09ᵐ 57ˢ   Prec. +   3ˢ·07
Dec. S   3° 03′·9   —— S 20″·03

Position 198°·5 (w 9)   Distance 21″·4 (w 5)   Epoch 1831·39
—— 198°·6 (w 9)   —— 20″·6 (w 9)   —— 1836·27

A fine double star, near the centre of Virgo's right wing, 3° due south of η Virginis, and one-third of the way from Spica to Regulus; formed by Nos. 32 and 33 of Piazzi's Hora XII., according to whom they are both 7½ magnitudes, though I cannot but say that B is certainly smaller than A. They are both of a silvery white tinge. It is also No. 22 of Ηℓ.'s list of 145 New Objects; but the first micrometric measures I meet with are those of H. and S., who obtained these results:

Pos. 197° 02′   Dist. 21″·01   Ep. 1823·33

This being an object which I had viewed with the amiable Piazzi at Palermo, in 1814, I was induced to bestow no small attention upon the observations at the first epoch here recorded; and being a fair and easy object, I am happy to say that to the *senses* they are every way satisfactory. The second epoch was from my desire to examine rigidly the space between this object, and 17 Piazzi, Hora XII., under the following circumstances.

In the month of September, 1835, I received a letter from M. Cacciatore, the successor of Piazzi at Palermo, of which this extract is literally translated: " One important thing I must communicate to you. In the month of May I was observing the stars that have proper motion; a labour that has employed me several years. Near the 17th star, 12th hour, of Piazzi's Catalogue, I saw another, also of the 7·8th magnitude, and noted the approximate distance between them. The weather not having permitted me to observe on the two following nights, it was not till the third night that I saw it again, when it had advanced a good deal, having gone further to the eastward and towards the equator. But clouds obliged me to trust to the following night. Then, up to the end of May, the weather was horrible; it seemed in Palermo, as if winter had returned; heavy rains and impetuous winds succeeded each other, so as to leave no opportunity of attempting anything. When, at last, the weather permitted observations at the end of a fortnight, the star was already in the evening twilight, and all my attempts to recover it were fruitless: stars of that magnitude being no longer visible. Meanwhile the estimated movement, in three days, was 10″ in Æ, and about a minute, or rather less, towards the north. So slow a motion would make me suspect the situation to be beyond Uranus. I was exceedingly grieved at not being able to follow up so important an examination."

Though this notification arrived after the apparition of Virgo had passed for the season, I lost no time in advising the astronomical world of its tenour; and bestirred myself, on the reappearance of the constellation, by making reticle diagrams of all the 7½-magnitude objects which I could find hereabouts. My endeavours proved fruitless; and after much good time lost in the search, I became convinced it would not be my fortune to rediscover a planet there.

An extract from my letter being read to the Academy of Sciences at Paris, 15th February, 1836, it was printed in the *Comptes Rendus* of that *séance*, with this sensible but severe animadversion by M. Arago:

"Il y a dans cette communication une circonstance que les astronomes auront beaucoup de peine à comprendre. Lorsque le tems redevint favorable à Palerme, à la fin de Mai, l'étoile mobile n'était plus visible, dit M. Cacciatore, à cause de la lumière crépusculaire du soir. L'explication est admissible lorsqu'il s'agit du passage de l'astre au méridien; mais deux, mais trois heures après le coucher du soleil, mais à nuit-close, rien ne pouvait empêcher de comparer la planète soupçonnée aux étoiles voisines, soit avec une machine parallactique, soit, à son défaut, *avec le grand cercle azimuthal qui occupe le premier rang parmi les instruments de l'Observatoire de Palerme.* Il nous paraît inconcevable qu'un observateur du mérite de M. Cacciatore, contrarié comme il l'était, comme il devait l'être, de ne pouvoir constater de réalité une découverte aussi capitale, ne se soit pas avisé de suivre l'astre hors du méridien."

CCCCXLI. 43 ♄. V. URSÆ MAJORIS.

Æ 12ʰ 11ᵐ 04ˢ	Prec. + 2ˢ·99
Dec. N 48° 11′·1	—— S 20″·02

Mean Epoch of the Observation . . . . 1837·26

A large white nebula, closely following the haunches of the Greater Bear, discovered by ♄. in 1788, and No. 1175 of his son's Catalogue.

It is a noble-sized oval, trending rather from the vertical in a direction *np* and *sf*, with a brightish nucleus in its southern portion; the lateral edges are better defined than the ends. It is preceded by two stars of the 10th magnitude, and followed by two others; and there are also some minute points of light in the field, seen occasionally by glimpses.

This object was carefully differentiated with Alkaid; and its place will be indicated by running a diagonal line across the square of Ursa Major, from *a* through γ, and carrying it 7½° into the south-east, that is, a little less than the distance between those stars.

CCCCXLII.   61 M. VIRGINIS.

Æ   12ʰ 13ᵐ 45ˢ   Prec. +   3ˢ·06
Dec. N   5° 21'·6   ——   S 20"·01

Mean Epoch of the Observation . . . . 1837·26

A large pale-white nebula, between the Virgin's shoulders. This is a well defined object, but so feeble as to excite surprise that Messier detected it with his 3½-foot telescope in 1779. Under the best action of my instrument it blazes towards the middle; but in H.'s reflector it is faintly seen to be bicentral, the nuclei 90" apart, and lying *sp* and *nf*. It is preceded by four telescopic stars, and followed by another. Differentiated with the following object, from which it bears about south by west, and is within a degree's distance.

This object is an outlier of a vast mass of discrete but neighbouring nebulæ, the spherical forms of which are indicative of compression.

CCCCXLIII.   17 VIRGINIS.

Æ   12ʰ 14ᵐ 24ˢ   Prec. +   3ˢ·06
Dec. N   6° 11'·8   ——   S 20"·01

Position 336°·2 (*w* 8)   Distance 19"·8 (*w* 6)   Epoch 1832·28

A neat double star, between the Virgin's shoulders; lying at nearly one-third of the distance from Denebola to Spica, and nearly north of η Virginis. A 6, light rose tint; B 9, dusky red. This object is 50 ℍ. IV., and from a difference between the first observations and those of H. and S., the change was suspected to be owing to the proper motions of the large star, the amount of which is thus assigned:

$$P.... \text{Æ} - 0''·25 \quad \text{Dec.} - 0''·02$$
$$B.... \quad - 0''·15 \quad\quad - 0''·07$$
$$A.... \quad - 0''·20 \quad\quad - 0''·08$$

But I am inclined to suppose that an error of 10° in the angular position may be imputed to the original entry at Slough, as the following measures of comparison would then be pretty coincident for a star of such disproportion:

Ḥ.	Pos. 328° 21′	Dist. 20″·15	Ep. 1782·10
H. and S.	339° 36′	20″·94	1823·20
Σ.	336° 45′	19″·32	1829·26

CCCCXLIV.   12 COMÆ BERENICIS.

Æ 12^h 14^m 27^s Prec. + 3^s·03

Dec. N 26° 44′·1   —— S 20″·00

Position 168°·2 (w 8)   Distance 66″·1 (w 4)   Epoch 1831·28

A bright star, with a distant companion, in the middle of the Tresses; about 1°¼ south-west of its lucida, and nearly mid-way between Cor Caroli and Denebola. A 5, straw-coloured yellow; B 8, rose-red; a third star of the same magnitude in the *sf* quadrant. This object is 121 Ḥ. v., first registered in the new-year's night of 1783; and its *comes* is alluded to by Piazzi's note to 59 Hora XII. The result of Ḥ.'s observations gave:

Pos. 163° 00′ Dist. 58″·91 Ep. 1783·00

It was next attacked by H. and S., who obtained:

Pos. 168° 47′ Dist. 65″·95 Ep. 1821·39

showing an increase both of angle and distance. A comparison of these, however, with my determinations, implies a relative fixity.

Berenice's Hair was intruded into the constellated host many ages ago, but was only confirmed between the time of the old 48 asterisms, and the gathering together of some clustered *amorphotæ* in the sixteenth century; for Ptolemy did not include it as a distinct asterism, but designates it only as πλόκαμος; and Ulugh Beigh enrols it as an *extra* of Leo, under the name of al ḍafírah, the tresses. Niebuhr heard it called *al-huzmeh*, at Cairo, the which signifies a bundle of wood, or corn; but the Arabs in general termed it *al-helba*, and the Trica of the Alphonsine Tables is recognised as being from τρίχες, a head of hair. It was anciently believed that the Tresses had been snatched into the heavens, because Conon, the astronomer, had so asserted, in order to console the lady for the loss of a lock of her hair, which she had dedicated to Venus, on account of a victory obtained by her husband, Ptolemy Evergetes; but it was Tycho Brahé who first fixed it, about 270 years ago. Old Thomas Hill, in his *Schoole of Skil*, 1599, calls these sacred tresses by the homely designation of Berenice's Bush; and there has been a name still homelier. It is readily found by running an imaginary line from Benetnasch, the outer horse of the wain, or tip of the Great Bear's tail, through Cor Caroli, and thence to Denebola, in the Lion's tail; midway between which two last, stands this fine though diffused cluster. The numbers have successively been:

Tycho Brahé . . . 14 stars	Flamsteed . . . 43 stars
Hevelius 21	Bode 117

CCCCXLV.   100 M. VIRGINIS.

ℛ 12ʰ 14ᵐ 52ˢ   PREC. + 3ˢ·04
DEC. N 16° 42'·6   —— S 20"·00

MEAN EPOCH OF THE OBSERVATION .... 1837·21

A round nebula, pearly white, off the upper part of the Virgin's left wing, and certainly at a great distance from Virgo's ear of corn, where the *Connaissance des Temps* places it: indeed, the true site will be hit upon just one-fifth of the way from β Leonis towards Arcturus. This is a large but pale object, of little character, though it brightens from its attenuated edges towards the centre; and is therefore proved to be globular. It was discovered by M. Méchain in 1781, and is accompanied by four small stars, at a little distance around it; besides minute points of light in the field, seen by occasional gleams.

We are now in the broad grand stratum of nebulæ, which lies in a direction almost perpendicular to the Galaxy, and passes from the south, through Virgo, Berenice's Hair, Canes Venatici, and the Great Bear, to the Pole, and beyond. This glorious but most mysterious zone of diffused spots, is an indisputable memorial to all future times, of the unwearied industry and indomitable scientific energy of Sir William Herschel. Yet has this unrivalled contributor to knowledge been disparagingly described, as a man indulging in "speculations of no great value to astronomy, rather than engage in computations by which the science can really be benefited." Save the mark! This is said of a philosopher of zeal and application hitherto unequalled: one whose contributions to the *Philosophical Transactions* prove the bold but circumspect grandeur of his conceptions, his consummate mechanical resources, and the exactness of his elaborate calculations. Herschel's labours, however, transcended those of the age in which he was cast, although he gave such animation and bias to sidereal astronomy that his mantle was caught at.

CCCCXLVI.   δ CORVI.

ℛ 12ʰ 21ᵐ 35ˢ   PREC. + 3ˢ·10
DEC. S 15° 37'·4   —— S 19"·96

POSITION 210°·9 (*w* 5)   DISTANCE 23"·5 (*w* 3)   EPOCH 1831·34

A fine double star, on the Raven's right wing. A 3, pale yellow; B 8½, purple. This object is 105 ℍ. IV., and was thus noticed by Piazzi, No. 101, Hora XII.: "Duplex. Socia summê exigua 0"·5 temporis præcedit, parumper ad austrum." The following are the measures with which I compared my own results, and discordant as the angle of position seems to be, I am more inclined to attribute the differences to instrumental and accidental oversights, than to orbital movement; moreover, there is a minus quantity of proper motion imputed to the large star by Piazzi, and the observations at so low an altitude are teased with variable refraction:

Ḥ.	Pos. 216° 00′	Dist. 23″·50	Ep. 1782·87
H. and S.	213° 33′	24″·00	1823·29

The movement in space attributed to this star, is diminishing under the recent rigorous comparisons, as will be seen by the values:

P....	Ɍ − 0″·07	Dec.	− 0″·20
Br...	− 0″·11		− 0″·23
B....	− 0″·02		− 0″·16

The Palermitan and other Catalogues have dubbed this star Algorab, from the Arabian *Al-ghoráb*, the raven, though the star is less brilliant than β. Wherefore α, which is usually the brightest star in an asterism, has here less brilliance than β, γ, or δ, and is recorded as *Minḳár-al-ghoráb*, the raven's beak; and it is also called *Al-khibá*, the tent, a name given by some of the Arabs to Corvus. With the addition of *al-yemáni*, the southern tent, the same star is signified, says 'Abdu-r-rahmán Ṣúfí. (*Hyde Syntag.*, pp. 79, 81.) In the old Alphonsine Tables, the name Algorab is applied to γ.

The alignment of Algorab is both easy and pleasing. To the west-south-west of Spica two stars of the 3rd magnitude, and 3° apart, will be seen prolonging the line. These are δ and γ Corvi; δ, the nearest to Spica, is 15° from it, and it forms with that star and the well-known γ Virginis, an exact equilateral triangle.

CCCCXLVII.   49 M. VIRGINIS.

Ɍ 12ʰ 21ᵐ 36ˢ		Prec. + 3ˢ·05	
Dec. N 8° 52′·9		— S 19″·95	

Mean Epoch of the Observation . . . . 1836·37

A bright, round, and well-defined nebula, on the Virgin's left shoulder; exactly on the line between δ Virginis and β Leonis, 8°, or less than half-way, from the former star. With an eyepiece magnifying 93 times, there are only two telescopic stars in the field, one of which is in the *sp* and the other in the *sf* quadrant; and the nebula has a very pearly aspect. This object was discovered by Oriani in 1771, and registered by Messier as a "faint nebula, not seen without difficulty," with a telescope 3½ feet in length. It is a pity that this active and very assiduous astronomer could not have been furnished with one of the giant telescopes of the present day. Had he possessed efficient means, there can be no doubt of the augmentation of his useful and, in its day, unique Catalogue: a collection of objects for which sidereal astronomy must ever remain indebted to him.

CCCCXLVIII.   88 M. VIRGINIS.

ℛ   12ʰ 23ᵐ 54ˢ   PREC. +   3ˢ·03
DEC. N 15° 18′·5   —— S 19″·94

MEAN EPOCH OF THE OBSERVATION . . . . 1836·37

A long elliptical nebula, on the outer side of Virgo's left wing. It is pale-white in colour, and trends in a line bearing *np* and *sf;* and with its attendant stars, forms a pretty pageant. The lower or northern part in the inverted field is brighter than the southern, a circumstance which, with its spindle figure, opens a large field for conjecture.

This is a wonderfully nebulous region, and the diffused matter occupies an extensive space, in which several of the finest objects of Messier and the Herschels will readily be picked up by the keen observer in extraordinary proximity. The following diagram exhibits the local disposition of the immediate nebulous neighbours north of 88 Messier; they being preceded by M., No. 84, and followed by M. 58, 89, 90, and 91, in the same zone; thus describing a spot only 2°½ from north to south, and 3° from east to west, as the micrometer shows it.   And it

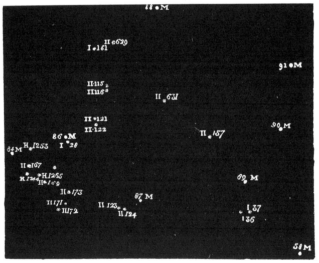

will be convenient to keep in mind, that the situation of the extraordinary conglomerate of nebulæ and compressed spherical clusters which crowd the Virgin's left wing and shoulder, is pretty well pointed out to the practised naked eye by ε, δ, γ, η, and β Virginis forming a semicircle to the east, whilst due north of the last-mentioned star, β Leonis marks the north-west boundary. Reasoning upon the Herschelian principle, this may reverently be assumed as the thinnest or shallowest part of our firmament; and the vast laboratory of the segregating mechanism by which compression and insulation are ripened, in the course of unfathomable ages. The theme, however imaginative, is solemn and sublime.

CCCCXLIX.   β CORVI.

$\mathcal{R}$ $12^h\ 26^m\ 00^s$ Prec. $+\ 3^s\!\cdot\!13$
Dec. S $22°\ 30'\!\cdot\!6$ —— S $19''\!\cdot\!92$

Position AB $119°\!\cdot\!5$ (w 1)   Difference $\mathcal{R} = 27^s\!\cdot\!4$ (w 1) $\Big\}$ Epoch $1831\!\cdot\!34$
—— AC $306°\!\cdot\!5$ (w 1)   ———————— $28^s\!\cdot\!0$ (w 1)

A Greenwich star of the second grade, elected in 1830. It is in the Raven's right claw, and lies nearly midway between the two distant companions, whose position and distance are here estimated. A $2\frac{1}{2}$, ruddy yellow; B 7, greenish yellow; C 8, dull grey. This is a fine star, and has unquestionably the precedence of lustre in Corvus, which could hardly have been the case in Bayer's time; and what is singular, it has no trivial Arabian designation. Ptolemy gave γ, or third degree of brightness, to a, β, γ, δ, and ε; but Tycho and Hevelius both rated a and ε of the 4th magnitude, and β has latterly been elevated to 2·3. Such discrepancies should be closely watched, for though the low altitude of the asterism may be against precision in this country, it must be recollected that Ptolemy, Ulugh Beigh, Alphonsus, and Piazzi, had a smaller south polar distance. The comparative lustre of the stars in Corvus, in the year 1796, was ably tabulated by ℍ., and is to be consulted in the 86th volume of the *Philosophical Transactions*, page 468.

Κόραξ, Corvus, is one of the constellated groups of the southern hemisphere, and though poor, is one of the ancient 48 asterisms. It is immediately to the east of Crater, and between Spica and Alphard, but considerably nearer to the former, where it is readily made out by a lozenge of four stars of the 3rd and 4th magnitudes. As it contains a part of the body of Hydra, on which the bird rests, it is sometimes designated *Hydra et Corvus*. It was piously regarded as Noah's raven; but this not being quite satisfactory to the Mosaicists, *Columba Noachi* was instituted by Royer, in 1679. (See a Hydræ.) The ancient Arabians called it 'Arsh al-simák al-a'zal, the throne of Spica; and with some it was also *Al-ajhmal*, the camels, *Al-khibá*, the tent, and 'Ajaz-al-Asad, the lion's rump, from an enormous constellation of which it was a part, without any reference to the Lion of the Zodiac, afterwards borrowed from the Greeks. Indeed these original Arab names relate to figures in the heavens imagined by the Bedawís, long before their descendants studied Hellenic astronomy. The constituent members of Corvus, have been thus numbered:

Ptolemy . . . . 7 stars		Hevelius . . . . 10 stars
Tycho Brahé . . 8		Flamsteed . . . 9
Kepler 7		Bode 61

A long occult line from Wega (a Lyræ) through Spica (a Virginis), and carried about 15° beyond, enters among the four principal and well-known stars of Corvus:

Mark in the space along the sky,   where Hydra's volumes are,
And 'twixt the Cup and Virgin's spike,   you'll find the Raven's square.

CCCCL. 8 CANUM VENATICORUM.

Æ 12ʰ 26ᵐ 08ˢ   Prec. + 2ˢ·93

Dec. N 42° 13′·7   —— S 19″·92

Position 228°·0 (*w* 1)   Distance 297″·0 (*w* 1)   Epoch 1835·24

A bright star with a distant companion, in the eye of Chara, the southern dog. A 4½, but suspected of variability, pale yellow; B 10, bluish; and there is another star in the *sp*, nearly on the parallel, at Δ Æ 28ˢ·5, as well as a very minute one in the *nf* quadrant. The large star is involved in a nebulous photosphere, as described by H., No. 1332 of Catalogue for 1830; but the nebulosity is no further apparent in my instrument, than in giving the object an apparent derangement of focal definition. A proper motion has been assigned to it as follows, and my observations countenance the largest value:

$$
\begin{array}{lll}
P.... \; Æ - 0''·02 & \text{Dec.} + 0''·33 \\
B.... \quad\;\; - 0''·98 & \quad\;\; + 0''·28 \\
A.... \quad\;\; - 0''·48 & \quad\;\; + 0''·28
\end{array}
$$

Here Piazzi, by comparison of his own observations with those of Flamsteed and Bradley, produced but a small value in Æ; but Mr. Baily has detected a serious amount.

CCCCLI. 24 COMÆ BERENICIS.

Æ 12ʰ 27ᵐ 06ˢ   Prec. + 3ˢ·01

Dec. N 19° 15′·5   —— S 19″·90

Position 272°·1 (*w* 9)   Distance 20″·5 (*w* 9)   Epoch 1836·38

A neat double star, between the Tresses and Virgo's left wing; lying at two-fifths of the distance from Arcturus to Regulus. A 5½, orange colour; B 7, emerald tint,—the colours very brilliant. This fine object is 27 ♓. iv., and Nos. 132 and 133, Hora XII., of the Palermo Catalogue. The following measures—in which reductions from Piazzi are also included—are those from which, compared with my own, we may venture to pronounce the relative orbital fixity of these stars; and the refinement of modern observations, and rigid scrutiny, is making its imputed proper motions disappear:

	Pos.	Dist.	Ep.
♓.	273° 28′	18″·44	1781·16
P.	273° 00′	20″·20	1800·00
H. and S.	272° 07′	20″·65	1822·24
Σ.	271° 57′	20″·42	1830·03

This seems to be one of the *informes* designated by Ptolemy as ἀμαυροὶ, *obscuræ*, but now in the distinct boundary of Berenice's Hair.

CCCCLII. 24 ♑. V. COMÆ BERENICIS.

Æ 12ʰ 28ᵐ 21ˢ   Prec. + 2ˢ·99
Dec. N 26° 52′·2   —— S 19″·89

Mean Epoch of the Observation . . . . 1837·25

A large white nebula, in the centre of the Tresses, and 2° south-east of the lucida, or 16 Comæ Berenicis; discovered by ♑. in April, 1784, and No. 1357 of his son's Catalogue. It is a curious, long, and streaky object, lying *np* and *sf* across the field, in somewhat of a weaver's shuttle shape, and preceded by four telescopic stars in a vertical curve. From the description which I received *vivâ voce* from H., my attention was intently fixed upon this nebula; and, after long and patient gazing, a parallel patch on the following limb was rather inferred than made out, by a peculiar glow on that part.

The parallel appendage to this nebula is a most extraordinary phenomenon, and is very beautifully figured in the *Philosophical Transactions* for 1833, by H.; who considers the two as constituting a flat annulus seen at a great obliquity, but having very unequal breadths and densities in its two opposite semicircles. " Or," asks he, "must we admit the appendage to be a separate and distant nebula, dependant, by some unknown physical relation, on its brighter neighbour?"

CCCCLIII. 143 P. XII. VIRGINIS.

Æ 12ʰ 30ᵐ 30ˢ   Prec. + 3ˢ·08
Dec. S 3° 29′·5   —— S 19″·87

Position 104°·5 (*w* 3)   Distance 50″·0 (*w* 3)   Epoch 1833·31

A wide double star, on the centre of Virgo's right wing; on the line and exactly two-thirds of the distance between Spica and η Virginis. A 6½, pale yellow; B 11, greenish; several small stars in the field, of which a pair in the *np* quadrant must be the " double star of the Vth class in view, preceding," recorded at 129 ♑. v. This object was registered in February, 1783, but without measures of position; yet the estimation 6° or 7° *sf* identifies it, as well as the distance = 46″·70. It appears to have been first micrometrically treated by S. with these results:

Pos. 105° 22′   Dist. 50″·55   Ep. 1825·36

143 P. is a borderer of the *locus lunæ* designated the XIIIth House; and also of a region of spherical nebular masses revealed by the telescope, of which it was well said, " Cœlique vias et sidera monstrat."

CCCCLIV.   68 M. HYDRÆ.

Æ 12ʰ 30ᵐ 56ˢ   Prec. +   3″·16
Dec. S   25° 51′·1   —— S 19‴·86

Mean Epoch of the Observation . . . . 1837·25

A large round nebula on Hydra's body, under Corvus, discovered in 1780 by Méchain. In 1786, Sir William Herschel's powerful 20-foot reflector resolved it into a rich cluster of small stars, so compressed that most of the components are blended together. It is about 3′ broad, and 4′ long; and he estimated that its profundity may be of the 344th order. It is posited nearly mid-way between two small stars, one in the *np* and the other in the *sf* quadrant, a line between which would bisect the nebula. It is very pale, but so mottled that a patient scrutiny leads to the inference, that it has assumed a spherical figure in obedience to attractive forces. Differentiated with β Corvi, from which it bears south by east, within 3° distance.

CCCCLV.   43 ♓. I. VIRGINIS.

Æ 12ʰ 31ᵐ 40ˢ   Prec. +   3ˢ·10
Dec. S   10° 43′·7   —— S 19″·85

Mean Epoch of the Observation . . . . 1837·25

A lucid white elliptical nebula, between the Virgin's right elbow and the Raven, in an elegant field of small stars; discovered by ♓. in May, 1784, and No. 1376 of his son's Catalogue. It lies nearly parallel to

the equatoreal line of the instrument, and on intense attention may be seen to blaze in the middle. The half dozen principal stars form a great Y, with the nebula as the centre. But it seems a mere wisp of subdued light, insomuch that my telescope does not afford me even the doubts inspired by the 20-foot reflector; for Herschel remarks that there is a faint, diffused oval light all about it, and that he is almost positive that there is a dark interval or stratum, separating the nucleus and the general mass of the nebula from the light above it. " Surely no illusion."

" The general form of elongated nebulæ is elliptic," says H., " and their condensation towards the centre is almost invariably such as would arise from the superposition of luminous elliptic strata, increasing in density towards the centre." This must be another of those vast flat rings seen very obliquely, already spoken of, and is an elegant example of that celestial perspective; it bears due west from Spica, and is 11° distant from that star, forming nearly a right angle with β Hydræ, which lies 12° to the southward.

CCCCLVI. γ VIRGINIS.

Ӕ 12ʰ 33ᵐ 33ˢ   PREC. + 3ˢ·07
DEC. S 0° 34′·3   S 19″·84

POSITION		DISTANCE		EPOCH
74°·9 (w 6)	DISTANCE	1″·6	(w 2)	EPOCH 1831·38
71°·4 (w 5)	——	1″·2	(w 3)	—— 1832·40
63°·6 (w 7)	——	not taken		—— 1833·23
62°·7 (w 8)	——	1″·3	(w 2)	—— 1833·44
48°·8 (w 6)	——	1″·0	(w 2)	—— 1834·20
45°·5 (w 5)	——	0″·8	(w 2)	—— 1834·39
15°·0 (w 5)	——	0″·5	(w 1)	—— 1835·40
round (w 9)	——	round	(w 9)	—— 1836·06
round (w 9)	——	round	(w 9)	—— 1836·15
blotty (w 8)	——	blotty	(·7 5)	—— 1836·25
350°·9 (w 4)	——	elongated (w 5)		—— 1836·30
348°·6 (w 4)	——	elongated (w 5)		—— 1836·39
265°·4 (w 3)	——	0″·6	(w 1)	—— 1837·21
235°·7 (w 3)	——	0″·8	(w 2)	—— 1838·28
217°·2 (w 5)	——	1″·0	(w 2)	—— 1839·40
192°·8 (w 5)	——	1″·9	(w 3)	—— 1843·08
191°·6 (w 8)	——	1″·9	(w 5)	—— 1843·33

A fine binary star, in Virgo's right side, heretofore known as *Porrima* and *Postvarta* by Calendar *savans*. A 4, silvery white; B 4, pale yellow, but though marked by Piazzi of equal magnitude with A, it has certainly less brilliance; and the colours are not always of the same intensity, but whether owing to atmospherical or other causes, remains undecided. They are followed by a minute star nearly on the parallel, and about 90″ off. With β, δ, and η, it formed the XIIIth Lunar Mansion, and was designated, from its position in the figure, *Záwiyah-al-'auwà*, the corner of the barkers. This most instructive star bears north-west of Spica, and is 15° distant, in the direction between Regulus and γ Leonis, which are already aligned. A very sensible proper motion in space has been detected in A, and there can be no doubt of B's standing on in the same course; the most rigorous comparisons of recent observations afford the following values:

P.... Ӕ - 0″·72   Dec. + 0″·10
B....   - 0″·50   - 0″·02
A....   - 0″·52   - 0″·02

It was with much gratification that I watched this very interesting physical object through a considerable portion of its superb ellipse, and I was fortunate enough to attack it during the most critical period of its march. It is rather singular that, brilliant as these two stars are,

T 2

various occultations of γ Virginis by the Moon have been recorded, without allusion to its being double. So lately as the 20th March, 1780, the phenomenon was watched by nine astronomers; yet at Paris only, on that occasion, is mention made of one star being occulted 10ˢ before the other. On the 21st January, 1794, the occultation was observed by four astronomers; yet no one mentions duplicity. This is passing strange, because Cassini had, in 1720, perceived and recorded the two stars, noting that the western disappeared 30″ before the other, behind the Moon's dark limb, but they emerged nearly together. He could not divide them with a telescope of 11 feet, but with one of 16 they were well severed, and of equal magnitudes. He watched the immersion, which was oblique, with great care, hoping by refraction or discoloration to detect a lunar atmosphere; but though the circumstances were favourable, he perceived no symptom. Yet the observation was held to be of importance, for, by enlisting that able astronomer and Bradley, Sir John Herschel considered that he gained some useful points in the orbital departure; and the results of more than a century, previous to my measures, may be thus shortly stated:

Bradley and Pound . . .	Pos. 160° 52′	Dist. *caret*	Ep. 1718·20
Cassini II.	*caret*	7″·49	1720·31
Mayer	144° 22′	6″·50	1756·00
Herschel I.	130° 44′	5″·70	1780·06
Herschel II. and South . .	103° 24′	3″·79	1822·25
Struve	98° 18′	2″·28	1825·42
Dawes	78° 15′	2″·01	1831·33

A mere inspection of the conditions here stated, shows the vast acceleration of the revolving star on approaching its periastre, and the retardation of its getting away again. These are the annual rates of retrograde angular progress:

Mayer . . .	0°·43	1756·00	Myself . . .	17°·37	1834·39
Herschel I. .	0°·57	1780·06	—— . . .	30°·20	1835·40
H. II. and Sth.	0°·64	1822·25	—— . . .	*round*	1836·06
Struve . . .	1°·59	1825·42	—— . . .	26°·78	1836·39
Dawes . . .	3°·39	1831·33	—— . . .	25°·55	1837·21
Myself . .	3°·43	1832·40	—— . . .	22°·01	1839·40
—— . . .	9°·40	1833·23	—— . . .	16°·52	1843·08
—— . . .	15°·25	1834·20	—— . . .	6°·63	1843·33

As the rigorous observations and computations of this object must be deemed a sort of *experimentum crucis* of the sidereal connected systems, I may be excused for entering into rather fuller details of the detection and establishment of so wonderful an elliptic motion, than I have yet indulged in among the binaries; and it will thereby serve as an example of the method of procedure with those interesting objects.

The various observations were most ably and zealously discussed by Sir John, and treated in a straight forward, geometrical mode, so as to be widely available; as will be seen on consulting the vth volume of the *Memoirs of the Royal Astronomical Society*. The method is equally novel and ingenious. Assuming that the motions of binary stars are governed by the universal law of gravitation, and that they describe conic sections about their common centre of gravity and about each other, he was bent on relieving their discussion from the analytical difficulties attending a rigorous solution of equations, where the data are uncertain, irregular, and embarrassing. Measures of position were to be the sheet-anchor; for distances, with the exception of the major semi-axis, were peremptorily excluded from any share of consideration in the investigation, because of their notorious looseness and insecurity.

The process, (said he,) by which I propose to accomplish this, is one essentially graphical; by which term I understand, not a mere substitution of geometrical construction and measurement for numerical calculation, but one which has for its object to perform that which no system of calculation can possibly do, by bringing in the aid of the eye and hand to guide the judgment, in a case where judgment only, and not calculation, can be of any avail.

Under the assumption, therefore, that gravitation governs, and one of the components revolves, while the other, though not necessarily in the focus, is at rest, the curve is constructed by means of the angles of position and the corresponding times of observation; and tangents to this curve, at stated intervals, yield the apparent distances at each angle, they being, by the known laws of elliptical motion, equal to the square roots of the apparent angular velocities.

Thus armed, Sir John proceeded with the orbit of γ Virginis. From the above positions and epochs, with interpolated intermediates, a set of polar co-ordinates were derived, and thence, for the apparent ellipse, the following elliptical elements:

Major semi-axis	5″·862
Position of major semi-axis	67° 20′
Excentricity	0·70332
Maximum of distance	9″·423
Position at the maximum distance	218° 55′
Minimum of distance	0″·514
Position at the minimum distance	1° 15′
Date of next arrival at minimum distance . .	1834·39
Greatest apparent angular velocity	− 68°·833
Least apparent angular velocity	− 0°·193

The next process was to obtain the elements of the *real* ellipse, and the whole consequent investigation is so succinctly described in the paper alluded to, that any zealous tyro may tread in the same steps, with a little attention. The results, together with a comparison of the elements and observations up to the period of the computation, and an ephemeris of the system for the years 1832, 1833, 1834, and 1835, were inserted in the Supplement to the *Nautical Almanac* for 1832. But finding a discrepancy between the measures then obtained and the places predicted, Herschel, nothing daunted, again took the field, and recalculated the orbit, as described in the vith volume of the *Astronomical Memoirs*. In this process, my measures of 1832 and 1833 were included, and the two conclusions stood thus:

	1831.		1833.
Major semi-axis	11″·830	...	12″·090
Perihelion projected . .	17° 51′	...	36° 40′
Excentricity	0·88717	...	0·8335
Inclination to plane of the heaven	67° 59′	...	67° 02′
Position of node . . .	87° 50′	...	97° 23′
Mean annual motion . . .	−0°·70137	...	−0°·57242
Period in tropical years . .	513·28	...	628·90
Perihelion passage . . .	1834·01	...	1834·63

In giving the first part of these remarkable elements to the astronomical world, Sir John said:

If they be correct, the latter end of the year 1833, or the beginning of the year

1834, will witness one of the most striking phenomena which sidereal astronomy has yet afforded, viz., the perihelion passage of one star round another, with the immense angular velocity of between 60° and 70° per annum, that is to say, of a degree in five days. As the two stars will then, however, be within little more than half a second of each other, and as they are both large, and nearly equal, none but the very finest telescopes will have any chance of showing this magnificent phenomenon. The prospect, however, of witnessing a visible and measurable change, in the state of an object so remote, in a time so short (for, in the mean of a very great number of careful measures with equal stars, a degree can hardly escape observation), may reasonably be expected to call into action the most powerful instrumental means which can be brought to bear on it.

And this was Sir John's projected ellipse:

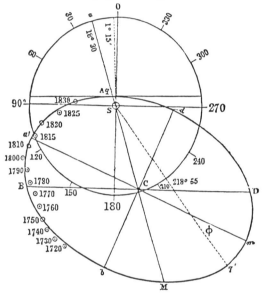

From the extreme delicacy of so novel a case, all the conditions were not yet met, so that this bold prediction was not circumstantially verified, although it was admirably correct in substance. Whilst rushing towards the nearest point of contact, or shortest distance of the revolving star from its primary, and the proximity became extreme, the field was left, as far as I know, to Sir John Herschel at the Cape of Good Hope, Professor Struve at Dorpat, and myself at Bedford. Our measures afforded unquestionable proofs of the wonderful movement under discussion; yet they certainly exhibited greater discrepancies than might have been expected, from the excellence of the instruments employed. But the increased angular velocity which so excentric a star acquired, when gaining its periastre, and the closeness of its junction, rendered the operations extremely difficult: added to which, the brightness of two such stars was sufficient to call forth that disadvantage, arising from the inflection of light, which the wire micrometer labours under, and which interferes in the exact contact between the line and the luminous body.

The accelerating velocity of angular change was thus vigilantly watched, until the commencement of the year 1836, when an unexpected phenomenon took place. Instead of the appulse which a careful projection, drawn from the above elements, had led me to expect, I was astonished, on gazing at its morning apparition in January, to find it a single star! In fact, whether the real discs were over each other or not, my whole powers, patiently worked from 240 to 1200, could only make the object round. I instantly announced this singular event to my astronomical friends, but the notice was received with less energy than such a case demanded; none of the powerful refractors in this country were pointed to it in time; and it is to be regretted, that we had not the benefit of the unexcelled Dorpat telescope's evidence, at the critical epoch in question. This state of apparent singleness may have existed during the latter part of 1835, for when I caught it, as may be seen in the observations above, it was very near a change. At length, about the beginning of June, 1836, a letter arrived from Sir John Herschel, addressed to Mr. Baily, wherein he detailed his observations on the single state of this star, at the villa of Feldhausen, Cape of Good Hope, in his 20-foot reflector. Under the date of February 27th, that unwearied astronomer says:

γ Virginis, at this time, is to all appearance a single star. I have tormented it under favourable circumstances, with the highest powers I can apply to my telescopes, consistently with seeing a well-defined disc, till my patience has been exhausted; and that lately, on several occasions, whenever the definition of the stars generally, in that quarter of the heavens, would allow of observing with any chance of success, but I have not been able to procure any decisive symptom of its consisting of two individuals.

The companion now took such a movement, as quite to confute a large predictive diagram I had constructed, showing that the orbit was extremely elongated, more like a comet's than a planet's; which gave me a suspicion that we had been looking at the ellipse the wrong way. Hereupon I returned to the Herschelian process to obtain the elements of the apparent and the true ellipse, with my new measures, but could neither accommodate the period, nor arrive at any satisfactory conclusions. When therefore M. Mädler's masterly computations appeared in the *Astronomische Nachrichten*, my views were greatly countenanced; but with a full value for the talent and zeal of that astronomer's process, I was still anxious for Sir John Herschel to return to his own field, and meet the apparently unaccountable informalities which still remained. Having made a request to this effect, he replied:

Maugre I cannot yet send you any finalities about γ Virginis, yet to prove that I have not been quite idle, I will state one or two general conclusions that a projection of all the observations has led me to, preparatory to exact numerical computation. 1. We are *all wrong*, Mädler and all of us, and it is the early observation of Bradley in 1718 which has misled us. That observation is totally incompatible with *any* reasonable ellipse, and must be absolutely rejected. Had it not been for my respect for that single observation, I should have got very near the true ellipse in my first approximation. 2. The period is short of 150 years. My conjecture, antecedent to *any* exact calculation from my projection, is 143, which is considerably less than the least of Mädler's, and beyond his assigned limits of error. 3. I suspect Mädler's perihelion to be half a year too early, and that the true perihelion passage took place at 1836·6, or thereabouts. We shall get on better now that we have found out the black sheep.

Thus duly authorized, I attacked the orbit again, rejecting, with some regret, Bradley, Pound, Cassini, and Mayer, and assuming Ḥ.'s observations of 1780 as the point of departure. Taking, therefore, the epochs from that date to 1843 for abscissæ, and the observed angles for

ordinates, a fresh set of periods was obtained, through which the inter-
polating curve was led, on a very large scale*. From the interpolated
positions corresponding to the *assumed dates* between 1780 and 1843,
the intervals being first decennial, then quinquennial, and afterwards
more rapid still, the angular velocities were concluded, and by their aid
the distances as radii vectores. These positions and distances were laid
down from the central star as an origin of polar co-ordinates. Now,
though this is a simple and merely graphic process of obtaining the
elements of both the apparent and true ellipse, and is liable to shakiness,
it undeniably shows the physical fact of a highly elongated orbit; and
several of the conditions prove that, notwithstanding the present anomalous differences, we are arriving near the mark. It is singular how all
the determinations of the excentricity have agreed, thus:

	First orbit.	Second orbit.			First orbit.	Second orbit.
Encke . . .	0·890	0·860		Mädler . . .	0·864	0·868
Herschel II. .	0·887	0·834		Myself . . .	0·883	0·872

As the ellipse projected by Sir John Herschel, under *all* the epochs,
has been given, the reader may like to see the figure produced by the
Bedford observations, which yields a period of about 180 years:

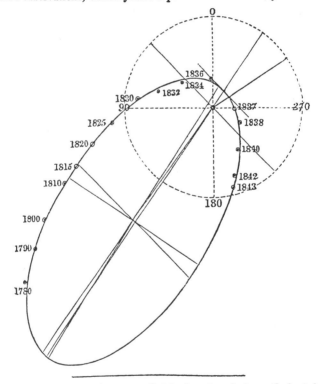

---

* Sir John Herschel informs me, that he has disused the method of drawing
tangents for the angular velocities. The substitute is a closer reading off of the curve,
equalizing the differences on paper, and thence deducing the angular velocities by first
and second differences (if needed); but first will generally suffice.

As the resulting elements, though better, were still unsatisfactory, I was about to take another point of departure, and try again, when I received a letter from Sir John Herschel, dated Collingwood, 9th July, 1843, of which the following is an extract:

I wrote to you last, that I could not make Bradley's observations agree with *any* ellipse consistent with the later observations, and that Mädler's elements, which assume the correctness of that observation, are inadmissible. I have now satisfied myself that this is really the case, and that Mädler's period admits of being yet reduced. But still it is necessary to suppose materially greater errors *in one direction* over the whole interval 1828, 1829, 1830, 1831, than I quite like. The mean of Dawes's and my own measures, however, is on the whole exceedingly well represented in all the critical and puzzling part of the orbit corresponding to 1830—1834 inclusive. Your observations of 1831, 1832, and 1833, offer discordancies of $+ 2°$, $+ 2°\frac{1}{4}$, and $+ 3°$, which are, considering the then considerable closeness of the stars, not more than might well be committed. But Struve's are quite inexplicable;—his errors, supposing the orbit correct, run thus:

1825	1828	1829	1831	1832	1833	1834
$+ 1°$	$+ 3°$	$+ 3°\frac{1}{4}$	$+ 4°\frac{1}{2}$	$+ 5°, + 7°$	$+ 6°$	$+ 7°\frac{1}{4}$

after which the deviation ceases.

On the whole I consider the proofs of gravitation afforded by this star quite satisfactory. It is true that I am forced to admit an error of $- 3°\frac{1}{4}$ in my father's measure of 1781, and an error exceeding $2°$ in the same direction in his subsequent mean result for 1803; but when I recollect what sort of micrometer and apparatus he used, I am not disposed to quarrel with these.

I am not satisfied with my inclination and node, and there is still a tendency in the curve of the star, if your measures of this year be correct, to run away from its proper course, *to bolt;* which leads me to believe that these elements are not yet so well determined as I hope to get them. Your ellipse from the Bedford observations is a very beautiful one, but I have not yet compared your elements with the observations. I am somewhat surprised at the length of your period, as I find 126 years represents the mean of all the observations (including Struve's) on the whole well. I have been chiefly attending to improving the *method* as a working one, and I am preparing a paper on the subject, in which the orbit of γ will occur in exemplification. What I aim at is, a *direct process* leading to the separate correction of each element, in place of a turmoil of calculus on the principle of least squares, which in cases of such discordant observations is, if not illusory, at least unnecessarily troublesome.

The inquirers into binary systems will yearn for the coming of this discussion; meantime, to use an expression of Pliny the Younger, I am fortunate in my heliacal rising, since what I have here stated may be of a little interest, before it shall be obscured and eclipsed in Herschel's brighter eminence.

One word more. To those who are earnest upon either of these topics, I submit a diagram of what I saw myself, which may render the above details more evident:

Such a phenomenon has had more discussers than beholders, so that astute doubts have been flung out of these stars being amenable to gravitation, whether their angular changes are reducible by the laws of elliptical motion, whether the period be a little longer or shorter, and all that. Nay, with such unquestionable instances before the world, and at the very time that admiration was incited in every reflecting mind, a blundering Zoilus, who, had he *flourished* at an earlier age, might have figured at Galileo's trial, was permitted to stain the *Church of England Quarterly Review*, April, 1837, p. 460, with the following Bœotian attempt at sprightliness:

We have forgotten the name of that Sidrophel who lately discovered that the fixed stars were not single stars, but appear in the heavens like soles at Billingsgate, in pairs; while a second astronomer, under the influence of that competition in trade which the political economists tell us is so advantageous to the public, professes to show us, through his superior telescope, that the apparently single stars are really three. Before such wondrous Mandarins of Science, how continually must *homunculi* like ourselves keep in the back-ground, lest we come between the wind and their nobility.

This plural unit must truly be, so far as education and intellect are concerned, the downright veritable *homunculus* he has written himself.

The would-be-wit, however, though quite as ignorant, is perhaps less malignant than a fellow reviewer, who must needs meddle with works beyond his ken. This stultified Bavius asserts, that it best suits with the knowledge we possess of our finite understanding, and with the purport and end of our being, to refrain from *silly* speculations which may perplex, but can never satisfy the mind. He holds it both vain and wicked to attempt to probe the infinity of space, and decries Sir William Herschel's estimate of the magnitude of the nebula in Orion, as a speculation to confuse rather than instruct the mind. This man is susceptible of very great improvement before his opinions command respect. So far from science being daring and proud, as he asserts, there are abundant reasons for it to feel humbled; but the effusion in question shows the proper nursery of those qualities,

For he that has but impudence,
To all things has a fair pretence.

CCCCLVII.   60 M. VIRGINIS.

Æ 12ʰ 35ᵐ 33ˢ   Prec. +   3ˢ·02
Dec. N 12° 26'·1   —— S 19"·80

Mean Epoch of the Observation . . . . 1837·22

A double nebula, in the centre of Virgo's left wing, lying *np* and *sf*, about 2' or 3' from centre to centre, the preceding one being extremely faint. The following, or brighter one, is that seen and imperfectly described by Messier in 1779, and is nearly between two telescopic stars vertically posited. A fine field is exhibited under the eye-piece, which magnifies 93 times, just as this object enters, because the bright little nebula 59 M. is quitting the *np* verge, and another small one is seen in the upper part, H. 1402: in fact, four nebulæ at once.

The hypothesis of Sir John Herschel, upon double nebulæ, is new and attracting. They may be stellar systems each revolving round the other: each a universe, according to ancient notions. But as these revolutionary principles of those vast and distant firmamental clusters cannot for ages yet be established, the mind lingers in admiration, rather than comprehension of such mysterious collocations. Meantime our clear

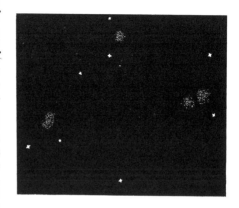

duty is, so industriously to collect facts, that much of what is now unintelligible, may become plain to our successors, and a portion of the grand mechanism now beyond our conception, revealed. "How much," exclaims Sir John Herschel, "how much is escaping us! How unworthy is it in them who call themselves philosophers, to let these great phenomena of nature, these slow but majestic manifestations of the power and the glory of God, glide by unnoticed, and drop out of memory beyond the reach of recovery, because we will not take the pains to note them in their unobtrusive and furtive passage, because we see them in their every-day dress, and mark no sudden change, and conclude that all is dead, because we will not look for signs of life; and that all is uninteresting, because we are not impressed and dazzled." . . . . . "To say, indeed, that every individual star in the Milky Way, to the amount of eight or ten millions, is to have its place determined, and its motion watched, would be extravagant; but at least let samples be taken, at least let monographs of parts be made with powerful telescopes and refined instruments, that we may know what is going on in that abyss of stars, where at present imagination wanders without a guide!" Such is the enthusiastic call of one, whose father cleared the road by which we are introduced to the grandest phenomena of the stellar universe.

This mysterious and shadowy doublet will be found 5° west of Vindemiatrix, in the direction of Regulus, where there is a very large and wonderful nebulous region.

CCCCLVIII. 196 P. XII. VIRGINIS.

Æ 12ʰ 43ᵐ 04ˢ   Prec. + 3ˢ·11
Dec. S   9° 28'·0   —— S 19"·69

Position 307°·9 (w 5)   Distance 33"·5 (w 3)   Epoch 1834·41

A neat but wide double star, between the Virgin's right arm and the tail of the Raven; about 8° west-half-north of Spica, and closely *sp* ψ Virginis, a star of the 5½ magnitude. A 6½, topaz yellow; B 9½, lucid purple, the colours finely contrasted. This was enrolled by the indefatigable Struve, No. 1682 of the Dorpat Catalogue, whose micro-metrical measures were:

Pos. 308° 46′   Dist. 33″·65   Ep. 1831·61

The rigorous comparisons of Messrs. Piazzi and Baily, with an interval of 40 years between them, produce an amount and direction of proper motion in the star A, which is almost identical; the values and signs being:

P.... Æ $-$ 0″·13   Dec. $-$ 0″·06

B.... $-$ 0″·11 $-$ 0″·07

CCCCLIX.   94 M. CANUM VENATICORUM.

Æ 12ʰ 43ᵐ 22ˢ   Prec. $+$ 2ˢ·84

Dec. N 41° 59′·7   —— S 19″·69

Mean Epoch of the Observation . . . . 1834·32

A large bright nebula discovered by Méchain, in 1781, immediately preceding the crown on Charles's Heart. It is a fine pale-white object, with evident symptoms of being a compressed cluster of small stars. It brightens towards the middle, and the gradual augmentation of intensity from the margin to the centre of this apparently orbicular object, is a direct proof of the real sphericity of the stellar mass. There are several small stars in the field, of which one in the *sf* quadrant is double. Differentiated with the bright star Cor Caroli, from which it is but 2°½ in the north-by-west.

CCCCLX.   202 P. XII. COMÆ BERENICIS.

Æ 12ʰ 43ᵐ 59ˢ   Prec. $+$ 2ˢ·98

Dec. N 20° 02′·6   —— S 19″·68

Position 201°·4 (w 9)   Distance 15″·9 (w 9)   Epoch 1831·39

——— 201°·9 (w 9)   ——— 16″·2 (w 9)   ——— 1838·28

A neat double star, between Berenice's Hair, and Virgo's left wing: it lies due west of Arcturus, or on its parallel, at the distance of 22°, where a line dropped south from Cor Caroli will intercept it. A 7½, and B 8, both white; other stars in the field, but small and distant. This object is 58 Ⱨ. iv.; and is also formed by Nos. 201 and 202, Hora XII., of the Palermo Catalogue. A comparison of my own with the following measures, afford presumptive proof of fixity:

Ⱨ.   Pos. 202° 03′   Dist. 15″·86   Ep. 1782·30
H. and S.   202° 11′   16″·96   1823·41
Σ.   200° 48′   15″·82   1829·87

CCCCLXI. 75 ♅. II. VIRGINIS.

Æ. 12ʰ 44ᵐ 50ˢ   Prec. + 3ˢ·01

Dec. N 12° 05'·9   —— S 19"·66

MEAN EPOCH OF THE OBSERVATION . . . . 1835·26

A pale elliptical nebula, in the middle of Virgo's left wing; discovered by ♅. in March, 1784, and No. 1466 of his son's Catalogue.

This is a fine object trending *sp* and *nf*, nearly in the vertical, but from its superior brightness at the south, or upper end, it rises while gazing from the dumpy egg-shape to that of a paper kite: over it is an arch formed by three telescopic stars, the symmetry of which is so peculiar as to add to that appearance. These stars trend, by two very faint ones, to a round nebula in the *np* quadrant, preceded by two stars of the 10th magnitude;

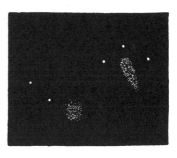

this is 74 ♅. II. The mean place of the *kite's* centre was obtained by differentiation with ε Virginis, from which known star it is only 2°¼ distant, on the western parallel.

CCCCLXII. 35 COMÆ BERENICIS.

Æ. 12ʰ 45ᵐ 25ˢ   Prec. + 2ˢ·96

Dec. N 22° 07'·0   —— S 19"·65

POSITION AB 30°·0 (*w* 1)   DISTANCE 1"·0 (*w* 1) ⎫
—— AC 126°·5 (*w* 8)   —— 28"·8 (*w* 6) ⎬ EPOCH 1834·38
—— AB 42°·0 (*w* 2)   —— 1"·5 (*w* 1) —— 1843·32 ⎭

A delicate triple star, between the Tresses and the Virgin's northern wing, about 7° south-south-east of its own lucida, and 20° west of Arcturus. A 5, pale yellow; B, indistinct; C 10, cobalt blue. Of this object A and C were classed as a double star, and registered 130 ♅. v.; but Σ. detected B, and rendered it a fine but extremely difficult triple. Indeed, the latter was so involved in the moulds and twirls of the primary, that but for A and B pointing directly upon a 12th-magnitude star in the *nf* quadrant, the estimation of angle and distance would have been hardly possible. The other two have been thus measured, and their fixity seems proved:

	Pos.	Dist.	Ep.
♅.	126° 51'	31"·29	1783·15
H. and S.	128° 18'	29"·49	1821·34
Σ.	124° 43'	28"·61	1830·13

CCCCLXIII. 221 P. XII. VIRGINIS.

Æ 12ʰ 47ᵐ 29ˢ   Prec. + 3ˢ·01
Dec. N 12° 22′·0   —— S 19″·62

Position 197°·9 (w 9)   Distance 29″·1 (w 9)   Epoch 1831·38

A neat double star, near the middle of Virgo's northern wing; preceding Vindemiatrix on the parallel by only 2°, and therefore readily fished up by the out-door gazer. A 7½, pale white; B 9, sky blue. This object is thus described by Piazzi: "Duplex. Socia 10ᵃ magnit. 0″·6 temporis præcedit, 20″ circiter ad austrum." Σ. has since entered it on his Catalogue; but the earliest micrometrical measures I find, are those of H. and S., as follows:

Pos. 196° 17′   Dist. 29″·17   Ep. 1823·40

CCCCLXIV. δ VIRGINIS.

Æ 12ʰ 47ᵐ 33ˢ   Prec. + 2ˢ·40
Dec. N 4° 16′·1   —— S 19″·62

Position 144°·2 (w 2)   Difference Æ = 5ˢ·8 (w 2)   Epoch 1832·31

A star with a distant companion, on Virgo's left side; it is readily seen by directing the eye 17° north-north-west of Spica, where it stands nearly mid-way between ε and γ Virginis. These stars, together with η, π, and several minor ones, assist in encircling the rich nebulous oasis which decks the damsel's left wing and shoulder; and which is so well delineated on Sir John Lubbock's map, published by the Society for the Diffusion of Useful Knowledge. A 3½, golden yellow; B 10½, reddish; several small stars in the field, but none nearer to the primary than B. My meridional observations confirm the fact of a proper motion of A, as compared with the epoch of the Palermo Catalogue, yet they are not sufficiently exact to pronounce upon a quantity for registry; but the amount has been ably looked to, and valued as follows:

P.... Æ − 0″·65   Dec. − 0″·02
B....   − 0″·40   − 0″·09
A....   − 0″·43   − 0″·06

so that the direction seems pretty well ascertained.

Ptolemy tells us, that those stars which Hipparchus placed on the Virgin's shoulder, were shifted by him to her side, " because their distances from the head were too great for the distance from the head to the shoulder;" hence the present situation of δ.

CCCCLXV.   232 P. XII. CAMELOPARDI.

ℛ 12ʰ 48ᵐ 08ˢ   Prec. +   0ˢ·29
Dec. N 84° 17′·0   —— S 19″·60

Position 327°·3 (w 9)   Distance 21″·8 (w 9)   Epoch 1833·97

A neat double star, near the animal's ear; it is nearly 7° south-by-east from Polaris, and forms the vertex of a nearly isosceles triangle with that star and δ Ursæ Minoris. A 6 and B 6½, both bright white; a third star in the *sp*, but only of the 11th magnitude. This was classed by Ⱨ. in 1780, as 15 ɪv.; but no angle was measured, and he merely gives a distance of 20″·01. It was afterwards measured by H. and S. thus:

Pos. 327° 00′ Dist. 21″·07 Ep. 1822·28

whence it appears to have had no motion either in angle or distance, in an interval of 12 years; and since the arrival of the Dorpat Catalogue, these results are confirmed by Professor Struve's observations. This object affords a fresh instance of the admirable correctness of the Palermo Catalogue, where its components are Nos. 230 and 232, Hora XII.; the mean places of which, on reduction, yield 329°·5 for the angle, and 21″·4 for the distance, for the epoch of 1800.

CCCCLXVI.   12 CANUM VENATICORUM.

ℛ 12ʰ 48ᵐ 32ˢ   Prec. +   2ˢ·84
Dec. N 39° 10′·9   —— S 19″·59

Position 227°·3 (w 7)   Distance 20″·2 (w 6)   Epoch 1830·64
——— 227°·0 (w 9)   ——— 19″·8 (w 9)   —— 1837·39

A fine double star, the lucida of the heart placed between Asterion and Chara. A 2½, flushed white; B 6½, pale lilac; a third star at a distance on the north vertical. This is a Greenwich star of the 2nd grade, and is a fine object, being 17 Ⱨ. ɪv., by whom it was thus measured:

Pos. 228° 33′ Dist. 20″·00± Ep. 1780·60

The companion is No. 226 of the Notes to Hora XII., in the Palermo Catalogue; and has been measured by H. and S. and Σ., showing a coincidence of results which sufficiently proves that there has been no appreciable alteration, in position or distance, during 57 years. A proper motion has been given to the large star, to the following amount:

P.... ℛ − 0″·34   Dec. + 0″·04
B....   − 0″·26   + 0″·04
A....   − 0″·30   + 0″·05

This star is the leader of Ptolemy's ἀμόρφωτοι to the Great Bear, and it appears on the Borgian globe. Ulugh Beigh records it by the Arabian designation *Kabd-al-asad*, liver of the lion, and there is abundance of evidence to prove that it was pretty notorious among astronomers. But it came to pass that it was named Cor Caroli by Halley, at the suggestion of Sir C. Scarborough, after a worthless man's heart. The popular story, or rather the vulgar one, runs,—how Scarborough, the court physician, gazed upon a star the very evening before the return of King Charles II. to London, the which, as in duty bound, appeared more visible and refulgent than heretofore; so the said star, which Hevelius had already made the lucida of Chara's collar, was thereupon extra-constellated within a sort of Valentine figure of a heart, with a royal crown upon it; and so the monarch, it would seem, by this extraction, remained heartless. Though this pretty symbol appears as a tail-piece to the preface of the *Atlas Cœlestis*, Flamsteed has not honoured it with a place on the plate of the Hounds.

Cor Caroli is readily found by alignment. With the three stars of the Greater Bear's tail as the head of a paper kite, it forms the tail end; or a line from the Pole-star over Alioth will point upon the heart. Another clue is obtained in the galley-rhyme:

When clear aloft, Boötes seek,	his brilliance leads the gaze,
And on each side its glitt'ring gems	the spacious arch displays;
Arcturus east to Wega join,	the Northern Crown you'll spy;
But west, to Ursa's second star,	he marks Cor Caroli.

The Canes Venatici form a new constellation, or pair of constellations, intruded into the heavens in recent times. Tycho Brahé, unfortunately less known by his excellence as a practical astronomer than as the author of an unmechanical system, had observed a couple of stars here; but Hevelius scraped together the *sporades* between the stern of the Bear and Boötes, and figured two Hounds, for the latter to chase the Bear with; that nearest the Pole he named Asterion, because the appellation is poetical, and it pleased him, and the southern one Chara: "Asterionis sociam, Charam appellavi, quòd fortè Booti, more venaticorum, canis illa fœmina, ob celeriorem ejus cursum, fuerit admodùm grata et chara." These dogs first appeared in the *Prodromus* of Hevelius, published by his widow, at Dantzig, in 1690. Since then, the number of its constituents has swelled very considerably, although, except Cor Caroli, the asterism offers little remarkable to the unassisted eye; they are thus registered:

| Hevelius . . . 23 stars | Piazzi . . . . 45 stars |
| Flamsteed . . 25 | Bode . . . . 139 |

CCCCLXVII.   64 M. COMÆ BERENICIS.

Æ 12ʰ 48ᵐ 52ˢ   Prec. + 2ˢ·95

Dec. N 22° 33'·2   —— S 19"·59

Mean Epoch of the Observation . . . . 1837·24

A conspicuous nebula, between Berenice's Hair, and the Virgin's left arm; discovered by M. in 1780, who, however, only saw it very faintly.

Yet it is magnificent both in size and brightness, being elongated in a line *np* and *sf*, and blazing to a nucleus. This is the object which Sir Charles Blagden, on being shown it by H̶l̶., likened to a *black eye*, which, though odd, is considered not an inapt comparison.

Sir John Herschel examined this nebula very minutely. He considers it resolvable, though not resolved; and adds, "I am much mistaken if the nucleus be not a double star, in the general direction of the nebula; 320 much increases this suspicion; 240 shows well a vacuity below the nucleus." My utmost endeavours only show it thus:

This nebula was fixed by differentiating it with ε Virginis, the bright star 11° south of it; and it lies between Arcturus and δ Leonis, about 20° west of the former bright star.

CCCCLXVIII.   44 VIRGINIS.

Æ   12ʰ 51ᵐ 30ˢ   Prec. +   3ˢ·08
Dec. S   2° 57′·4   —— S 19″·54

Position 55°·0 *(w 1)*   Distance 20″·0 *(w 1)*   Epoch 1834·21

A delicate double star, on the lower part of Virgo's right or southern wing; lying 10° north-west of Spica, in the direction of Denebola. A 6, bright white; B 13, blue. This object is so difficult, that my estimations are merely registered as confirmatory of its general details. It is 51 H̶l̶. ɪᴠ., and has been thus measured:

H̶l̶.   Pos. 57° 30′   Dist. 22″·29   Ep. 1782·10
Σ.   53° 01′   21″·29   1830·63

The discrepancies here observable are such as must be rather attributed to the errors of observation incident to so delicate a test, than to any perceptible motion in the individuals as to angle of position, or alteration of distance. Yet my meridional reductions warrant the conclusion of a proper motion in space to the primary, at least equal to that which Piazzi, referring to Mayer, has assigned it; although it does not find such a value in Mr. Baily's list. The amounts stated are:

P.... Æ − 0″·40   Dec. −′0″·06
B....   + 0″·03   + 0″·03

My own operations, referred to Piazzi's, would give even a greater quantity; but as they are not sufficiently directed to that object, I merely mention it to call attention to a somewhat slighted, though gravely interesting branch of astronomy. 'Tis true, the amount in question is so small as to require the nicest practice to detect it precisely; but these aberrations from the common law of precession, however secondary and minute the quantities may appear, are matters of high import as leading to ulterior abstruse discoveries. They should therefore be most rigidly and vigilantly watched; for though the present age of astronomy cannot arrive at any great or final conclusion on them or their causes, posterity probably will, if exact uranographical details are left them. We must not, therefore, any longer neglect studying a motion so perceptibly carrying stars through space, towards some unknown point in the firmament. Proper motion is not only in itself an object of curious research, but, as Mr. Baily has forcibly expressed it, "more especially as it frequently interferes with and deranges those general rules which are laid down, with so much accuracy, for the determination of places of stars at any distant period." We shall know more about it by-and-bye.

CCCCLXIX. ε VIRGINIS.

$$\text{Æ } 12^h\ 54^m\ 13^s \qquad \text{Prec. } +\quad 3^s\cdot00$$
$$\text{Dec. N } 11°\ 49'\cdot3 \qquad \text{——} \quad \text{S } 19''\cdot49$$

Position 123°·8 (w 4)   Distance 229''·0 (w 2)   Epoch 1838·42

A star with a minute distant companion, on the upper extreme of the Virgin's left wing. A 3½, bright yellow; B 15, intense blue,—this last colour on so small an object is very striking, and an astronomical friend, who examined it at my request, with powerful means, confirms both the tint and its intensity. The large star, which I suspect to be slightly variable, has a sensible proper motion, of which the following value has been given:

$$
\begin{array}{llll}
P.... & \text{Æ} - 0''\cdot37 & \text{Dec.} + 0''\cdot07 \\
B.... & \quad - 0''\cdot18 & \quad + 0''\cdot03 \\
A.... & \quad - 0''\cdot25 & \quad + 0''\cdot03 \\
\end{array}
$$

This star appears, in various treatises on astronomy, as symbolizing the gatherer of grapes; but *Vindemiatrix*, in the Alphonsine Tables, is an adaptation of the longer word *Provindemiator* (Vitruv. ix. 4), a translation of προτρυγητὴρ, given to ε Virginis, because it rises in the morning, just before the vintage. (*Schol. in Arat.*, 138.) Hence it became *Mukdim-al-kitáf*, the forerunner of the vintage, among the Arabians, some of whom included it as a boundary of the XIIIth Lunar Mansion, in their *Manázil-al-Kamar*, implying the Moon's Zodiac.

On completing my examination of this star, I made a second search at Mr. Baily's request, for 42 Virginis, which was missed by H. in 1828; but could find nothing in the place indicated in the Astronomical Society's Catalogue, No. 1490, except a star of the 10th magnitude. This answers to the Æ pretty well, and differs in declination only within blundering distance: it may, therefore, be the lost 7th-magnitude one retiring, and such retreats should be well watched with the Mammoth telescopes. It is, however, singular, that Mr. Baily could find no observation of 42 Virginis of the British Catalogue in Flamsteed's *Historia Cœlestis;* "nor," adds he, "can I discover that such a star exists." That fixed upon by Baron de Zach, and enrolled in his *Tabulæ Speciales*, as 42, differs 3′ in Æ, and, what is remarkable, it is now also missing. Neither Piazzi, Lalande, nor Bessel has any star that can be mistaken for it. It may often happen that such anomalies arise from erroneous entries, but here at least De Zach's star was regularly observed and registered, and yet has probably disappeared from the visible heavens, for an indefinite period.

In star-gazing, Vindemiatrix may be identified by drawing an ideal line from Dubhe to Spica, which will pass it at about one quarter of the distance between those fine stars. Or it may be found nearly midway between Arcturus and β Virginis. Spica forms a remarkable triangle with Arcturus and Denebola; and of the bright stars in this triangle, Vindemiatrix is the one nearest to the line joining Arcturus and the Lion's tail. Though only a secondary kind of star, it has been deemed of sufficient importance to meet the notice of the galley rhymester:

Would you the star of Bacchus find,   on noble Virgo's wing,
A lengthy ray from Hydra's heart   unto Arcturus bring;
Two-thirds along that fancied line,   direct th' inquiring eye,
And there the jewel will be seen   south of Cor Caroli.

CCCCLXX. θ VIRGINIS.

Æ 13ʰ 01ᵐ 40ˢ   Prec. + 3ˢ·10
Dec. S 4° 41′·0   —— S 19″·32

Position	AB 345°·2 (w 5)	Distance	7″·1 (w 5)	Epoch 1831·15
——	AC 295°·0 (w 3)	——	65″·0 (w 1)	
——	AB 350°·0 (w 2)	——	7″·4 (w 2)	—— 1833·40
——	AB 344°·2 (w 9)	——	7″·2 (w 9)	—— 1837·07

A triple star on the lower part of the Virgin's southern wing; and 7° to the north-west of Spica. A 4½, pale white; B 9, violet; C 10, dusky. This star was considered as having a considerable orbital motion, which appeared to be countenanced by my two first epochs; but as the measures of 1833 were made with a new spherical crystal micrometer, of which the zero point was not rigorously exact, little reliance was placed on them, and my last observations were under so favourable a train of circumstances, as to leave no doubt of the object's fixity. The results of other astronomers were certainly not so coincident, as a star bearing illumination admirably would lead us to expect; for the following were the points of comparison:

	Pos.	Dist.	Ep.
H.	339° 18′	7″·13	1782·10
H. and S.	347° 08′	8″·30	1821·24
D.	345° 07′	7″·52	1832·25

Since these angles of position were discussed, the Dorpat Catalogue arrived, in which Σ.'s results justify the above conclusion. A slight proper motion is traced to the principal constituent, which will bear

U 2

upon some future discussion of the question; Mayer had made the value in $Æ = -0''\cdot30$, but the best registers are conflicting, being:

$$P.... \quad Æ - 0''\cdot15 \qquad \text{Dec.} - 0''\cdot05$$
$$B.... \quad\quad\ + 0''\cdot02 \qquad\qquad - 0''\cdot05$$

CCCCLXXI. 42 COMÆ BERENICIS.

$$Æ \quad 13^h\ 02^m\ 12^s \qquad \text{Prec.} + \quad 2^s\cdot95$$
$$\text{Dec. N}\ 18°\ 22'\cdot6 \qquad\qquad - \ \text{S}\ 19''\cdot31$$

Position	round (w 8)	Distance	round (w 8)	Epoch 1832·38
———	10°·0 (w 2)	———	elongated (w 1)	——— 1839·41
———	5°·0 (w 1)	———	0''·3 (w 1)	——— 1842·50

A close double star, between Berenice's Hair and the Virgin's left hand. A 4½, and B 5, both pale yellow. It was discovered by Σ., No. 1728, and recorded as a high class " vicinissimæ;" and his first measures prove to have been:

$$\text{Pos. } 9°\cdot5 \quad \text{Dist. } 0''\cdot64 \quad \text{Ep. } 1827\cdot83$$

When I first attacked this object in 1832, it appeared quite round; and I several times returned to it with similar results. At the middle epoch above recorded, however, though I could not raise a vacancy between the individuals, or even palpably notch them, the elongation was so distinct, under a power magnifying 850 times, as to be capable of a tolerable estimation: the last, which was obtained at Hartwell, is little to be relied on, as the object was at times quite round. In the *np* quadrant is a 10th-magnitude grey star, at an angle $= 315°$, and $\Delta\ Æ = 7^s$. The primary appears to be subject to a proper motion, the amount of which has been thus severally recorded:

$$P....Æ - 0''\cdot45 \qquad \text{Dec.} + 0''\cdot15$$
$$B.... \quad\ - 0''\cdot41 \qquad\qquad + 0''\cdot14$$
$$A.... \quad\ - 0''\cdot44 \qquad\qquad + 0''\cdot16$$

42 Comæ is placed fortunately for the out-door gazer, being midway between Arcturus and Denebola on the parallel, and vertically halfway from Spica to Cor Caroli.

CCCCLXXII. 53 VIRGINIS.

$$Æ \quad 13^h\ 03^m\ 33^s \qquad \text{Prec.} + \quad 3^s\cdot17$$
$$\text{Dec. S}\ 15°\ 20'\cdot0 \qquad\qquad - \ \text{S}\ 19''\cdot28$$

Position 35°·0 (w 1)   Distance 45''·0 (w 1)   Epoch 1833·40

A wide and very delicate double star, in the space between the Virgin's spike and Hydra; at 5½ from the former, on a south-south-west line towards β Corvi. A 5$\frac{1}{2}$, yellowish white; B 15, bluish,—there is also a bluish 10th-magnitude star in the *np* quadrant, and a dusky one of the 12th, nearly on the *s* vertical. As I could only catch a sight of B by gleams, with the equatoreal clock driving the telescope, the above results are but estimations. It was first registered by H., in one of his 20-foot Sweeps, No. 2645; and is a good test for an instrument. A proper motion is given to the star A, to this amount:

$$P.... \quad \text{Æ} - 0''\text{·}09 \qquad \text{Dec.} - 0''\text{·}36$$
$$B.... \quad + 0''\text{·}14 \qquad\qquad - 0''\text{·}29$$
$$A.... \quad - 0''\text{·}10 \qquad\qquad - 0''\text{·}27$$

CCCCLXXIII. 54 VIRGINIS.

$$\text{Æ} \quad 13^h \, 04^m \, 55^s \qquad \text{Prec.} + 3^s\text{·}19$$
$$\text{Dec. S } 17° \, 58'\text{·}5 \qquad \text{——— S } 19''\text{·}24$$

Position 33°·5 (*w* 6)   Distance 5''·9 (*w* 4)   Epoch 1831·41

——— 33°·5 (*w* 9)   ——— 5''·7 (*w* 9)   ——— 1839·30

A neat double star, between the Virgin's right hand and Hydra; where a line from Arcturus carried through Spica, and 8° beyond, will hit it. A 7, and B 7$\frac{1}{2}$, both white. This object is 45 ₥. ɪɪ., and is called "duplex" by Piazzi. The micrometrical measures previous to my attack were:

₥.   Pos. 33° 00'   Dist. 4''·00   Ep. 1782·26
H. and S.   33° 43'   6''·77   1823·28

whence it was inferred, that though the angle of position was nearly stationary, the distance was on the increase. This surmise is not confirmed by my results; and it must be remembered that ₥.'s interval between the stars, was only estimated on the apparent diameter of the larger star. Mr. Baily utterly differs from Piazzi in the value and direction of 54's proper motions:

$$P.... \quad \text{Æ} - 0''\text{·}02 \qquad \text{Dec.} - 0''\text{·}07$$
$$B.... \quad + 0''\text{·}18 \qquad\qquad + 0''\text{·}02$$

CCCCLXXIV. 53 M. COMÆ BERENICIS.

$$\text{Æ} \quad 13^h \, 05^m \, 03^s \qquad \text{Prec.} + 2^s\text{·}94$$
$$\text{Dec. N } 19° \, 01'\text{·}3 \qquad \text{——— S } 19''\text{·}24$$

Mean Epoch of the Observation . . . . 1835·41

A globular cluster, between Berenice's tresses and the Virgin's left hand, with a coarse pair of telescopic stars in the *sf* quadrant, and a single one in the *sp*. This is a brilliant mass of minute stars, from the 11th to the 15th magnitude, and from thence to gleams of star-dust, with stragglers to the *np*, and pretty diffused edges. From the blaze at the centre, it is evidently a highly compressed ball of stars, whose law of aggregation into so dense and compact a mass, is utterly hidden from our imperfect senses. It was enrolled by Messier in 1774 as No. 53, and resolved into stars by Sir W. Herschel.

The contemplation of so beautiful an object, cannot but set imagination to work, though the mind may be soon lost in astonishment at the stellar dispositions of the great CREATOR and MAINTAINER. Thus, in reasoning by analogy, these compressed globes of stars confound conjecture as to the modes in which the mutual attractions are prevented from causing the universal destruction of their system. Sir John Herschel thinks, that no pressure can be propagated through a cluster of discrete stars; whence it would follow, that the permanence of its form must be maintained in a way totally different from that which our reasonings suggest.

Before quitting this interesting ball of innumerable worlds, I may mention that it was examined by Sir John Herschel, with Mr. Baily, in the 20-foot reflector; and that powerful instrument showed the cluster with curved appendages of stars, like the short claws of a crab running out from the main body. A line through δ and ε Virginis, northward, meeting another drawn from Arcturus over η Boötis, unite upon this wonderful assemblage; or it is also easily found by its being about 1° northeast of 42 Comæ Berenicis, the alignment of which is already given.

CCCCLXXV. 25 P. XIII. VIRGINIS.

Æ 13ʰ 06ᵐ 34ˢ   PREC. + 3ˢ·14
DEC. S 10° 30′·4   —— S 19″·20

POSITION 62°·8 (w 7)   DISTANCE 42″·4 (w 5)   EPOCH 1831·38

A wide double star, preceding the Virgin's right hand, at 2°½ to the west of Spica. A 7½, and B 8½, both bluish. This object is composed of Nos. 25 and 26 of Piazzi's Hora XIII., and though coarse is not without interest, especially in a moderate telescope. The only micrometrical measures I am aware of, previous to my own, are those of

H. and S. Pos. 61° 39′ Dist. 44″·84 Ep. 1823·35

CCCCLXXVI. 63 M. CANUM VENATICORUM.

Æ 13ʰ 08ᵐ 38ˢ   PREC. + 2ˢ·70
DEC. N 42° 52′·7   —— S 19″·15

MEAN EPOCH OF THE OBSERVATION . . . . 1836·63

An oval nebula, on the chest of Asterion, the northern dog; discovered by M. Méchain in 1779. This object is of a milky-white tint, and brightens in the centre, where the nucleus resembles a small star.

It is placed between two telescopic stars which cross the parallel vertically, while closer to it in the *sp* is a third. Sir W. Herschel figured this object in the *Philosophical Transactions* for 1811, and described it as very bright, extending from *np* to *sf*, 9′ or 10′ long, and near 4′ broad, with a very brilliant nucleus. The mean apparent place was differentiated from Cor Caroli, from which it bears north-north-east $5°\frac{1}{2}$, on the line indicated from Denebola through Charles's Heart.

CCCCLXXVII.   61 VIRGINIS.

Æ. 13^h 10^m 03^s Prec. + 3^s·19

Dec. S 17° 25′·1   —— S 19″·12

Position 340°·6 (*w* 2)   Difference Æ = 2^s·8 (*w* 1)   Epoch 1832·31

A star with a distant companion, in a barren field, between the Virgin's right hand and Hydra's back, and 7° south-south-west of Spica. A $4\frac{1}{2}$, light straw colour; B $10\frac{1}{2}$, pale blue; a distant star in the *sp* quadrant. This was merely looked at as being 90 ♅. vi., thus registered:

Pos. 345° 00′   Dist. 73″·25   Ep. 1782·26

The primary has considerable proper motions, which my meridional observations were sensible of, and which are thus valued:

P.... Æ − 1″·30	Dec. − 1″·08	
B.... − 1″·04	− 1″·03	
A.... − 1″·09	− 1″·02	

In Mr. Baily's *Address to Astronomical Observers*, in May, 1837, he cites the Æ of this star, thus:

Robinson	13^h 9^m 31^s·28 from	4 observations
Argelander	13^h 9^m 31^s·49	12
Johnson	13^h 9^m 31^s·53	10
Taylor	13^h 9^m 31^s·97	14
Pond	13^h 9^m 31^s·93	11

So that we have here a difference of no less than 0^s·65 between two of our best modern astronomers. "Has *personal equation*," asks Mr. Baily, "anything to do with these anomalies?" But, after all, the coincidences are so admirable, that the shade of Römer may be delighted with the products of the transit instrument.

CCCCLXXVIII.   63 P. XIII. COMÆ BERENICIS.

Æ. 13^h 14^m 01^s Prec. + 2^s·93

Dec. N 18° 36′·4   —— S 19″·01

Position 225°·0 (*w* 2)   Distance 15″·0 (*w* 1)   Epoch 1834·33

A delicate double star, preceding the right foot of Boötes. A 8, white; B 13, blue; and a line through them passes near a brightish distant star in the *nf* quadrant. *Σ.*, in the Catalogue of 1827, mistook his No. 1736, for 63 P. xiii., which he has therefore since rejected; but it is, no doubt, his No. 1737, though the identity of the apparent place is not quite exact. It lies 14° west-by-south from Arcturus, in the line projected from that star towards Denebola and Regulus; or rather less than half-way between the two first.

CCCCLXXIX.   *α* VIRGINIS.

Æ 13ʰ 16ᵐ 47ˢ   Prec. + 3ˢ·15

Dec. S 10° 19′·5   —— S 18″·93

Position 57°·3 (*w* 2)   Difference Æ = 19ˢ·4 (*w* 2)   Epoch 1833·28

A standard Greenwich star, in the Virgin's right hand. A 1, brilliant flushed white; B 10, bluish tinge. This beautiful bright star is in a clear dark field, and, in a manner, insulated, for it has no companion nearer than the one here described; and it is subject to a very slight proper motion, the value of which has been thus registered:

P.... Æ − 0″·09   Dec. − 0″·03
B....   0″·00   − 0″·05
A.... − 0″·04   − 0″·03

α Virginis is the *Στάχυς*, Spica, As-Sumbuleh, or ear of corn, of the Greeks, Latins, and Arabians *. It is also designated *As-Simák-al-a'zal*, the unarmed or defenceless Simák; and Chrysococcas calls it, in reference to Arcturus, *Μικρὸς κονταράτος*, the little lance-bearer. The true meaning of Simák is uncertain, but it appears to have been a leg of an enormous asterism of the ancient Arabs, called the Lion, without any reference to that of the zodiac, and as such forms the XIVth Lunar Mansion. Fírúzábádí, in the Ḳámús, mentions another name for Simák and Al'Auwa, *Al-anharán*, the two rivers, on account of their rising being accompanied by rains; and Riccioli asserts that its Nubian name was *Eleazelet*. The last seems like the Arab *El'azelat*.

This star has strong claims to regard, as affording presumptive evidence that Hipparchus, the first astronomer on record who really made systematic observations, was acquainted with the fact of the precession of the stars, or rather the retrogradation of the equinoctial points. The argument which supports this opinion, is the comparison which this celebrated philosopher made of the places of Spica, determined by himself, with those assigned to it by Aristyllus and Timocharis, about 170 years previously. This lover of truth, as Ptolemy styles him,

---

While the fussy to-do was making as to whether Orion's name should be changed for that of Nelson, or Napoleon, a poet stepped forth for another enrolment of Spica Virginis:

> The star which crowns the golden sheaf,
> And wants a name, O glory of the skies!
> And shall not justice dignify thy sphere
> With the great name of Newton? Be at least
> To *me* for ever the *Newtonian Star*.

bestowed intense application to both the theoretical and practical branches of astronomy. From a difference which he detected between some early observations of this star, and the place which he determined for it by two lunar eclipses, he entertained a suspicion that there existed an inequality in the length of the solar year. It is therefore probable that, in order to ascertain this point, he made comparisons with the ancient registers of celestial phenomena, especially that of a solstice which had been made by Aristarchus, or Archimedes, at the end of the 50th year of the first Calippic period, B.C. 281, or 145 years before his own observation. The latter circumstance was decisive, for there appeared a difference of 12 hours between the calculation and the observation, on the supposition that the year consisted of $365\frac{1}{4}$ days; but $\frac{0\cdot5}{145}$ is $= \cdot00345$, or nearly $\frac{1}{300}$, therefore that supposition seemed to be in excess about $\frac{1}{300}$ of a day, and Hipparchus concluded that the number of days in a tropical year was $= 365 + \frac{1}{4} - \frac{1}{300}$, or to $365\cdot24655$. This value is greater than the truth by $6'\ 13''$ only; since, according to La Place, the length of the tropical year at that time must have been equal to $365\cdot242215$ days, or about $4''\cdot2$ shorter than in the present age. By such a result much was gained; but Hipparchus, conscious of the uncertainty attending the observations of the solstices, from the smallness of the variations in the lengths of the shadows cast by the gnomon, employed the method of the equinoxes, by observations made with the equatoreal armillæ. Under these means, with the lapse-epoch of 33 years afforded by his own results, his expanded mind approximated to the exact length of the tropical year; a grand step in the solar theory, not only on account of its utility in the regulation of the calendar, but also because upon it, depend the elements of the apparent solar orbit. See α Leonis, No. CCCLXXIV.

Virgo, παρθένος, is one of the old 48 constellations; being the sixth sign in zodiacal order, and the last of the summer signs. According to mythology, the lady represents Ceres, or Isis, or Parthenos, or Erigone, or the Singing Sibyl, or some one else, who wore a stern but majestic countenance; though the scales at her feet seemed to fix her as Astrea or Justitia. She is considered as symbolizing the Earth, the producer of fruits and animals; and Dr. Hyde, observing that the Orientals call Spica Sumbuleh, says that this was the Σίβυλλα of the ancients. The Arabians designated her 'Adhrá neḍhífah, the pure virgin; and among Christians she has been recognised as the Blessed Mother, "but how wisely," observes Hood, "any child may judge:"

O Virgo felix! O Virgo significata
Per stellas, ubi Spica nitet.

We are told, that in Ogygian ages and among the Orientals, she was represented as a sun-burnt damsel, with an ear of corn in her hand, like a gleaner of the fields; but the Greeks, Romans, and moderns, have concurred in depicting her as a winged angel, holding wheat ears, typical of the harvest, which came on in the time of the Greeks as the sun approached this star. She forms a conspicuous and extensive asterism, replete with astronomical interest; but astrologers, nothing daunted by classic attributes, stigmatized it as a barren sign, and the illuminated manuscript Almanack of 1386, tells us that whoever is born under the

dominance of its earthy triplicity, he shall "wythowten gylt be blamed."
The constituents have been thus numbered:

Ptolemy	. . . 32 stars	Bullialdus	. . . 43 stars
Copernicus	. . . 32	Hevelius	 50
Tycho Brahé	. . 39	Flamsteed	. . . 110
Bayer	 42	Bode	 411

and there are moreover 323 nebulæ enrolled within its boundaries, by the
unrivalled scrutiny of the elder Herschel.

To find the lucida of this constellation by alignment, is easy enough.
A long line through the conspicuous stars a and γ Ursæ Majoris, will
pass close to Spica, which makes nearly an equilateral triangle with
Arcturus and Denebola, in the Lion's tail. Or a line from Polaris
through ζ Ursæ Majoris, the 6th of the large stars, or middle of the tail,
passes, at 70° distance, through Spica:

From the Pole-star through Mizar glide   with long and rapid flight,
Descend, and see the Virgin's spike   diffuse its vernal light.
And mark what glorious forms are made   by the gold harvest's ears,
With Deneb west, Arcturus north,   a triangle appears;
While to the east a larger still,   th' observant eye will start,
From Virgo's spike to Gemma bright,   and thence to Scorpio's heart.

CCCCLXXX.   ζ URSÆ MAJORIS.

Æ 13ʰ 17ᵐ 28ˢ   Prec. + 2ˢ·42
Dec. N 55° 45′·8   —— S 18″·91

Position 147°·0 (w 7)   Distance 14″·6 (w 5)   Epoch 1830·85
—— 147°·4 (w 9)   —— 14″·4 (w 9)   —— 1839·32

A splendid double star, conspicuous in the middle of Ursa Major's
tail, but was rejected from the Greenwich list in 1830. A 3, brilliant
white; B 5, pale emerald; a distant bluish star of the 8th magnitude,
with minute companions, in the *sf* quadrant; and Alcor, of the 5th,
away down in the *nf*, at a Δ of Æ = 77ˢ·5. This truly fine object is
2 H. III., and is formed by Nos. 78 and 79 Piazzi, Hora XIII. It has
an ascertained proper motion which, though questioned by the Bishop
of Cloyne, is thus registered:

P....Æ − 0″·08   Dec. − 0″·01
B....   + 0″·30   − 0″·04
A....   + 0″·28   − 0″·04

As this proper motion has been pronounced peculiar to both stars,
they would appear to be connected, or their apparent motion is parallactic; but H. thought he detected a retrograde change of position
= 5° 32′ in 20 years 319 days, on which he remarks: "this cannot be
accounted for by a parallactic motion of ζ, which would have occasioned
a contrary change of the angle." His earlier observations suggested to
him an idea that the distance was also rapidly increasing; but both
these opinions have been dissipated by the late observations, and cer-

tainly my own possess most gratifying coincidence. The data previous
to my first epoch were:

Bradley	Pos. 143° 05′	Dist. 13″·88	Ep. 1755·00
Herschel I.	146° 46′	14″·50	1779·63
Piazzi	146° 01′	15″·91	1800·00
Struve	145° 20′	14″·24	1819·70
Herchel II. and South	147° 46′	14″·45	1822·24

Professor Struve made some elaborate observations on these stars,
in 1814 and 1815, for the investigation of their parallax, and the aberration of light. The results, however, effected little more than disproving
the hypothesis of MM. Fuss and Soldner, in the Berlin Ephemerides
for 1785 and 1803.

In 1723, a German astronomer thought he had discovered a new
wandering star near ζ, and not remarking that it was a strange location
for a planet, immediately dubbed it *Sidus Ludovicianum.* in honour of
his sovereign, Louis V., landgrave of Hesse Darmstadt. This was
probably the 8th-magnitude star to the southward of Alcor, which was
first noticed by D. Einmart, the Nuremberg astronomer, in 1691. About
sixty years afterwards, M. Flaugergues was wont to try his telescopes
on ζ, without ever noticing its being double; but in August, 1787, he
was astonished to find that it was composed of two stars. On continuing
to observe them closely, he found a continual augmentation in the distance,
and that the smaller component had increased in size and brightness:
" Ce progrès est actuellement bien sensible, et il y a au moins quinze
secondes de distance entre elles, c'est-à-dire, trois ou quatre fois plus
que lorsque je fis cette observation." This must have been merely the
effect of becoming better acquainted with the object before him.

But there is no end of mistakes respecting ζ Ursæ Majoris, for it
has since been frequently observed by continental astronomers, as a
single star. Hence it is the supposed cause of several discrepancies, in
results of movement, more especially in those of M. Méchain, at Barce-
lona, in 1792. M. Nicollet, in discussing the operations of the French
meridian, states that the telescopes attached to the repeating-circles used
by Delambre and Méchain, were unable to separate Mizar. This was
so limited a performance for instruments on such an important service—
for it requires but little optical aid to divorce the components—that I
was not at all surprised on receiving a letter from Mr. Airy, our Astronomer Royal, in which he says: "About seeing ζ Ursæ Majoris with
the telescope of Méchain's circle, I can only tell you that I saw it in the
beginning of September, 1829, at Milan, and that I made it double
perfectly well, but of course, rather close, the power of the telescope
being low. I did not inquire for Delambre's telescope. I asked for
La Caille's sector, but could not hear of it."

ζ Ursæ Majoris is familiarly known as Mízár, which means a waist
cloth or apron, a name unknown to the Arabians. It was introduced
into the celestial maps in consequence of a conjecture of Scaliger's, who
substituted it for Mirák, the name of β Ursæ Majorís, also given in the
old Tables to ζ instead of *Al-'anák-al-benát*, the goat of the mourners.
Mizar occurs as a proper name in the 42nd Psalm.

Mizar must not be quitted without a notice of Alcor, its more distant
companion, usually called the Rider, since it gave rise to the Arabic

proverb, applied to one who in searching for a mote overlooks a beam: "Thou canst see Alcor, yet canst not perceive the full Moon." But they are wrong who pronounce the name to be an Arabian word importing sharp-sightedness: it is a supposed corruption of *al-jaún*, a courser, incorrectly written *al-jat*, whence probably the *Alioth* of the Alphonsine Tables came in, and was assigned to ε Ursæ Majoris, the "thill-horse" of Charles's Wain. This little fellow was also familiarly termed *Suhà*, and implored to guard its viewers against scorpions and snakes, and was the theme of a world of wit in the shape of saws: "I show her Suhà, and she shows me the Moon," said one Arab wag; while another asks, "How should Suheïl (Canopus) and Suhà ever come in each other's way?" and a third observes, "Truly, as soon as the Sun appears, what can Suhà do further than hide himself?" In the Latin version of the *Almagest*, *Alcor* is written *Aliore*,

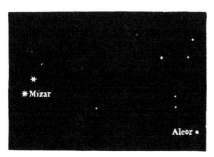

whence Joseph Scaliger conjectured the word ought to be *alyah* in the original, meaning the tail of the broad-tailed sheep; and the name Saddak is also applied to it. This star has led many reckless assertors to declare, that they could see Mizar double with the naked eye; but this is another of the errors alluded to, as the accompanying diagram of the whole group will testify.

From a presumed identity of proper motion, Mizar and Alcor, the *eques stellula* of old, though upwards of 700″ apart, have been suspected of having a physical connexion, albeit under an *annus magnus* of 190,000 of our years; but this may only prove an additional error. However, to assist a watch upon them, I will add their position and distance from each other, and from a third star—the Sidus Ludovicianum—at a vertex between them, as above shown:

	Pos.	Dist.	
Mizar and Alcor.	71°·7	11′ 30″	Ep. 1839·32
Mizar and third star.	102°·6	8′ 45″	

CCCCLXXXI. μ HYDRÆ.

Æ. 13ʰ 20ᵐ 59ˢ   Prec. +   3ˢ·26
Dec. S 22° 27′·1   ——   S 18″·81

Position 80°·9 (*w* 2)   Difference Æ = 19ˢ·5 (*w* 2)   Epoch 1834·38

A variable star, in the caudine portion of Hydra, with a distant companion: it is about 12° south, a little easterly, from Spica, in the line towards Cor Caroli, and is the third of three equidistant stars, ψ, γ, and μ, on the same parallel. A, at the time of observation, was 5½, pale orange-yellow; B 8, greenish, with a small one *sp* it, near the vertical, both companions being the preceding outliers of a following group. Both the magnitude and colour of the primary here given, are liable to the uncertainty created by low altitude, refraction, and vapours. Piazzi remarked it in May, 1805, as "5ˣ magnitudinis, et rubei coloris."

Montanari had called attention to the changes of this star in 1670, and in 1704 Maraldi closely observed it, continuing to examine it at intervals till 1712, when he concluded it variable under a period of about two years. The conditions were investigated by Pigott, who made the time of the star's passing through all its gradations of light and magnitude to be 494 days, by a mean of Maraldi's best observations; but only 487 by his own, under the following conditions:

1. When at its full brightness it is of the 4th magnitude, and has no perceptible change for about a fortnight.
2. It is about six months in increasing from the 10th magnitude, and returning to the same.
3. Therefore it may be considered as invisible also during six months.
4. It is considerably quicker in increasing than decreasing, perhaps by half.

In several recent publications, this star has been designated γ Hydræ, which is the adjacent greenish-yellow star with a minute purple *comes* following by about 11ˢ. But though γ, from its low altitude, has been variously rated, it has never been indistinct, let alone invisible. Ptolemy marked it δ, or 4; Ulugh Beigh and Hevelius, 3; Flamsteed, 3·4; Mayer, 4; and Piazzi, 4·5. I certainly saw it considerably brighter than its neighbour ψ, which is also rated 4·5 in the Palermo Catalogue.

Mr. Samuel Dunn, of Chelsea, in a paper read to the Royal Society in February, 1762, thinks there may be a gross atmosphere interposed between us and the varying stars: such an ethereal medium he deems sufficient to account for the appearance of new stars, and the disappearance of others.

CCCCLXXXII.   72 VIRGINIS.

Æ 13ʰ 22ᵐ 05ˢ   Prec. + 3ˢ·11
Dec. S 5° 38'·5   —— S 18″·77

Position 18°·5 (w 3)   Distance 25″·0 (w 1)   Epoch 1832·26

A very delicate double star, on the right side of Virgo's lower garment; and about 4°½ north by east of Spica, just preceding the line produced between that star and Arcturus. A 7½, yellowish white; B 13, violet tint; a third star in the *sp* quadrant. This is No. 27 of ℍ.'s 145 New Double Stars, which was registered in March, 1785; no measures were, however, obtained, and it is merely noted, "Extremely unequal. Position about 30° *nf*. Large white. Small red." Its detected spacial movement has been thus opposingly valued:

$$P.... \text{Æ} - 0″·12 \quad \text{Dec.} - 0″·01$$
$$B.... \quad + 0″·11 \quad\quad + 0″·02$$

CCCCLXXXIII. 113 P. XIII. URSÆ MAJORIS.

Æ 13ʰ 23ᵐ 00ˢ   Prec. + 2ˢ·22
Dec. N 60° 45'·5   —— S 18"·75

Position 151°·0 (w 3)   Distance 1"·8 (w 2)   Epoch 1835·38

A close double star between the Dragon's and the Bear's tails, 5°
north-by-east of Mizar, and exactly in midway of Alioth and Thuban.
A 8½, and B 11, both bluish; with three stars stretching across the
south part of the field, in a line east and west; the whole seen during a
lively Aurora Borealis. This fine object was discovered by Σ., and is
No. 1752 of the Dorpat Catalogue, where it is thus registered:

Pos. 149°·43   Dist. 1"·63   Ep. 1832·17

CCCCLXXXIV. 51 M. CANUM VENATICORUM.

Æ 13ʰ 23ᵐ 06ˢ   Prec. + 2ˢ·54
Dec. N 48° 01'·7   —— S 18"·74

Mean Epoch of the Observation . . . . 1836·69

A pair of lucid white nebulæ, each with an apparent nucleus, with
their nebulosities running into each other, as if under the influence of a
condensing power. They are near the ear of Asterion, the northern
hound; and the smaller nebula, or northern one, having the brightest
nucleus, was differentiated by the wire micrometer; they are 3° south-
west of Alkaid, where the place is indicated by a line from Dubhe
through Megrez, extended nearly twice that distance into the south-
east beyond. There are three telescopic stars following, and a bright
7th-magnitude about as far beyond them as they are from the nebulæ,
but the preceding part of the field is quite clear. Sir John Herschel has
given a very beautiful representation of this extraordinary object, No.
25, in the illustrations to his Catalogue of 1830.

This fine field was discovered by Messier in 1772, and described as a
faint double nebula whose centres are 4' 35" apart, but with "the
borders in contact." The southern object is truly singular, having a
bright centre surrounded with luminosity, resembling a ghost of Saturn,
with his ring in a vertical position. They form Nos. 1622 and 1623 of
H.'s Catalogue, who terms the southern, or halo nebula, a most astonishing object, probably a similar system to our own, the halo representing
the Galaxy. "Supposing it," he remarks, "to consist of stars, the
appearance it would present to a spectator placed on a planet attendant
on one of them, excentrically situated towards the *np* quarter of the
central mass, would be exactly similar to that of our Milky Way, tra-
versing, in a manner precisely analogous, the firmament of large stars,

into which the central cluster would be seen projected, and (owing to its greater distance) appearing, like it, to consist of stars much smaller than those in other parts of the heavens. Can it then be that we have here a brother-system, bearing a real physical resemblance and strong analogy of structure to our own?"

We have then an object presenting an amazing display of the uncontrollable energies of OMNIPOTENCE, the contemplation of which compels reason and admiration to yield to awe. On the outermost verge of telescopic reach we perceive a stellar universe similar to that to which we belong, whose vast amplitudes no doubt are peopled with countless numbers of percipient beings; for those beautiful orbs cannot be considered as mere masses of inert matter. And it is interesting to know that, if there be intelligent existence, an astronomer gazing at our distant universe, will see it, with a good telescope, precisely under the lateral aspect which theirs presents to us. But after all what do we see? Both that wonderful universe, our own, and all which optical assistance has revealed to us, may be only the outliers of a cluster immensely more numerous. The millions of suns we perceive cannot comprise the Creator's Universe. There are no bounds to infinitude; and the boldest views of the elder Herschel *only* placed us as commanding a ken whose radius is some 35,000 times longer than the distance of Sirius from us. Well might the dying Laplace exclaim: " That which we know is little; that which we know not is immense."

CCCCLXXXV.   75 VIRGINIS.

ℛ 13ʰ 24ᵐ 19ˢ   PREC. +   3ˢ·19

DEC. S 14° 32'·3   ——   S 18''·70

POSITION 112°·0 (*w* 1)   DISTANCE 93''·0 (*w* 1)   EPOCH 1835·32

A star with a minute distant companion, on the tip of the wheat-ears in Virgo's right hand; and nearly 5° south-south-east of Spica. A 6, pale white; B 14, dusky; two other small stars at a distance in the *np* quadrant. This object was merely looked at from being among H.'s Sweeps, No. 2658; otherwise it is too difficult to measure, and too wide for tolerable estimation,—indeed the *comes* was best seen by averting the eye to another part of the field of view. Careful comparisons have produced these values for the spacial movement of No. 75:

P.... ℛ - 0''·10   Dec.   0''·00
B....   - 0''·02   - 0''·10

The minute companions occasionally registered in this Cycle, are the nearest objects to be seen with my telescope, and therefore will be tests for the general ones; but no doubt the grand instruments will bring out some still smaller, and nearer to those lettered A.

CCCCLXXXVI. 127 P. XIII. VIRGINIS.

Æ 13ʰ 26ᵐ 07ˢ   Prec. + 3ˢ·06
Dec. N 0° 30'·4   —— S 18"·65

Position 24°·1 (w 5)   Distance 1"·5 (w 3)   Epoch 1832·39
———— 31°·0 (w 8)   ———— 1"·7 (w 4)   —— 1838·48
———— 37°·9 (w 8)   ———— 1"·7 (w 6)   —— 1842·52

A close binary star in Virgo's lower garment; it is 11° north-by-east of Spica, just preceding the line between that lucida and Arcturus, and close to ζ Virginis, a star of the 4th magnitude. A 8, pale white; B 9, yellowish; and the two point to a telescopic star at a distance in the *nf* quadrant. This was discovered by Σ., and is No. 1757 of the Dorpat Catalogue, where the earliest measures for a starting point, by which the direct angular motion is seen, are:

Pos. 10°·0 Dist. 1"·60 Ep. 1825·37

From comparing this early measure with his and my subsequent ones, embracing a period of 17 years, it might be inferred that during the first part the angular progress was at the rate of 2° per annum; that it then diminished to 1°, and is now on the increase, amounting to 1°¼, whilst the distance appears to have continued the same. We may hence conclude that we are gazing at this binary couple nearly full face, in which case its revolution, assuming its present rate as a mean, will be accomplished in about 240 years.

CCCCLXXXVII. 81 VIRGINIS.

Æ 13ʰ 29ᵐ 13ˢ   Prec. + 3ˢ·13
Dec. S 7° 03'·2   —— S 18"·54

Position 39°·8 (w 7)   Distance 2"·8 (w 4)   Epoch 1832·36

A close double star, on the right side of the lower garment, and just 4°½ north-north-east of Spica. A 7½ (rated 6 by Flamsteed), bright white; B 8, yellowish; a minute blue star in the *np* quadrant. This fine object is 80 ℍ. ɪ., whose register, allowing a distance upon the star's diameter, will stand thus:

Pos. 48° 48' Dist. 1"·75± Ep. 1783·10

The subsequent measures, previous to my own, were:

H. and S. Pos. 42° 24' Dist. 4"·02 Ep. 1822·94
Σ.   39° 01'   2"·69   1830·34

whence, on a comparison of all the results, we may infer a slow retrograde change in the angle of position, though requiring confirmation.

CCCCLXXXVIII. Prec. 3 M. Can. Venaticorum.

Æ. 13ʰ 30ᵐ 28ˢ   Prec. + 2ˢ·78

Dec. N 29° 08′·6   —— S 18″·50

Position 191°·5 (w 3)   Distance 1″·0 (w 1)   Epoch 1835·48

A close double star, on the flank of Chara, the southern Hound; lying nearly in mid-distance between Arcturus and Cor Caroli, and a little to the west of the large cluster, No. 3 Messier. A 9½ and B 10½, both white, with a pale blue telescopic companion in the *nf* quadrant. I first noticed this beautiful object while viewing the outliers of Messier's gorgeous mass of stars, No. 3, with Sir John Herschel, at Slough, in his 20-foot reflector; and it is thus entered in his admirable Catalogue, No. 1663:

> Observed with Captain Smyth, who " saw something remarkable" in a small star 2′ or 3′ preceding the cluster, which proved on closer examination to be a fine first-class double star.

This entry is here copied to show that my telescope, with its 5₁₀⁹ths inches aperture, had no small task inflicted upon its performance, to attack an object which had been thus picked up with an instrument of 18 inches aperture. It accordingly required much attention and coaxing to gain a fair division in its elongation, and the result of my several estimations is drawn from fitful gleams. In 1839, I never divided it, yet it had so *long* a disc as to give an impression that it ought to be, under favourable conditions. I therefore afterwards begged my friends, the Rev. J. Challis of Cambridge, and the Rev. W. R. Dawes, to make an examination of its state and condition. Though several circumstances prevented Mr. Dawes from observing it with the accuracy which so delicate an object requires, he yet made it double on the 6th April, 1842, but was unable to verify it afterwards. Mr. Challis applied the Northumberland equatoreal to it, and on the 22nd June, 1842, wrote me word: "I looked a long time last night at the star pr. 3 Messier, and am convinced I saw it double, though the distance was certainly not more than 0″·6. I looked at it with a power of 715, and took two measures of position, the mean of which gave 13° for the angle of position. I then desired Mr. Glaisher, in my absence, to observe it, and say whether he perceived it double. Without knowing my result he obtained 3° for the angle of position, which, considering the proximity of the stars, is as near as might be expected. He said he had no doubt it was double, and the larger star appeared to him *preceding*. I could not decide on this point, but I thought you would be interested in the details, from your having observed this star when the individuals were much more separate." Since this, on the 24th of April 1843, I examined it with Mr. Dawes, from Mr. Bishop's observatory in Regent's Park, when it was decidedly elongated.

CCCCLXXXIX. 156 P. XIII. URSÆ MAJORIS.

$\mathbb{R}$ 13ʰ 31ᵐ 18ˢ　　Prec. + 2ˢ·42

Dec. N 51° 31'·9　　—— S 18"·48

Position 119°·9 (w 9)　Distance 1"·9 (w 6)　Epoch 1832·81

A close double star, towards the tip of the Bear's tail. A 6, topaz yellow; B 8, livid, followed within 2ˢ by a third star similar in lustre with B; and they precede another pair in the *nf* quadrant, probably Piazzi's No. 157. This exquisite object is one of Σ.'s " aureæ vicinæ," being No. 1770 of the Great Dorpat Catalogue. The other registered measures are:

	Pos.	Dist.	Ep.
H.	116° 32'	1"·59	1830·22
Σ.	120° 57'	1"·79	1831·80
D.	116° 47'	*caret*	1832·25

156 P is easily found, lying but 2° to the north-north-west of Alkaid, in the direction of Megrez, δ Ursæ Majoris.

CCCCXC. 1 BOOTIS.

$\mathbb{R}$ 13ʰ 33ᵐ 02ˢ　　Prec. + 2ˢ·86

Dec. N 20° 46'·0　　—— S 18"·42

Position 147°·1 (w 5)　Distance 4"·9 (w 3)　Epoch 1832·23

A fine double star, preceding the right shin of Boötes, where it is 8°¼ preceding Arcturus on a west-half-north line, which prolonged 33° would pass between δ and θ Leonis. A 6, sapphire blue; B 10, smalt blue; and this beautiful object is the more remarkable not only in these stars being thus coloured, but in there being two others in the field, one *np* and the other *sp*, which are also bluish. This is No. 1772 of the Dorpat Catalogue, and thus registered:

Pos. 148°·72　Dist. 4"·84　Ep. 1831·23

A trifling proper motion is attributed to this star, which Piazzi gives to the declination only, and Mr. Baily to the right ascension.

CCCCXCI. 163 P. XIII. CANUM VENATICORUM.

$\mathbb{R}$ 13ʰ 33ᵐ 16ˢ　　Prec. + 2ˢ·78

Dec. N 28° 52'·6　　—— S 18"·41

Position 215°·0 (w 2)　Distance 68"·0 (w 2)　Epoch 1831·26

A star with a companion, preceding the right knee of Boötes, among the outliers of 3 Messier, the next described object. A 6½, light orange;

B 13, ash coloured; two other small stars in the *sp* quadrant. A is undoubtedly Piazzi's No. 163, Hora XIII., and it is classed from the Palermo Catalogue, otherwise it would hardly have been rated at more than the 8th magnitude.

This being an object to which light was inadmissible, the non-illuminating principle was necessary. It was therefore selected for a trial of Mr. Dollond's spherical crystal micrometer on the angle of position, by getting the double image of A in a line towards B. The distance, being too great for the value of the scale, was obtained by $\triangle$ Æ.

CCCCXCII. 3 M. CANUM VENATICORUM.

Æ 13^h 34^m 45^s Prec. + 2^s·77
Dec. N 29° 10'·6   —— S 18"·35

MEAN EPOCH OF THE OBSERVATION . . . . 1831·26

A brilliant and beautiful globular congregation of not less than 1000 small stars, between the southern Hound and the knee of Boötes; it

blazes splendidly towards the centre, and has outliers in all directions, except the *sf*, where it is so compressed that, with its stragglers, it has something of the figure of the luminous oceanic creature called *Medusa pellucens*. This noble object is situated in a triangle formed by three small stars in the *np*, *nf*, and *sf* quadrants, which, by their comparative brightness, add to the beauty of the field. It is nearly in mid-distance between Arcturus and Cor Caroli, at 11° north-west of the former star.

This mass is one of those balls of compact and wedged stars, whose laws of aggregation it is so impossible to assign; but the rotundity of figure gives full indication of some general attractive bond of union. It was discovered in 1764 by Messier, who described it as "a nebula without a star, brilliant and round:" his instrument must have been rather moderate not to resolve this object, and it is matter of regret, that the exertions of such a man were straitened to such means. It was next pronounced to be a "mottled nebulosity;" but in 1784, Sir W. Herschel attacked it with his 20-foot reflector, and resolved it into a "beautiful cluster of stars, about 5' or 6' in diameter." By the gauging process, which he has fully described, he estimated its profundity to be of the 243rd order.

CCCCXCIII. 84 VIRGINIS.

Æ 13^h 35^m 02^s Prec. +   3^s·03
Dec. N   4° 21'·0   —— S 18"·35

Position 232°·9 (w 7)   Distance 3"·7 (w 5)   Epoch 1831·19
—— 231°·8 (w 5)   —— 3"·6 (w 5)   —— 1836·35
—— 233°·4 (w 9)   —— 3"·5 (w 9)   —— 1839·37

A close double star, on the tip of Virgo's left wing; it is distant 10°½ north by east from Spica, and points from that star towards Arcturus. A 6, yellowish; B 9, smalt blue, a third star of the 9th magnitude in the *sf* quadrant. I am inclined to attribute a small minus proper motion, in declination, to the primary.

The registered observations, previous to my measures, had led to the suspicion of a considerable orbital change, which my own operations did not confirm; and I may add that, as far as the *senses* are concerned, my results in 1839 were perfect. The previous details were:

H.   Pos. 240° 55'   Dist. 4"·00 ±   Ep. 1782·12
H. and S.   227° 51'   3"·91   1821·37

This angular discrepancy being a larger amount than could be assigned to instrumental error, induced a belief that a mean annual motion of − 0°·288 existed: but the lapse of eight years between my measures is a basis for the contrary opinion. The object, however, except with the best instruments, and under favouring circumstances, is rather difficult to handle.

CCCCXCIV. 171 P. XIII. VIRGINIS.

Æ 13^h 35^m 13^s Prec. +   3^s·12
Dec. S   3° 27'·9   —— S 18"·34

Position 336°·3 (w 4)   Distance 30"·0 (w 2)   Epoch 1830·99

A delicate double star, on the middle of the lower garment; and about 8° north-north-east of Spica. A 8, light orange-tint; B 10½, pale lilac. This was discovered by Σ., and is No. 1775 of the Dorpat Catalogue, and in the publication of 1837 appears thus:

Pos. 335°·69 Dist. 27"·76 Ep. 1829·35

It may be here noticed, that in Sir John Herschel's observations with the 7-foot equatoreal, he has mistaken the synonyme of the star which he measured as 171 P. XIII.; that being 1776 of Σ., which is not in Piazzi's Catalogue.

CCCCXCV. 85 VIRGINIS.

Æ 13ʰ 36ᵐ 59ˢ   Prec. + 3ˢ·21
Dec. S 14° 57'·6   —— S 18"·28

Position 320°·0 (w 1)   Distance 30"·0 (w 1)   Epoch 1834·28

A most delicate double star, between the Virgin's skirt and the tail of Hydra; and 7° south-east of Spica. A 6, white; B 16, faint, with a distant companion in the *sp* quadrant, and another away in the *np*. This most difficult object was merely examined from being one of H.'s Sweeps, No. 2677, of the Fifth Series of observations with the 20-foot reflector. It is the *minimum visibile* of my instrument, and therefore impossible to measure, being only caught by evanescent glimpses, aided by the smooth motion of the equatoreal clock. The direction of B is nearly pointed out by the distant 10th-magnitude star in the *np*; and as the estimations were pretty coincident *inter se*, they are, perhaps, entitled to the weight assigned them.

CCCCXCVI. η URSÆ MAJORIS.

Æ 13ʰ 41ᵐ 14ˢ   Prec. + 2ˢ·39
Dec. N 50° 06'·5   —— S 18"·12

Position 323°·0 (w 2)   Difference Æ = 29ˢ·3 (w 1)   Epoch 1835·37

A standard Greenwich star, with a distant companion, at the tip of the Greater Bear's tail; and that animal being itself the very corner stone of alignment, it need only be added, that the worst time for observing η is when the constellation is sub-polar, in autumn, for the tail then, in these latitudes, almost trails along the horizon. A 2½, brilliant white; B 9, dusky; and the primary has had the following proper motions assigned to it, viz.:

P....	Æ − 0"·50	Dec.	− 0"·00
B....	− 0"·14		− 0"·03
A....	− 0"·13		− 0"·03

In the VIIIth volume of Baron de Zach's *Correspondance Astronomique*, p. 516, it is remarked: "M. Struve observe que le P. Piazzi marque dans son dernier Catalogue plusieurs étoiles doubles, qui ne le sont pas, comme, par exemple, l'étoile η de la Grande Ourse." I have myself examined most of the stars to which Piazzi assigns companions, as herein often instanced, and have invariably found him accurate. In the case before us, among the Notæ, Hora XIII., p. 94 and No. 209, are these words: "Duplex. Comes 9ᵃ magnitud. in eodem verticali paullisper ad boream;" and I will venture to assert, that the A and B I have here given thirty-five years afterwards, were as fairly described as the notes to such a meridian Catalogue required.

This star is designated Alkaid, or Benetnasch, both of which are taken from its Arabian denomination, *Al káyid-al benát-al-na'sh*, the governor of the mourners, in allusion to the fancied figure of a bier: the stadtholder of Ideler. It forms a fine termination to the Bear's long tail, which queer appendage is thus accounted for, by old Thomas Hood, a Fellow of Trinity College, Cambridge, who wrote on the celestial globe in 1590:

"*Scholar.* I marvell why (seeing she hath the forme of a beare) her taile should be so long.

"*Master.* Imagine that Jupiter, fearing to come too nigh unto her teeth, layde holde on her tayle, and thereby drewe her up into the heaven; so that shee of herself being very weightie, and the distance from the earth to the heavens very great, there was great likelihood that her taile must stretch. Other reason know I none."

The tailed bear was, however, an important asterism to Lilly, Dee, Alasco, Hemenga, and other astrologers; and, as Recorde says, "its motion is so evident, that every child may mark it." The star in question has gained location in Hudibras:

> Cardan believed great states depend
> Upon the tip o' th' Bear's tail's end,
> That, as she whisk'd it towards the Sun,
> Strow'd mighty empires up and down;
> Which others say must needs be false,
> Because your *true bears have no tails.*

Cardan, however, with all his astronomy and subtlety and mathematics and physics, was no conjuror; for, if the author of *Sidrophel Vapulans* is to be believed, he starved himself to death to verify *his own* prediction of *his own* decease. It is not yet settled, whether he discovered the general theory of equations of the third degree; the claims of Tartaglia, his rival, being admitted in Italy. But the Hindú algebraists may, perhaps, spurn both pretensions.

CCCCXCVII.   220 P. XIII. BOOTIS.

Æ 13ʰ 42ᵐ 51ˢ   Prec. + 2ˢ·83
Dec. N 22° 04'·4   — S 18"·06

Position 208°·5 *(w 6)*   Distance 85"·8 *(w 4)*   Epoch 1831·14

A wide pair of stars, on the right shin of Boötes, 6° west-by-north of Arcturus, in the direction of the lucida of Coma Berenices. A 7½, and B 8, both flushed white, between two stars nearly on the parallel. B is Piazzi's No. 219, Hora XIII., who also noticed the companion following; but the object was first micrometrically measured by Sir James South, whose results may thus be compared with data obtained from the mean places in the Palermo Catalogue,—a severe trial of the altitude and azimuth circle:

P.   Pos. 210° 30'   Dist. 88"·50   Ep. 1800·00
S.   208° 10'   86"·03   1825·20

CCCCXCVIII. 238 P. XIII. VIRGINIS.

Æ 13ʰ 46ᵐ 35ˢ   PREC. + 3ˢ·14

DEC. S 7° 16'·1   —— S 17"·92

POSITION 55°·0 (w 6)   DISTANCE 2"·5 (w 5)   EPOCH 1834·29

A close double star, nearly in the middle of the lower garment. A ray from Algorab to the east-north-east, through Spica, and prolonged 8° beyond that star, catches it up; or another from Denebola in the Lion's tail, passed close under δ Virginis, and carried as far again into the south-east, also hits it. A 7, and B 8½, both white, and they are preceded by two small dusky stars, one in the *sp* quadrant, and the other in the *np*. The previous measures of this pretty object are:

S.	Pos. 51° 44'	Dist. 2"·76	Ep. 1825·39
H.	48° 38'	2"·57	1830·27
Σ.	54° 01'	2"·37	1831·38

And by M. Struve's numbers which produced this mean after four years, the great difficulty of precise observation in this intimate pair, is sufficiently evinced.

CCCCXCIX. η BOOTIS.

Æ 13ʰ 47ᵐ 04ˢ   PREC. + 2ˢ·86

DEC. N 19° 12'·0   —— S 17"·90

POSITION 117°·6 (w 4)   DISTANCE 123"·7 (w 2)   EPOCH 1832·42

A second-rate Greenwich star, with a distant companion, on the right leg of Boötes, and 5°½ west-by-south of Arcturus. A 3, pale yellow; B 10½, lilac, and a telescopic star in the *sp* quadrant. This object is 95 Ⱨ. VI., but its entry in 1782 is unaccompanied by measures, a mere estimation being made that its angle of position was about 25° or 30° *sf*, and its distance about 90". H. and S. gained these results:

Pos. 119° 27'   Dist. 126"·20   Ep. 1822·66

This star is called Muphrid on the Palermo and other Catalogues, from the Arabic *al-mufrid-al rámih*, the single, or solitary star of the lancer; and though its place in the figure was denoted by the word *saák*, shin-bone, it was called the lance of 'Auwá. It has a sensible proper motion, especially in declination, the amount of which has, on a very rigid scrutiny, been thus given:

P....	Æ − 0"·10	Dec.	− 0"·40
B....	+ 0"·02		− 0"·35
A....	− 0"·02		− 0"·35

D. 187 ⯎. I. CANUM VENATICORUM.

ℛ 13ʰ 49ᵐ 55ˢ   Prec. + 2ˢ·39
Dec. N 48° 02'·1   —— S 17'''·78

Mean Epoch of the Observation . . . . 1837·28

A small round nebula, of an oval shape, and pale-white tint, preceding the right arm of Boötes. Its major-axis trends *sp* and *nf*, and there are several small stars in the field, of which three form a triangle near the north vertical. It was discovered by ⯎. in May, 1787; and is No. 1712 of his son's Catalogue for 1830. The mean apparent place was obtained from η Ursæ Majoris, by differentiation; being a couple of degrees to the south-south-east of that star, in the direction of γ on the right shoulder of Boötes.

DI. 277 P. XIII. URSÆ MAJORIS.

ℛ 13ʰ 53ᵐ 03ˢ   Prec. + 2ˢ·20
Dec. N 53° 53'·1   —— S 17'''·65

Position 4°·9 (*w* 2)   Distance 6''·8 (*w* 1)   Epoch 1835·36

A neat but very delicate double star, between the Bear's tail and the right hand of Boötes. A 7½, bright white; B 12, pale blue, and, for its magnitude, singularly distinct, hence forming a fine test object. It was discovered by Σ., and is No. 1795 of the Dorpat Catalogue, where it is thus registered:

Pos. 3° 15'   Dist. 7''·61   Ep. 1832·13

This object is easily fished up by alignment, since it is but 4° north-north-east of Alkaid, and in the line with ζ, ε, and δ Ursæ Majoris.

DII. τ VIRGINIS.

ℛ 13ʰ 53ᵐ 31ˢ   Prec. + 3ˢ·04
Dec. N 2° 19'·3   —— S 17'''·64

Position 291°·4 (*w* 5)   Distance 78''·6 (*w* 3)   Epoch 1831·24

A star with a distant companion, a little to the north of Virgo's knee; it is 15°¼ north-north-east of Spica, midway towards the vertical stars of the left leg of Boötes. A 4½, bright white; B 8½, lilac. This was merely examined as being 77 ⯎. vi., enrolled in February, 1782, as

68″·36 apart; and found by H. and S. to have apparently increased the distance by 11″ in forty-one years. A comparison, however, of all the measures, brings me to the conclusion, that an error exists in ⬡.'s original entry; and that these stars are not subject to any change sensible in short periods.

DIII. 101 M. BOOTIS.

Æ 13ʰ 57ᵐ 31ˢ   Prec. + 2ˢ·13

Dec. N 55° 08′·3   —— S 17″·47

MEAN EPOCH OF THE OBSERVATION . . . . 1837·26

A pale white nebula, in the nebulous field *np* the right hand of Boötes; it is 5° north-north-east of Alkaid, and a similar distance east-half-south from Mizar. This object was discovered by Mechain in 1781, in whose instrument it was very obscure; and it only exhibited a mottled nebulosity to ⬡. Under a favourable view it is large and well spread, though somewhat faint except towards the centre, where it brightens. There are several telescopic stars in the field, one of which is very close to the nebula.

From the nature of this neighbourhood, and a trifling uncertainty in the earlier data, this object may be 214 ⬡. ɪ.; but that astronomer does not appear to have been aware of the identity. It is one of those globular nebulæ that seem to be caused by a vast agglomeration of stars, rather than by a mass of diffused luminous matter; and though the idea of too dense a crowd may intrude, yet the paleness tells of its inconceivable distance, and probable discreteness.

DIV. α DRACONIS.

Æ 14ʰ 00ᵐ 03ˢ   Prec. + 1ˢ·63

Dec. N 65° 08′·4   —— S 17″·36

POSITION 45°·8 (*w* 2)   DIFFERENCE Æ = 23″·9   EPOCH 1834·41

A bright star with a distant companion, in the middle of Draco's body. A 3½, pale yellow; B 8, dusky; two other telescopic stars in the following part of the field. The primary is suspected of variability, for Ptolemy, Ulugh Beigh, and La Caille, mark it as of 3rd magnitude; and Pigott as a bright 4th; Tycho Brahé, Hevelius, and Bradley, rank it of the 2nd; and though marked of the latter size in the British Catalogue, Mr. Baily found that in the original entries it is designated once of the 3rd and once of the 4th. I have had it in view many times, and always looking like a small 3rd; though Baron de Zach, but shortly before, classed it 2·3.   It has a slight appreciable proper motion, which is thus registered:

P....	ℛ − 0″·15	Dec. + 0″·09
B....	− 0″·14	− 0″·02
A....	− 0″·15	+ 0″·01

This star, though not the *lucida* of the asterism, unless we admit its variability, has been lettered *a*, and was once rated of the 2nd magnitude: it is named Thuban, from the Arabian *al-Thu'bán*, the dragon, a word synonymous with *tinnín* and *azhdehá*. Upwards of 4600 years ago, it was the pole-star of the Chaldeans, being then within 10′ of the polar point; a point which will not be approached by *a* Ursæ Minoris nearer than 26′ 30″. *a* Draconis, in that remote age, must have seemed stationary during the apparent revolution of the celestial sphere about the northern extremity of the polar axis; though now it has, by the slow movement to which the stellar host is subject, deviated from the pole as much as 24° 52′.

Δράκων, Draco, is figured as a strange bird-headed reptile meandering around the north pole of the ecliptic, passing its tail between the two Bears and its head under the right foot of Hercules, and extending over so many hours of ℛ as to be quite confusing. Virgil, and other old writers, constantly place the constellation between the two Bears; which location hardly suits its present place, since the principal stars are between Ursa Minor, Cepheus, Cygnus, and Hercules. In a rare volume in my possession, printed at Venice, *anno salutifere incarnationis* 1448, an edition which escaped the industrious gleaning of La Lande, are some very taking figures of the constellations, and among others the two Bears are regularly enfolded in the embrace of Draco; while Virgil, *viâ* Dryden, says:

> Around our Poles the spiry Dragon glides,
> And like a wand'ring stream the Bears divides.

By mythologists it was viewed as *Hesperidum Custos*, the grim dragon which guarded the golden fruit of those far-famed gardens, though the Mosaicists insist that it represents the wily serpent that tempted Eve with an apple; and a third class esteemed it merely the emblem of vigilance: while Olaus Rudbeck, in the spirit of a true Swede, tells us that Draco symbolized the Baltic Sea. This extensive and convoluted *Maximus Anguis* was, however, one of the original forty-eight asterisms, and has thus gradually increased its components, as practical optics have advanced:

Ptolemy . . . 31 stars		Hevelius . . . 40 stars
Tycho Brahé . . 32		Flamsteed . . 80
Bayer 33		Bode 255

An imaginary line projected northward from Cor Caroli through Mizar, passes by it; or it may be looked for about mid-way between Mizar and Pherkad. Thuban had gained admission to the Greenwich roll, but was discarded by the *Nautical Almanac* reformers of 1830, whence the brackish rhymes:

> Though long the captain of the stars,   which Draco's body grace,
> Thuban has given up the *pas*,   and *beta's* in the place.

DV. ϰ BOOTIS.

Æ 14ʰ 07ᵐ 44ˢ   Prec. + 2ˢ·15

Dec. N 52° 32′·5   —— S 17″·01

Position 237°·9 (w 7)   Distance 12″·5 (w 5)   Epoch 1830·93

—— 238°·1 (w 8)   ——— 12″·7 (w 8)   —— 1838·78

A neat double star in the herdsman's right hand, where, with θ and ι, and λ on the upper arm, it forms *Aulád al-dhibá'*, the young of the hyæna. A 5½, pale white; B 8, bluish. This fine object is formed by Nos. 30 and 31 P. Hora XIV.; and it is 11 Ḥ. III. The standard measures previous to my own were, including a reduction of Piazzi's mean apparent places:

	Pos.	Dist.	Ep.
Ḥ.	240° 00′	12″·50	1779·74
P.	239° 30′	12″·80	1800·00
H. and S.	238° 45′	13″·14	1822·62
Σ.	237° 42′	12″·42	1828·77

These results led me to infer that a slight angular change had occurred, in direction *np sf*, or retrograde; but my second series of measures, made under the best circumstances, discountenance the inference. It may be discerned as forming the north-western point of a little triangle with its own ι and θ; or carrying a line from ϵ Ursæ Majoris through ζ, the two inner stars of the tail, and 6° further to the eastward, will show it.

DVI. α BOOTIS.

Æ 14ʰ 08ᵐ 22ˢ   Prec. + 2ˢ·81

Dec. N 20° 00′·9   —— S 16″·98

Position 49°·3 (w 2)   Difference Æ = 15ˢ·1 (w 1)   Epoch 1835·47

Arcturus, a standard Greenwich star, between the legs of Boötes, with a distant companion in the *nf* quadrant. A 1. reddish yellow; B 11, lilac; and at the extreme verge of the field, under moderate magnifying power, is the star *infra Arcturum*, noticed by Flamsteed on the 14th of February, 1690, then *preceding* the primary, but now, from the large proper motions of Arcturus, *following* it. These motions have been well watched, and the best registers of their quantity and direction give the following values:

		Dec.
P....	Æ − 1″·17	− 1″·96
B....	− 1″·11	− 1″·98
A....	− 1″·18	− 1″·96

Mayer of Manheim, who had some odd notions in astrognosy, considered Arcturus as a cluster; in the which he was mistaken. Various crotchets have been started respecting its insulation, while from its brilliance and proper motions, a conclusion was drawn that it was the nearest star to our system, but this has been abandoned since the still greater motion of 61 Cyni, μ Cassiopeia, and other stars, has been ascertained. Dr. Hornsby pronounced upon its proximity, because the variation of its place seemed more remarkable than that of any other of the stars; and by comparing a variety of observations respecting its motions, he inferred that the obliquity of the ecliptic decreases at the rate of 58″ in one hundred years; a quantity, he observes, " which will be found nearly at a mean of the computations framed by Mr. Euler and M. de la Lande, upon the principles of attraction." Sanguine hopes of arriving at its distance were entertained, but every exertion met with disappointment; and Piazzi dismissed the case thus: " Pluries inquisi-vimus in hujus stellæ parallaxim, sed nihil profecimus." The star has, however, been extensively useful in modern astronomy; and it is remarkable as being the body by which the fine discovery was made, that stars and planets may be advantageously observed during the Sun's presence. The Abbé Picard imagined himself to be the discoverer of the feasibility of so doing, from observing the meridional altitude of this star on the 13th of July, 1669, while the Sun was elevated 17°; but he who observes should also read, or he may " make many inventions," for the fact of the principal stars being visible in broad day-light had been announced by the enthusiastic M. Morin, in 1635, and, curiously enough, his first star was also Arcturus.

Sir William Herschel supposed the true diameter of this star to be about one-tenth of a second, having detected for the apparent, two-tenths. This would give as the diameter of Arcturus four millions of leagues, or eleven times the diameter of our Sun.

Arcturus is compounded of ἄρκτου, and οὐρὰ, bear's tail, from its proximity to the latter, being somewhat in a direction pointed by a line drawn through the two hinder stars, ζ and η, of the tail. Being one of the most brilliant of the stellar bodies, it was noticed by very early star-gazers; but though, as I have said under η Tauri, the name is dragged into our translation of Job as a synonyme of 'Aïsh, it must be inferred that the Seventy wished rather to express a brilliant emblem of Majesty, than to be critically exact: like the exclamation of Job, "Oh that my words were *printed* in a book!" It is first mentioned by Hesiod, whose æra is nearly approximated by the passage in the *Works and Days*, in which the star is mentioned. From this passage it appears that there is a difference of 40 days in the achronical rising of that lucida, since the time of the poet; whence, by allowing 50″¼ annually as the recession of the equinoxes, we obtain about 2800 years since the days of Hesiod.

Aratus—of course from Eudoxus—and Hyginus, place Arcturus in the herdsman's girdle: " In zona unā clarius cæteris lucentē, hæc stella Arcturus appellatur;" but it is now usually marked between the calves of his legs, or as others have it, in *fimbriâ*, and on the "skirts of his coat." Vitruvius, lib. IX., expressly says of it, " stella media genuorum custodis Arcti." The Arabians call it *Simák-al-rámih*, the prop or leg of the lancer, but the true meaning of *simák* is very uncertain, as was noticed under a Virginis, the other simák. From this designation, however, came the *Aramech*, or, as the Alphonsine Tables have it, *Azimech*, which have been applied to this fine star; but it is difficult to guess where Riccioli picked up his designation, *Kolauza*. Ulugh Beigh, in his Catalogue, places Arcturus, "extra hanc figuram," expressly noting its position, "inter femora figuræ:" whence probably R. Recorde, in 1556, derived his description, "Boötes hath 22 starres, beside one very bryghte starre called Arcturus, which standeth between Boötes his legges." From the days of Evander it was a noted star among ancient mariners, but its influences were reckoned ungenial; and the change between the summer and autumnal Etesian winds, being preceded by eight or ten days of squally weather, the *prodromi* of old, they were ascribed to the direct power of Arcturus, instead of the alteration consequent upon the solar march. We learn from Demosthenes, that a sum of money was lent at Athens on bottomry, on a vessel going to the Crimea and back, at $22\frac{1}{2}$ per cent. on the voyage out and home; but unless they returned before the rising of Arcturus, 30 per cent. was to be paid. Virgil repeatedly spurns this paranatellon of Virgo, as a cold star; while Horace shows that a contented man is neither anxious about the tempestuous sea, nor the malign aspects of stars, thus:

> Nec sævus Arcturi cadentis
> Impetus, aut orientis Hædi.

Arcturus opens the *Rudens* of Plautus in person, by delivering the prologue; and the act is curious, inasmuch as it is one of the early opinions of the presence of invisible agents amongst mankind. This fine introduction has long been admired for its train of beautiful and religious sentiments; while Bonnel Thornton's English version of it may be cited as one of the very rare instances of a translation even exceeding the original.

To pick up Arcturus by alignment is very easy. A ray from the Pole-star through Alkaid, the first horse of the wain, and carried about 30° beyond it to the southward, will pass *a* Boötis; which bright object may be further identified by forming an equilateral triangle with Spica and Deneb. Arcturus and Polaris also make nearly a right angle with Wega; whence the galley-rhymes:

From staid Polaris cast a glance,	to beauteous Lyra's lines,
'Twill guide, rectangular from these,	to where Arcturus shines:
Or lead a line from two bright stars,	in Ursa's tail the last,
The same prolonged thrice ten degrees,	will on that gem be cast.

ΒΟΩΤΗΣ, Boötes, is one of the old forty-eight constellations, and the name appears to be from *βοῦς*, an ox, alluding to the herdsman; but the ancients as frequently called it *Arctophylax*, or bear-keeper. Aratus designates it by both names, as shown by his translators, Cicero and Festus Avienus; while Germanicus, in common with several others, called it Icarus, and the translators of Ptolemy rendered it Vociferator. The asterism is usually figured as a robust man walking, with one hand upraised, and the other holding a club, spear, pastoral staff (*pedum*), or sickle (*merga*), for he is represented at various epochs with each of these symbols. The attitude, especially in early representations, well countenanced the title of *Clamator*; but Hevelius having introduced the two hounds in 1690, they were given as attributes to Boötes, and the cords which held them passed into his upraised hand. The figure was adopted from elsewhere by the Greeks, no doubt, since they give no certain account of its origin; their stories making a yaw between Icarius, the father of Virgo of the zodiac, and the farming son of Calisto. Those who considered Ursa as an animal, dubbed Boötes the bear-keeper; but numbers who saw in the disposition of those stars a waggon, or wain, made him the driver. Hence Claudian's

> Boötes with his Wain the north unfolds;
> The southern gate Orion holds.

Among the several offices assigned him, however, the majority are for that of herdsman, and to make him truly Arcadian, Hevel placed Mount Mænalus under his feet, with 11 component stars: such at least is the story, but it were better to call it Menelaus, after the Alexandrian astronomer referred to by Ptolemy and Plutarch. The worthy Pole, however, is jealous of his prerogative, and insists that his having first observed and registered the constituents, *debitè et accuratè*, gave him the exclusive privilege of naming them: "hinc etiam nomina illis stellis imponendi facultas mihi soli competit." While the Mosaicists and Papists were contending that the figure symbolized Nimrod and St. Sylvester, Weigelius stepped in and formed the asterism into the three Swedish crowns.

The Arabians were equally unsettled in their epithets for this figure, it being designated *Al-rámih*, the lancer; *Al-'auwà*, the shouter; *Háris-as-semá*, keeper of heaven; *Nikkar*, a corruption from *al bakkár*, the herdsman; and *Al-kalurops*, which is conjectured by Grotius to be the proper reading for *inkalúrus* in the Alphonsine Tables, which is evidently taken from καλαύροψ, a shepherd's crook. But there is no end of names to this polyonymous asterism, of which the principal may be thus quoted:

Al Kalurops	Ceginus	Nekkar
Al Rámih	Canis latrans	Nimrod
Al 'Auwà	Clamator	Plorans
Arcas	Háris-al-simák	Plaustri-Custos
Arctophylax	Háris-as-semá	Philomelus
Arcturus Minor	Icarius	Septentrio
Boötes	Icarus	Thegius
Bubulcus	Lanceator	Venator
Bubulus	Lycaon	Vociferator

This constellation has some remarkable stars besides Arcturus, though of inferior brilliance; and Izár was a great favourite, ages before its beauty and delicacy had been revealed to man by the telescope. The components have been thus successively enumerated in the best Catalogues:

Ptolemy	. . . . 23 stars	Bullialdus	. . . 29 stars
Tycho Brahé	. . . 28	Hevelius	 52
Kepler	 29	Flamsteed	. . . 54
Bayer	 34	Bode	 319

Of these upwards of 30 are double stars, some exceedingly fine and interesting; and there are several clusters, and nearly 100 nebulæ, of various classes. So that Boötes is a truly rich asterism.

DVII. 418 ♉. II. BOOTIS.

Æ 14ʰ 09ᵐ 19ˢ   Prec. + 2ˢ·52
Dec. N 36° 05′·7   —— S 16″·94

Mean Epoch of the Observation . . . . 1837·56

A faint nebula preceding the right side of Boötes, which was first registered by ♉. in 1785, and is No. 1766 of his son's Catalogue. This object is small, round, and pale, but perfectly distinct. It has a nucleus, or at least is brightest at the centre, and its edges so attenuated as to give it the appearance of a star in a burr. It is attended by a coarse group of small stars in the *nf* quadrant, followed by a conspicuous one of the 6·7 magnitude. Uniting Cor Caroli to Arcturus as the base of a triangle, the spot now treated of will form the north-east apex, at an equal distance, 16° from each.

DVIII. ι BOOTIS.

Æ 14ʰ 10ᵐ 30ˢ   Prec. + 2ˢ·14
Dec. N 52° 06′·4   —— S 16″·88

Position AC 33°·7 (w 7)   Distance 38″·1 (w 9)   Epoch 1831·66
—— AB 195°·0 (w 1)   —— 0″·5 (w 1)⎱
—— AC 33°·4 (w 9)   —— 37″·9 (w 9)⎰ —— 1838·18

A singularly delicate triple star, in the herdsman's right hand, where, with θ and κ, and λ on the arm, it forms one of the *Aulád al-dhiba'*, the young of the hyæna. A and B 4½, pale yellow; C 8, creamy white; and they will be picked up by the alignment already given at κ, being the southern vertex of the triangle there mentioned. Measured as a double star, my first observations were made to compare with the following registered details:

♉.   Pos. 37° 09′   Dist. 37″·56   Ep. 1779·74
H and S.   33° 24′   38″·05   1822·24

This difference of 3°½ in forty-three years, though easily attributable to instrumental errors, also gave rise to a suspicion of a slow retrograde orbital motion, which my measures did not confirm; and there, as far as my own views were concerned, the matter had rested. But when Professor Struve sent me his magnificent volume, at the close of 1837, I found that he had succeeded in dividing A and B with his powerful instrument, and that his register stood thus:

A B Pos. 149°·0 Dist. 0″·3
A C Pos. 33°·9 Dist. 38″·06 Ep. 1836·28

On receiving this stimulus, I sought the star at an early apparition; but it was not till after long and patient gazing that I became persuaded of its being rather oblong than round. Even then, however, nothing was certain, and the estimations I made, are mere guesses. A has a proper motion, to the following amount:

$$P.... \mathcal{R} - 0''{\cdot}34 \quad Dec. - 0''{\cdot}03$$
$$B.... \quad - 0''{\cdot}21 \quad \quad + 0''{\cdot}07$$
$$A.... \quad - 0''{\cdot}24 \quad \quad + 0''{\cdot}08$$

DIX. 99 ♓. I. BOOTIS.

$$\mathcal{R} \ 14^h \ 11^m \ 44^s \quad \text{Prec.} + 2^s{\cdot}52$$
$$\text{Dec. N } 37° \ 14'{\cdot}4 \quad \text{------} \ \text{S} \ 16''{\cdot}82$$

Mean Epoch of the Observation . . . . 1838·51

A white round nebula, preceding the right shoulder of Boötes. It is very pale except on the centre, and is amidst some scattered telescopic stars, of which the closest is one of the 10th magnitude *np*. This object was discovered by ♓. on the 1st May, 1785, and is No. 1776 of his son's Catalogue. The mean apparent place is differentiated from ε Boötis, from which it lies 10° to the north-north-west, in the direction of Alkaid, at the tip of the Bear's tail.

DX. 62 P. XIV. LIBRÆ.

$$\mathcal{R} \ 14^h \ 14^m \ 11^s \quad \text{Prec.} + 3^s{\cdot}16$$
$$\text{Dec. S } 7° \ 01''{\cdot}7 \quad \text{------} \ \text{S} \ 16''{\cdot}71$$

Position 166°·8 (*w* 9)   Distance 5''·2 (*w* 9)   Epoch 1836·44

A fine double star, in a strange boundary nook of Libra, but on Virgo's garment, 15° east-by-north of Spica. A and B, both 8th magnitude, and both silvery white; a line led through them into the *np* quadrant leads close to an ash-coloured telescopic star. This is an easy object, which bears illumination well, and is of considerable brightness. It was classed by Piazzi: "Duplex. Comes 4'' circiter ad boream, 0''·2 temporis praecedit;" but the first micrometrical measures I find, are those of H. and S., as follows:

Pos. 347° 06'   Dist. 5''·88   Ep. 1823·44

which, by altering the quadrant, as both stars are equal in magnitude, differ so little from what I obtained thirteen years afterwards, that their relative fixity is established. But there is a sensible proper motion brought home to A, which in time will prove whether it is in connexion with B or not; the stated values are:

$$P.... \mathcal{R} - 0''{\cdot}23 \quad Dec. - 0''{\cdot}12$$
$$B.... \quad - 0''{\cdot}11 \quad \quad - 0''{\cdot}12$$

DXI.   69 P. XIV. BOOTIS.

Æ 14^h 15^m 31^s Prec. +   2^s·95
Dec. N 9° 10'·7   —— S 16''·64

Position 186°·2 (w 9)   Distance 6''·3 (w 9)   Epoch 1835·44

A very neat double star, on Mons Mænalus, between the left foot of Boötes and Virgo; it is 11° south-half-east from Arcturus, and on the line between Spica and ζ Boötis. A 6, flushed white; B 7½, smalt blue. This pretty pair was first classed by Piazzi: "Duplex. Comes 8·9æ magn. et in eodem verticali." Here that excellent astronomer has certainly under-rated the brightness of the companion, unless it be variable; a comparison of the following observations with my own, will prove its fixity, but shows the verticality to be rather an inaccurate phrase:

H. and S.   Pos. 186° 36'   Dist. 7''·18   Ep. 1823·42
Σ.   186° 03'   6''·26   1825·40

DXII.   70 P. XIV. LIBRÆ.

Æ 14^h 16^m 06^s Prec. +   3^s·21
Dec. S 10° 56'·3   —— S 16''·61

Position 325°·8 (w 4)   Distance 1''·6 (w 2)   Epoch 1833·36

A close double star, in the precincts of Libra, but hard upon Virgo's right heel: it is 15° east of Spica, where it is preceded by a star of the 6th magnitude. A 7½, pale yellow; B 9½, greenish. This very beautiful difficult object is one of the "pervicinæ" of Σ.'s Catalogue of 1827; and he has since measured it with these results:

Pos. 326°·87   Dist. 1''·41   Ep. 1829·83

The proper motions of A in space, have been thus watched down:

P.... Æ − 0''·32   Dec. − 0''·03
B....   − 0''·12   − 0''·01

DXIII.   φ VIRGINIS.

Æ 14^h 19^m 58^s Prec. +   3^s·09
Dec. S 1° 30'·4   —— S 16''·42

Position 115°·0 (w 1)   Distance 5''·0 (w 1)   Epoch 1837·42

A most delicate double star in the *nf* corner of Virgo's skirt, discovered at Dorpat by Σ.: it will be found about 18° north-east-by-east of

Spica, where it is the apex of a triangle of which a and β Libræ form
the base. A 5, pale yellow; B 13, fine blue. The little acolyte is clear
and distinct, but being too minute for illumination, its position and
distance were carefully estimated under a knowledge of the actual lines
of parallel and perpendicular. Since this was recorded, the results of Σ.
have been published, and the mean presents:

Pos. 108°·32   Dist. 3″·73   Ep. 1829·74

This star forms a kind of vertical curve with ι and κ, and is the skirt
of the garment, given as *chimar* in the *Almagest*. But the Latin translator, by the omission of the point over the first letter, made it *ḥimár*,
an ass; so that ι, the σύρμα of Ptolemy, and the central one of the
three, is designated " in asino." These stars form the XVth room of the
Manázil al ḳamar, or Lunar Mansions; giving a faint light, whence
they were called *Al-ghafr*, the covering, because the beauty of the earth is
hidden when they rise on the 18th *Tishrín*, or 1st of November; others
say on account of the shining of the stars being lessened, as if covered.

ϕ Virginis is charged with a spacial movement, to these amounts:

P.... ℞ − 0″·13   Dec. − 0″·02
B....   − 0″·03   − 0″·02

DXIV. 70 ♅. I. VIRGINIS.

℞ 14ʰ 21ᵐ 13ˢ   Prec. + 3ˢ·14
Dec. S 5° 15′·1   —— S 16″·36

Mean Epoch of the Observation . . . . 1838·41

A pale white nebula, over Virgo's left ancle, in a line between a
9th-magnitude star in the *sf* quadrant, and an 11th a little nearer in the
np; it is 4° south, a little easterly, from the above described object.
This object was first registered in March, 1785, and is No. 1813 of
H.'s Catalogue; though small it is very distinct, and its *candied* aspect
betokens a wonderfully remote globular cluster. Indeed the powerful
instrument of H. has resolved it; and he describes it as being composed
of stars of the 19th magnitude. So that here we find another universe
in the plenitude of space!

DXV. 95 P. XIV. VIRGINIS.

℞ 14ʰ 21ᵐ 41ˢ   Prec. + 3ˢ·12
Dec. S 3° 31′·7   —— S 16″·33

Position 266°·5 (w 2)   Distance 40″·0 (w 1)   Epoch 1836·49

A wide double star, on the lower left side of Virgo's skirt, where a
ray from Spica, carried about 18° into the east-north-east, will pick it

up a little to the south of ϕ. A $7\frac{1}{2}$, and B 12, both bluish. This is a truly difficult object, the position of which was approximated by the spherical rock-crystal micrometer; but the distance being too wide for its range, and the faintness of B precluding any illumination, it was merely estimated. Σ. entered this upon his list, but did not measure it. The small star is certainly best seen on averting the eye to another part of the field of view, when its gleams become strong.

DXVI. 5 URSÆ MINORIS.

Æ 14h 27m 56s   PREC. −   0s·27
DEC. N 76° 24'·5   —— S 16"·01

POSITION 135°·5 (w 2)   DISTANCE 45"·0 (w 1)   EPOCH 1833·61

A wide double star, under the Lesser Bear's belly. A 4, fine yellow; B 11, plum colour; and the two point precisely to a distant telescopic star in the *sf* quadrant: it may easily be found, since a ray through the Guards, carried about two-thirds as far again into the north-west, will reach it. The reduction of my meridian results, seem to indicate a sensible proper motion to this star, though unnoticed by Piazzi; yet my observations are not of a nature to decide. Argelander gives a quantity, which I cannot but consider as too small.

DXVII. π BOOTIS.

Æ 14h 33m 12s   PREC. +   2s·81
DEC. N 17° 06'·5   —— S 15"·73

POSITION 99°·3 (w 9)   DISTANCE 6"·0 (w 9)   EPOCH 1836·51

A neat double star, on the herdsman's left leg, being one of four bright stars in that limb, and the nearest to Arcturus, lying east-south-east of that splendid gem, 7° distant. A 3½, and B 6, both white. This fine object is 8 IιI. III., and was thus registered by Piazzi, Nota 147, Hora XIV.: "Duplex. Comes in eodem parallelo sequitur 0"·4 temporis." From this expression, and the results of the following measures:

IιI.   Pos. 96° 28'   Dist. 6"·17   Ep. 1779·72
H. and S.   97° 53'   6"·90   1822·05

I was led to infer that there existed a slight direct orbital motion. This suspicion would have been confirmed by my observations, but that Σ., No. 461 of his first Catalogue, found the angle 9° 50' *sf*, in 1819·61; and ten years afterwards he concluded 9° 12' *sf* to be the mean position. (*Mensuræ Micrometricæ*, No. 1864.)

DXVIII. ζ BOOTIS.

℞ 14ʰ 33ᵐ 31ˢ   Prec. + 2ˢ·85
Dec. N 14° 25′·1   —— S 15″·71

Position 129°·9 (w 5) Distance 1″·3 (w 5) Epoch 1833·39
—— 129°·8 (w 4)   —— 1″·4 (w 3)   —— 1834·38
—— 128°·6 (w 8)   —— 1″·3 (w 6)   —— 1838·45
—— 127°·3 (w 9)   —— 1″·2 (w 6)   —— 1842·43

A close double star on the herdsman's left heel, being the southern-most of the four stars above mentioned, and bearing from Arcturus east-south-east, 8° distant. A 3½, bright white; B 4½, bluish white, and supposed to be variable. This is a fine object, and not difficult of mea-surement, for I have operated upon it in full day-light. The close duplicity was overlooked by Ⱨ. in 1782, when from noting the dusky *comes* which precedes ζ, nearly on the parallel at about 90″ distance, he entered it as No. 104 of Class VI., nor was B detected by him till 1796. Though I have stated that the stars are easy to handle, it must be admitted that the results in position are not so coincident as might have been expected; but perhaps the closeness and oblique position contribute to this discordance in angle. A comparison of the whole leads me to pronounce, that no orbital variation has occurred. Yet, as strong asser-tions have been made to the contrary, I here submit the details and epochs with which mine were discussed:

	Pos.	Dist.	Ep.
Ⱨ.	131° 59′	1″·00±	1796·60
H. and S.	126° 58′	1″·68	1823·27
Σ.	129° 17′	1″·19	1830·47
D.	128° 17′	1″·32	1832·47

It has been stated that each constituent presents, alternately, a clear and a dim face to us; but though I frequently examined them in a dark field, it is a point which I cannot confirm.

DXIX. 10 HYDRÆ.

℞ 14ʰ 36ᵐ 46ˢ   Prec. + 3ˢ·46
Dec. S 24° 45′·5   —— S 15″·54

Position 138°·4 (w 5) Distance 9″·8 (w 3) Epoch 1831·49

A neat double star, about 9° south-by-west of a Libræ, and 5° due west of Zuban-al-kravi (20 Libræ), a star of the 3½ magnitude. It is close to the boundary of *Turdus Solitarius*, an insignificant asterism intruded by Le Monnier, in 1776, to commemorate the Hermit-bird of India, between Hydra, Virgo, and Libra. A 5½, pale orange; B 7½, violet tint. This is a very beautiful object for a moderate telescope; but its identity has been jeopardized by its synonymes, being 54 Hydræ of ♃., 30 Turdi Solitarii of H. and S., and 73 Hydræ in Bode. It is, however, the star which Piazzi thus describes, at No. 163 of the Notæ to Hora XIV.: "Duplex. Comes 7·8ᵐ magnit. 0″·3 temporis sequitur 7″ ad austrum." It had been previously registered 97 ♃. III. under these measures:

Pos. 128° 15′ Dist. 11″·29 Ep. 1783·03

Nearly forty years afterwards H. and S. found it:

Pos. 136° 40′ Dist. 9″·95 Ep. 1822·87

showing a considerable direct motion in the elapsed time, which my measures appear to substantiate. The apparent diminution in distance, may be merely an instrumental anomaly. But there is a considerable proper motion in space detected, to these values:

$$P.... \ \mathit{R} - 0'''·33 \quad Dec. - 0'''·13$$
$$B.... \quad - 0'''·20 \quad\quad - 0'''·08$$

DXX. ε BOOTIS.

Æ 14ʰ 38ᵐ 00ˢ	Prec. + 2ˢ·62
Dec. N 27° 45′·1	—— S 15″·47

Position 321°·6 (w 5)	Distance 3″·2 (w 3)	Epoch 1831·46
—— 323°·8 (w 5)	—— 3″·8 (w 2)	—— 1833·53
—— 321°·2 (w 7)	—— 2″·9 (w 5)	—— 1838·68

A standard Greenwich star, on the herdsman's left hip; where it is readily shown by a line from Mizar in the Great Bear's tail through Alkaid at its tip, carried away into the south-south-east till it meets a bright star in mid-distance between Arcturus and Gemma. A 3, pale orange; B 7, sea green; the colours being distinct and strongly contrasted. This lovely object, which Σ. calls "pulcherrima," is 1 ♃. I., or *par excellence* A¹, as they would say at Lloyd's; and being in a manner insulated, afforded grounds to its discoverer's concluding the *comes* to be a binary partner, and not merely a star at a vast distance nearly in a line behind it, as may be the case in those parts of the heavens where small stars are profusely scattered. Though subsequent observations appeared to confirm this, among which are my own of 1833, showing a direct rotation of the companion, my last set of measures were under such favourable circumstances, that I attribute the former want of coincidence rather to the proximity and brightness, as well as obliquity of position, which render them of somewhat difficult measurement, than to a supposed elliptical orbit, amenable to the laws of gravitation. Nor can proper motion be cited, since Piazzi's value of that element is disappearing under more recent observations. But as a skilful discussion by H. of the changes he observed, indicated a slow angular motion, forming a periodical revolution of not less than 980 years; and as the progression has since been pronounced "indisputable," and its amount computed $+ 0°\cdot4378$, it will be necessary to submit the details and epoch which have led me to consider the question to be, as yet, unestablished:

	Pos.	Dist.	Ep.
H.	301° 34′	4″·00 $+$	1779·67
H. and S.	322° 59′	3″·93	1822·55
Σ.	320° 58′	2″·64	1829·39
D.	321° 35′	...	1831·36

ϵ Boötis is the $\pi\epsilon\rho\iota\zeta\omega\mu\alpha$ of Ptolemy; but appears as Izár, in various Catalogues, from the Arabian word signifying a zone, or girdle. It is also designated *Mizár*, a waist-cloth, or apron; and when ill-written in the Arabic, Mirár, whence the Merer and Meirer of the Alphonsine Tables, subsequently changed into Mirac and Micar. Tízíní calls it *Mintakah 'al-auwá*, the belt of Auwá.

DXXI. α^2 LIBRÆ.

Æ 14^h 42^m 02^s PREC. $+$ 3^s·31

DEC. S 15° 22′·3   —— S 15″·24

POSITION 314°·3 (w 2)   DISTANCE 229″·0 (w 2)   EPOCH 1836·47

A standard Greenwich star with a distant companion. A 3, pale yellow; B 6, light grey. Though this object is the leader of Libra, and one of its two lucidæ, it is located on the southern chelate hand-claw of Scorpio; yet it is called Kiffa Australis, from the Arabian *al kiffah-al-jenúbíyah*, the southern scale. It may be found by carrying an occult line from Arcturus to Spica, and from thence a rectangular one, led about 22° to the eastward, passes nearly over α^2 Libræ. Or a readier reference is found in the galley rhymes:

Where yon gaunt Bear disports a tail,	seek Alkaid at its tip,—
From thence a ray athwart the space	to south-south-east must dip;
And when Arcturus has been pass'd,	prolong th' imagin'd line,
'Twill mark a star, as far again,	the first in Libra's sign.

The star B of this pair is a^1 of Bayer, or No. 186 P. XIV.; and the two form a fine though wide object, which was measured in full twilight without artificial illumination. It is No. 186 of H. and S., and a comparison of our measures shows it to have undergone no appreciable alteration in an interval of thirteen years. The primary is subject to a slight proper motion, of which the amount has been thus stated:

	Æ	Dec.
P....	$- 0″·20$	$- 0″·08$
B....	$- 0″·03$	$- 0″·07$
A....	$- 0″·07$	$- 0″·03$

Libra, the Balance, is the 7th in order of the twelve signs of the zodiac, and the first of the autumnal ones. The integrity of its boundaries has been largely encroached upon by Scorpio; but the numbers of its component stars have thus progressively increased, with the improvement of optical means:

Ptolemy	. . . . 17 stars		Hevelius	. . . 21 stars
Tycho Brahé	. . 18		Flamsteed	. . 51
Griemberger	. . . 20		Bode	 180

Servius assures us, that the original Chaldean zodiac consisted of but eleven constellations; and this may be explained by assuming that the body of Scorpio occupies one sign, and the pincers, or χῆλαι, another; for Aratus mentions them as being distinct. Libra is supposed to have been introduced by the astronomers of Alexandria, though another story says that the Roman *savans* added it in honour of Julius Cæsar; and, in either case, its being met with on the famous zodiac of Tenterah, destroys the claim of remote antiquity set up for that performance. The *chelæ* have, however, long been drawn back to make room for the scales; for the latter are named by Manilius, and Cicero rather sneeringly mentions the *jugum*, whence the Arabians derived their *al-mizán*, or balance. Vitruvius, Pliny, and Columella, recognise the sign Libra, while Ovid and Germanicus hoist the colours of Chelæ; and others attempted to smooth the matter over by a mezzo-termine varnish, as on the Farnese globe, where Scorpio's claws carry the Scales. But Ruæus is in error when he accuses Virgil, in common with other ancients, of being ignorant of Libra; for though in flattering Augustus in the first *Georgic* he makes mention of the *Chelæ*, yet, in the very same book, he distinctly says,

Libra die somnique pares ubi fecerit horas.

It is true, that a century afterwards Ptolemy describes the Chelæ as an asterism in his Sidereal Catalogue, giving it 8 regular stars and 9 *amorpholæ;* but he mentions Libra in his text. Petavius has shown that Servius mistook his author, when he pulled in Virgil to certify that Libra was of Augustus's day; and some set no bounds to its antiquity, telling us that it is mentioned in the sacred books of India, 1200 years before that time, nay, even the 15th verse of the xlth chapter of Isaiah has been pressed into its service. The Egyptian Institute have recently pronounced that Libra was once the Coptic *Faramour*, whence the Greeks made their φαρμουθε, *mensura, regula confecta temporis.*

Libra is considered as typifying the equality of the days and nights in autumn, as well as the uniform temperature of the air at that season. According to Kircher, who nicknames it *Wezn*, weight, it was the emblem painted on the standard of Asher; but this man of "immense but undigested learning" was greatly given to twaddle on these subjects. Some astrologers, not keeping the proximity of the accursed Scorpion before their eyes, considered it a happy sign, and represented it, as indicative of abundance, by "vir utraque manu spiculum tenens:" yet this aërial trigon is harshly denounced in the illuminated Almanack for 1386, for it is there asserted that "whoso es born in yat syne sal be an ille doar and a traytor." But notwithstanding the struggle made by Dupuis for its Ogygian antiquity, Libra is often deemed an interloper upon the *Scorpionis forceps;* and since that intrusion, Le Monnier, in 1776, formed some stars between the southern saucer of the Balance, and the tail of Hydra, into the *Turdus Solitarius;* the property having belonged to those asterisms. Vulcan, the tutelar guardian of Libra, ought to have looked to it.

DXXII. ξ BOOTIS.

| Æ 14^h 44^m 00^s | Prec. + 2^s·75 |
| Dec. N 19° 46′·1 | —— S 15″·13 |

Position 332°·1 (w 6)	Distance 7″·3 (w 3)	Epoch 1831·53
—— 327°·4 (w 5)	—— 7″·0 (w 5)	—— 1837·49
—— 324°·8 (w 9)	—— 7″·1 (w 9)	—— 1839·61
—— 322°·9 (w 9)	—— 6″·9 (w 9)	—— 1842·42

A binary star, in the left knee of Boötes; being the northernmost of the four stars forming his leg, and 10° east of Arcturus. A 3½, orange; B 6½, purple; the colours in fine contrast. I have designated this a binary system, from the striking orbital retrogression, although the connexion between the two components was not held to be fully established until Sir John Herschel's last investigation; as it was asserted to be within possibility that the movement observed might be merely the difference in their proper motions. "If the relative path of the small star," said H., "be really the straight line it appears to be, the angle of position will never reach 50° *np*, and the angular velocity will diminish continually from the present moment (1823). On the other hand, if the stars form a binary system, the present angular velocity of about 1° per annum, will continue for some time nearly uniform, and in 15 or 20 years the limit of 50° *np* will be attained or passed." Now it will be seen by my last series of measures, that this binary proof was in a fair way to be accomplished; for the companion has so decidedly weathered or rounded the north end of the ellipse, as to have completed, speaking in round numbers, a fourth of its apparent orbit, since it has been under measurement. In the same *round* manner, assuming the primary to be near the *sf* portion of the orbit, the axis major will appear at least twice the length of the transverse; so that the ellipse described is highly elongated, and very oblique to the line of sight. Sir John Herschel, from a grouping of all the observations, computed the orbit in 1833, and made the period equal, in tropical years, to 117·14, with an excentricity of 0·59374, and a mean motion of — 3°·0733. The physical connexion of the components may therefore be deemed "fully proven," and that fact alone is a gratification to the contemplative mind.

This fine object is 18 ♓. II., and was discovered in April, 1780, the companion then being 24° from the vertical in the *nf* quadrant. Piazzi, who probably observed it in 1792, makes a remark which is highly important to the discussion: at Note 197, Hora XIV., he says, "Duplex. Comes in eodem verticali ad boream," and in 1814 he showed it to me in the *np* quadrant. The principal micrometric results on register, when I took the field, were:

♓.	Pos. 24° 07′	Dist. 3″·42	Ep. 1780·28
	353° 54′	6″·00	1804·25
H. and S.	340° 54′	8″·70	1822·63
Σ.	334° 11′	7″·22	1829·46

The proper motions in space to which I alluded, may be registered under the following values, those of Piazzi being deduced between Bradley's observations and his own:

$$P.... \text{Æ} - 0''{\cdot}23 \qquad \text{Dec.} - 0''{\cdot}18$$
$$B.... \quad + 0''{\cdot}18 \qquad - 0''{\cdot}14$$

DXXIII.   39 BOOTIS.

Æ 14^h 44^m 16^s Prec. $+$ $2^s{\cdot}04$
Dec. N $49°$ $22'{\cdot}8$ —— S $15''{\cdot}12$

Position $43°{\cdot}9$ (w 8)   Distance $3''{\cdot}9$ (w 6)   Epoch $1834{\cdot}51$
——— $44°{\cdot}7$ (w 9)   ——— $3''{\cdot}8$ (w 6)   ——— $1839{\cdot}00$

A neat double star, following the herdsman's right wrist; and it may be seen by running a line from Polaris through β, the preceding guard, and prolonging it just as far again. A $5\frac{1}{2}$, white; B $6\frac{1}{2}$, lilac. This pretty object is 79 H. II., and may, from a comparison of the several results, have a slow retrograde motion. These are the selected results of micrometrical measures, previous to my own:

Ħ.   Pos. $51°$ $39'$ Dist. $4''{\cdot}00 \pm$ Ep. $1783{\cdot}02$
H. and S. $45°$ $05'$ $4''{\cdot}63$ $1822{\cdot}93$
Σ. $44°$ $12'$ $3''{\cdot}71$ $1830{\cdot}02$

This double star is an outlier of the thirty-two sporades which La Lande scraped together in 1795, to form the new asterism Quadrans Muralis, in commemoration of his nephew's *Histoire Céleste Française;* which work enrolled no fewer than 50,000 stars, and constitutes, as Olbers justly remarked, one of the most important productions of the eighteenth century. The instrument with which the observations were made was therefore placed aloft, as a candidate for immortality; and it certainly merited celestial honours, more than some of the recent intruders. The little star about $2°\frac{1}{2}$ directly south of it, 38 Boötis, is the *Merga*, or corn-fork, of Bayer.

DXXIV.   212 P. XIV. LIBRÆ.

Æ 14^h 48^m 05^s Prec. $+$ $3^s{\cdot}40$
Dec. S $20°$ $40'{\cdot}9$ —— S $14''{\cdot}95$

Position AB $272°{\cdot}6$ (w 6)   Distance $10''{\cdot}3$ (w 4) ⎫
——— AC $320°{\cdot}0$ (w 1)   ——— $20''{\cdot}0$ (w 1) ⎬ Epoch $1833{\cdot}44$
                                              ⎭

A most delicate triple star, preceding the southern Lanx Libræ; and nearly in mid-distance between a and Flamsteed's No. 20 Libræ. A 6, straw-coloured; B 8, orpiment yellow; C 16, pale red; and several minute stars in the field. Piazzi, speaking of this object, remarks: "Eadem duplex, et ipsius comes 0″·7 temporis præcedit parumper ad austrum;" and the companion thus mentioned, is registered by H. and S.:

Pos. 270° 09′   Dist. 10″·82   Ep. 1823·32

A and B were also seen by ⟨H⟩. in 1785, and form No. 28 of his list of 145. But C must have escaped mortal ken until detected by H., No. 2755; and it is no easy matter to get it even by occasional glimpses, under equatoreal clock motion, and attentive gazing. My details of A C are therefore a mere estimation, for the *comes* is the very *minimum visibile* of my refractor; and not to be caught at all times.

DXXV. 18 LIBRÆ.

Æ 14ʰ 50ᵐ 15ˢ   Prec. + 3ˢ·24
Dec. S 10° 29′·8   —— S 14″·76

Position 38°·9 (*w* 8)   Distance 20″·0 (*w* 3)   Epoch 1834·27

A delicate double star, under the centre of the Balance beam; where it is nearly the last of a little group, at about 5° north-north-east of *a* Libræ. A 7, straw colour; B 11, grape-red; and an occult line through the two leads to a third star in the *nf* quadrant, of the 12th magnitude, and ruddy. This neat object is 56 ⟨H⟩. iv., the measures of which show larger discrepancies than are merely attributable to its difficulty. The former results stand thus:

⟨H⟩.   Pos. 45° 15′   Dist. 17″·98   Ep. 1782·26
H. and S.   35° 52′   26″·61   1823·30
Σ.   38° 45′   19″·45   1831·09

The earlier reductions and comparisons of this star, indicated a very sensible spacial movement, but the late rigorous investigations have abolished that for Æ; the results of Piazzi, Brioschi, and Baily, being:

P.... Æ − 0″·29   Dec. − 0″·10
Br...   − 0″·32   − 0″·13
B....   0″·00   − 0″·12

DXXVI. β URSÆ MINORIS.

Æ 14ʰ 51ᵐ 14ˢ   Prec. − 0ˢ·29
Dec. N 74° 48′·2   —— S 14″·71

Position 5°·5 (*w* 1)   Distance 165″·0 (*w* 1)   Epoch 1833·64

A standard Greenwich star, with a distant *comes* on the Lesser Bear's left shoulder. A 3, reddish tint; B 11, pale grey; several small stars in

the field. A sensible movement in space is detected, of which the
following are the values:

$$P\ldots \mathbb{R} - 0''{\cdot}30 \quad \text{Dec.} - 0''{\cdot}18$$
$$A\ldots \quad - 0''{\cdot}12 \quad\qquad - 0''{\cdot}04$$
$$B\ldots \quad - 0''{\cdot}11 \quad\qquad - 0''{\cdot}06$$

This star is designated Kocab, from the Arabian *kaúkab-al-shemáli*,
the north star, it having been nearer to the Pole than *a*, in Ptolemy's
time: hence it became the *Reicchabba* of the Alphonsine Tables. In
Ulugh Beigh it appears as *Anwar-al-ferkadeïn*, the brightest of the two
calves; the other being γ. It is within the Arctic Circle, which has
been through all ages esteemed the vertex of the heavens, and anciently
was a variable distance from the polar point, always equal to the latitude
of the place. The first person who fixed the polar circles to a constant
distance, was our countryman Sacro-bosco, *anglicè* Holywood. From
such circumstances Kocab, which is still a useful star, had its day; and
a line from Arcturus through Alkaid, and by Thuban, will identify it;
or resorting to the rhymester, its vicinity to the present Polaris may be
thus stated:

Kocab, one bright, and two faint stars,   grace Lesser Ursa's side,
In oblong square; trace her bent tail,   and to the Pole you'll glide.

β Ursæ Minoris and γ were stars of no small utility and renown
among those who craved the use of a natural and never-failing nocturnal
clock; and if in our day navigators are indebted to lunar observations,
so the old mariners were beholden to the heavens for various professional
facilities. Thus they shifted their tides by the Moon's bearing, and told
the hour by the Sun's rhumb. It was also readily seen how smoothly
the Little Bear was every day swung round the Pole-star as about an
axis; and as it never sets to Europe, the circle described is universally
available in these latitudes. The two southernmost of the stars—making
nearly a right line to the direction of the Pointers—with Polaris, are the

most conspicuous objects of the
constellation, though all the
principal constituents are clearly
traceable by the naked eye, forming a miniature of the Great
Bear, only with the tail more
curved. β, on the animal's
shoulder, and γ, in its ear, are
designated the Guards—" of the
Spanish word *guardare*," saith
Hood, " which is to beholde,
because they are diligently to be
looked unto, in regard of the
singular use which they have in
navigation." In that rare old
work, the *Safegard of Saylers,*

there is a chapter " Howe to knowe the houre of the night by the Guards,
by knowing on what point of the compass they shall be at midnight,
every fifteenth day throughout the whole yeare." * * " Now, when you
know on what point of the compasse the Guardes are alwaies at mid-

night, then may yee by it also knowe the houres before or after midnight, alway reckoning for every point that they shall lack of the midnight point three quarters of an houre." But Richard Eden, the worthy friend of Sebastian Cabot, bestirred himself beyond most of his competitors in the improvement of hydrography, and was the cause of much attention to nautical objects. In the *Arte of Navigation* which he " Englished out of the Spanyshe," in 1561, the xviith chapter treats of the " composition and use of an instrument generall for the houres of the night," by the circle which the " two starres called the Guardians, or the mouth of the horne," describe. The base-work being:

19th April.   Guard perpendicular over the Pole-star } at Midnight.
12th October.   Guard perpendicular under the Pole-star }

This instrument consisted of a fixed circle, and a moveable one: "When you desire to know the houre, you shall turn the index of the less rundell in which is written 𝕋𝕚𝕞𝕖, to that part of the great rundell where is marked the day in which you desire to know the houre, and directing your face toward the north you shall turn the head toward the height of heaven." You are then directed to look at the Pole-star through the centre, and turn the concentric circle till the two Guards are seen in their respective holes, " and all three with one eye," and the hour shall be shown on the smaller circle. There is still more upon the Guards, in Tap's *Seaman's Grammer*, 1609; all tending to show the important utility of the asterism. "How often," meditates Hervey, " has this star beamed bright intelligence on the sailor, and conducted the keel to its destined haven." In the preceding diagram the changes visible to the eye are shown when the Bear is on the meridian, and at six hours, midnight, and eighteen hours.

DXXVII.   756 ♅. II. BOOTIS.

Æ 14ʰ 53ᵐ 53ˢ   Prec. + 1ˢ·78
Dec. N 54° 32′·7   —— S 14″·55

Mean Epoch of the Observation . . . . 1837·46

A small pale-white nebula, with a bright centre, in the space between the right hand of Boötes and Draco's belly. It precedes a fine though wide double star, of the 7th and 8th magnitudes, from the following of which, four equidistant very minute stars extend in a line *sf* = 150°. This object was discovered by ♅. in May, 1788, and is No. 1898 of his son's Catalogue. The place is derived from a differentiated observation with θ Draconis; and if a curved line is drawn from thence to Alkaid, at the tip of the Great Bear's tail, passing through the three stars in the hand of Boötes, it will pass the nebula's site in mid-distance. A ray from Polaris, dropping closely to the east of β Ursæ Minoris, the preceding Guard, also leads to it.

DXXVIII. β BOOTIS.

Æ 14^h 55^m 55^s PREC. + $2^s\cdot26$

DEC. N $41°$ $01'\cdot5$ —— S $14''\cdot42$

POSITION $310°\cdot9$ (w 3)   DISTANCE $593''\cdot0$ (w 2)   EPOCH $1835\cdot61$

A bright star with a very distant companion in the herdsman's eye; where a ray from Polaris between the Guards, and $2\frac{1}{4}$ times as far again, will meet it, or from the southward, a line from Spica to the west of Arcturus, and as much further, reaches β Boötis. A 3, golden yellow; B 11, pale grey; and there are two very minute stars near the south vertical. This star, together with γ, δ, and μ, in the head, shoulders, and staff, form the trapezium which the Arabian astronomers term *al-dhiba'*, the hyæna; but β is designated Nekkar in the Catalogues, which might appear to be from *al-nakkár*, the digger, but that in Arabic it may so easily be an error of transcription for *al-bakkár*, the herdsman, as Ibn Yúnis has given it. As this object appeared to have at least as much proper motion as that assigned to it by Piazzi, though it has been altered by Mr. Baily, the companion was measured with a view of watching its progress; the conclusions hitherto arrived at are:

P.... Æ − $0''\cdot20$ Dec. − $0''\cdot02$
B....   + $0''\cdot02$ − $0''\cdot05$

It has since been examined, at my request, by the Rev. Mr. Challis, with the Northumberland equatoreal, and these results obtained:

Pos. $313°$ $31'$ Dist. $600''\cdot83$ Ep. $1841\cdot69$

DXXIX. 44 BOOTIS.

Æ 14^h 58^m 31^s PREC. + $2^s\cdot01$

DEC. N $48°$ $16'\cdot8$ —— S $14''\cdot27$

POSITION $233°\cdot8$ (w 5)	DISTANCE $2''\cdot9$ (w 3)	EPOCH $1830\cdot82$
—— $234°\cdot7$ (w 5)	—— $3''\cdot1$ (w 3)	—— $1831\cdot42$
—— $235°\cdot1$ (w 5)	—— $3''\cdot3$ (w 7)	—— $1834\cdot55$
—— $234°\cdot9$ (w 7)	—— $3''\cdot6$ (w 5)	—— $1836\cdot71$
—— $235°\cdot3$ (w 9)	—— $3''\cdot5$ (w 7)	—— $1839\cdot62$
—— $235°\cdot9$ (w 9)	—— $3''\cdot7$ (w 7)	—— $1842\cdot58$

A close double star, in the space following the right arm of Boötes; where a line from the Pole-star, carried between the two Guards, and continued nearly as far again to the southward, will meet it about $7°$ to the north of the above-described object. A 5, pale white; B 6, lucid grey; a distant star in the *np* quadrant. This fine physical object was discovered by IꞪ., and registered No. 15 of Class I.; and is, as he remarked, a miniature of Castor. As, for reasons which will presently appear, this star demanded much of my very best application, it is proper to show the recorded amount of its proper motions:

$$P.... \text{Æ} - 0''\cdot91 \quad \text{Dec.} \quad 0''\cdot00$$
$$B.... \quad - 0''\cdot59 \quad \quad + 0''\cdot03$$
$$A.... \quad - 0''\cdot62 \quad \quad + 0''\cdot03$$

44 Boötis, notwithstanding some apparent contradiction in ⩜.'s results, is a remarkable and highly interesting star. It is quite palpable that the distance has augmented since 1781; and that the stars are still separating, though it is difficult to establish the exact annual increase of such increment. There is also strong presumptive evidence of a slow orbital change in the same period; but, except there exist some conspiring errors in the angles or readings, the case is beset with difficulties. The angles show that the apparent path is not entirely rectilinear, and if all the observations be retained, that the orbital plane of this system is very nearly parallel with that of our own, since the companion first appeared in a direction opposite to the present one. " It is true," says H., " that an orbit passing nearly through the eye, and so situated that the longer axis of the ellipse into which it is projected shall form an angle of about 6° with the meridian, would account for the remarkable *jump* the small star seems to have made from one side to the other of the large one." At all events there are now sufficient observations enrolled to destroy M. Struve's paradox of several entire revolutions having been performed in 38 years; but the idea had already been abandoned.

The standard results for bringing the history of this double star up to the epoch of my observations, are:

	Pos.	Dist.	Ep.
⩜.	60° 06′	1″·50±	1781·62
H. and S.	229° 07′	2″·28	1821·33
Σ.	234° 01′	2″·86	1832·24
D.	235° 39′	3″·28	1833·39

DXXX. 279 P. XIV. BOOTIS.

Æ 14ʰ 59ᵐ 49ˢ   Prec. + 2ˢ·90

Dec. N 9° 50′·6 ——— S 14″·19

Position 209°·7 (w 8)   Distance 4″·0 (w 5)   Epoch 1835·39

A very neat double star, in the space between the left foot of Boötes and the snake of Ophiuchus. A and B, each 7½, and each pale white. This fine object is No. 1910 of the Dorpat Catalogue, and a comparison of all the observations stamps the relative fixity of the components. The measures for comparison are these:

	Pos.	Dist.	Ep.
H. and S.	209° 10′	4″·78	1823·42
H.	209° 05′	4″·29	1830·32
Σ.	209° 09′	3″·80	1832·08

The optical star before us, is preceded at a quarter of a minute's time by 277 P. xiv., by which it may be identified in *fishing;* and it may be readily picked up by running an imaginary line from Arcturus over ζ Boötis, and continuing it rather more than as far again into the south-east region.

DXXXI. 219 ♓. I. DRACONIS.

Æ 15ʰ 02ᵐ 03ˢ   Prec. + 1ˢ·63
Dec. N 56° 23′·0   —— S 14″·05

Mean Epoch of the Observation . . . . 1837·60

A bright-class oval nebula, under the body of Draco, with its major axis trending towards the vertical of the *np* and *sf* quadrants. It is rather faint at the edges, though not so as to obscure the form. It was discovered by ♓. in March, 1789, and is No. 1909 of his son's Catalogue. There is a small star nearly above it, and three larger more distant, of which the preceding is coarsely double. The mean place is differentiated with θ Draconis; from which it lies about 7° to the west-south-west, or one-third of the distance towards Alkaid.

DXXXII. *ι*¹ LIBRÆ.

Æ 15ʰ 03ᵐ 07ˢ   Prec. + 3ˢ·40
Dec. S 19° 10′·8   —— S 13″·98

Position 110°·5 (*w* 4)   Distance 51″·3 (*w* 3)   Epoch 1837·43

A wide double star, on the southern claw of Scorpio, preceding the group of which β Scorpii is the principal, by 15° on the parallel. A 5½, pale yellow; B 9½, purple; there is a third star of the 10th magnitude in the *sf* quadrant, which in 1822 was noted by H. and S. as " precisely in a line," but the angles at my epoch were from A to B 110°·5, and from A to C 107°·4. This object is 44 ♓. VI., and has been thus measured:

♓.	Pos. 112° 31′ Dist. 59″·05 Ep. 1781·40	
H. and S.	111° 39′ 50″·63 1822·84	

Although this star is unnoticed by Argelander, my results are symptomatic of a small proper motion. Piazzi has assigned it a small amount, and Baily still smaller:

P.... Æ − 0″·10   Dec. − 0″·15
B....   + 0″·03   − 0″·02

DXXXIII. 22 M. (?) DRACONIS.

Æ 15ʰ 04ᵐ 50ˢ   Prec. + 1ˢ·55
Dec. N 57° 36′·1   —— S 13″·87

Mean Epoch of the Observation . . . . 1835·52

A small but brightish nebula, on the belly of Draco, with four small stars spreading across the field, north of it. There may be a doubt as to whether this is the nebula discovered by Méchain in 1781, since Messier merely describes it as "very faint," and situated between o Boötis and ι Draconis. But there must be some mistake here; the one being on the herdsman's leg, and the other in the coil of the Dragon far above the head of Boötes, having 22° of declination and 44' of time between them, a space full of all descriptions of celestial objects. But as the θ in the raised right hand of Boötes, if badly made, might be mistaken for an *omicron*, this is probably the object seen by Méchain, and H.'s 1910; it being the brightest nebula of five in that vicinity. A line from κ in Draco's tail, led to the south-east through Thuban, and prolonged as far again, strikes upon its site.

DXXXIV. 14 P. XV. LIBRÆ.

Æ 15ʰ 05ᵐ 26ˢ   Prec. + 3ˢ·38
Dec. S 17° 49'·5   ——— S 13"·83

Position 141°·7 (*w* 5)   Distance 48"·4 (*w* 3)   Epoch 1835·52

A coarse double star, on the lower hand-claw of Scorpio; it is north-north-east of ι¹ Libræ, and 9° south of β, in the same asterism. A 8, silvery white; B 9, pale grey; a third star follows in the north. This object is composed of Piazzi's Nos. 14 and 15, Hora XV.; and is also 131 ℍ. v., imperfectly registered in 1783 as being in a line parallel to β and π Scorpii, with a distance of 47"·77: it is pretty evident that it must also be his No. 27 of the same class, noted in 1781. It was well measured by H. and S., who thus record it:

Pos. 140° 58'   Dist. 49"·04   Ep. 1823·28

DXXXV. 19 ℍ. VI. LIBRÆ.

Æ 15ʰ 08ᵐ 06ˢ   Prec. + 3ˢ·43
Dec. S 20° 26'·7   ——— S 13"·67

Mean Epoch of the Observation . . . . 1837·43

A large compressed cluster of very minute stars, in the Scorpion's southern pincer, with eight brightish stars vertically curved in the following part of the field, and half a dozen duller preceding. Discovered in March, 1785. It is faint and pale, but owing to the fineness of the night, steadiness of gaze, and excellent action of the telescope, was as well seen as so low and so awfully remote an object could be expected to be. There was much interest attached to the observation, because this cluster is one of the gradations from the palpable congeries of stars, in the Herschelian system, towards the distant nebulæ. The illustrious astronomer thus treats of it, in the LXXXI.st vol. of the *Philosophical Transactions:*

"When I pursued these researches, I was in the situation of a natural philosopher who follows the various species of animals and insects from the height of their perfection down to the lowest ebb of life; when, arriving at the vegetable kingdom, he can scarcely point out to us the precise boundary where the animal ceases, and the plant begins; and may even go so far as to suspect them to be not essentially different. But recollecting himself, he compares, for instance, one of the human species to a tree, and all doubt upon the subject vanishes before him.

"In the same manner we pass through gentle steps from a coarse cluster of stars, such as the Pleiades, the Præsepe, the Milky Way, the cluster in the Crab, the nebula in Hercules, that near the preceding hip of Boötes, &c. &c., without any hesitation, till we find ourselves brought to such an object as the nebula in Orion, where we are still inclined to remain in the once adopted idea, of stars exceedingly remote and inconceivably crowded, as being the occasion of that remarkable appearance. It seems, therefore, to require a more dissimilar object to set us right again. A glance like that of the naturalist, who casts his eye from the perfect animal to the perfect vegetable, is wanting to remove the veil from the mind of the astronomer. The object I have mentioned above (see 69 $\underline{H}$. IV., $\cancel{R}$ 3^h 59^m), is the phenomenon that was wanting for this purpose. View, for instance, this cluster, and afterwards cast your eye on that cloudy star, and the result will be no less decisive than that of the naturalist we have alluded to. Our judgment, I may venture to say, will be, that *the nebulosity about the star is not of a starry nature.*"

This wondrous but difficult object was differentiated with β Libræ, to obtain its mean apparent place; and it bears 9° due south of that star.

DXXXVI. β LIBRÆ.

$\cancel{R}$ 15^h 08^m 24^s PREC. + 3^s·22
DEC. S 8° 47'·4 —— S 13"·64

POSITION 85°·3 (w 2)   DISTANCE 570"·0 (w 1)   EPOCH 1836·34

A second-grade Greenwich star, with a distant companion, on the Scorpion's northern hand-claw; it will be identified by projecting an occult curve from Antares through β Scorpii, and carrying it to twice the interval between those two stars. A 2½, pale emerald; B 12, light blue; and there are two other stars in the north part of the field, forming the base of a triangle whose vertex is A. The same bright object is also, in alignment, the apex of a grander triangle, of which Arcturus and Spica form the base; and we learn:

Two stars from Scorpio's heart, will form	a westward rising line,
This Scorpio's second star, and *that*	the same in Libra's sign.

A Chaldean observation of the approach of Mars to this star is recorded, which appulse was observed in the 476th year of Narbonassar, or 271 B.C. It was a star of some note of old, and was not *en bonne odeur* with astrologers. It appears on Catalogues under the name of Kiffa Borealis, from the Arabian *al-kiffah-al-shemáliyah*, the northern scale. β and α Libræ form the XVIth Lunar Mansion, which was called *Al-zubáná*, the claws or pincers, and corrupted to *Zuban-al-kravi*, for *Zubán-al'akrab*, and applied to 20 Libræ, the γ Scorpii of Bayer.

β Libræ was ranked as a star of sensible proper motions, but the last, and perhaps most rigorous, investigation, has almost annihilated the amount, as may be thus shown:

P....	℞ − 0″·30	Dec. −	0″·07
Br...	− 0″·26	−	0″·10
B....	− 0″·05		0″·00

DXXXVII. δ BOOTIS.

℞ 15ʰ 09ᵐ 03ˢ   Prec. + 2ˢ·41

Dec. N 33° 54′·9   —— S 13″·60

Position 75°·0 (w 1)   Distance 110″·0 (w 1)   Epoch 1835·49

A star with a distant companion, on the left shoulder of Boötes. A 3½, pale yellow; B 8½, light blue. This object is 16 H. vi., and is thus described by Piazzi, Note 29, Hora XV.: "Videtur duplex, 10″ circiter temporis 1′ ad boream sequitur alia vix fere visibilis." This description perfectly identifies the *comes*, though the Δ ℞ is sufficiently large, and it is rather too bright to be called hardly visible. The measures have not that accordance which so easy an object, under slight proper motion, would promise, the other results being:

	Pos.	Dist.	Ep.
H.	84° 14′	135″·00 ±	1780·56
H. and S.	79° 39′	105″·33	1822·80

δ Boötis is the south-east component of the trapezium formed with β, γ, and μ, which was designated *Al-dhibá'*, the hyæna, in Arabian astronomy: a line from Arcturus through ε Boötis, carried as far again beyond, reaches it.

DXXXVIII. 5 M. LIBRÆ.

℞ 15ʰ 10ᵐ 26ˢ   Prec. + 3ˢ·02

Dec. N 2° 41′·3   —— S 13″·51

Mean Epoch of the Observation .... 1838·38

A close cluster of stars, over the beam of the Balance, in a narrow channel led into Serpens, and close to No. 5 of that asterism, which is double, and the next described. This superb object is a noble mass, refreshing to the senses after searching for faint objects; with outliers in all directions, and a bright central blaze, which even exceeds 3 M. in concentration. Messier, who registered this in 1764, describes it as a beautiful round nebula, adding, " et je me suis assuré qu'elle ne contient aucune étoile." This is curious, as the mass is so easily resolvable; though its laws of aggregation into so dense and compact a ball, are at present beyond reach. In May, 1791, Sir William Herschel directed his grand 40-foot reflector to this object, and counted about 200 stars; though the middle of it

was so compressed, that it was impossible to distinguish the components.

In his description of this object, Sir William remarks on the difficulty of observing with so large an instrument, as well from variable temperature as from alteration in the mirror's lustre; " but," he adds, " if we will have superior views of the heavens, we must submit to circumstances that cannot easily be altered."

DXXXIX.   5 SERPENTIS.

Æ 15ʰ 11ᵐ 08ˢ   Prec. +   3ˢ·03
Dec. N   2° 22'·6   —— S 13"·47

Position 40°·7 (w 5)   Distance 10"·5 (w 3)   Epoch 1832·41
—— 39°·8 (w 8)   —— 10"·3 (w 4)   —— 1838·36

A delicate double star, among the marginal stragglers of the preceding object, No. 5 Messier: it is 9° to the south-west of α Serpentis, and 24° to the south-east of Arcturus. A 5½, pale yellow; B 10½, light grey. This fine object is 106 ₤. iii., and was discovered in May, 1783; but no angle or distance are given in the original entry, Sir William merely noting that its position was 30° or 40° *nf,* and that it was excessively unequal, too obscure for measures. Those of Sir James South are, therefore, the first I met with:

Pos. 39° 03'   Dist. 10"·70   Ep. 1825·45

My results being as nearly coincident as could be expected in so delicate an object, I considered this star to be " done with;" but finding, on the arrival of the Dorpat Catalogue in 1837, that Professor Struve considers A to be oblong, and consequently double, I took the first good opportunity of apparition to scrutinize it closely, and catch another set of measurements. The night was beautiful, the instrument in its best action, and the stars distinct and clear; but no effort could elongate the primary.

This star has a proper motion which escaped Piazzi; but which I

Z 2

should have suspected the existence of in declination, though my observations were too general for deciding such a point. The quantity assigned by other astronomers is:

$$B.... \quad \text{Æ} - 0''\cdot01 \qquad \text{Dec.} - 0''\cdot53$$
$$A.... \qquad + 0''\cdot13 \qquad \qquad - 0''\cdot53$$

DXL. 759 ♉. II. BOOTIS.

Æ 15ʰ 11ᵐ 40ˢ	Prec. + 1ˢ·55
Dec. N 56° 54'·3	—— S 13''·43

MEAN EPOCH OF THE OBSERVATION . . . . 1838·61

A long pale-white nebula, which though classed in Boötes is on the belly of Draco. It was first registered by ♉. in 1788, and is No. 1917 of his son's Catalogue, where it is designated "a superb ray nebula." In my instrument it is a faint streak, trending *np* and *sf*; it requires the eye to *settle* before its outline is well seen. There is a little group of five stars in the *np* quadrant, and one in each of the others, all about 11th magnitude. In H.'s 20-foot telescope, it was 7½ in length. The mean place is derived, by differentiating from θ Draconis, from which it bears south-south-east, nearly 3°; being exactly on the following parallel of the distant Alioth.

DXLI. η CORONÆ BOREALIS.

Æ 15ʰ 16ᵐ 36ˢ	Prec. + 2ˢ·46
Dec. N 30° 52'·2	—— S 13''·12

Position	57°·2 (w 3)	Distance 0''·8 (w 1)	Epoch 1832·63
——	61°·9 (w 2)	—— 0''·8 (w 1)	—— 1833·57
——	68°·1 (w 5)	—— 0''·6 (w 1)	—— 1834·60
——	75°·2 (w 5)	—— 0''·6 (w 1)	—— 1835·65
——	89°·2 (w 4)	—— 0''·5 (w 1)	—— 1836·59
——	102°·3 (w 2)	—— 0''·5 (w 1)	—— 1837·68
——	109°·9 (w 2)	—— 0''·5 (w 1)	—— 1838·19
——	120°·1 (w 6)	—— 0''·5 (w 1)	—— 1839·67
——	151°·3 (w 5)	—— 0''·5 (w 1)	—— 1842·58

A binary star, midway between the Northern Crown and the club of Boötes; where a north-north-west ray from α Coronæ carried through β, and half as far again, will hit it. A 6, white; B 6½, golden yellow; a third and very minute star seen by glimpses in the *nf* quadrant. This wondrous physical object is 16 Ⅰ. ɪ.; and the discoverer recounts that he could not have detected the duplicity with a power of 227, adding, " with 460 they are as fine a miniature of i Boötis (44) as that is of a Geminorum." Under this note of preparation I commenced my attack, and difficult enough it proved, for the observations of position were mostly unsatisfactory, and those of distance were estimations. The result of 1837 is almost a guess, for the measures were made in the midst of sheet-lightning. Perhaps my last observations are the best, being taken under excellent circumstances just before the dismantling of my observatory; the angular velocity being then under rapid and direct acceleration, while the distance was diminishing, so that the fine black division seen between the stars in 1832, had not only disappeared, but the object was not always elongated. As a preparation for examining this star, I usually tested the telescope upon that beautiful adjacent object, Coronæ 1 (No. 1932 of Σ.), the components of which being small, white, and only 1″·5 apart, form an excellent preliminary. The registered results of η, for comparison, stood thus:

Ⅰ.	Pos. 30° 41′	Dist. 1″·00	Ep. 1781·69
H. and S.	25° 57′	1″·57	1823·27
Σ.	35° 16′	1″·08	1826·77
D.	63° 31′	0″·70	1833·39

From the measures of 1823, as compared with those of his father, Sir John Herschel did not infer greater motion than 4° or 5° of retrograde position; but some observations which he made in 1830, led to the remarkable conclusion, that this object had actually undergone more than a revolution, direct, since its discovery in 1781. "If this reasoning, and all the measures can be relied on," he adds, " η Coronæ is the most remarkable binary star known, being the only one which has completed a whole revolution; and this it has done in a period of 43·2 years, and with a mean value of + 8°·34 per annum." The subsequent observations of Struve, Dawes, and myself, substantially confirm this bold assumption; and as the orbit is elliptic, the velocity was increasing rapidly when my measures were discontinued. It may also be added, in corroboration, that the general mean drawn from a comparison of my own and other observations was = + 9°·41, and the period about 44 years. The excentricity, by the graphic process, is 0·3561. The connexion of the components is therefore "fully proven," and that fact alone is a gratification to the contemplative mind, for such instances of actual development in the heavenly motions cannot be studied without inspiring an increase of veneration for the Almighty Disposer of the Universe, and of zeal for the progress of astronomical inquiry.

From the reductions of my observations, I entertain little doubt of a slight proper motion in this star, although such is not noticed by Piazzi. It appears, however, upon Argelander's list, with an allowance in $Æ$ = + 0″·16, and in Dec. = − 0″·17; which value is countenanced by Mr. Baily's recent and rigorous scrutiny. Under the term " this star," both the components must be included.

DXLII. μ^1 BOOTIS.

$$\text{Æ } 15^h\ 18^m\ 27^s \qquad \text{Prec. } +\ 2^s\text{·}28$$
$$\text{Dec. N } 37°\ 56'\text{·}5 \qquad\text{——}\qquad \text{S } 12''\text{·}99$$

Position AB 171°·8 (w 8) Distance 109''·0 (w 9) Epoch 1832·31
—— BC 321°·4 (w 5) —— 1''·3 (w 2)

A triple star, on the upper part of the staff or club of Boötes; and one of the trapezium called *Al-dhibá'*, the hyæna, by the Arabians, the other members being β, γ, and δ,—μ being 4° south-east of β. A 4, flushed white; B 8, and C $8\frac{1}{2}$, both greenish white. This object is formed by 17 ʜ. vɪ. and 17 ʜ. ɪ.; and the latter so decidedly binary, that it will be entered as a distinct system in the next article. The large star appears as Alkalurops on the Catalogues, from *al-kalaúrops*, or καλαύροψ, a shepherd's crook or herdsman's staff, written in the Arabic characters; and not very happily conjectured by Grotius to be the proper reading for *inkalunis* in the Alphonsine Tables*.

This star's proper motion in space is a point of very great interest, in deciding upon its physical connexion with μ^2. The most standard authorities give the values to these amounts:

P....	Æ − 0''·30	Dec. + 0''·16
B....	− 0''·15	+ 0''·09
A....	− 0''·17	+ 0''·10

Now, even if the lowest quantity here assigned be at all correct, μ^2 must be similarly affected, for the measures are surprisingly coincident, as is here shown:

	Pos.	Dist.	Ep.
ʜ.	170° 25'	128''·00	1781·58
H. and S.	170° 25'	108''·54	1821·35
Σ.	172° 36'	108''·73	1821·78

to which results, though gained by a widely different process, those of Piazzi may be added:

Pos. 171°·5 Dist. 112''·0 Ep. 1800

Now, in discussing the position and distance of A and B, as found by H. and S. in 1821, and which are almost identical with my own, Sir John Herschel considers their fixity of importance as a palpable reference in establishing, by indisputable evidence, the fact of the close pair's rotation. In 1781, it was remarked by Sir William Herschel, that the small star followed the line joining the large one to μ (μ^1 and μ^2), and in 1802 that it had changed sides, and preceded the same line. The following object will show that this change is fully confirmed; and a distant period may ascertain whether the whole system has a common motion in space.

---

* Some have *wondered* why the Arabic names should have been so barbarized in the Alphonsine Tables, as Isaac Hazen is held to have been a Moor; but it seems he was Isaac the *Khazzan*, or chief of the synagogue, and therefore a Jew. Yet as he must have been familiar with Arabic, the errors of those Tables are probably due to transcribers, rather than to the learned Isaac.

DXLIII. μ^1 BOOTIS. (74 P. XV.)

Æ 15^h 18^m 28^s Prec. + 2^s·28
Dec. N 37° 54'·7   —— S 12"·99

Position	Distance	Epoch
321°·4 (w 5)	1"·3 (w 2)	1832·31
319°·9 (w 7)	1"·2 (w 2)	1834·56
314°·8 (w 4)	1"·0 (w 1)	1837·29
310°·6 (w 7)	0"·9 (w 2)	1839·32
306°·1 (w 6)	0"·8 (w 2)	1842·52

A binary star, on the tip of the herdsman's staff, where, absurdly enough, there is a third μ, viz. that of Corona Borealis, on some of our best delineations. A 8, and B 8½, both greenish white; these are the second and third components of the preceding object, and there registered also, for identity, as B and C, or 17 ʬ. I. The retrograde orbital motion of this pair is completely established, as, besides the mark of reference above stated, the following epochs of comparison will show:

	Pos.	Dist.	Ep.
ʬ.	357° 14'	1"·50 ±	1782·68
H. and S.	333° 42'	1"·65	1823·41
Σ.	324° 01'	1"·25	1829·73
D.	319° 47'	1"·15	1833·38

In discussing this interesting star in 1823, Sir John Herschel has said, " If this double star be a binary system, of which there can be little doubt, its period is about 622 years, and the most probable mean annual motion is − 0°·5783, in the direction *np sf*, or retrograde." All the subsequent observations prove that the motion has greatly accelerated; and the object is yearly becoming more difficult of scrutiny, from the diminution in its distance. From the earliest epoch here registered down to my latest, an annual mean movement appears = − 0"·85; but from Herschel Junior and Sir James South's period it averages − 1"·44, so that the period may be within 460 years; but the annual rates are as yet distressingly irregular.

DXLIV. 76 P. XV. SERPENTIS.

Æ 15^h 18^m 52^s Prec. + 2^s·72
Dec. N 18° 44'·2   —— S 12"·96

Position 326°·3 (w 3) Distance 1"·7 (w 1) Epoch 1834·41

A close and delicate double star, preceding the head of Serpens. A 8 and B 9½, both white. This most difficult object is one of Σ.'s " vicinæ," being No. 1940 of the great Dorpat Catalogue, thus registered by its discoverer:

Pos. 325°·47 Dist. 1"·477 Ep. 1830·35

This star is 8° south-by-west from Gemma, the lucida of the Northern Crown, where it is intercepted by a line drawn east from Arcturus and passed under ξ Boötis; it is so closely *nf* 874 ℍ. ii., that Sir John Herschel assumes it as a pointer whereby to identify that globular object, in the terms, "a * 7·8 m, 6′ n."

DXLV. ι DRACONIS.

AR 15ʰ 21ᵐ 23ˢ   PREC. + 1ˢ·32
DEC. N 59° 31′·7   —— S 12″·79

POSITION 46°·2 (*w 2*)   DISTANCE 117″·0 (*w 1*)   EPOCH 1833·60

A bright star with a distant companion, in the middle of Draco's body. A 3, orange tint; B 9, pale yellow; several other stars in the field. This object precedes several of ℍ.'s nebulæ, and may be readily found by running a line from the Pole-star through Pherkad Major (γ Ursæ Minoris), and carrying it nearly as far again to the south. Some discordant anomalies have attended the early entries of this star in the Catalogues. Mr. Baily suggests the correction of Flamsteed's AR; but adds, "I cannot discern how the declination was determined; as the first star called ι Draconis in the *Historia Cœlestis*, vol. i., p. 72, is ι Herculis." It was designated *Al-dhiba'*, the hyæna, by the Arabians, and appears thus on the Dresden globes, and in Ulugh Beigh's Catalogue; but others of the same school call it *Al-dhikh*, the wolf, or hyæna.

DXLVI. 91 P. XV. LIBRÆ.

AR 15ʰ 22ᵐ 32ˢ   PREC. + 3ˢ·44
DEC. S 19° 36′·8   —— S 12″·71

POSITION 282°·6 (*w 6*)   DISTANCE 11″·8 (*w 5*)   EPOCH 1832·48

A neat double star, in the space between the *chelæ* of Scorpio, and 9° due west of β Scorpii. A 7½, bluish white; B 9, smalt blue; with two or three minute stars in the field. This object is Sir James South's No. 672, having been discovered in the observatory which he erected at Passy, near Paris; and he thus first registered it:

Pos. 283° 14′   Dist. 11″·47   Ep. 1825·35

whence it seems to have remained nearly constant, during an interval of seven years. The slight proper motion in AR imputed to it by Piazzi, has entirely disappeared under Mr. Baily's inquiry, but the movement in declination seems to be − 0″·09, at present; which may also vanish under the next exact series of meridional observations.

DXLVII. δ SERPENTIS.

ℛ 15ʰ 27ᵐ 10ˢ Prec. + 2ˢ·86
Dec. N 11° 04′·7 —— S 12″·40

Position 196°·5 (w 9) Distance 2″·9 (w 9) Epoch 1831·43
—— 197°·3 (w 8) —— 2″·7 (w 8) —— 1838·38
—— 196°·2 (w 9) —— 2″·8 (w 8) —— 1842·35

An elegant double star, in the bend of the Serpent's neck; where it forms a component of the Arabian *Nasak Yemáni*, or southern boundary of the pasturage, cited under a Serpentis, from which it bears north-by-west, distant 5°. A 3, bright white; B 5, bluish white; but under the very best vision, both have a bluish tinge, which, in such a pair, is rather against the theory of contrast. This fine object is 42 ♓. ɪ., and the earliest results for comparison are:

♓. Pos. 227° 12′ Dist. 3″·00+ Ep. 1782·99
H. and S. 199° 13′ 3″·05 1821·33

The change of angle indicated by these observations, gave H. reason to assume a mean annual retrograde motion of − 0°·726; but the subsequent measures of Σ., D., and myself, do not bear out so large a quantity. Nor has the distance between them increased, as once suspected by Σ. and H.; which, coupled with the diminution of angular velocity apparent on a comparison of the several epochs, induced the latter to conclude the orbit to be elliptic, and so situated as to allow of its ellipticity being visible without distortion.

It must not be omitted that a small movement in space has been detected in A, which, when surer known, will afford further demonstration of its physical connexion with B; the present values are:

P.... ℛ − 0″·06 Dec. + 0″·07
B..... + 0″·02 + 0″·05

DXLVIII. α CORONÆ BOREALIS.

ℛ 15ʰ 27ᵐ 54ˢ Prec. + 2ˢ·53
Dec. N 27° 15′·2 —— S 12″·35

Position 214°·5 (w 2) Difference ℛ = 11ˢ·6 (w 1) Epoch 1834·51

A standard Greenwich star, with a distant companion, in the front centre of the Northern Crown; where it is known as Gemma, or the precious stone of the diadem. A 2, brilliant white; B 8, pale violet; and there is a very neat little double star nearly preceding A by 48ˢ·3, on the angle = 275°, between which and A, in nearly mid-distance, is a red star of the 10th magnitude. The primary of this group has had the following proper motion assigned to it; and though I detected nothing in Æ, I am inclined to corroborate a spacial change in the declination:

$$P.... Æ \quad - \; 0''{\cdot}10 \qquad \text{Dec.} \; - \; 0''{\cdot}10$$
$$B.... \quad + \; 0''{\cdot}17 \qquad\qquad - \; 0''{\cdot}07$$
$$A.... \quad + \; 0''{\cdot}15 \qquad\qquad - \; 0''{\cdot}06$$

Gemma does not seem to have been a name known to the ancients, though the astral genealogists are fain to derive it from the " Gemmasque novem transformat in ignes" of Ovid. The star, however, was honoured as *Ashtaroth*, the Syrian Venus, as *Ariadne*, as *Gnossia stella*, as *Clara stella*, and other designations, from the earliest times.   When the Arabians boarded Ptolemy, and clouded his λαμπρὸς, there was no end of the epithets given to *a* Coronæ Borealis; and the whole tribe of commentators, glossators, scholiasts, etymologists, and lexicographers, have been thrown into fault.   It appears as *Alfeta*, in the Almagest and Alphonsine Tables; as *Alfeta*, and *Foca*, in other books; as *Káshehi Dervíshán*, or *Scutella pauperum*, from a fancied resemblance of the asterism to a beggar's patin; and as *Kásheh shekesteh*, or *Scutella fracta*, because the asterism appears as an incomplete circle.   But its most usual name on the Catalogues is Alphecca, from *al fekkah*, the dervish's cup or platter, from the said break in the ring of stars; and it was *Naïr-al-fekkah*, the lucida or bright star of the pauper's platter.   To find this by alignment, place ζ and η, the two last stars in the Greater Bear's tail, in a direction following, and they will point upon the seven stars of the Northern Crown in the south-east, where *a* shines conspicuous enough; or it may be picked up at one third of the distance, and on the line, from Arcturus to Wega.   And the salt-water poet points out another method:

From *epsilon* in Virgo's side	Arcturus seek, and stem,
And just as far again you'll spy	Corona's beauteous gem:
There no mistake can well befal	e'en him who little knows,
For bright and circular the Crown	conspicuously glows.

Στέφανος βόρειος, the Northern Crown, is one of the old forty-eight constellations, and though small, possesses some very interesting double stars and nebulæ.   A table of the comparative lustre of the components will be found in the *Philosophical Transactions* for 1797, p. 315; and their number has thus risen under increased optical application:

Ptolemy . . . 8 stars		Flamsteed . . . 21 stars
Bayer 20		Bode 87

DXLIX. ζ CORONÆ BOREALIS.

Æ 15ʰ 33ᵐ 21ˢ   Prec. + 2ˢ·26

Dec. N 37° 09′·6   —— S 11″·97

Position 301°·2 (w 9)	Distance 6″·4 (w 5)	Epoch 1831·61
—— 300°·9 (w 9)	—— 6″·5 (w 9)	—— 1839·50
—— 301°·2 (w 9)	—— 6″·1 (w 9)	—— 1842·57

A fine double star, in the middle of the space over the wreath, and 10° north, a little easterly, from Gemma.   A 5, bluish white; B 6, smalt blue. This beautiful object is 8 Ḥ. II., and its measures previous to my observations stood thus:

Ḥ.   Pos. 295° 51′   Dist. 5″·468   Ep. 1779·76
H. and S.   300° 57′   7″·168   1822·30

In discussing the latter results H. states that his father's distance is a mean of two observations; and Ḥ. seems to imply that the position is a mean of two years' measures. The star was therefore considered as well established at the first epoch; whence the second tended to show a slow direct orbital variation, with a slight increase of distance. My operations in August, 1831, inclined me to think there might be a movement *nf sp;* but my second series, being to the *senses* the most satisfactory measures possible, establish its fixity; and this is also confirmed by Professor Struve's observations.

Since the above was written, I have obtained a splendid set of measures at Hartwell House, by which it is rendered still more evident, that these stars are optical, and relatively at rest. Indeed, there are few of these objects whose details come out so satisfactorily.

DL. 764 Ḥ. II. DRACONIS.

Æ   15ʰ 35ᵐ 53ˢ   Prec. + 1ˢ·21
Dec. N 59° 52′·0   —— S 11″·79

Mean Epoch of the Observation . . . . 1838·50

A small round nebula, in the centre of Draco's body, being one of those mentioned *supra*, as following *ι* Draconis. This object is pale but distinct, and was first registered in May, 1788. There are several stars in the field, of which one of the 10th magnitude precedes pretty close to the nebula, and a brighter one follows near the parallel. The mean apparent place was differentially obtained from *ι* Draconis, from which it bears east-by-north, and is about a degree and a half distant.

DLI. γ CORONÆ BOREALIS.

Æ   15ʰ 36ᵐ 01ˢ   Prec. + 2ˢ·52
Dec. N 26° 48′·4   —— S 11″·78

Position AB *round* (w 8)   Distance . . *round* (w 8)   Epoch 1834·66
—— AB 225°·0 (w 1)   —— . . 0″·3 (w 1) ⎫
—— AC 88°·7 (w 2)   Difference Æ = 14ˢ·5 (w 2) ⎬ —— 1839·69
—— AB *round* (w 8)   Distance . . *round* (w 8) ⎭ —— 1842·58

A most difficult binary star, with a distant companion, in the middle of the lower side of the wreath; where it follows Gemma a little south of the parallel, at about $2°\frac{1}{2}$ distance. A 6 (but certainly looking very much brighter), flushed white; B, uncertain; C 10, pale lilac. This is the præses of Σ.'s " vicinissimæ," being pronounced by him as closer and more troublesome than ζ Herculis. In 1832, Sir John Herschel examined it with the 20-foot reflector and eye-pieces magnifying 320, 480, and 600 times, and under each saw but a round disc without any companion. My attacks, therefore, under various powers, were attended with a similar result. But previous to dismantling my equatoreal, I took advantage of a superb night, 7th September, 1839, and after much gazing, became *almost* assured that it was really elongated in a direction *sp* and *nf;* the which was so different from what Σ. had registered, that I ventured at once to receive it as a physical object, and a binary system. The data of the Dorpat astronomer, upon which I drew this conclusion, are:

Pos. 111°·05   Dist. 0″·725   Ep. 1826·75
simplex   *simplex*   1835·40

On my repairing to Hartwell House in 1842, the observing of this interesting star was among the *agenda* which I took with me, and though it was still a dumpy mis-shaped object, with an axis major perhaps *np* and *sf,* I could only record it round. While writing this, however (June 23rd, 1843), Mr. Dawes tells me that he has just detected it again double, or, as he expresses it, " with the companion coming out again."

As usual, my magnitude of γ Coronæ is received from the Palermo Catalogue, otherwise its brilliance would be placed considerably higher. Ptolemy rated it δ, a size followed by Tycho Brahé, Hevelius, and Flamsteed. But may not the apparent differences be owing to the actual duplicity and orbital action of the star? A proper motion in space has been assigned it, to the following amount:

P.... Æ − 0″·23   Dec. + 0″·10
B....   − 0″·76   + 0″·08
A....   − 0″·08   + 0″·09

DLII. α SERPENTIS.

Æ 15ʰ 36ᵐ 23ˢ   Prec. + 2ˢ·94
Dec. N 6° 55′·9   ——— S 11″·75

Position 1°·5 (*w* 1)   Distance 50″·0 (*w* 1)   Epoch 1837·32

A standard Greenwich star with a minute *comes*, on the heart of the Serpent. A 2½, pale yellow; B 15, fine blue, followed a little south of the parallel, at about 25ˢ Δ Æ, by a telescopic star. This extremely delicate object was discovered by H., in his 20-foot Sweeps; and the primary has a proper motion assigned to it in space thus:

P.... Æ − 0″·10   Dec. + 0″·05
B....   + 0″·21   + 0″·06
A....   + 0″·18   + 0″·07

This star appears on the Catalogues as Unukalhay, from '*unḳ-al-*

hayyah, the serpent's neck. It has also been designated *Alyah*, and supposed to be from the "broad sheep's tail" of some Oriental figure for this asterism; but no such name is known to the Arabian astronomers, although *Alangue* and *Ras Alaugue* appear on the Alphonsine Tables. It was substituted for *Alioth*, found in some works, by an ingenious but erroneous conjecture of Joseph Scaliger's. It is the Cor Serpentis of astrologers; and doubtless the *lucidus anguis*, one of the stormy warnings of the *Georgics*. It may be found by being looked for at nearly mid-distance between Gemma and Antares; or our friend the galley-poet may be called in:

To strike th' insidious Serpent's heart,	a line from Altair wield,
From thence below Ras Alague,	across th' Arabian Field;
And when as far again you've reached,	as those two stars may be,
The middle one of three fair gems,	Serpentis Cor you'll see.

The Serpent is one of the ancient forty-eight constellated groups, and is large, as its windings bring it in contact with Ophiuchus, Aquila, Libra, and Hercules; the head and neck, which are directly under the crown, are marked by some remarkable stars, and the tail terminates in a splashy part of the Galaxy. The Greeks distinguished the Dragon, the Snake, and the Water-snake, by the distinctive epithets Δράκων, Ὄφις, and Ὕδρη, but the Latins used only *Anguis* and *Serpens* for the whole three. The Arabians termed it *Huweyyah*, the snake; but they had also a great configuration around this spot. Kazwíni states that β and γ Serpentis, with the stars of the same rank and letter on the arm of Hercules, were the *Nasak shámí;* while δ, a, and ε Serpentis, with δ, ε, ζ, and η Ophiuchi, were the *Nasak Yemaní;* and the north and south lines thus formed were considered as the boundaries of a vast pasturage, or garden. Serpens is astronomically, though not mythologically, distinguished from its bearer; and is held to be emblematic of prudence and vigilance. Its constituents have been thus enumerated:

Ptolemy	. . . 18 stars		Bayer	37 stars
Copernicus	. . 18		Hevelius . . .	22
Tycho Brahé	. . 13		Flamsteed . .	64
Kepler .	. . . 26		Bode	187

DLIII. β SERPENTIS.

Ꞁ 15ʰ 38ᵐ 48ˢ   Prec. + 2ˢ·76
Dec. N 15° 55'·7   —— S 11"·57

Position 264°·5 (w 3)   Distance 31"·0 (w 2)   Epoch 1836·53

A delicate double star, on the Serpent's under-jaw; being one of the components of the *Nasak shámí*, or northern boundary wall of the Arabian Field above mentioned; and it is 12° south-by-east of Gemma. A 3½, and B 10, both of a pale blue tint, and there is a distant telescopic star, also blue, in the *sp* quadrant. This object is 36 Ḥ. IV., and being considered too obscure for measuring at its discovery, it was "pretty exactly estimated." The results may be stated thus:

Pos. 266°·5 Dist. 24"·00 Ep. 1781·62

By Σ., the next who examined it, it stood thus:

Pos. 265°·0   Dist. 30″·64   Ep. 1832·14

whence its fixity may be considered as established, notwithstanding the apparent anomaly in the distance.

DLIV.   39 SERPENTIS.

Æ   15ʰ 45ᵐ 45ˢ   Prec. +   2ˢ·80

Dec. N 13° 42′·2   —— S 11″·08

Position 355°·0 (*w* 1)   Distance 12″·0 (*w* 1)   Epoch 1835·57

A most delicate double star, at the back of the Serpent's head, and about 7° distant from its own lucida, on a north-north-east line. A 7½, white; B 15, bluish, and there is a telescopic star in the *sf* quadrant, which points through A to a still smaller one in the *np*. This object is 25 Ħ. III., and was thus registered by its discoverer:

Pos. 357° 14′   Dist. 10″·0   Ep. 1780·65

39 Serpentis has been suspected of variability, and was even mistaken for Harding's variable star, which is 2° north of it. It is marked as of the 6th magnitude in the British Catalogue; "but," says Mr. Baily, "in the *original* entry, it is designated of the 6½, which I have therefore retained." It has been thus charged with proper motion:

P.... Æ – 0″·16   Dec. – 0″·16
B....   – 0″·11   – 0″·51
A....   – 0″·13   – 0″·51

DLV.   220 P. XV. SERPENTIS.

Æ   15ʰ 49ᵐ 15ˢ   Prec. +   2ˢ·99

Dec. N   3° 52′·4   —— S 10″·83

Position 324°·7 (*w* 8)   Distance 10″·5 (*w* 8)   Epoch 1834·35

A neat double star, on the Serpent's back; it is about 5° from *a*, on a south-east-by-south line, which passes ε in nearly mid-distance. A 8, white; B 9, grey; this appeared red to Ħ., who remarks that probably a "dry fog" at the time of observation had tinged it. This object is 103 Ħ. III., and was described by Piazzi: "Duplex. Comes 10 magnitudinis parumper ad boream, 0″·6 temporis præcedit." A comparison of the several series of measures of this star, shows a slight increase of angle, which may as likely be owing to *personal equation* as to motion; but still the results are remarkably coincident. They are:

Ħ.   Pos. 320° 12′   Dist. 12″·46   Ep. 1783·18
H. and S.   323° 04′   10″·66   1823·46
Σ.   324° 01′   10″·27   1831·91

DLVI. ζ URSÆ MINORIS.

Æ 15ʰ 49ᵐ 52ˢ   Prec. − 2ˢ·36
Dec. N 78° 16′·7   —— S 10″·78

Position 210°·5 (w 2)   Distance 310″·0 (w 1)   Epoch 1834·64

A second-rank Greenwich star, in the middle of the animal's body, and the most northern member of its square. A 4, flushed white; B 11, bluish; a yellow star, of the 9th magnitude, in the *sp* quadrant, and a minute *comes* to B was seen by Mr. Challis, who examined this star at my request, in September, 1841, with the great Northumberland equatoreal. There is every inference that A has a motion through space, and now that it is placed on the Greenwich list, it will probably be so watched as to get the value detected. The present registered amount is thus:

$$P.... \qquad 0''·00 \qquad \text{Dec.} - 0''·14$$
$$B.... \qquad + 0''·16 \qquad \qquad 0''·00$$
$$A.... \quad Æ + 0''·21 \qquad \qquad + 0''·01$$

This star is erroneously termed *Anpha al Pherkadein*, and η Ursæ Minoris *Anwer al Pherkadein*, in several respectable authorities. But Ulugh Beigh and other astronomers distinctly apply these names to β and γ Ursæ Minoris, the first being *Anwar-al-ᶠerkadeïn*, the brightest of the two calves, and the second *Akhfa-al-ferkadeïn*, the dimmest of the two calves; and it will be recollected that *a* of this asterism was called *Juddah*, the kid, by the Arabs. The two stars of γ were latinized into *Pherkad major* and *Pherkad minor*.

Before quitting this object, and while the telescope and roof aperture are pointed in this direction, a couple of good double stars may be looked at, which were accidentally omitted from my working list, by not being in the Palermo Catalogue. To the *np* of ζ, and in the mid-distance of a line drawn between *a* and γ of the same asterism, are π¹ and π² Ursæ Minoris, upon the flank of the animal. The first was discovered in 1782, and classed 90 Ⱨ. ɪv.; the components being of the 6th and 7th magnitudes, contrasted yellow and blue in colour, and 30″ apart. The second is an interesting object, being No. 1989 of the Dorpat Catalogue; and it is so close as to be in contact. In my instrument it was only wedged, the axis-major seeming from *sp* to *nf;* but in this there was no certainty.

DLVII. ϱ CORONÆ BOREALIS.

Æ 15ʰ 54ᵐ 56ˢ   Prec. + 2ˢ·30
Dec. N 33° 47′·5   —— S 10″·40

Position 126°·4 (w 4)   Distance 80″·0 (w 2)   Epoch 1833·61

A star with a distant companion, over the *nf* portion of the wreath; it is 7°½ distant from Gemma, on a north-east bearing. A 6, creamy white; B 11, blue. This poor object is 93 ♓. vi., and from the data given by its discoverer, as compared with those of Sir James South, an inference of a great angular change had obtained, which my observations do not confirm. The former measures were:

$$\begin{array}{llll} \text{♓.} & \text{Pos. } 144^\circ\ 27' & \text{Dist. } 87''{\cdot}73\pm & \text{Ep. } 1782{\cdot}55 \\ \text{S.} & 125^\circ\ 06' & 79''{\cdot}20 & 1825{\cdot}48 \end{array}$$

Although Piazzi assigns no proper motion to this star, my reductions indicate a sensible one in declination. It has been thus registered since that astronomer's epoch:

$$\begin{array}{lll} B.... & \mathbb{R} - 0''{\cdot}24 & \text{Dec. } - 0''{\cdot}77 \\ A.... & - 0''{\cdot}20 & - 0''{\cdot}75 \end{array}$$

DLVIII.　51 LIBRÆ.

$$\begin{array}{ll} \mathbb{R}\ 15^\text{h}\ 55^\text{m}\ 35^\text{s} & \text{Prec. } +\ 3^\text{s}{\cdot}29 \\ \text{Dec. S } 10^\circ\ 55'{\cdot}6 & \text{——— S } 10''{\cdot}36 \end{array}$$

Position AB 6°·6 (w 5)	Distance 1″·4 (w 3)	} Epoch 1834·42
——— AC 76°·1 (w 8)	——— 7″·2 (w 5)	
——— AB 13°·3 (w 8)	——— 1″·1 (w 5)	} ——— 1838·60
——— AC 74°·2 (w 8)	——— 7″·2 (w 6)	
——— AB 23°·5 (w 8)	——— 1″·2 (w 5)	——— 1842·56

A fine triple star, between the upper *lanx* Libræ and the right leg of Ophiuchus, above the upper or left chela of Scorpio; and it is 16° from Antares, on a line running north-north-westward from that star to α Serpentis. In this I am particular, because Flamsteed designated it ξ Libræ, thinking he thereby followed Bayer; whereas in fact this star is ξ Scorpii; which he says is "in origine forficis, seu forficulæ, Barbari graffias vocant:" it consequently ought not to have been jumbled into the Balance. The components are, A 4¼, bright white; B 5, pale yellow; and C 7½, grey; and there is a neat small pair of stars in the *np* quadrant, whence ♓., who did not at first perceive the duplicity of A, called the whole a double-double star.

This beautiful object is formed by 33 ♓. i. and 20 ♓. ii.; but there is some remarkable anomaly respecting A and B, which it is difficult to account for, since the comparative steadiness of the distance indicates a circular orbit, constantly showing both components to us, at a similar distance. Now H. and S. observing this object on the 18th June, 1822, when their measures of A and C "were extremely satisfactory," and guarded before-hand by ♓.'s remarks, could not see the large star double. Hence the "videtur triplex" of Piazzi's Note 245, Hora XV., must have been AC, and the small pair in the *np*, for his telescope was certainly inferior to that of the Huddart equatoreal. The first measures were those of Sir William Herschel:

$$\begin{array}{lll} \text{Pos. AC } 88^\circ\ 37' & \text{Dist. } 6''{\cdot}38 & \text{Ep. } 1780{\cdot}39 \\ \text{AB } 7^\circ\ 58' & 1''{\cdot}50 & 1782{\cdot}36 \end{array}$$

When Sir James South's observations of 1825·49 were summed up, the result appeared so identical with the above—merely displacing a quadrantal n for s—that it was stated, "Not the slightest alteration appears to have taken place in the relative position or distance of the close stars." But this conclusion of its having remained perfectly unaltered, was shaken by the subsequent observations of Σ., D., and H.; and it is shown by Σ., that if H⨍. made an error in the quadrant of the star, which the nearly equal magnitudes will easily admit of, and if, from similar causes, we add 180° to S.'s deductions, it will show a direct motion of 182° in fifty-five years; giving about a century as its *annus magnus*. The stars A and C, however, are evidently retrograding at about − 0°·2 per annum, which is not accountable on proper-motion conditions, the latter element's estimated values being:

$$P.... \mathbb{R} - 0''·23 \qquad Dec. - 0''·10$$
$$B.... \quad\;\; - 0''·03 \qquad\qquad - 0''·02$$

In conclusion I may add, that my measures in 1838, despite of certain teasings from the variable refractions of such a low altitude, and the closeness of A and B, were so satisfactory, that I place great confidence in the results. The mean annual motion yielded by my own observations, under proper allowances for the respective weights, is + 1°·85.

DLIX. β SCORPII.

$$\mathbb{R}\quad 15^h\; 56^m\; 08^s \qquad Prec. +\; 3^s·47$$
$$Dec.\; S\; 19°\; 21'·7 \qquad\text{———}\quad S\; 10''·31$$

Position 24°·9 (w 9)   Distance 13''·1 (w 9)   Epoch 1835·39

A second-rate Greenwich star with a companion, at the root of the upper hand-claw. A 2, pale white; B 5½, lilac tinge; and the two point nearly to a 5th-magnitude star in the *nf* quadrant. This fine object, which was measured under the natural illumination of full daylight, is 7 H⨍. III., and Nos. 251 and 252 of Piazzi's Hora XV., and both stars have a slight proper motion assigned to them in the Palermo Catalogue. They are sometimes marked β¹ and β²; and the measures at the epochs compared with mine, are thus registered:

H⨍.	Pos. 25° 09'	Dist. 14''·37	Ep. 1779·72
H. and S.	26° 30'	13''·65	1823·28

This comparison, embracing an interval of fifty-five years, shows the components to have been pretty constant, and the object must therefore be assigned to the optical class.

Though only the second star of the asterism, β Scorpii has obtained the designation of Acrab, or Scorpion, in the Catalogues. It is also called *Iklíl-al-Jebhah*, the crown of the brow, and with δ and π forms the XVIIth Lunar Mansion; these three stand in a nearly vertical line, and to them Tízíní adds ρ, and Ulugh Beigh ν, in continuation of that line. It may be readily found, by running a line from α Lyræ southwards between α Herculis and α Ophiuchi, and equally far beyond, where it will hit upon β Scorpii, which is the northern of the three stars just named, and whence it may also be recognised; or resorting to the rhymer's advice, his rule runs thus:

| From Virgo's spike to east-south-east, | direct th' inquiring eye; |
| You'll pass between the Scale's bright stars | to where Iklil doth lie. |

DLX. $\varkappa^1$ HERCULIS.

Æ 16ʰ 00ᵐ 51ˢ   Prec. $+ 2^s \cdot 70$
Dec. N 17° 28'·7   —— S 9"·96

Position 9°·7 (*w* 9)   Distance 31"·4 (*w* 9)   Epoch 1835·45

A neat double star, on the Hero's left elbow; about 30° distant from Wega, in the west-south-west, where it is nearly mid-way between γ Herculis and β Serpentis. A 5½, light yellow; B 7, pale garnet, being $\varkappa^3$ of Piazzi, No. 285, Hora XV. This object is 8 Ħ. v., and its measures, at the epochs previous to my observations, are thus registered, including deductions from the Æs and Decs. of the Palermo Catalogue:

	Pos.	Dist.	Ep.
Ħ.	10° 23'	39"·98	1779·72
P.	13° 12'	32"·70	1800·00
H. and S.	9° 35'	31"·17	1821·39
Σ.	9° 35'	31"·21	1832·60

These results, compared with my own, afford strong presumption that the angle of position has remained stationary. There may, however, have been a diminution of distance, especially if the operations of Flamsteed, 132 years before mine, are to be relied on. From M. Argelander, under the able discussion of Professor Struve, the interesting details of this question are:

Pos. 14° 35'   Dist. 56"·48   Ep. 1703·31

Struve, however, doubts the decrease of distance in so large a ratio. Future observers must decide this point; meantime the near coincidence of direction in the proper motions may be noted:

$$
A \begin{cases} \varkappa^1 \; Æ \; - 0"·09 & \text{Dec.} - 0"·01 \\ \varkappa^3 \quad - 0"·12 & \quad - 0"·21 \end{cases}
$$
$$
B \begin{cases} \varkappa^1 \quad - 0"·75 & \quad 0"·00 \\ \varkappa^3 \quad - 0"·02 & \quad + 0"·01 \end{cases}
$$

DLXI. ν SCORPII.

Æ 16ʰ 02ᵐ 42ˢ   Prec. $+ 3^s \cdot 47$
Dec. S 19° 02'·3   —— S 9"·82

Position 338°·5 (*w* 6)   Distance 40"·0 (*w* 4)   Epoch 1831·50

A neat double star, at the root of the upper or northern handclaw; included by Ulugh Beigh in the XVIIth Lunar Mansion, called

Al-Iklíl al-Jebhah, crown of the forehead. It lies east-by-north from β Scorpii, at about 2° distance. A 4, bright white; B 7, pale lilac. This object is 6 Ḥ. v., and, including the Palermo reductions, may be thus exhibited:

	Pos.	Dist.	Ep.
Ḥ.	339° 28′	38″·33	1779·72
P.	341° 00′	41″·80	1800·00
H. and S.	338° 12′	40″·81	1821·37

In Sir John Herschel's discussion of these results, he has taken a position from his father's MSS. for 1782; but the above epoch has been copied from the LXXII.nd vol. of the *Philosophical Transactions*, where they are evidently entered with great regularity. Piazzi's deductions are inserted, because they neutralize the notion of an increase of distance; and the full discussion requires the proper motions to be stated, although those registered are neither large nor accordant:

	Æ	Dec.
P....	− 0″·06	− 0″·08
Br...	− 0″·09	+ 0″·01
B....	+ 0″·06	+ 0″·03

DLXII. 49 SERPENTIS.

Æ 16h 05m 51s   PREC. + 2s·78

DEC. N 13° 57′·6   —— S 9″·58

POSITION 317°·8 (*w* 8)   DISTANCE 3″·7 (*w* 5)   EPOCH 1832·43

—— 318°·1 (*w* 9)   —— 3″·3 (*w* 9)   —— 1839·29

A close double star, which, though pertaining to the Serpent, is absurdly placed on the left arm of Hercules. A 7, pale white; B 7½, yellowish. This is 82 Ḥ. I., and is a fine and tolerably easy object, therefore its measures at the several epochs are entitled to confidence. By a comparison of the two first, it appeared to have a direct angular velocity of + 0°·510 per annum; but subsequent observations greatly diminish this value. These are the periods of comparison:

	Pos.	Dist.	Ep.
Ḥ.	291° 33′	2″·50±	1783·18
H. and S.	311° 57′	4″·21	1823·28
D.	314° 50′	3″·17	1831·40
Σ.	316° 41′	3″·20	1832·70

Still a rough investigation of the whole, gives above 600 years for the orbital revolution of the *satellite* about its primary,—or rather, of one sun around the other. More observations at longer epochs are, however, necessary, before it can actually be pronounced a binary system. Indeed, as improved methods and means are progressively applied, various suspected movements will be decisively confirmed or rejected, results more exact obtained, and the vague allowances for *personal equation* destroyed. Meantime he who wishes to fish the object up, may find it at the distance of 10°½ to the north-east of *a* Serpentis, on the line leading from thence to Wega.

DLXIII. δ OPHIUCHI.

$\mathbb{R}$ 16ʰ 05ᵐ 58ˢ   Prec. + 3ˢ·14
Dec. S 3° 16′·7   —— S 9″·57

Position 131°·5 (w ı) Distance 319″·0 (w ı) Epoch 1834·45

A standard Greenwich star, with a distant companion, on the right hand of the Serpent-bearer; where it constitutes one of the stations in the *Nasak Yeменí*, or southern boundary of the Arabian garden, mentioned under a Serpentis. A 3, deep yellow; B 10, pale lilac; a third minute star in the *sf* quadrant. I should certainly join with Piazzi in imputing an appreciable proper motion in $\mathbb{R}$ to this fine star, although it is greatly reduced by the scrutiny of Mr. Baily:

P.... $\mathbb{R}$ – 0″·15   Dec. – 0″·08
B....   – 0″·02   – 0″·11

This is the preceding of two bright stars on the right hand of Ophiuchus, δ and ε, which are called *Yed prior* and *Yed posterior*, from the Arabic word *yed*, the hand. A line drawn from Gemma to Antares, passes these stars, at about two-thirds of the distance down. Or they may be picked up thus:

See bright Altair and Virgin's spike,   and mid-way cast your eye,
Grasping the huge and tortuous snake,   the Bearer's hand you'll spy.

DLXIV. 80 M. SCORPII.

$\mathbb{R}$ 16ʰ 07ᵐ 28ˢ   Prec. + 3ˢ·56
Dec. S 22° 35′·4   —— S 9″·45

Mean Epoch of the Observation .... 1837·36

A compressed globular cluster of very minute stars, on the right foot of Ophiuchus, which is on Scorpio's back. This fine and bright object was registered by Messier in 1780, who described it as resembling the nucleus of a comet; and indeed, from the blazing centre and attenuated disc, it has a very cometary aspect. There are some small stars both above and below its following parallel, of which three of those in the *nf* form a coarse triangle; but the field and the vicinity are otherwise barren. An early star of Ophiuchus, No. 17 P. xvi., slightly precedes this splendid conglomerate, about half a degree to the northward, and though only of the 8th magnitude, is a convenient index of approach to the out-door gazer. Such particulars are not needed by the man with fixed instruments, but will greatly facilitate the operations of those who are more remarkable for intellectual energy than for means.

The mean apparent place is differentiated from δ Scorpii, from which it lies east, at 4° distance; and it is mid-way between a and β Scorpii.

This is a very important object when nebulæ are considered in their relations to the surrounding spaces, which spaces, Sir William Herschel found, generally contain very few stars: so much so, that whenever it happened, after a short lapse of time, that no star came into the field of his instrument, he was accustomed to say to his assistant, "Make ready to write, Nebulæ are just approaching." Now our present object is located on the western edge of a vast obscure opening, or space of 4° in breadth, in which no stars are to be seen; and Sir William pronounces 80 Messier, albeit it had been registered as *nébuleuse sans étoiles*, to be the richest and most condensed mass of stars which the firmament can offer to the contemplation of astronomers. See also 4 Messier, No. DLXIX.

DLXV. σ CORONÆ BOREALIS.

$$\text{Æ} \quad 16^h \ 08^m \ 41^s \qquad \text{Prec.} \ + \ 2^s{\cdot}26$$
$$\text{Dec. N } 34° \ 16'{\cdot}1 \qquad \underline{\qquad} \ \text{S } 9''{\cdot}36$$

Position AB 107°·6 (w 5)	Distance 1″·3 (w 2)	Epoch 1830·76
———— AC 90°·0 (w 5)	———— 43″·3 (w 3)	
———— AB 114°·9 (w 6)	———— 1″·4 (w 2)	———— 1832·37
———— AC 88°·7 (w 7)	———— 44″·1 (w 3)	
———— AB 120°·7 (w 3)	———— 1″·2 (w 2)	———— 1833·58
———— AB 130°·9 (w 5)	———— 1″·4 (w 3)	———— 1835·50
———— AC 89°·3 (w 8)	———— 44″·0 (w 4)	
———— AB 145°·1 (w 8)	———— 1″·6 (w 3)	———— 1839·67
———— AC 88°·9 (w 9)	———— 44″·2 (w 5)	
———— AB 155°·9 (w 9)	———— 1″·8 (w 5)	———— 1843·35

A binary and triple star, between the Wreath, and the left thigh of Hercules, 10° distant from and north-east of Gemma, and a little more than half-way between that star and η Herculis. A 6, creamy white; B 6½, smalt blue; C 11, dusky blue; and a fourth star, of the 17th magnitude, was seen by H. in the 20-foot reflector, in the *sp* quadrant, at about 20″ distance.

This very interesting object is 3 ♅. I., and the close pair is physical, being found to have a rapid direct orbital motion, which has been accelerating since the year 1800. When H. and S. examined it in 1821, this accelerated velocity was detected, since the first interval of twenty-one years gave a mean of 1°·139 per annum; and the following interval of seventeen years, showed an augmentation to $2^{\circ}\cdot298$; and from thence, by an almost sudden start, it was supposed to have increased to $6^{\circ}\cdot982$ per annum. This, however, was not verified by Sir James South's observations of 1825; but the accelerated velocity from that date may be seen by inspecting the various measures,—and it should be added, that the rotatory motion continues in the direction assigned by H., and supports his hypothesis. In 1828 the *comes* changed its quadrant from *nf* to *sf*; and the angular mutations appear to have been accompanied by a sensible diminution of distance, since about 1830 it was a miniature of ξ Ursæ Majoris, which, in 1780, was considered by ☿. to be the closest of the two. The following are my selected sources of comparison:

		Pos.	Dist.	Ep.
☿.	{	347° 32′	1″·8±	1780·60
		11° 24′	...	1802·74
H. and S.		71° 33′	1″·46	1822·83
Σ.		89° 21′	1″·31	1827·02
D.		120° 37′	1″·30	1833·36

Having thus submitted the recorded statements respecting this binary object, to illustrate this beautiful phenomenon still further, I will add, my measures afford presumptive evidence that the components are again separating; and presuming its orbit to be elliptic, with an excentricity of 0·6988, it must occupy a period of not less than 560 years; with its motion performed in a plane passing nearly through the eye, at these direct annual values:

Herschel I.	.+1″·08	1802·74	Dawes . .	+4″·92	1833·36
H. II. and Sth.	3″·00	1822·83	Myself . .	5″·11	1835·50
Struve . . .	4″·27	1827·02	—— . .	3″·40	1839·67
Myself . .	4″·53	1832·37	—— . .	2″·93	1843·35

From the steadiness of the little star C, it may be concluded to have only an optical affinity with A and B; and ☿. says, " The great number of small stars in the neighbourhood, is not favourable to a supposed connection between any of them and σ Coronæ." Besides its orbital march, this star has a proper motion through space, to which the following values have been assigned:

$$\text{B.... } R - 0''\cdot35 \quad \text{Dec. } - 0''\cdot04$$
$$\text{A.... } - 0''\cdot37 \quad\quad\quad - 0''\cdot02$$

DLXVI. υ CORONÆ BOREALIS.

R 16ʰ 10ᵐ 20ˢ Prec. + 2ˢ·40
Dec. N 29° 33′·0 —— S 9″·23

Position AB 24°·2 (*w* 6) Distance 86″·9 (*w* 4) ⎫
—— AC 55°·2 (*w* 8) —— 128″·0 (*w* 5) ⎬ Epoch 1831·59
—— BD 221°·5 (*w* 2) —— 10″·0 (*w* 1) ⎭

A quadruple star, in the space between the Wreath and the back of Hercules; it is about $9^{\circ}\frac{1}{4}$ distance from Gemma on an east-by-north bearing, and in a line between a Serpentis and η Herculis. A 6, B 10,

C 9, and D 13, all bluish. This is 37 ɪɟ. v., and was registered in 1781. It was measured as triple by H. and S., No. 223; but B, which is the apex of the obtuse-angled triangle formed by the group, has a little companion overlooked by those astrometers, although duly notified by the eagle-eyed ɪɟ. The results of H. and S. were:

$$\begin{array}{llll} \text{A B} & \text{Pos. } 24° \ 27' & \text{Dist. } 83''\!\cdot\!69 \\ \text{A C} & \quad\ \ 54° \ 51' & \quad\ \ 126''\!\cdot\!42 \end{array} \right\} \ \text{Ep. } 1823\!\cdot\!36$$

DLXVII. 48 P. XVI. SCORPII.

ᴿ 16ʰ 11ᵐ 09ˢ   Pʀᴇᴄ. + 3ˢ·49

Dᴇᴄ. S 19° 43'·5   —— S 9"·17

Pᴏsɪᴛɪᴏɴ 21°·5 (w 3)   Dɪsᴛᴀɴᴄᴇ 13"·9 (w 3)   Eᴘᴏᴄʜ 1831·39

A neat double star, above the Scorpion's head. A 8, dull white; B 9, flushed; and there is a wide pair of stars in the *np* quadrant of similar magnitudes with these. This object is 124 ɪɟ. ɪᴠ., and Piazzi's Nos. 48 and 49 of Hora XVI.; and it lies just 4° distance to the eastward of β Scorpii. The micrometrical measures were:

$$\begin{array}{llll} \text{ɪɟ.} & \text{Pos. } 27° \ 06' & \text{Dist. } 15''\!\cdot\!40 & \text{Ep. } 1783\!\cdot\!22 \\ \text{H. and S.} & \quad\ \ 20° \ 31' & \quad\ \ 13''\!\cdot\!28 & \quad\ \ 1823\!\cdot\!45 \end{array}$$

From these results, a perceptible change in the angle, and a very sensible diminution of distance, were inferred; but a comparison of the latter with my observations, imply a relative fixity. Mr. Taylor, who observed them meridionally at Madras, assigns to them a slight proper motion in space, which is, however, unnoticed by other astronomers.

The other pair above alluded to are Nos. 44 and 45 P. Hora XVI.; they bear from this object on an angle of 124°, about 7' distant, lying in a position of 330°, and 50" apart.

DLXVIII. σ SCORPII.

ᴿ 16ʰ 11ᵐ 28ˢ   Pʀᴇᴄ. + 3ˢ·63

Dᴇᴄ. S 25° 12'·2   —— S 9"·14

Pᴏsɪᴛɪᴏɴ 271°·4 (w 6)   Dɪsᴛᴀɴᴄᴇ 20"·3 (w 3)   Eᴘᴏᴄʜ 1831·39

——— 271°·6 (w 8)   ——— 20"·5 (w 4)   —— 1838·32

A delicate double star, in the middle of Scorpio's body, and about 2° distant west-by-north from Antares. A 4, creamy white; B 9½, lilac tint. This object is 121 ɪɟ. ɪᴠ., and was entered in March, 1783, as in position 0, " or perhaps a single degree *np*;" and distance = 21"·667. H. and S. procured good measures, from both of which, compared with my own, I deduce that these stars are only optically connected:

Pos. 271° 11'   Dist. 20"·60   Ep. 1822·43

σ Scorpii precedes Antares, which is followed at nearly the same distance by τ, whence the Arabians termed σ and τ, *Al-niyât*, the præcordia, or out-works of the heart. They will be readily distinguished by taking up Antares. The whole of these have had considerable proper motions imputed to them, which are rapidly disappearing under able treatment; proximity to the horizon may have been to blame.

P....	Æ $- 0''\cdot 06$	Dec.	$- 0''\cdot 04$
Br...	$- 0''\cdot 10$		$- 0''\cdot 05$
B....	$+ 0''\cdot 03$		$0''\cdot 00$

DLXIX.   4 M. SCORPII.

Æ $16^h 13^m 51^s$ Prec. $+ 3^s\cdot 65$

Dec. S $26° 07'\cdot 5$ —— S $8''\cdot 95$

Mean Epoch of the Observation . . . . $1837\cdot 36$

A compressed mass of very small stars, in the middle of the creature's body, with outliers and a few small stellar companions in the field. The place is carefully differentiated with Antares; from which it is only $1°\frac{1}{2}$ distant to the west. This object is elongated vertically, and has the aspect of a large, pale, granulated nebula, running up to a blaze in the centre. It was discovered by Messier in the year 1764, and duly reported in the *Connaissance des Temps*. In 1783, Sir William Herschel resolved this object into stars; and gauging it by a modification of the method which he applied to fathom the Galaxy, he concluded that his 10-foot reflector, having a power to show stars exceeding that of the eye 28·67 times, gave the profundity of this cluster of the 344th order. He describes it as having a ridge of eight or ten pretty bright stars, running from the middle to the *nf;* a description which I found to be very correct.

Under the head of 80 Messier (which see, No. DLXIV.), a slight allusion was made to nebulæ considered in their relations to the surrounding spaces. Like that singular mass, the group before us is also situated on the western edge of an area which contains no stars, *i. e.*, none which we can descry; and in such spaces invariably, according to the testimony of Sir William Herschel, are nebulæ found. "Let us," says M. Arago, "connect these facts with the observation which has shown that the stars are greatly condensed towards the centre of spherical nebulæ, and with that which has afforded the proof that these stars sensibly obey a certain power of condensation (or clustering power), and we shall feel disposed to admit with Herschel, that nebulæ are sometimes formed by the incessant operation of a great number of ages, at the expense of the scattered stars (*étoiles dispersées*) which originally occupied the surrounding regions; and the existence of empty, or *ravaged* spaces, to use the picturesque expression of the great astronomer, will no longer present anything which ought to confound our imagination."

DLXX. γ HERCULIS.

ℛ 16ʰ 14ᵐ 53ˢ   Prec. + 2ˢ·64
Dec. N 19° 32'·0   — S 8"·87

Position 242°·3 (w 6)   Distance 38"·7 (w 3)   Epoch 1831·48

An open double star in a dark field, on the hero's left arm. A 3½, silvery white; B 10, lilac; and it points nearly upon a third star at a distance in the *sp* quadrant. This is 19 Ⱨ. v., and when discovered was thus registered:

Pos. 250° 30'   Dist. 41"·81   Ep. 1780·68

It was then examined by H. and S., with these results:

Pos. 243° 46'   Dist. 38"·32   Ep. 1821·85

Their apparent angular retrogression and diminution of distance here observable in a period of forty-one years, were not confirmed by my observations; and as the object is difficult under the micrometer, the discrepancies may be imputed to errors in the first observation.

γ Herculis is a portion of the *Nasak shámí*, or northern boundary wall of the Arabian garden, described by Ḳazwíní. It may be readily seen in mid-distance of a line produced between Rasalague and Gemma, passing also over Rasalgeti. An almost imperceptible movement in space is attributed to γ, of which the following are the most accurately investigated values:

P.... ℛ − 0"·04   Dec. + 0"·09
B....   − 0"·02   + 0"·05

DLXXI. ꝑ OPHIUCHI.

ℛ 16ʰ 16ᵐ 00ˢ   Prec. + 3ˢ·58
Dec. S 23° 04'·3   — S 8"·79

Position 3°·1 (w 6)   Distance 3"·8 (w 4)   Epoch 1832·38

A fine neat double star, on the Serpent-bearer's foot which rests upon Scorpio. A 5, pale topaz yellow; B 7½, blue; and they directly point upon No. 72 P. Hora XVI., about 2½ in the *nf*. There are two other companions in the field, the whole forming a pretty group, to the north-by-west of Antares, at 3° distance. This is 19 Ⱨ. ii., and it is registered with a couple of errors, which should be noticed, though one of them scarcely deserves it. Sir William Herschel says it is at the angular point of the three telescopic *g's*, making a rectangle on the body of Cancer, whereas he meant Scorpio; and he has placed the *comes* in the *sp* quadrant instead of the *nf*; but Sir John Herschel, on reference to his father's original observations, and the diagram made at the time, has found that it is indisputably an error of the press. By this correction, the former measures will be:

H̷.   Pos. 7° 50'   Dist. 4"·00±   Ep. 1780·34
H. and S.   2° 30'   4"·06   1822·46

On the whole, the angle may have undergone a slight change. Piazzi assigns a trifling proper motion to this star, but it has not been confirmed. In the Palermo Catalogue it is erroneously designated 5 g Scorpii; and it has inadvertently had g applied to it instead of ρ, in the British Catalogue, which letter was assigned to ξ, but Mr. Baily, finding that Bayer had mistaken *north* latitude for *south*, has respectively restored them. These are the conflicting values of the proper motions of ρ by Piazzi and Baily:

P.... Æ − 0"·12   Dec. − 0"·12
B....   + 0"·03   + 0"·02

DLXXII. ν^2 CORONÆ BOREALIS.

Æ 16ʰ 16ᵐ 27ˢ   Prec. + 2ˢ·26
Dec. N 34° 04'·8   ——— S 8"·75

Position 17°·9 (w 2)   Distance 137"·0 (w 1)   Epoch 1838·57

A star with a minute distant companion, on the following boundary of the asterism, where it bears north-east from Gemma, at 11° distance. A 5, pale yellow; B 12, garnet. This is 18 H̷. vi., and was first registered in July, 1780; but there are no measures of distance, and we are only told that the position is "about 80° *nf*." Outside the *comes*, A is the apex of a fine scalene triangle, the *np* member of which is ν^1, and its distance to ν^2 is rather more than 6' on an angle of 165°. The *nf* individual is the star alluded to in Piazzi's notes to Hora XVI., as having been observed at Toulouse by Darquier, and erroneously given by Bode. ν^2 has also caused a world of mistakes, for the Rev. F. Wollaston and Sir William Herschel were at issue about it; and even Flamsteed inserted it under a false position in the British Catalogue, owing, says Mr. Baily, to "a singular confusion of the figures," from which Miss Herschel rescued it.

DLXXIII.   23 HERCULIS.

Æ 16ʰ 16ᵐ 48ˢ   Prec. + 2"·29
Dec. N 32° 42'·6   ——— S 8"·72

Position 20°·1 (w 6)   Distance 36"·2 (w 4)   Epoch 1830·72

A double star in a dark field, on the boundary between Hercules and the Northern Crown; it is 1°½ distant from the preceding object, a little to the eastward of south. A 6, white; B 9, violet; a star at a distance in the *sp*, and another nearly following. This is 38 ♅. v., the " largest of a telescopic triangle;" it was first noticed in September, 1781, when 36″·48 was given as a " rather narrow distance," and no angle was registered. It was first measured by Sir James South, as follows, but is erroneously entered in his Catalogue as 88 ♅. v.:

Pos. 20° 22′ Dist. 36″·84 Ep. 1825·46

DLXXIV. α SCORPII.

Æ. 16ʰ 19ᵐ 36ˢ   Prec. + 3ˢ·66
Dec. S 26° 04′·3   —— S 8″·50

Position 129°·5 (*w* 1) Difference Æ = 33ˢ·1 (*w* 1) Epoch 1837·35

A standard Greenwich star, with a companion on the Scorpion's heart. A 1, fiery red; B 8, pale; and A is followed by a telescopic star on the parallel. A slight proper motion is assigned to this object, to the following values, which are so small that it may be needless to add they are not indicated by my results:

P.... Æ – 0″·05   Dec. – 0″·10
B....   + 0″·06   – 0″·03
A....   + 0″·01   – 0″·01

This star is the noted Antares, said to have been so called from Ἀντάρης, *i. e.* rivalling Mars in colour, a phrase first given in Ptolemy, and probably invented by very early astrologers: such is the generally received opinion, but Grotius shows, from an obscure note of Suidas, that Antares signifies a bat. " Utrum horum mavis accipe." It was named Καρδία Σκορπίου by the Greeks, *Cor Scorpionis* by the Latins, and *Kalb-al-'akrab*, also the Scorpion's heart, by the Arabians; which last is written *Calbalacrab* by the framers of the Alphonsine Tables. It forms the XVIIIth Lunar Mansion, and may be readily known by alignment. With Spica and Arcturus it forms a conspicuous triangle; and it may be picked up by running a long line from the Pole-star, by Rasalgeti, down to where Antares will be seen, with four stars forming an arc over him, in the shape of a paper kite. Should clouds in the polar regions obstruct this, shoot a ray from Regulus through Spica, and at 45° further to the eastward, it will pass below Antares; also, a line from Wega, carried by Rasalague, and extended as far beyond, will reach it:

Through Ras Alague, Wega's beams   direct th' inquiring eye,
Where Scorpio's heart, Antares, decks   the southern summer sky.

Σκορπίος, Scorpio, or Scorpius, the reputed slayer of the giant Orion, is one of the ancient 48 asterisms, being the 8th sign in order in the zodiac, and the 2nd of the southern signs; and it is seen, says Sherburne, to crawl towards our meridian at midnight, about the end of May. Cicero and Manilius apply the Latin word Nepa to this asterism; and little as its size and figure correspond with the Megatherium so ably described by my friend Dr. Buckland, it was called $M\acute{\epsilon}\gamma a \; \theta\eta\rho\acute{\iota}o\nu$ by Aratus. Hood, in his use of the *Celestial Globe in plano*, says, "Novidius, writing of this constellation, is verie childish in his conceite, and for ought that I can see, doth falsifie the word of God; for he sayth, that this was the Scorpion or Serpent, whereby Pharaoh, king of Ægypt, was enforced to let the children of Israel depart out of his countrey, whereas there is no such thing in the Scripture."

Scorpio is not large, but so brilliant as to have attracted much notice from the corps of astrologers, with whom it was "the accursed constellation," and the baleful source of war and discord; for, besides its being accompanied by tempests when setting, it was of the watery triplicity, and the stinging symbol of autumnal diseases, as it winds along with its receding tail. But though stigmatized as *signum falsitatis* by seers of every degree, the redoubtable Gadbury, at whose birth it ascended, broke many a lance in its defence, and stoutly contended for its beneficial influences; and the alchymists were well assured, that the transmutation of iron into gold could only be performed when the sun was in that sign.

From some confusion of the Greeks, this constellation formerly occupied two signs of their zodiac; and the pulling back of the chelæ to make room for the scales, by the Romans, is mentioned under a Libræ. The subject has been warmly contested among authors; meantime astronomers have successively numbered their components, thus:

Ptolemy	. . . 24 stars	Hevelius	. . . 20 stars
Tycho Brahé	. . 10	Halley	 29
Kepler	 27	Flamsteed	. . . 44
Bayer	 29	Bode	 200

DLXXV.   88 P. XVI. OPHIUCHI.

$\mathbb{R}$ 16^h 20^m 10^s PREC. $+ 3^s \cdot 23$
DEC. S 7° $45' \cdot 9$ —— S $8'' \cdot 46$

POSITION $305^\circ \cdot 0$ (w 2)   DISTANCE $5'' \cdot 0$ (w 1)   EPOCH $1833 \cdot 47$

A very delicate double star, close upon the Serpent-bearer's right thigh; it is 25° distant from a Herculis, on a line south-south-west towards β Scorpii, where it is also pointed out by a line through δ and ϵ, the two stars of the 3rd magnitude in the hand of Ophiuchus. A $7\frac{1}{2}$, yellow; B 12, dusky; other stars in the field, particularly one of a deep orange tinge in the np. Nor ought I to omit mentioning that 6 Hevelii, or v, precedes it to the southward by about a minute of time, since it is of the 5th magnitude, and forms a sort of pointer to the delicate pair, on an angle of 133°. This is a most difficult object, though Σ., its discoverer, gives the *comes* a 9th magnitude only, whereas, by comparison, I cannot rate it higher than I have. His measures are:

Pos. $302^\circ \cdot 73$ Dist. $4'' \cdot 687$ Ep. $1831 \cdot 48$

DLXXVI. η DRACONIS.

Æ 16ʰ 21ᵐ 48ˢ   PREC. + 0ˢ·79
DEC. N 61° 52'·4   — S 8"·33

POSITION 31°·0 (ᵥ ₁)   DISTANCE 190"·0 (ᵥ ₁)   EPOCH 1833·62

A second rate Greenwich star, with a companion, in the middle of the Dragon's body. A 3, deep yellow; B 11, pale grey. The Arabians designated η and ζ *Al-dhibaïn*, the two jackals, and the adjacent dim or *dark* pair, ω and *f*, *Adhfár al-dhib*, the jackal's claws; but the glossator of Ulugh Beigh cites them as *Al-'auhakán*, black cattle, or crows. They will be at once recognised from being nearly in mid-distance from *Al-'awáïd* towards *Alferkadeïn*, or between the head of the Dragon and the two bright stars of the Little Bear's fore-body, called the Guards. Though this object is in a barren field, the neighbourhood to the *sp* offers some brilliant groups of small stars; and directly north of it is the fine double star of Σ.'s First Class "vicinæ," No. 2054 of the great Dorpat Catalogue. It is a capital telescope-test.

Since this was written, Professor Struve has applied the gigantic Poulkova refractor to this object, and has detected a minute *comes* to A, at about 4" distance. I have had no opportunity of looking for it since the intelligence reached me.

DLXXVII. λ OPHIUCHI.

Æ 16ʰ 22ᵐ 51ˢ   PREC. + 3ˢ·02
DEC. N 2° 20'·4   — S 8"·25

POSITION 351°·2 (ᵥ ₃)   DISTANCE 1"·0 (ᵥ ₁)   EPOCH 1834·48
———— 352°·9 (ᵥ ₅)   ———— 1"·1 (ᵥ ₂)   —— 1836·51
———— 356°·5 (ᵥ ₅)   ———— 1"·0 (ᵤ ₂)   —— 1839·67
———— 1°·4 (ᵥ ₈)   ———— 1"·1 (ᵥ ₃)   —— 1842·50

A fine binary star, in the bend of the Serpent-bearer's right arm. A 4, yellowish white; B 6, smalt blue. This physical object is 83 ℍ. ɪ., and is truly, as its discoverer said, very beautiful and close, for the measures are attended with great difficulty except when, as H. expresses it, "the star will be quiet." The measures of comparison are:

ℍ. Pos. 75° 30' Dist. 0"·50 Ep. 1783·18
Σ.   331° 48'   0"·84   1825·51
D.   342° 20'   *wedged*   1831·33

Hence it is seen, that this star has a great progressive angular motion; but my observations are not indicative of the acceleration which has been spoken of by other astrometers. By adding 180° to ℍ.'s angle, the mean rate will be about + 1°·70 per annum. From the shown course and velocity, it is evidently making an elliptical and rapid orbit, of which the *annus magnus* may be between eighty and ninety years. The inconsiderable amount of proper motion is, as yet, too variously stated to render it admissible into the discussion, being:

$$P.... \text{ÆR} - 0''{\cdot}06 \quad \text{Dec.} - 0''{\cdot}06$$
$$Br... \quad - 0''{\cdot}05 \quad \quad + 0''{\cdot}02$$
$$B.... \quad + 0''{\cdot}09 \quad \quad - 0''{\cdot}06$$

The Arabians designated λ Ophiuchi *Marfik*, meaning the elbow; and it may be found south-west of its own *a*, distant about 18°, rather more than half-way towards μ Serpentis.

DLXXVIII.   β HERCULIS.

ÆR 16^h 23^m 21^s Prec. $+$ 2^s·58
Dec. N 21° 50'·6   —— S 8''·12

Position 276°·0 (*w 2*)   Distance 278''·0 (*w 1*)   Epoch 1835·66

A fine bright star in a barren field, on the hero's left shoulder; forming the *nf* terminus of the *Nasak shámi*, or northern boundary wall of the Arabian pasturage, mentioned under *a* Serpentis. A 2½, pale yellow; B 11, lilac tint. The proper motions in space are thus registered:

$$P.... \text{ÆR} - 0''{\cdot}22 \quad \text{Dec.} - 0''{\cdot}04$$
$$B.... \quad - 0''{\cdot}09 \quad \quad + 0''{\cdot}01$$
$$A.... \quad - 0''{\cdot}12 \quad \quad - 0''{\cdot}01$$

This star appears on the Catalogues as Korneforos, a corruption of the Greek κορυνηφόρος, *i. e.* Claviger, one of the numerous epithets of Hercules. It may be found nearly half-way between Rasalgeti and Gemma, where it is sufficiently insulated to render it pretty conspicuous; or, agreeably to the brackish rhymes:

Where in the heav'ns they strangely paint	Alcides on his knees,
His *beta* should you wish to find,	adopt such rules as these :
Bright Gemma and Rasalgeti	mark *beta* in mid-way,
And far to the north of Scorpio's heart	does Kornefóros stay.

DLXXIX.   40 ℍ. VI. OPHIUCHI.

ÆR 16^h 23^m 35^s Prec. $+$ 3^s·34
Dec. S 12° 41'·1   —— S 8''·19

Mean Epoch of the Observation . . . . 1837·38

A large but pale *granulated* cluster of small stars, on the Serpent-bearer's right leg. There are five telescopic stars around it, so placed as to form a crucifix, when the cluster is high in the field; but the region immediately beyond is a comparative desert. After long gazing, this object becomes more compressed in the centre, and perplexes the mind by so wonderful an aggregation. It was discovered by ꜫ. in May, 1793, and was registered 5' or 6' in diameter. The mean place was obtained by differentiation with ζ Ophiuchi, from which it is distant 3° to the south-south-west, in the line between β Scorpii and β Ophiuchi.

DLXXX. 136 P. XVI. HERCULIS.

Ⱥ 16^h 30^m 26^s Prec. + 2^s·76
Dec. N 14° 01'·0   —— S 7"·64

Position 310°·0 (*w* 3)   Distance 25"·0 (*w* 1)   Epoch 1834·59

A delicate double star in a barren field, in the space between the left arm and the head of Hercules; and 9° distant from the latter, on the western parallel. A 7, pale yellow; B 12, bluish,—this star bearing no illumination, the position was obtained by the spherical crystal-micrometer, and the distance carefully estimated. The object was first classed by Σ. as one of his Third Class, No. 2071; and in the following Catalogue of 1837, his measures were:

Pos. 311°·60 Dist. 25"·117 Ep. 1830·14

DLXXXI. 17 DRACONIS.

Ⱥ 16^h 32^m 28^s Prec. + 1^s·41
Dec. N 53° 14'·9   —— S 7"·48

Position AB 115°·7 (*w* 9) Distance 3"·8 (*w* 8)⎫
——— AC 194°·6 (*w* 9) ——— 90"·5 (*w* 9)⎭ Epoch 1832·87

A triple star, preceding the Dragon's jaws, where a line from γ Draconis carried over β, and prolonged twice their distance, reaches it. A 6, pale yellow; B 6½, faint lilac; C 6, white; four other small stars in view. This object is 4 ꜫ. i., where A is designated 16 Draconis, which is the Flamsteed number of C; but there was no small confusion respecting them in the British Catalogue, as may be seen in the Notes 2301 and 2302 of Mr. Baily's edition. M. Argelander, who took some trouble with their identity, assigns 17 Draconis a slight motion in space to the following value:

Ⱥ − 0"·10   Dec. + 0"·02

The close star has been repeatedly measured, and the great coincidence of all the results, sufficiently stamps its fixity as an optical object. These are the results for comparison:

	Pos.	Dist.	Ep.
ꜫ.	114° 00'	4"·00 ±	1781·88
H. and S.	115° 26'	4"·51	1823·33
H.	115° 56'	3"·38	1828·64
Σ.	116° 27'	3"·74	1831·91

DLXXXII.   37 HERCULIS.

Æ. 16ʰ 32ᵐ 42ˢ   Prec. + 2ˢ·97
Dec. N 4° 32'·2   —— S 7"·45

Position 229°·9 (w 9)   Distance 69"·2 (w 9)   Epoch 1835·48

A pair of stars in a barren field, which, though pertaining to
Hercules, are placed upon the right arm of Serpentarius, due west of
his β, and 16° distant from it. A 6½, pale blue; B 7½, blue, and just
preceding this star is a minute *comes*, incapable of bearing the smallest
illumination. This object is 72 Ⅱ. v., and Nos. 147 and 149 of Piazzi,
Hora XVI.; and we are told in his notes, that Herschel suspected some
change in these stars, " but neither from Bradley's observations, nor from
ours at the distance of ten years from one another, is anything like it
to be obtained." This conclusion, derived from the reduction of meridian observations, is fully confirmed by the micrometrical measures,
especially since the first distance is corrected by H., from his father's
manuscripts; deductions from Piazzi are added:

Ⅱ.   Pos. 233° 03'   Dist. 67"·77   Ep. 1782·38
P.   230° 30'   68"·30   1800·00
H. and S.   230° 23'   68"·83   1821·39

The direction of the spacial movement of star A is not yet settled,
and in giving the following values, it may be stated that Piazzi assigns
nearly the same amount to B:

P.... Æ − 0"·10   Dec. + 0"·03
B....   + 0"·08   − 0"·03

DLXXXIII.   42 HERCULIS.

Æ. 16ʰ 34ᵐ 24ˢ   Prec. + 1ˢ·63
Dec. N 49° 14'·6   —— S 7"·32

Position 93°·5 (w 3)   Distance 20"·0 (w 1)   Epoch 1835·57

A very delicate triple star between the left knee of Hercules and
Draco's head, of which only the two forming 63 Ⅱ. iv. are here measured. A 6, orange; B 12, blue; the third star, which is still more
minute, makes a neat triangle of the object, in a rich field; and it may
be found to the west-south-west of β Draconis, at 6°¼ distance: the
small components are caught by averting the eye to other parts of the
field of view. These are preceded by a 7th-magnitude star on an angle
of 211° and 4' 20" distant. A comparison of the above results with
the former measures, indicates fixity:

Ⅱ.   Pos. 93° 42'   Dist. 21"·52   Ep. 1782·61
Σ.   92° 21'   22"·39   1828·43

DLXXXIV. ζ HERCULIS.

Æ 16ʰ 35ᵐ 15ˢ   Prec. + 2ˢ·30
Dec. N 31° 53'·7   —— S 7"·26

Position	Distance	Epoch
190°·0 (w 1)	0"·5 (w 1)	1835·68
176°·3 (w 1)	0"·7 (w 1)	1836·73
169°·0 (w 4)	1"·2 (w 1)	1838·65
136°·9 (w 6)	1"·2 (w 2)	1842·57

A close binary star, over the left hip of Hercules, where with ε, its companion in magnitude, it is rather conspicuous between Wega and Gemma. A 3, yellowish white; B 6, orange tint; a third star of the 9th magnitude in the *nf* quadrant, with Δ Æ 23ˢ·6. This wondrous object is 36 ♓. I., and as everything connected with its movements is most important, I may cite the following values for its proper motion in space:

	Æ	Dec.
P....	− 0"·70	+ 0"·47
B....	− 0"·45	+ 0"·43
A....	− 0"·47	+ 0"·38

The duplicity of ζ Herculis was first detected under the eagle-gaze of Sir William Herschel, in July, 1782. In October, 1795, he again beheld the *comes*, but it afterwards disappeared, or, under the most delicate treatment, was only wedged; and the able astronomer remarked, "My observations of this star furnish us with a phenomenon which is new in astronomy; it is the occultation of one star by another. This epoch, whatever be the cause of it, will be equally remarkable, whether owing to solar parallax, proper motion, or motion in an orbit whose plane is nearly coincident with the visual ray."

In this state it remained unobserved for some years; and during 1821, 1822, 1823, and 1825, baffled all the endeavours of H. and S. to divide, or even elongate. At length Σ. caught it double in 1826, though it again became single in two years, and remained so to the Dorpat instrument till 1832, when that persevering observer again measured it. It has since become comparatively of easy vision, for when my friend, the Rev. W. R. Dawes, sent me his results, he informed me that, in 1840, he saw both the stars yellow; and that with a magnifying power of 400, they were readily separable with his 5-foot telescope. The several epochs of comparison with my results, by which the *np sf* direction, or retrograde ellipticity of movement, is proved, may be thus given:

		Ep.
♓.	{Pos. 69° 18' Dist. 1"·00 ±	1782·55
	{The star single	1802
Σ.	{Pos. 23° 24' Dist. 0"·91	1826·63
	{The star single	1828·77
	{Pos. 222° 30' Dist. 0"·81	1832·75

In addition to these published observations, Mr. Dawes has obligingly furnished me with the results obtained by him during my residence at Cardiff, which afford beautiful corroborations of the indicated movement; and a momentous hiatus in the new series of angles is filled up:

Pos. 161° 55' Dist. 1"·221 Ep. 1839·758
150° 40'   1"·230   1840·655
142° 58'   1"·239   1841·651

On the erection of the large telescope at Bedford, this was one of the first objects of my attention; but from Professor Struve's description,— " Magnitudinis constanter notari (3) et (7), colorem majoris album sub-flavum, minoris subrubrum. Difficultas in stella hac duplici videnda ex splendore oritur majoris,"—I did not expect to notch it. Following ꜧ.'s method, I first got my eye and instrument into order by scruti-nizing η Coronæ; and then turning upon ζ Herculis, felt confident that I saw, and that readily, a red spot on its disc, which, from the above-quoted words, I took for the *comes*. It may, however, have been a spurious image or colour; for on the following apparition of Hercules, in 1831, wishing to show the same phenomenon to Captain Kater and Mr. Maclear, I could no longer receive the same impression. But notwith-standing this disappointment, I watched it occasionally under various powers; and, at length, in 1835, became satisfied that it was positively elongated towards the vertical, and at times could trace the deeper-coloured point of the wedge upwards, or towards the south, in an inverted field of view. But all attempts to notch it failed till the summer of 1838, when, though still deserving of Σ.'s epithet " vicinis-simæ," the distance had palpably increased, and the stars were occa-sionally fairly divorced. As the question was of deep interest, unusual pains were taken with the measures, although, from the difficulty of observation, they could not be stamped with high weights. Indeed, all the evils of a double-wire micrometer had to be encountered; and what with the necessity of using great magnifying powers upon an object, the components of which were so close as occasionally to flow into each other, there were the threads hanging, dragging, and fiddling, proving by jerks, imperceptible under most objects, the inaccuracies of the screw. And yet my micrometer is among the very best of the age; and my telescope was smoothly carried by the equatoreal clock, which had been put into excellent gear. At the last epoch, the stars had widened, and were much less fatiguing to observe; the weights are therefore given with a greater degree of confidence than heretofore.

This wonderful object ought to be narrowly looked after by all the new giant telescopes, since it offers, according to M. Struve's conclu-sions, the astounding velocity of an apparent and very elliptical orbit revolving in little more than fourteen years! What a motion! Bacon little knew the force of his own expression, when he exclaimed, " Hea-venly bodies have much veneration, but no rest." My own views, however, do not quite square with this velocity, although they acknow-ledge one about as astonishing: a scrutiny of the observations leads me to suppose an orbit with an excentricity of 0·4186, and a period of about thirty-five years. This, of course, concludes rather more than one entire revolution to have occurred, between ꜧ.'s epoch of 1782 and Σ.'s result in 1826.

DLXXXV. 13 M. HERCULIS.

Æ 16ʰ 35ᵐ 58ˢ   Prec. + 2ˢ·14
Dec. N 36° 45'·8   —— S 7"·19

Mean Epoch of the Observation . . . . 1836·62

A large cluster, or rather ball of stars, on the left buttock of Hercules, between ζ and η; the place of which is differentiated from η Herculis, from which it lies south, a little westerly, and 3°¼ distant.

This superb object blazes up in the centre, and has numerous outliers around its attenuated disc. It was accidentally hit upon by Halley, who says, "This is but a little patch, but it shows itself to the naked eye, when the sky is serene, and the moon absent." The same paper, in describing this as the sixth and last of the nebulæ known in 1716, wisely admits that "there are undoubtedly more of these which have not yet come to our knowledge:" ere half a century had passed, Messier contributed his 80 or 90 in the Catalogue of 103; and before the close of that century Ħ. alone had added to the above 6, no fewer than 2500; and his son, in re-examining these, added 520 more! In my own refractor its appearance was something like the annexed diagram; but I agree with Dr. Nichol, that no *plate* can give a fitting representation of this magnificent cluster. It is indeed truly glorious, and enlarges on the eye by studious gazing. "Perhaps," adds the Doctor, "no one ever saw it for the first time through a telescope, without uttering a shout of wonder.

This brilliant cluster was discovered by Halley in 1714; and fifty years afterwards it was examined by M. Messier, with his 4-foot Newtonian, under a power of 60, and described as round, beautiful, and brilliant; but, "ferret" as he was in these matters, he adds, "Je me suis assuré qu'elle ne contient aucune étoile." This is rather startling, since the slightest optical aid enables the eye to resolve it into an extensive and magnificent mass of stars, with the most compressed part densely compacted and wedged together under unknown laws of aggregation. In 1787, Sir William Herschel pronounced it "a most beautiful cluster of stars, exceedingly compressed in the middle, and very rich." It has been recently viewed in the Earl of Rosse's new and powerful telescope, when the components were more distinctly separated, and brighter, than had been anticipated; and there were singular fringed appendages to the globular figure, branching out into the surrounding space, so as to form distinct marks among the general outliers.

DLXXXVI.   η HERCULIS.

$\mathbb{R}$ 16ʰ 37ᵐ 25ˢ   Prec. + 2ˢ·05

Dec. N 39° 13'·8   —— S 7"·07

Position AB *round* (ω 9)   Distance *round* (ω 9)⎱
—— AC 265°·0 (ω 2)   —— 141"·0 (ω 1)⎰   Epoch 1835·65
—— AB 150°·0 (ω 1)   —— 0"·3 (ω 1)   —— 1842·53

A bright star with a distant companion, on the left thigh of Hercules, and nearly in a line with the last two objects. A 3, pale yellow; B, only inferred; C 10, dusky. The proper motion in space of the principal is thus valued:

P.... $\mathbb{R}$ − 0"·24   Dec. − 0"·09
B....   + 0"·08   − 0"·07
A....   + 0"·07   − 0"·07

Here A forms No. 2093 of the Dorpat Catalogue, and was described in 1827 as a first-class "vicinissimæ," like ζ Herculis and ν Coronæ; and its components were registered of the 4th and 8th magnitudes. Many were the efforts I made at distinguishing a proximate *comes*, but without effect; and when Σ.'s measures arrived in 1837, finding it was styled "simplex," I relinquished the attack. Having, however, heard of its subsequent elongation, I re-examined it at Hartwell, and think I may safely give the above details, as tolerable estimations of an egg-shaped object.

This star is sufficiently conspicuous to the north-east of Gemma, at about 16° distance; and it is also 19° from Wega, on its western parallel; it forms an equilateral triangle with its own ζ and π.

DLXXXVII.   Σ. 5 N. HERCULIS.

$\mathbb{R}$ 16ʰ 37ᵐ 46ˢ   Prec. + 2ˢ·51

Dec. N 24° 05'·8   —— S 7"·04

Mean Epoch of the Observation . . . . 1836·61

A small planetary nebula, between the hero's shoulders, which H. aptly compares to a star out of focus. There are four stars in the field, of which that in the *sf* quadrant is of the 6th magnitude and brightly reddish, affording a fair test of comparison with the pale blue nebula. This curious object is Professor Struve's fifth nebula in the list at the end of the Dorpat Catalogue for 1827; and as it presents a visible disc of 8" in diameter, at a distance probably equal to that of the star near it, the vastness of its dimensions is within the range of reasonable conjecture, however it may stagger the comprehension.

This nebula is situated at about one-third the distance on an east-

south-east line from Gemma to Altair, and is to the north-east of
γ and β, in the left arm of Hercules, at a distance similar to that
between those two stars.

DLXXXVIII. 43 HERCULIS.

Æ 16ʰ 38ᵐ 09ˢ   Prec. + 2ˢ·87
Dec. N 8° 52′·8   —— S 7″·01

Position 230°·7 (w 8)   Distance 79″·5 (w 5)   Epoch 1832·59

A wide pair of stars, in the asterism of Hercules, but on the shoulder
of Serpentarius; it is south-west of Rasalgeti, at 8°½ distance. A 5,
rose tint; B 9, light blue; a third and very minute glimpse-star in the
nf quadrant. This object is 116 ♅. vi., and not 41 ♅. iii.   Under
the latter head I should have concluded Sir William Herschel to have
meant A, and the point of light in the *nf*, but that he has registered the
components " both equal;" from which, and the result of the measures,
it seems evident that 100 Herculis was the star observed in 1781, by
Mr. Bryant, of Bath. The mistake occasioned this object to be placed
on the working lists, and secured its being well attended to; and the
following are the previous results:

♅.	Pos. 231° 12′	Dist. 74″·62	Ep. 1783·44
Σ.	230° 18′	83″·70	1819·63
H. and S.	230° 51′	80″·09	1821·42

DLXXXIX. 46 HERCULIS.

Æ 16ʰ 38ᵐ 43ˢ   Prec. + 2ˢ·38
Dec. N 28° 39′·3   —— S 6″·96

Position 163°·8 (w 9)   Distance 5″·1 (w 5)   Epoch 1834·50

A neat double star, on the hero's back, and 7° distant north-by-
east from β. A 7½, pale white; B 10, sky blue. This fine object is
79 ♅. i., and from its class would seem to have been closer at its
discovery than it is now; whilst the interval between the discs, as
estimated by ♅., cannot be assumed at more than 2″·5. When S.
measured it, an inference was drawn that the distance had increased
materially; and there were symptoms of an angular movement also.
The great coincidence, however, between S. Σ. and myself, invalidates
the conjecture, and stamps the fixity of both stars; these are the results
for comparison:

♅.	Pos. 156° 36′	Dist. 2″·50 ±	Ep. 1783·10
S.	163° 51′	5″·39	1825·05
Σ.	163° 56′	4″·06	1830·57

The investigations for the proper motions of this star, assign an almost imperceptible amount to the Æ; but they are unanimous in giving it none in declination. The next rigid series of meridional observations will perhaps clear it off.

DXC.   12 M. OPHIUCHI.

Æ   16ʰ 38ᵐ 56ˢ   Prec. + 3ˢ·11
Dec. S   1° 40'·3   —— S 6"·94

Mean Epoch of the Observation . . . . 1837·61

A fine rich globular cluster, between the right hip and the elbow of Ophiuchus, with a *cortège* of bright stars, and many minute straggling outliers. This resolvable mass is greatly condensed towards the centre, with several very bright spots: it was discovered by Messier in 1764, but, probably from the imperfection of his means, was registered as "a round nebula, unaccompanied by any star." Its place was obtained by differentiating with ε Ophiuchi, from which it is 8°¼ distant, on a north-west-by-west line, leading nearly upon β.

Sir William Herschel resolved this object in 1783; and in the following year his 20-foot reflector made it "a brilliant cluster, 7' or 8' in diameter; the most compressed parts about 2'. By the gauging process, he held its profundity to be of the 186th order.

DXCI.   19 OPHIUCHI.

Æ   16ʰ 39ᵐ 06ˢ   Prec. + 3ˢ·02
Dec. N   2° 21'·5   —— S 6"·93

Position 92°·9 (w 5)   Distance 21"·8 (w 2)   Epoch 1834·36

A delicate double star, under the Serpent-bearer's right axilla, 15° distant from α Ophiuchi, on a south-west line towards the following of the two bright stars in the hand; 19 being nearly in mid-distance. A 6½, pale white; B 10, livid. This difficult object is 123 H. IV., and marked "the most south of the two;" and Piazzi's note, No. 180, Hora XVI., says, "Binæ præcedunt, et una sequitur," but from his further details, neither of these is B. Indeed, there appears to be no little confusion in the several Catalogues, as to the identity of the several members of the little group on this spot, of which 19 is the largest. These are the measures prior to mine:

H.   Pos. 93° 09'   Dist. 20"·47   Ep. 1783·19
Σ.   92° 35'   22"·25   1832·14

DXCII.   50 ♅. IV. HERCULIS.

Æ 16^h 42^m 23^s Prec. + 1^s·68
Dec. N 47° 49'·0   —— S 6"·66

Mean Epoch of the Observation . . . . 1835·33

A fine planetary nebula, between the left heel and the right knee of Hercules; the mean apparent place of which was carefully differentiated from τ Herculis; from which it bears east-by-north, 4° distant, and from γ Draconis west-south-west, about 10°. This object, which was discovered in May, 1787, is large, round, and of a lucid pale-blue hue; but its definition and distinctness are encroached upon by the brilliance of the 6th-magnitude stars near it; one of which precedes the nebula by 22^s, nearly on the parallel. It was offered as a "prize comet" to Maria Louisa of Lucca, in 1819; and the *dénouement* made by Baron de Zach, led to the appointment of M. Pons as her Majesty's *astroscoper* in the new observatory at La Marlia, where he was to receive 100 dollars for every comet that might be discovered. But the establishment, though commenced under considerable pomp and circumstance, only lingered about four years, and was then formally abolished. This is one of the mortifying instances, wherein the poverty of a queen's treasury prevented her manifesting that reverence for science, which she really felt.

DXCIII.   236 P. XVI. SCORPII.

Æ 16^h 47^m 40^s Prec. + 3^s·51
Dec. S 19° 16'·8   —— S 6"·22

Position 230°·6 (w 2)   Distance 5"·8 (w 2)   Epoch 1833·38

A neat double star, over the Scorpion's back, but in an absurd demesnal *hook* between the legs of Ophiuchus; it is 10° distant from Antares, to the north-north-east, which line leads upon η Ophiuchi, at about 5° further. A 6½, yellowish white; B 8, pale green,—another small star follows in the *sf* by about 9^s, and there is a telescopic coarse pair on the north vertical. This object was thus pointed out by Piazzi: "Duplex. Comes vix ferê visibilis, proxime praecedit paulisper ad austrum." It then became No. 240 of H. and S., who gathered these results:

Pos. 227° 16'   Dist. 5"·64   Ep. 1823·44

There is considerable coincidence in the proper motions detected in A, and there can be little doubt, if the change of angle be correct, that B also partakes in them:

P.... Æ − 0"·07   Dec. − 0"·02
B....   − 0"·09   − 0"·03

DXCIV.   56 HERCULIS.

ℛ 16ʰ 48ᵐ 29ˢ   Prec. + 2ˢ·45
Dec. N 25° 59′·6   —— S 6″·15

Position AB 96°·1 (*w 3*) Distance 15″·0 (*w 3*)⎫
—————— AC 170°·0 (*w 2*)   ——————— 540″·0 (*w 1*)⎬ Epoch 1838·71

A most delicate double star, with several companions, forming a small group inclining from *np* to *sf*, on the hero's right shoulder, at about 6° distance east-north-east of β Herculis, on a line leading upon Wega. A 6½, light yellow; B 13, pale red; C 11, greenish,—and between A and C are three minute stars nearly on the parallel with each other. This group forms so severe a test for a telescope, that I requested the Rev. James Challis to examine it with the great Northumberland equatoreal, in the autumn of 1841; and at the same time asked the Rev. W. R. Dawes to do the same with Mr. Bishop's large refractor. It was a gratification to find that their diagrams were in very nice agreement with my own, showing the utility of careful estimation in those cases where the delicacy defies metrical observations. It appears pretty certain that this star has a slight movement in ℛ; but Mr. Baily has extinguished that in declination. These are the registered values:

P.... ℛ - 0″·26   Dec. - 0″·05
B....   - 0″·08   0″·00

DXCV.   10 M. OPHIUCHI.

ℛ 16ʰ 48ᵐ 45ˢ   Prec. + 3ˢ·15
Dec. S 3° 51′·8   —— S 6″·13

Mean Epoch of the Observation . . . . 1835·54

A rich globular cluster of compressed stars, on the Serpent-bearer's right hip. This noble phenomenon is of a lucid white tint, somewhat attenuated at the margin, and clustering to a blaze in the centre. It is so easily resolvable with very moderate means, that we are surprised at Messier's remark, on registering it in 1764: "A beautiful round nebula. It may be seen, with attention, by a telescope three feet in length." The mean apparent place of the central mass, was differentiated with ε Ophiuchi, which it follows nearly on the eastern parallel, at about 8° distance; being nearly midway between β Libræ and α Aquilæ, and about a degree preceding 30 Ophiuchi, a star of the 6th magnitude, with a smaller one preceding it. Sir William Herschel resolved this object; in 1784 he applied his 20-foot reflector, and made it a beautiful cluster of extremely compressed stars, resembling Messier's No. 53. He estimated its profundity to be of the 243rd order.

DXCVI. 62 M. SCORPII.

Æ 16ʰ 51ᵐ 04ˢ   Prec. + 3ˢ·80
Dec. S 29° 50'·6   —— S 5"·94

Mean Epoch of the Observation .... 1837·46

A fine large resolvable nebula, at the root of the creature's tail, and in the preceding part of the Galaxy. It is an aggregated mass of small stars running up to a blaze in the centre, which renders the differentiating comparatively easy and satisfactory; and in this instance it was referred to its neighbour, 26 Ophiuchi, which is 5° distant to the north: and it lies only about 7° from Antares, on the south-east. This was registered in 1779, and Messier described it as "a very pretty nebula, resembling a little comet, the centre bright, and surrounded by a faint light." Sir William Herschel, who first resolved it, pronounced it a miniature of Messier's No. 3, and adds, "By the 20-feet telescope, which at the time of these observations was of the Newtonian construction, the profundity of this cluster is of the 734th order." To my annoyance, it was started as a comet a few years ago, by a gentleman who ought to have known better.

DXCVII. 19 M. OPHIUCHI.

Æ 16ʰ 52ᵐ 44ˢ   Prec. + 3ˢ·69
Dec. S 26° 02'·2   —— S 5"·82

Mean Epoch of the Observation .... 1837·46

A fine insulated globular cluster, of small and very compressed stars, between the Scorpion's back and the left foot of Ophiuchus; and nearly midway between two telescopic stars, in the preceding branch of the Via Lactea. It is of a creamy white tinge, and is slightly lustrous in the centre; but H. tells us, that even in the 20-foot reflector it did not brighten to a blaze, or to a confusion of the stars with one another. It was discovered by M. in 1764, and described as a nebula without stars, of a round form, and seen well with a 3½-foot telescope; but in 1784, Sir William Herschel resolved it, and pronounced its profundity to be of the 344th order. The mean apparent place is obtained by differentiation with 36 Ophiuchi, from which it is 2°¼ distant on a line west-by-north; and it is 7°¼ due east from Antares.

The above nebulæ, and the whole vicinity, afford a grand conception of the grandeur and richness even of the exterior creation; and indicate the beauteous gradation and variety of the heaven of heavens. Truly has it been said, "Stars teach as well as shine." This is near the large opening or hole, about 4° broad, in the Scorpion's body, which ℍ. found almost destitute of stars.

DXCVIII.   270 P. XVI. OPHIUCHI.

Æ 16^h 54^m 18^s Prec. + 2^s·87
Dec. N 8° 41'·3   —— S 5"·67

Position 137°·0 (*w* 4)   Distance 1"·5 (*w* 2)   Epoch 1832·41

A close double star, on the Serpent-bearer's right shoulder; it is to the west-south-west of *a* Ophiuchi, at 9° distance, and is the *sf* of three stars, viz. 270, *κ*, and *ι*. A 7, and B 8, both white. This was discovered by *Σ*., and entered among his "vicinæ," and was recommended to me by H., as an object more difficult to measure than *σ* Coronæ Borealis, which it greatly resembles. He kindly forwarded me his results; and when the Dorpat Catalogue arrived at the close of 1837, we gained another comparison. The previous epochs, therefore, are:

Σ. Pos. 135° 40'   Dist. 1"·34   Ep. 1830·97
H.   134° 03'   0"·84   1832·41

Notwithstanding the flattering regular progression of angle and distance here exhibited, it is too delicate a test for speculation yet awhile.

DXCIX.   20 DRACONIS.

Æ 16^h 55^m 38^s Prec. + 0^s·28
Dec. N 65° 17'·0   —— S 5"·56

Position 245°·0 (*w* 4)   Distance 0"·8 (*w* 1)   Epoch 1832·41
—— 243°·7 (*w* 6)   —— 0"·7 (*w* 2)   —— 1839·72

A close double star, in the middle of the creature's body; it is nearly in mid-distance between *β* Ursæ Minoris and *γ* Draconis, where it precedes the bright star *ζ*. A 7, and B 7½, both white. This very difficult object is 19 ᚻ. I., and the above star, 270 P. XVI., was a good preparative for attacking this. Nothing exhibits the comparative ease of observing at present, with the labour and exposure undergone by the zealous discoverer, more than his directions for this star; whose system, though so much smaller than that of *η* Coronæ, can be managed well enough under my rotatory roof. After stating that it is one of the most minute of all the double stars he had found, and too small for any micrometer in his possession, ᚻ. adds, " It is in vain to look for them if every circumstance is not favourable. The observer as well as the instrument must have been long enough out in the open air to acquire the same temperature. In very cold weather, an hour at least will be required; but in a moderate temperature, half an hour will be sufficient."

While quoting from ᚻ., a slight inadvertence in the entry may be noticed; the star is designated "*h* Draconis, near F. l. 19,"—but it should have been "near *h* Draconis, F. l. 19." And there is another point which demands attention: he says the components of this object were "considerably unequal," whereas they are now so nearly of the same magnitude, that it was only after much comparison that I felt inclined to place the *comes* in the *sp* quadrant. Can it be variable?

The observations of this refractory star, still leave us in doubt as to its being in motion or at rest—a physical or an optical object. "Aut itaque nil in his stellis mutatum est inde ex 1781, aut motus perficitur in plano valde ut videtur inclinato, et in 55 annis tota revolutio est absoluta," concludes Σ. Indeed, it was seeing the apparent increase of angle in the Dorpat Catalogue, which induced me to take extraordinary pains to confirm it, and form an epoch, previous to the dismantling of my observatory; but the angle proved smaller than that which I had obtained seven years before. The following is a list of the standard results for comparison:

Ḥ.	Pos. 243° 00'	Dist. 0″·50±	Ep. 1781·70
H.	242° 35'	0″·63	1830·32
Σ. {	246° 23'	0″·85	1832·30
	247° 01'	0″·71	1836·75

DC. 60 HERCULIS.

Æ 16ʰ 57ᵐ 58ˢ   Prec. + 2ˢ·77

Dec. N 12° 58'·0   —— S 5″·36

Position 310°·0 (*w* 2)   Distance 45″·0 (*w* 1)   Epoch 1838·53

A wide double star in a poor field, between the heads of Hercules and Ophiuchus, exactly on the western parallel of *a* Ophiuchi, which it precedes by 7°. A 5, silvery white; B 12, lilac; and at a distance *sf* are the two stars mentioned by Piazzi, Note 293, Hora XVI., in these terms: "Alia 8ᵃ magnit. 34″ temporis sequitur 6' ad austrum; et alia 39″ temporis 9 ad austrum." This object, which, from the minuteness of the *comes*, is here entitled to but small weight, is 133 Ḥ. v., and was thus registered:

Pos. 300° 0'   Dist. 48″·68   Ep. 1783·19

DCI. η OPHIUCHI.

Æ 17ʰ 01ᵐ 13ˢ   Prec. + 3ˢ·43

Dec. S 15° 31'·3   —— S 5″·09

Position 260°·5 (*w* 1)   Distance 269″·0 (*w* 1)   Epoch 1833·61

A brilliant star, with a distant companion, on the Serpent-bearer's left knee; and on the margin of the Milky Way. A 2½, pale yellow; B 13, blue; and they lie nearly in mid-distance between Kelb-al-Raï and Antares. A proper motion in space has been assigned to this star, of which the following are the registered values:

$$P.... \ \text{Æ} - 0''\!\cdot\!03 \qquad \text{Dec.} + 0''\!\cdot\!09$$
$$B.... \qquad + 0''\!\cdot\!08 \qquad \qquad + 0''\!\cdot\!12$$
$$A... \qquad + 0''\!\cdot\!03 \qquad \qquad + 0''\!\cdot\!13$$

η Ophiuchi is the *sf* terminus of the *Nasak Yemení*, or southern boundary wall of the Arabian pasturage so repeatedly mentioned. It is also termed *Sábik*, preceding, by Tízíní; but Beigel deems the word to be a corruption of *sáik*, successor: be that as it may, the star can be identified by a line carried from Antares into the north-north-east, to a distance of 16°. Again,

From Scorpio's deadly heart to trace   the Serpent-bearer's knee,
Look for the Eagle's tail, and then   one third that way 'twill be.

DCII. μ DRACONIS.

Æ 17ʰ 02ᵐ 02ˢ   Prec. + 1ˢ·24
Dec. N 54° 41′·2   —— S 5″·02

Position 206°·7 (*w* 8)   Distance 3″·6 (*w* 4)   Epoch 1830·79
—— 200°·3 (*w* 8)   —— 3″·3 (*w* 5)   —— 1839·53

A very neat binary star, on the tip of the Dragon's tongue; it is to the north-west of β, and 3°½ distant from it. A 4 and B 4½, both white. This object is 13 Ḥ. ɪɪ., and it has been considered a miniature of Castor, but the stars are too nearly equal to bear out the resemblance accurately. It is the Arrakis of the Catalogues, from the Arabian *al-rákis*, the trotter, *i. e.* the trotting camel, in reference to the neighbouring stars, called *al-'awáyid*, the sucking camels: the principal components of the herd being β, γ, μ, ν, and ξ, in Draco's head. In the middle of the latter is the small star of the Borgian globe, called *Al-ruba'*, the camel's foal; but though described as being of the 6th magnitude, it is not in Piazzi's list. The previous measures in proof of its retrograde orbital movement, are:

	Pos.	Dist.	Ep.
Ḥ.	232° 22′	4″·35	1781·73
H. and S.	208° 21′	3″·91	1821·38
Σ.	205° 06′	3″·23	1832·22
D.	204° 38′	3″·36	1832·61

My own observations were very satisfactory, since the results, *inter se*, were as accordant as in operations of this nature can be expected. A geometrical rough-cast of the whole, yields a period of about 600 years for the orbital revolution; since the velocity has appeared to decrease to − 0°·3 per annum, and then to accelerate to − 0°·7, during this small south-west portion of its orbit. But the proper motion introduced into the discussion has been greatly pared down by Mr. Baily, for the registered values now stand:

$$P.... \ \text{Æ} - 0''\!\cdot\!20 \qquad \text{Dec.} + 0''\!\cdot\!16$$
$$B.... \qquad - 0''\!\cdot\!12 \qquad \qquad + 0''\!\cdot\!03$$

DCIII. ε URSÆ MINORIS.

$\mathcal{R}$ 17^h 02^m 37^s Prec. − 6^s·53

Dec. N 82° 17′·1   —— S 4″·98

Position 358°·5 (w 1)   Distance 41″·0 (w 1)   Epoch 1835·55

A second-grade Greenwich star, with a minute *comes*, at the root of the Lesser Bear's tail, where it is readily identified, being the third star from Polaris. A 4, bright yellow; B 12, pale blue; three other telescopic stars in attendance, the nearest of which is about 2′ distant. It is about half way on the line between the Pole-star and η Ursæ Minoris; and its proper motion in space has been thus valued:

P.... $\mathcal{R}$ − 0″·82   Dec. + 0″·01

B....   + 0″·45   0″·00

DCIV. 36 OPHIUCHI.

$\mathcal{R}$ 17^h 05^m 29^s Prec. + 3^s·71

Dec. S 26° 21′·5   —— S 4″·73

Position AB 226°·1 (w 7)	Distance	5″·2 (w 5)	Epoch 1831·57
—— AC 289°·9 (w 4)	——	193″·8 (w 2)	
—— AB 221°·4 (w 6)	——	5″·0 (w 4)	—— 1835·33
—— AB 219°·5 (w 6)	——	5″·3 (w 5)	—— 1839·28
—— AB 216°·6 (w 8)	——	4″·9 (w 8)	—— 1842·46

A triple, or rather multiple star, of which our principal business is with A and B, since the above results prove B to be on a retrograde march. Mayer made the two stars to be exactly on the same meridian, with a Δ of declination = 13″; this accidental statement was the cause of considerable error, since, assuming as points of departure the following epochs,

Mayer.	Pos. 360° 00′	Dist. 13″·00	Ep. 1780·00
H. and S.	227° 19′	5″·55	1822·52
S.	228° 28′	5″·20	1825·17

there appeared to be an appreciable direct motion; and this was not a little strengthened by one of H.'s Sweeps, No. 1297, giving 235° as the value of the angle about five years after the last epoch. My measures, however, show a motion exactly contrary, and while writing this, on requesting the Astronomer Royal to give a casting vote on the subject, he kindly forwarded me the following results:

Pos. 213° 20′   Dist. 5″·32   Ep. 1843·52

As there is some confusion in assigning the stars about here to their proper asterisms, we may be particular. 36 Ophiuchi is on the margin of the gap made in Scorpio's precincts, in carrying down those of

Ophiuchus; by which arrangement or mis-arrangement it is artificially divided from 30 Scorpii, with which, as will be presently shown, it has an occult affinity. Both are between the Serpent-bearer's left foot and the root of the Scorpion's tail; and No. 36 is 10° due east of Antares. Of the individuals first measured, A is 4½, and ruddy; B 6½, pale yellow; and C 7½, greyish; the latter being double, with a most minute *comes* near the *sf* vertical, whose existence Sir John Herschel first pointed out to me. It would appear that Piazzi did not see A divided, as his note, "Duplex. 15″·5 temporis alia 7·8 magnit. præcedit, 1′ ad boream," evidently alludes to A and C. The principal star is thought to be variable, though I have always seen it as now registered.

36 Ophiuchi is upwards of 12′ distant from 30 Scorpii, on an $\angle = 70°$. As the parallax on wires so widely separated made the micrometer questionable, these results were obtained by the Lee Circle:

$$\Delta \, \mathbb{R} = 13′ \; 11″·4 \quad \Delta \text{ Dec.} = 3′ \; 03″·63 \quad \text{Ep. } 1831·57$$
$$= 13′ \; 10″·6 \qquad\quad = 3′ \; 04″·41 \qquad\quad 1839·28$$

M. Bessel first pointed out that 36 Ophiuchi and 30 Scorpii have a common proper motion, in these words (*Fundamenta Astronomiæ*, 30 Scorpii): "Ex observationibus hæc proficiscitur differentia inter stellam hanc atque A Ophiuchi:

Flamsteedii . .	1690 + 13′ 32″·4	+ 2′ 56″·0
Bradleii . . .	1755 ...	+ 3′ 02″·7
Mayeri . . .	1756 + 13′ 13″·1	
Piazzi . . .	1800 + 13′ 07″·0	+ 3′ 04″·2

unde perspicuum fit, huic stellæ et illi duplici stellæ communem esse gravem motum." The movements of these stars through space have pretty nearly the same values; those for 36 Ophiuchi are:

$$P.... \, \mathbb{R} - 0″·59 \qquad \text{Dec.} - 1″·25$$
$$B.... \quad - 0″·48 \qquad\qquad - 1″·14$$

And my own observations afford $\mathbb{R} - 0″·53$, and Dec. $- 1″·23$; so that while in itself a singular revolving binary system, it is accompanying another and a most distant object in an *annus magnus*, to contemplate the period of which makes imagination quail. This is a curious example of two stars transferred, by a progressive uniform motion common to both, in a curve so vast as to appear a straight line, towards some unknown region, or, as Mr. Baily expresses it, "journeying together through space, and leaving the neighbouring stars behind." In the present state of human knowledge it is impossible to ascertain whether this is a real motion arising from gravitation, or an apparent one, owing to the actual progression of the solar system towards some pre-ordained point in the heavens, between which and us they lie: at all events, such a march indicates that these stars cannot be only optically, or accidentally, connected.

While making observations towards a future determination of this question, I perceived a star of the 14th magnitude, nearly between 36 Ophiuchi and 30 Scorpii, a little to the southward of a line joining them; and which, though in the range of the micrometer run, had escaped the gaze of preceding observers. This star being so placed as to become of interest, either as a point of departure, or as partaking of the motions of its neighbours, is of greater importance than from its size it would otherwise be: it was therefore carefully diagrammed in 1835.

Wishing to know the action of the great Northumberland equatoreal upon this group, I requested the Rev. James Challis, in 1839, to give it a rigid scrutiny. This was kindly complied with, and an arduous examination was rewarded with no fewer than four new stars; but they

were most minute ones, as the largest *sp*, 36 Ophiuchi, is barely of the 16th magnitude. This was to be expected from an instrument of double the dimensions of mine, it having 11½ inches clear aperture and 19½ feet focal length; whence the objective part is in greater proportion to the pupil of the eye, than in my smaller one. Its field is therefore never without little specks of light; as was, from like causes, the case on looking into Sir John Herschel's 20-foot reflector at Slough,—I found it was never without company. Mr. Challis thus described the four additional stars: "*a* is a star nearly as easily seen as the small one of your coarse double, and is situated nearly at the point where lines passing through the two stars of 36 Ophiuchi and the two stars of your coarse double meet; *b* is about the same size; and *c* was seen with more difficulty, but very decidedly. *d* is the faintest star of all; it was as much as the telescope would do to make it visible, but I think I may be sure of its existence. I am greatly interested in giving you these particulars, as I am desirous of knowing what the light-transmitting power of the great Northumberland is, as compared with other telescopes."

DCV. α HERCULIS.

Æ 17ʰ 07ᵐ 21ˢ	Prec. + 2ˢ·73
Dec. N 14° 34′·5	—— S 4″·57

Position 119°·4 (*w* 8)	Distance 4″·6 (*w* 5)	Epoch 1832·51
—— 118°·9 (*w* 9)	—— 4″·8 (*w* 7)	—— 1838·71
—— 118°·7 (*w* 9)	—— 4″·5 (*w* 9)	—— 1842·57

A standard Greenwich star with a companion, on the head of Hercules. A 3½, orange; B 5½, emerald, or bluish green; and there are two distant stars of 10th and 12th magnitudes in the *nf* quadrant, which are remarkable for their lilac tinge. I have here registered the medium brightness, for A was found to be variable by ♅., who compared it with κ Ophiuchi, changing from maximum 3 to minimum 4 in a period of 60¼ days. Σ. has since suggested that B also varies from 5 to 7. And it has a proper motion in space, which, however slight, becomes of singular interest in studying its conditions; these are the best investigated values:

$$
\begin{array}{llll}
P.... & \mathbb{R} - 0''\!\cdot\!11 & \text{Dec.} & + 0''\!\cdot\!12 \\
B.... & + 0''\!\cdot\!03 & & + 0''\!\cdot\!05 \\
A.... & - 0''\!\cdot\!002 & & + 0''\!\cdot\!06 \\
\end{array}
$$

This lovely object, one of the finest in the heavens, is 2 ♅. ii., and was described to be double by Piazzi, though not always easily seen so. "Duplex," ait, "comes sequitur ad austrum; et non semper nec facile distinguitur. Aptius ad id tempus Septemb. initium paulo post solis occasum." From the observations of its discoverer it was considered to have undergone an orbital increase of 11°½, in little more than twenty-three years; therefore when Σ. attacked it in 1819, he expected to find the angle amount to about 130°. But the result was an actual retrogradation from ♅.'s determination, and as the Dorpat astronomer was convinced that he was within 1° of the truth, and indeed his last mean is drawn from five years' measures, it was concluded, either that the former observations were uncertain, or that one of the stars had *rebroussé chemin*. But all the subsequent measures, however they differ *inter se*, coincide in establishing the fixity of this object, thereby adding another instance to that of γ Andromedæ, that highly-coloured stars are not necessarily in motion. The following are the previous results which I compared with my own, in arriving at this conclusion:

	Pos.	Dist.	Ep.
♅. {	111° 28'	4''·74	1779·66
	121° 57'	5''·05	1803·40
Σ.	116° 36'	5''·61	1819·60
H. and S.	119° 33'	5''·29	1821·74
Σ.	118° 28'	4''·65	1829·63
D.	120° 23'	4''·85	1830·62

A discussion of the delicate observations of ♅. and Σ. led Baron de Zach to exhort those Uranian amateurs who wish to be useful, to work in the rich field of double stars: it is, he says, "un vaste et un très-fertile champ à défricher, que nous recommandons aux soins des amateurs qui voudront se rendre utiles, et faire encore autres choses que des observations banales qu'on répète par-tout."

The principal star is called Rasalgeti, from the Arabian *rás al-játhí*, the kneeler's head; and the casual gazer may pick it up by noting that Altair, Wega, and Rasalague, form a triangle nearly equilateral, the latter being the preceding star, and having Rasalgeti about 5° before it; the heads both of Hercules and Serpentarius lie between Lyra and Scorpio. The galley rhymes afford another clue:

Amid yon glorious starry host,   that feeds both sight and mind,
Would you the Serpent-bearer's head,   and that of Herc'les find,
From Altair west direct a ray   to where Arcturus glows,
One-third that distance, by the eye, will both those heads disclose.

Rasalgeti is the *lucida* of Hercules, one of the old forty-eight asterisms, called 'Εν γόνασιν, Ingeniculus, Genuflexus, Saltator, and Incumbens Genubus, by the ancients; and represented as a man kneeling, weary, and sad. It was probably therefore not originally figured for the Theban;

Eudoxus and Aratus, speaking by the well-known verses of Cicero, merely allude to his sorrow, and tell us:

Engonasin vocitant, genibus quod nixa feratur.

Manilius, speaking through Sherburne, says:

Next the cold Bears—the cause t' himself best known—
Shines forth a *kneeling* constellation.

This kneeling posture has given rise to momentous discussion; and whether it represents Lycaon lamenting his daughter's transformation, or Prometheus sentenced, or Ixion ditto, or Thamyras mourning his broken fiddle, remains still unsettled. But in process of time, this figure became a hero, and Hyginus mentions both the lion's skin and the club; while the right foot's being just over the head of Draco, satisfied the mythologists that he was crushing the Lernæan hydra. But this is a matter upon which much twaddle has been raked together by the "learned,"who fancy they see in this position, as well as a similar one of the Indian Krishna, a bruising of the serpent's head, in illustration of the Mosaic record. The Arabians called it *El-rákis*, the dancer or leaper, and *El-játhi 'alá rukbeteïhi*, the man who kneels on both his knees; an epithet which Bayer, who sadly worries the Orientals, has brought to *Elgezi ale rulxbachei*. The early Venetian editions of Hyginus figure Hercules as going to attack a snake coiled round the trunk of an apple tree; and Bayer depicted a mystic apple-branch in the Theban's hand. Hevelius transformed it into a bunch of snakes, under the name of Cerberus, from the watch-dog of the infernal portals; with the fox carrying a goose for his breakfast, as shown in the *Prodromus Astronomiæ*. Some have considered the emblem as typifying the serpent which infested the vicinity of Cape Tænarus, whence a sub-genus of Ophidians still derives its name. At all events a poet, indignant at the heathen exaltations of Hevelius, has said:

To Cerberus too a place is given—
His home of old was far from heaven.

This symbol of the "tricapitem canem infernalem voracem" figures among the new constellations which follow Hevelius, in his homage to Urania and the great astronomers, in the elaborate frontispiece to his *Uranographia*. Bode has adopted both the apple-branch and the snakes, in his Atlas, under the style and title of *Cerberus et Ramus*.

This constellation is of great extent and importance, notwithstanding it boasts of no star larger than the 3rd magnitude: yet several of that and the 4th size decorate the head, back, shoulders, hips, thighs and right ankle of the figure. But though this asterism is not very remarkable to the eye, its double stars, nebulæ, and clusters, render it telescopically interesting and glorious. The components have increased as optical means have been enlarged, and the registered numbers are thus:

Ptolemy . . . . 30 stars	Hevelius . . . . 45 stars
Tycho Brahé . . 28	Flamsteed . . . 113
Clavius 31	Bode 451

Hevelius, in his *Prodromus Astronomiæ*, reported a nebula on the top of Hercules' head, close to Rasalgeti, which Messier searched for in vain. The nearest nebula to this star appears to be 901 ℍ. II., but that being too small and faint for the power of their telescopes, the object must have been a comet.

DCVI. 31 SCORPII.

Æ 17ʰ 07ᵐ 44ˢ   Prec. + 3ˢ·71
Dec. S 26° 26'·7   —— S 4"·53

Position 330°·4 (w 6)   Distance 6"·8 (w 4)   Epoch 1835·64

A delicate double star, between the left foot of Ophiuchus and Scorpio's back, and closely following 36 Ophiuchi, already treated of. A 6½, pale white; B 11, ash coloured. This object is 35 ♓. ɪ., and has generally been registered as 38 Ophiuchi ; but the two may be considered as identical, since Flamsteed's remark, that three or four telescopic stars follow it, shows that it is 31 Scorpii of the British Catalogue. The preceding measures were:

♓.   Pos. 330° 48'   Dist. 4"·00±   Ep. 1782·45
S.   330° 50'   7"·14   1825·53

These results led to a belief that the stars had greatly receded from each other, in an interval of forty-three years, though the angle of position had remained constant. The assumption was owing to ♓.'s having estimated the distance by the inaccurate method of diameters of the large star, for my observations go far to show its fixity as an optical object. It is, however, difficult to handle, from the variable refractions of so low an altitude, and the smallness of the *comes*.

This object is about 10° to the north of the *karazat*, or end lump in the Scorpion's tail, from which *al-shaúlah*, the sting, issues; and which is the XIXth Lunar Mansion. The stars which mark this moon-station, λ and υ, are among the lowest objects in our hemisphere: the first is named Shaula, and the second Lesath, from *les'ah*, a sting, formed by Scaliger's conjecture from Alascha, which is a corruption of *al-shaúlah*. Lesath, therefore, is not a term used by the Arabs, who designate all these bumps which form the tail, *Al-fikrah*, vertebrated twirls; they are formed by ε, μ, ζ, η, θ, ι, κ, λ, and υ, and it is supposed that the sting, *punctura scorpionis*, was formerly carried to the following star, γ, marked nebulous by Ptolemy.

DCVII. 39 OPHIUCHI.

Æ 17ʰ 08ᵐ 16ˢ   Prec. + 3ˢ·65
Dec. S 24° 06'·3   —— S 4"·49

Position 355°·6 (w 7)   Distance 11"·6 (w 5)   Epoch 1830·63
——— 356°·2 (w 9)   ——— 12"·1 (w 9)   ——— 1838·52

A neat double star, on the toes of the Serpent-bearer's left foot, and 1° distant to the north-west of the bright star θ. A 5½, pale orange;

B 7½, blue. This is a very fine object, and was measured at my first epoch in full twilight. It is 25 ♄. III., also Nos. 31 and 32 of Piazzi's Hora XVII.; and was thus micrometrically measured at Slough:

Pos. 357° 14′   Dist. 10″·03   Ep. 1782·46

The subsequent measures of H. S. and myself, are so coincident as to indicate the fixity of this optical object. Yet a reduction of Piazzi's mean apparent places, except that turning Æs and Decs. into the metric desiderata is not sufficiently exact to dwell upon, would imply a direct orbital movement, for there is no appreciable proper motion:

Pos. 348° 0′   Dist. 12″·0   Ep. 1800

About the end of September, 1604, the scholars of Kepler discovered a very remarkable star near this. They were examining the planets Mars, Jupiter, and Saturn, which were then sufficiently close to each other in that quarter, to engage the attention of astronomers, and Mæstlin very quickly detected the interloper. At first it surpassed Jupiter in magnitude, and its brilliancy even rivalled that of Venus; but it afterwards became as small as Regulus, and as dull as Saturn. It was white near the horizon, but as it rose, assumed alternately the varying colours of the rainbow; it had no parallax, and was exactly round. When in this state, it had the honour of being particularly observed by Galileo; and the lovers of salad will recollect Kepler's attack on the Epicureans, in his account of its advent. It gradually diminished in splendour till October, 1605, when it was very small, and about the beginning of 1606 it entirely disappeared, nor has it been seen since. Mr. Pigott searched in vain for it in 1782; but the vicinity should be constantly watched.

DCVIII. δ HERCULIS.

Æ 17ʰ 08ᵐ 28ˢ   PREC. + 2ˢ·46
DEC. N 25° 01′·9   —— S 4″·47

POSITION 173°·9 (w 9)   DISTANCE 25″·9 (w 7)   EPOCH 1830·71
———— 174°·9 (w 9)   ———— 24″·7 (w 5)   ———— 1837·49
———— 175°·1 (w 9)   ———— 24″·5 (w 9)   ———— 1839·62

A binary star, on the hero's right shoulder, and due north of its lucida about 11°, forming a nearly equilateral triangle with it and β. A 4, greenish white; B 8½, grape red. A proper motion is assigned to the primary, which I here submit, although my own observations would dispense with that in Æ:

P.... Æ − 0″·20   Dec. − 0″·14
B....   − 0″·05   − 0″·15
A....   − 0″·10   − 0″·13

This neat object is 1 ♄. V., and as the movement suspected by the re-examination of H. and S. is confirmed, it may be recognised as a

2 C 2

physical system. The previous measures and epochs which I compare with mine are these:

$$\begin{array}{llll}
\text{H.} & \text{Pos. } 162°\ 28' & \text{Dist. } 33''\cdot75 & \text{Ep. } 1779\cdot61 \\
\text{H. and S.} & 172°\ 10' & 28''\cdot87 & 1821\cdot37 \\
\Sigma. & 173°\ 42' & 26''\cdot11 & 1829\cdot77
\end{array}$$

The results show a very appreciable decrease of distance, and a direct angular increase *sp nf;* the object, therefore, as H. observes, merits particular attention, "as the change is contrary to what the presumed proper motion of the large star would alone produce." My last epoch was under the very best atmospheric and instrumental circumstances; and on the whole I am led to infer, that if all the series could be depended on, B had lately passed its apastron in the south-east portion of its orbit, and that it is slackening its march as it recedes from the extremity of the ellipse, now barely moving a degree in ten years.

DCIX. 9 M. OPHIUCHI.

$$\begin{array}{ll}
\text{Æ } 17^h\ 09^m\ 42^s & \text{Prec. } + 3^s\cdot50 \\
\text{Dec. S } 18°\ 20'\cdot7 & \text{——— S } 4''\cdot37
\end{array}$$

MEAN EPOCH OF THE OBSERVATION . . . . 1834·54

A globular galaxy-cluster, on the Serpent-bearer's left leg, with a coarse telescopic double star in the *np* quadrant. This fine object is composed of a myriad of minute stars, clustering into a blaze in the centre, and wonderfully aggregated, with numerous outliers seen by glimpses. It was registered by Messier in 1764; and described by him as a nebula, "unaccompanied by any star." The mean apparent place was carefully differentiated from *η* Ophiuchi. Sir William Herschel resolved it with his 20-foot reflector, in 1784; and he estimated its profundity as, at least, of the 344th order. He thinks it a miniature of No. 53 Messier; and it is one of those which forms a capital object, for proving the space-penetrating power of a telescope. It lies 3° to the south-east of *η*, and rather more than a quarter of the way from Antares to Altair.

DCX. ν SERPENTIS.

$$\begin{array}{ll}
\text{Æ } 17^h\ 11^m\ 49^s & \text{Prec. } + 3^s\cdot36 \\
\text{Dec. S } 12°\ 40'\cdot7 & \text{——— S } 4''\cdot18
\end{array}$$

POSITION 31°·3 (w 4) DISTANCE 50''·8 (w 2) EPOCH 1832·60

A wide double star, in the middle of the Serpent; it lies 4° to the north-east of *η*, and consequently in the line between Antares and the Eagle's tail. A 4½, pale sea-green; B 9, lilac; and there is a third star, of a dusky tint, at a distance in the *np* quadrant. This object was first classed by ♃. as No. 29 v., but he merely says the distance is about 35″. It was, however, submitted to measurement by H. and S., No. 247, who obtained these results:

Pos. 30° 47′ Dist. 50″·21 Ep. 1821·97

ν Serpentis has movement in space, but not to the amount that had been suspected by some astronomers. The most recent and thoroughly investigated values are:

P....	Æ + 0″·25	Dec.	+ 0″·08
Br...	+ 0″·17		+ 0″·06
B....	+ 0″·09		+ 0″·03

DCXI. 92 M. HERCULIS.

Æ 17ʰ 12ᵐ 14ˢ   Prec. + 1ˢ·84
Dec. N 43° 18′·4   —— S 4″·15

Mean Epoch of the Observation .... 1835·56

A globular cluster of minute stars, preceding the right leg of Hercules. This object is large, bright, and resolvable, with a very luminous centre; and, under the best vision, has irregular streamy edges. It is immediately preceded by a 12th-magnitude star, distinct from the outliers, and there are several other stars in the field, of which the brightest is of the 7th magnitude in the *nf*, with a Δ Æ = 28ˢ. Messier, who enrolled it in 1781, remarks that "it is easily seen with a telescope of one foot;" and it really demands very little optical aid to render it visible. Messier's own instrument did not, it seems, re-solve it, for he compares the shining centre, with its attendants, to the nu-cleus of a comet surrounded by nebu-lous matter; but, of course, it rose into a brilliant cluster, of 7′ or 8′ in diameter, before the reflectors of Sir W. Herschel in 1783. The mean place was obtained by carefully differentiating the cluster  with η Herculis, from which it bears north-by-east, 1°¼ distant; bearing to the north of *a* Herculis, and west of Wega.

DCXII. 94 P. XVII. OPHIUCHI.

Æ 17ʰ 17ᵐ 21ˢ   Prec. + 2ˢ·70
Dec. N 15° 45′·4   —— S 3″·71

Position 65°·0 (*w* 2)   Distance 5″·0 (*w* 1)   Epoch 1835·52

A very delicate double star, between the heads of Ophiuchus and Hercules; it is 2° north-east of Rasalgeti, on the line towards β Cygni.

A 7, brilliant white; B 13, violet tint: A is followed by a ruddy star of nearly the same magnitude at a △ ℛ = 16ˢ·5, which, with another in the *nf*, forms a neat isosceles triangle. This severe test was discovered by Σ., and measurement being precluded by the minuteness of the *comes*, the results are from a mean of several estimations made near the meridian, but amid occasional glares of sheet lightning. On the arrival of the Dorpat Catalogue, this was among the first objects I referred to, in order to see the degree of weight assignable to my estimations, when placed in juxtaposition with actual and trusty measurements; these are Professor Struve's results:

Pos. 61°·93   Dist. 4″·073   Ep. 1830·23

DCXIII. ρ HERCULIS.

ℛ 17ʰ 18ᵐ 10ˢ   Prec. + 2ˢ·07
Dec. N 37° 17′·9   —— S 3″·63

Position 308°·5 (*w* 7)   Distance 3″·6 (*w* 7)   Epoch 1831·60
——— 308°·9 (*w* 9)   ——— 3″·7 (*w* 9)   ——— 1839·74

A beautiful optical double star, in the middle of the hero's right thigh; it is 2° to the eastward of π, on the line towards Wega. A 4, bluish white; B 5½, pale emerald. This object is 3 ♓. II., and was entered double by Piazzi, Note 105, Hora XVII., "Duplex, minor præcedit." From the earliest strict measures, it was concluded to have made a direct orbital change of nearly 9° in little more than half a century, and the distance to have increased materially; but all the subsequent observations tend to prove its fixity. These are the several measures previous to mine:

	Pos.	Dist.	Ep.
♓.	300° 21′	2″·97	1779·66
H. and S.	307° 53′	4″·46	1821·38
Σ.	307° 22′	3″·60	1830·35
D.	308° 35′	3″·96	1830·63

This star is circumstanced similarly to δ Herculis, No. DCVIII., since if all the epochs could be strictly depended upon, the orbital progress would appear to be equally slow.

DCXIV. λ HERCULIS.

ℛ 17ʰ 24ᵐ 16ˢ   Prec. + 2ˢ·42
Dec. N 26° 14′·2   —— S 3″·11

Position 295°·0 (*w* 1)   Difference ℛ = 23ˢ·4 (*w* 1)   Epoch 1834·58

A star with a distant companion, on the right arm of Hercules, and 14°½ due north of *a* Ophiuchi. A 4½, deep yellow; B 10, light blue,—

several other stars in the field. This star is Masym of various Catalogues, from the Arabian *mi'sam*, the wrist; and it is notable as the
point in the heavens to which ⩊. considers the whole solar system is
slowly journeying, an hypothesis which the researches of M. Argelander
strengthen. Both Cacciatore and Brioschi considered λ to be a star of
very considerable proper motion; but the most rigid examinations only
show this statement:

$$P.... \text{Æ} - 0''\cdot05 \quad \text{Dec.} \quad 0''\cdot00$$
$$B.... \quad + 0''\cdot02 \quad \quad + 0''\cdot07$$

DCXV. 147 P. XVII. DRACONIS.

$$\text{Æ} \quad 17^h \ 25^m \ 07^s \quad \quad \text{Prec.} \ + \ 1^s\cdot44$$
$$\text{Dec. N } 50^\circ \ 59'\cdot9 \quad \quad - \ \text{S } 3''\cdot03$$

Position 266°·2 (w 8)   Distance 3''·2 (w 6)   Epoch 1836·53

A fine double star, between the right foot of Hercules and the
Dragon's eye; from the latter of which, Alwaid, it is about 1°¼ south.
A 8, pale white; B 8½, ruddy. This pretty object is 66 ⩊. i., and
though bearing a physical aspect, seems to have undergone no alteration
since its discovery. The following are the previous epochs of comparison:

	Pos.	Dist.	Ep.
⩊.	267° 36'	3''·50±	1782·84
H.	265° 18'	3''·26	1828·48
Σ.	265° 28'	3''·17	1831·29

DCXVI. β DRACONIS.

$$\text{Æ} \quad 17^h \ 26^m \ 48^s \quad \quad \text{Prec.} \ + \ 1^s\cdot35$$
$$\text{Dec. N } 52^\circ \ 25'\cdot2 \quad \quad - \ \text{S } 2''\cdot90$$

Position 100°·6 (w 2)   Difference Æ = 29ˢ·8 (w 1)   Epoch 1834·59

A standard Greenwich star, with a very distant companion, in the
Dragon's eye. A 2, yellow; B 10, bluish; with a coarse telescopic
double star *nf* of B, by about 30ˢ, and several other stars in the field.
This object is the Alwaid of the Catalogues, from the Arabian
al-'awáyid, the suckling camels; and it was corrupted to Asvia in the
Middle Ages. The commentator on Ulugh Beigh, quoted by Hyde
(*Syntagma*, Dissert. I., p. 18), says that the four stars in the head of
Draco are called *'Awáyid* and *El salib wáķi*; the latter words signify
the Falling Cross, evidently meaning β, γ, ξ, and μ; of which β and
ξ are supposed to be joined by the perpendicular, and γ and μ by the
transverse beam of the cross. But the earlier Arabs seem to have
patronized *'Awáyid*, and our star is the principal of the herd already
alluded to under μ and γ; glossators, however, are not wanting who

overthrow the whole story of the camel, the she-camels, and the camel-foal, by making out μ to be *al-rákis*, the dancer, and *al-'awad*, the lute-player. Leaving that as a point for grave consideration, we look to β as having displaced a both as a Greenwich standard, and a *lucida*. It is now generally rated of the 2nd magnitude, whereas Ptolemy marked it γ in brightness; and Ulugh Beigh, Tycho Brahé, Hevelius, Bradley, Zach, and Mayer, followed him in stamping it of the 3rd size. Flamsteed, however, made it $2\frac{1}{2}$, and Piazzi raised it to 2. A future day may restore the rights of a, for they may both be variable at long periods. β and γ Draconis, in the head of Draco, may be readily known from being nearly in the curved line joining it to Arided and Arcturus; and with two stars, μ and ξ, just to the north of them, they form an irregular trapezium. The poetaster has a word upon it:

> From Alkaid on the Great Bear's tail, to Cygnus cast your eye;
> Midway between the Bird and Beast the Dragon's head you'll spy.

DCXVII. 54 OPHIUCHI.

℞ 17ʰ 26ᵐ 59ˢ   Prec. + 2ˢ·76
Dec. N 13° 16'·6   —— S 2"·88

Position 75°·0 (w 3)   Distance 20"·0 (w 1)   Epoch 1833·56

A most delicate double star, on the crown of the Serpent-bearer's head, and closely due north of the lucida. A 6, pale straw-colour; B 14, blue; several other stars in the field. This object is 35 Ͱͱ. iii., and was discovered in August, 1781. In the original register it is merely described as excessively unequal, and about 8" apart. Now, with all my gazing, as I could only see the glimpse point of light noted B, which I could not estimate at less than 20", I concluded that there was a still smaller companion beyond my reach. On the arrival, however, of the Dorpat Catalogue, I found that Professor Struve had measured my *comes*, and had seen no other with his then unequalled refractor; so that I cannot but think Ͱͱ. had written 18" on his original entry. The Dorpat results for a first epoch are:

Pos. 76° 77' Dist. 21"·42 Ep. 1830·19

DCXVIII. 53 OPHIUCHI.

℞ 17ʰ 27ᵐ 01ˢ   Prec. + 2ˢ·84
Dec. N 9° 42'·1   —— S 2"·87

Position 192°·5 (w 9)   Distance 41"·3 (w 9)   Epoch 1836·51

A wide double star, closely following the Serpent-bearer's neck, and 3° due south of the lucida. A 6, and B 8, both bluish. This object is 30 ℍ. v., and Piazzi's Nos. 149 and 150, Hora XVII.; and it is followed by the two 8th-magnitude stars mentioned in his Note 150. The measures on my list, including a reduction which I made from the mean apparent places given in the Palermo Catalogue, when I took it in hand, were:

ℍ.	Pos. 192° 48′	Dist. 32″·35	Ep. 1782·38
P.	195° 58′	42″·60	1800
H. and S.	191° 19′	41″·66	1821·47

From these results, the position was evidently stationary; but though ℍ. calls his a "narrow distance," the difference on such an object was so great as to demand attention. My observations were therefore specially conducted under the best circumstances, and I am able to place my highest weights on the conclusion.

DCXIX. α OPHIUCHI.

$$\text{Æ}\ 17^h\ 27^m\ 30^s \qquad \text{Prec.} + 2^s·77$$
$$\text{Dec. N } 12°\ 40'·8 \qquad\qquad — \text{ S } 2''·83$$

Position 187°·2 (w 2)   Difference Æ = 1ˢ·4 (w 1)   Epoch 1833·54

A standard Greenwich star, with a minute companion, at the back of the Serpent-bearer's head. A 2, sapphire; B 9, pale grey; and there is a coarse triplet of small stars preceding, nearly on the parallel. A is designated Rasalague, from the Arabian *rás-al-hawwá*, the serpent-charmer's head: an alteration traceable to the *el-hauwé* of the Moors, descendants of the Psylli, being spelt *el-hague* by the Spaniards. But it was variously corrupted in astronomical writings. It was once called *Al-ráyi*, from *ar-ra'i*, the shepherd; and the *Afeichus* of the Alphonsine Tables is the Arabo-Latin for Ophiuchus: *Ras-al-hangue* is taken from the Almagest, *Azalange* follows from *ras-al-hangue*, and *Al-hayro* from *al-hawwá*. This, however, is merely mentioned to identify the several references; in observation it may be easily found, as it lies nearly in mid-distance of a line drawn from Wega down to Antares; and it is the preceding point of a nearly equilateral triangle, which it forms with Wega and Altair. In starting from the last-named star, our galley-poet gives his sage advice, and thus it runs:

> From *Altair* let a ray be cast,   where we Arcturus view,
> One-third that distance will reveal   the star *Ras-al-hague*.

This star may have undergone a slight change of brilliance, since Ptolemy rated it γ, and he was followed by Ulugh Beigh and Tycho Brahé; but Hevelius, Flamsteed, Bradley, Mayer, Zach, and Piazzi, are unanimous in designating it of the 2nd magnitude. Though my observations hardly warrant the conclusion of this star's having a proper motion in space, it may possibly be the case, the following values being already on record:

P....	Æ + 0″·09	Dec. − 0″·18
B....	+ 0″·12	− 0″·19
A....	+ 0″·10	− 0″·19

Ὀφιοῦχος, Ophiuchus, is a large, though not a very conspicuous, constellation, most uncouthly figured as respects Hercules: it has no proper name, but is merely designated from holding a serpent. Yet he is not without appellations enough, being the Anguitenens of Cicero, the Anguifer of Columella, and the Serpentarius of the Tables: and he also appears as Serpentis lator, Effæminatus, and even as Cacus, the bad man. While some think this figure typifies one of the exploits of Hercules, others deem it a tribute to Esculapius, and a third class implicate it with the absurd heresy of the Ophites, by which the serpent that seduced Eve was elevated to divine honours. Novidius differs from all these, and has an opinion of his own: he thinks it clear enough that Ophiuchus and his snake are but types of St. Paul and the Maltese viper. Whatever may have been the original intention in placing this figure in the heavens, it is one of the old forty-eight asterisms, and its constituents have been thus numbered:

Ptolemy	29 stars	Hevelius	44 stars
Ulugh Beigh	29	Flamsteed	74
Tycho Brahé	37	Bode	289

In 1604, the scholars of Kepler discovered a new star in the eastern foot of Ophiuchus, which appeared brighter than one of the 1st magnitude; but in a few months it again became invisible. Such reports call for attention to the prospect of a re-appearance.

DCXX.   ν¹ DRACONIS.

Æ. 17ʰ 29ᵐ 01ˢ   PREC. + 1ˢ·16

DEC. N 55° 17′·7   —— S 2″·70

POSITION 311°·8 (w 9)   DISTANCE 61″·9 (w 9)   EPOCH 1837·51

A wide pair of stars, in Draco's mouth; 3° to the north of β, on a line towards Polaris. A and B, both 5, and both pale grey. This object is 11 Ⅲ. v., and Piazzi's Nos. 168 and 169 of Hora XVII., or ν¹ and ν² Draconis; to both of which are attributed proper motions, to the following amount:

$$A \begin{cases} \nu^1 \, \text{Æ} + 0''\cdot 30 & \text{Dec.} + 0''\cdot 05 \\ \nu^2 \quad + 0''\cdot 26 & \quad + 0''\cdot 03 \end{cases} \quad B \begin{cases} \nu^1 \, \text{Æ} + 0''\cdot 32 & \text{Dec.} + 0''\cdot 03 \\ \nu^2 \quad + 0''\cdot 30 & \quad + 0''\cdot 01 \end{cases}$$

These values, though slight, are important, for being under similar signs, they indicate a physical connexion of the two stars. The verification of this point requires another epoch of exact observation.

When Ⅲ. registered this object, he remarked: "From the right ascension and declination of these stars in Flamsteed's Catalogue, we gather that in his time their distance was 1′ 11″·418; their position 44° 23′ n. preceding; their magnitude equal or nearly so. The difference in the distance of the two stars is so considerable, that we can hardly account for it otherwise than by admitting a proper motion in either one or the other of the stars, or in our solar system; most probably, neither of the three is at rest." Yet, a reconsideration of the case leads me to infer that they must have remained very nearly constant; and so far from the anomalies of the results being in the way, I am only surprised at the coincidences of such different means of arriving at the general conclusion. They may be thus stated:

BY $\triangle$ ℞ AND DEC.

Flamsteed.	Pos. 314° 23′	Dist. 71″·42	Ep. 1690
Piazzi.	314° 12′	61″·41	1800

BY MICROMETER.

I⬩H.	Pos. 314° 49′	Dist. 54‴·80	Ep. 1779·80
H. and S.	312° 23′	62″·24	1822·44

This accordance may be deemed satisfactory, on considering that deductions from ℞s and Decs. at so early a period as Flamsteed's, can hardly merit reliance upon such delicate questions. To find this star by alignment, run an occult ray from γ to η, and it will pass over the two ν's at about one-third the distance.

DCXXI.   14 M. OPHIUCHI.

℞	17ʰ 29ᵐ 13ˢ	PREC. + 3ˢ·14
DEC. S	3° 09′·1	—— S 2″·68

MEAN EPOCH OF THE OBSERVATION . . . . 1835·54

A large globular cluster of compressed minute stars, on the Serpent-bearer's left arm. This fine object is of a lucid white colour, and very nebulous in aspect; which may be partly owing to its being situated in a splendid field of stars, the lustre of which interferes with it. By diminishing the field under high powers, some of the brightest of these attendants are excluded, but the cluster loses in definition. It was discovered by Messier in 1764, and thus described: "A small nebula, no star; light faint; form round, and may be seen with a telescope $3\frac{1}{2}$ feet long." The mean apparent place is obtained by differentiation from γ Ophiuchi, from which it is south-

by-west about 6°¼, being nearly midway between β Scorpii and the tail of Aquila, and 16° due south of Rasalague. Sir William Herschel resolved this object in 1783, with his 20-foot reflector, and he thus entered it: "Extremely bright, round, easily resolvable; with 300 I can see the stars. The heavens are pretty rich in stars of a certain size, but they are larger than those in the cluster, and easily to be distinguished from them. This cluster is considerably behind the scattered stars, as some of them are projected upon it." He afterwards added: "From the observations of the 20-feet telescope, which in 1791 and 1799 had the power of discerning stars 75·08 times as far as the eye, the profundity of this cluster must be of the 900th order." "It resembles the 10th *Connaissance des temps*, which probably would put on the same appearance as this, were it removed half its distance farther from us." For this 10 M., see No. DXCV.

DCXXII.   200 P. XVII. HERCULIS.

Æ 17^h 34^m 31^s PREC. $+$ $2^s \cdot 46$
DEC. N $24°$ $35' \cdot 8$ —— S $2'' \cdot 23$

POSITION $9° \cdot 5$ (w 8)   DISTANCE $16'' \cdot 3$ (w 6)   EPOCH $1830 \cdot 71$

A neat double star, in the space south of the hero's right arm, where it lies in a nest of half a dozen stars, two-thirds of the way from Wega towards Rasalgeti. A $6\frac{1}{2}$, topaz yellow; B 9, purple; a third star at a distance in the *sf* quadrant. This object is 104 $\mathbb{H}$. III., and was thus registered under a position of $83° 48'$ *np*, reducing thus:

Pos. $353° 48'$ Dist. $14'' \cdot 33$ Ep. $1783 \cdot 23$

The next epoch, that of Sir James South, gave these results:

Pos. $8° 58'$ Dist. $17'' \cdot 21$ Ep. $1825 \cdot 00$

which would imply a great orbital change to have taken place in forty-two years. But there is reason to suppose that $\mathbb{H}$. may have intended to note *nf* position, instead of *np;* which error of quadrant makes a direct difference of $12° 24'$ in the angle.

DCXXIII.   61 OPHIUCHI.

Æ 17^h 36^m 32^s PREC. $+$ $3^s \cdot 01$
DEC. N $2°$ $39^\wedge 2$ —— S $2'' \cdot 05$

POSITION $93° \cdot 9$ (w 8)   DISTANCE $20'' \cdot 7$ (w 8)   EPOCH $1833 \cdot 53$

A neat double star below β, on the Serpent-bearer's left shoulder, where it is $2°$ south of the bright star β, Celbalrai, which lies about $7°$ south-by-east of a Ophiuchi. A and B, both $7\frac{1}{2}$, and both silvery white. This object is 32 $\mathbb{H}$. IV.; and Nos. 215 and 216 of Piazzi's Hora XVII. In the first registry of the measures, $\mathbb{H}$. mentions the position as a mere estimation, "almost exactly following;" but this suffices to indicate fixity. The following are the measures previous to my epochs, and though a severe trial to meridian deductions, the results obtained from the Palermo Catalogue are included:

	Pos.	Dist.	Ep.
$\mathbb{H}$.	$90° 00'$	$19'' \cdot 07$	$1781 \cdot 55$
P.	$90° 00'$	$26'' \cdot 10$	$1800 \cdot 00$
H. and S.	$93° 33'$	$20'' \cdot 52$	$1821 \cdot 77$
Σ.	$94° 01'$	$20'' \cdot 55$	$1827 \cdot 37$

This being one of the four stars so satisfactorily measured by the Rev. R. Sheepshanks, in the summer of 1834, with the great equatoreal at Campden Hill, the obtained results will here be acceptable:

	Pos.	Dist.	
ϵ Boötis.	322° 11′	2″·92	
ξ Boötis.	329° 50′	7″·35	Ep. 1834·58
a Herculis.	119° 06′	4″·94	
61 Ophiuchi.	93° 37′	20″·45	

In forwarding these standard measures, Mr. Sheepshanks tells me that he relies more upon the accuracy of the distances than of the positions, because less pains were taken in observing the latter. He moreover says:

I have elsewhere described the mode of making the observation, but I will here say a few words. The star having been found and placed in the middle of the field, the clock was put in gear. The stars were then made to lie between two close parallel wires, or on one, and the vernier read off. When this had been repeated five or six times, or oftener, the mean was taken, the vernier set 90° forward, and one star split on one wire, while the other star was bisected by the other wire. The first star was then bisected by the second wire (using the screws of the slipping-piece), and the first wire brought down to the second star, when it had moved over twice the distance. In this way five double distances were taken, and then the zero for position got by making the star run along the wire.

The distances were particularly attended to, as they were said to be the most difficult to measure. With our means we found them the most easy, the truth being that position is measured almost as well without clockwork as with, but distance can only be measured satisfactorily with clockwork. Our apparatus was, on the whole, the most convenient I have yet seen, as the slipping-piece, on which this mode chiefly depends, was, I think, lighter than yours. The clock went admirably, and carried that large instrument so easily that, if I remember rightly, putting it in and out of gear only altered the rate a second in ten minutes.

DCXXIV. μ HERCULIS.

Æ 17ʰ 40ᵐ 12ˢ   Prec. + 2ˢ·37

Dec. N 27° 49′·0   ⸺ S 1″·73

Position 241°·8 (*w* 9)   Distance 30″·1 (*w* 5)   Epoch 1837·67

A delicate double star, in the bend of the Theban's right arm; 14° distant from Wega, to the south-west, and preceding β Cygni by about 26°, exactly on the parallel. A 4, pale straw-colour; B 10, cerulean blue. This is 41 Ⱨ. iv., and difficult to measure, especially in distance, from its bearing illumination badly. Still the results are surprisingly coincident, except that Ⱨ.'s distance, though marked 18″, "by pretty exact estimation," must have been erroneous, or a misprint for 28″. The following are my epochs of comparison:

	Pos.	Dist.	Ep.
Ⱨ.	240° 00′	18″·00	1781·78
S.	240° 46′	29″·30	1825·50
Σ.	241° 21′	29″·88	1831·60

μ Herculis has a very sensible proper motion, and well deserves attention from those who may be investigating this lamentably deficient department of astronomical knowledge. The assigned values are these:

	Æ	Dec.
P....	− 0″·29	− 0″·84
B....	− 0″·36	− 0″·72
A....	− 0″·39	− 0″·72

DCXXV.   ψ¹ DRACONIS.

Æ 17ʰ 44ᵐ 47ˢ   Prec. − 1ˢ·09
Dec. N 72° 13′·6   —— S 1″·33

Position 15°·2 (w 9)   Distance 30″·9 (w 9)   Epoch 1833·41
—— 14°·9 (w 9)   ———— 31″·3 (w 9)   —— 1838·37

A neat double star, near the middle of the Dragon's back; it is easily identified, being on the same parallel with γ Ursæ Minoris, the following of the two Guards, and about one-third of the distance from that star towards β Cephei. A 5½, and B 6, both pearly white. This object is 7 Hͪ. ɪᴠ., and Nos. 286 and 287 of Piazzi's Hora XVII., B being his ψ¹ seq. and not ψ³ as erroneously stated, which last follows at Δ Æ of 13ᵐ. Piazzi gave both A and B the same amount of proper motion, but Argelander, who has given very strict attention to the subject, finds that they differ; the following are the values:

$$P.... \begin{cases} \psi^1 \, pr. \; Æ + 0''{\cdot}25 & Dec. - 0''{\cdot}30 \\ \psi^2 \, seq. \quad\; + 0''{\cdot}25 & \quad\;\; - 0''{\cdot}30 \end{cases}$$

$$A.... \begin{cases} \psi^1 \, pr. \; Æ - 0''{\cdot}05 & Dec. - 0''{\cdot}26 \\ \psi^2 \, seq. \quad\; - 0''{\cdot}02 & \quad\;\; - 0''{\cdot}27 \end{cases}$$

$$B.... \begin{cases} \psi^1 \, pr. \; Æ - 0''{\cdot}09 & Dec. - 0''{\cdot}25 \\ \psi^2 \, seq. \quad\; - 0''{\cdot}03 & \quad\;\; - 0''{\cdot}26 \end{cases}$$

Hͪ., who registered this object in October, 1779, did not observe the angle of position, and made the distance 28″·23; but on comparing my results with the following micrometric measures, its constancy may be inferred. With these, I cannot but include the results obtained by reducing Piazzi's admirable observations:

	Pos.	Dist.	Ep.
P.	14° 00′	32″·00	1800
H. and S.	14° 46′	31″·78	1821·46
Σ.	15° 08′	30″·89	1832·34

DCXXVI.   23 M. OPHIUCHI.

Æ 17ʰ 47ᵐ 32ˢ   Prec. + 3ˢ·53
Dec. S 18° 58′·2   —— S 1″·09

Mean Epoch of the Observation . . . . 1835·54

A loose cluster in the space between Ophiuchus's left leg and the bow of Sagittarius. This is an elegant sprinkling of telescopic stars over the whole field, under a moderate magnifying power; the most clustering portion is oblique, in a direction *sp* and *nf*, with a 7th-magnitude star in the latter portion. The place registered is that of a neat pair, of the 9th and 10th magnitudes, of a lilac hue, and about 12″ apart. This object was discovered by Messier in 1764, and it precedes a rich out-cropping of the Milky Way. The place is gained by differentiating the cluster with μ Sagittarii, from which it bears north-west, distant about 5°, the spot being directed to by a line from σ on the shoulder, through μ at the tip of the bow.

After having examined this object, I lowered the telescope a couple of degrees, and gazed for the curious trifid nebula, 41 H̶. IV.; but though I could make out the delicate triple star in the centre of its opening, the nebulous matter resisted the light of my telescope, so that its presence was only indicated by a peculiar glow. Pretty closely preceding this is No. 20 M., an elegant cruciform group of stars, discovered in 1764, which he considered to be surrounded with nebulosity.

DCXXVII. 300 P. XVII. HERCULIS.

Æ 17^h 49^m 23^s Prec. + 2^s·63
Dec. N 18° 21'·3   —— S 0''·93

Position 114°·9 (w 9)   Distance 2''·5 (w 6)   Epoch 1835·61

A close double star, in the space between the hero's head and the Eagle's tail; it is 7°½ north-north-east of a Ophiuchi, or one-quarter of the distance from that star towards γ Lyræ. A 7½, and B 8, both lucid white. This exquisite object was discovered by Σ., and is No. 2245 of the Dorpat Catalogue. Both Σ. and H. make the components equal in magnitude, but on a very careful comparison I cannot but think B the smallest. The following are the results of the previous observations:

Σ.   Pos. 113° 57'   Dist. 2''·63   Ep. 1829·18
H.   115° 21'   2''·99   1830·59

DCXXVIII. 67 OPHIUCHI.

Æ 17^h 52^m 38^s Prec. + 3^s·00
Dec. N 2° 56'·6   —— S 0''·64

Position 143°·6 (w 6)   Distance 54''·7 (w 4)   Epoch 1830·63

A wide double star, in the space between Ophiuchus and Taurus Poniatowski, at the distance of 4°½ to the east-south-east of β Ophiuchi. A 4, straw-colour; B 8, purple. This object is 2 H̶. VI., registered in 1779, but without measures. Piazzi described it, " Duplex. Comes 2'' temporis sequitur 50'' circiter ad austrum." The results obtained at the re-examination by H. and S., were:

Pos. 143° 04'   Dist. 55''·23   Ep. 1823·41

This star is designated by the Greek letter o in the British and other Catalogues; " but," says Mr. Baily, " there is no such star in Bayer's Map; I have therefore rejected it."

DCXXIX. γ DRACONIS.

Æ 17ʰ 52ᵐ 53ˢ   Prec. + 1ˢ·39
Dec. N 51° 30′·6   —— S 0″·62

Position 123°·5 (w 1)   Difference Æ = 9ˢ·7 (w 1)   Epoch 1832·63

A standard Greenwich star, with a telescopic companion, on the crown of the Dragon's head. A 2, orange tint; B 12, pale lilac, a third star in the *nf* quadrant, making with A and B, a neat triangle. A is the Etamin of the Catalogues, from *rás-el-tannín*, the dragon's head; it may be readily found by its being nearly in mid-distance between Polaris and Rasalague. A line from Wega to Pherkad Major also passes through it, and points γ as the following star to β Draconis; the two latter, with its immediate northern neighbours, forming an irregular square, which constitute the Dragon's head; and which once included the contested herd of camels, alluded to under μ and β. The Alphonsine Tables term this star Rasaben, which some have viewed as a corruption of Etamin; but Scaliger rightly points it out as *Rás-al-thu'bán*, head of the devouring basilisk. The rhymester shows the monster's extent:

A line from Dubhe, in the Bear,   sent right the Guards between,
The stars which form the Dragon's tail   in midway will be seen.
Far to the east the body winds,   where Lyra's lustres glow,
A ray from Wega to the Pole,   its lozenge-head will show.

This star affords another proof of the defective state of the degrees of brightness; for Ptolemy registered it γ in magnitude, and has been followed by Ulugh Beigh, Tycho, and La Caille; Hevelius, De Zach, and Groombridge called it 2½; and Flamsteed, Bradley, Mayer, and Piazzi, elevated it to the 2nd rank. In my own comparisons, it appears small for its class. The imputed proper motion has been thus valued, though even its existence may be doubted:

P.... Æ − 0″·31   Dec. − 0″·07
B....   + 0″·06   − 0″·03
A....   + 0″·04   − 0″·04

γ Draconis is a valuable star, from passing very near the zenith of the south of England; and it is celebrated as being the one by which the important discovery of the aberration of light was made. The exertions of Bradley and his friend Molyneux to find its distance from us, are too well known to need repetition; but in the course of them, the fact was established, that " if light is propagated in time, the apparent place of a fixed object will not be the same when the eye is at rest, as when it is moving in any other direction than that of the line passing through the eye and the object; and, that when the eye is moving in different directions, the apparent place of the object will be different." It is recorded that these *savans* were embarrassed beyond measure when, instead of an indication of parallax, they found a regular motion directly opposite to what they expected, which baffled both theory and conjecture. At length one day, when Dr. Bradley was enjoying the then usual and laudable feat of sailing about on the Thames, he observed every time the boat tacked, the direction of the wind, estimated by the direction of the vane, seemed to change. Here was another εὕρηκα, and one even more deserving the sacrifice of a hecatomb than that of Archimedes; the perplexity vanished, and the phenomenon was found to be an optical illusion occasioned by a combination of the motion of light with the motion of his telescope while observing the polar stars. In a word, he enriched Astronomy by the weighty announcement, that "all the phenomena proceeded from the progressive motion of light and the Earth's annual motion in its orbit," or, as he afterwards called it, aberration of light. Having thus detected the existence of this effect, he also determined its constant at 20″, whence it followed that the interval of time in which light travels from the Sun to the Earth is $= 8^m\ 7^s$. The investigations of the constant of aberration, stand thus:

Bradley	20″·00	Brinkley	20″·37
Delambre	20″·25	Struve	20″·35
Bessel	20″·68	Richardson	20″·50

It is very interesting, among other circumstances connected with the important discovery of ABERRATION, that the original entry of the first night's observation at Kew, which confirmed the fact of an unexplained motion in the star before us, is preserved in Bradley's own hand-writing. It is written on an old scrap of paper, and dated 21st December, 1725, exactly in the annexed form and terms; and an excellent facsimile of it is given by Professor Rigaud, in his volume on the *Miscellaneous Works and Correspondence* of Bradley.

Being a Greenwich zenith-star, and therefore little affected by refraction, γ Draconis was employed by our astronomers to ascertain the parallax of the Earth's orbit, and thus determine our distance from the fixed stars. Dr. Hook attacked it at Gresham College, with a 36-foot telescope, in 1669; and twenty years afterwards Flamsteed opened trenches in the same cause. From the united results thus obtained, Whiston concluded the parallax to be 47″, and that a cannon-ball could not have reached the star in 160,000 years, though moving 500 miles an hour. But we now know pretty well that the said ball would not have got over the fortieth part of its journey. The celebrated operations of Molyneux and Bradley followed; but though the observations were of the most rigorous exactness, on a base of 190 millions of miles, they proved in the result, that the parallax was a quantity not cognisable by

Dec 21st Tuesday 5h 40′ sider. time
Adjusted ye mark to ye Plumb Line
& then ye Index stood at 8
5h 48′ 22″ ye star entred
49 52½ Star at ye Cross
51 24 Star went out
s   could
At soon as I let go ye course
screw I perceived ye Star too
much to ye right hand &
so it continued till it passed
ye Cross thread and within a quarter
            was
of a minute after it had passed
            graduat
I turned ye fine screw till I saw
ye light of ye star perfectly
    bissected, and after ye obser
        vation I found ye index
    at 11¾. so that by this
        observation ye
        mark is about 3″¾
        too much south.
        but adjusting
        ye mark and plumbline
I found ye Index at 8½

any astronomical instrument then used, however accurately constructed. Had the parallax amounted to a single second, Bradley considered he should have detected it; his conclusion therefore was, that it did not amount to so much, and consequently, that γ Draconis is above 400,000 times farther from us than the Sun. Such were the approaches towards a barrier which has now been passed, and the apparently insurmountable obstacles to ascertaining the wonderful distance of the stars, are now prostrate before observation and computation. But this, as we shall presently see, was not effected by Bradley's method; zenith distances are so charged with errors of nutation, aberration, and instrumental irregularities, as to make an angle difficult to pronounce upon within a second. See 61 Cygni.

The term Zenith-star, which γ Draconis has obtained at Greenwich, is rather relative than real; for no star has yet been actually observed in the zenith of any observatory, the most interesting of all the points in the apparent concavity of the visible hemisphere. If the Earth had no annual or diurnal motion, nor any nutation of its axis, the zenith of each place on the Earth's surface would be so many fixed points in the heavens; but as we cannot control either of these elements, the actual zenith of every place, is continually changing, so that the true zenith must be singled out from the succession of apparent ones generated in the heavens by the Earth's diurnal motion on its axis. Means might be taken under the equator to reduce the problem to a simple condition.

While on this subject, it may be noted that Al-dhib, or ζ Draconis, is the perpetual zenith-star of Jupiter, whence, from the vast flattened expanse of that planet, it meets Sir A. Hunt's verses better than does our own Polaris:

> Where in the zenith shines the polar star,
> And the cold sun looks dimly from afar,
> Obliquely skims the drear horizon round,
> And flings Periscian shadows on the ground.

DCXXX.   τ OPHIUCHI.

Æ 17ʰ 54ᵐ 22ˢ	Prec. + 3ˢ·26
Dec. S 8° 10′·4	—— S 0″·49

Position AB *round* (w 9)	Distance *round* (w 9)	Epoch 1832·55
—— AC 114°·5 (w 2)	—— 83″·1 (w 1)	
—— AB 214°·0 (w 2)	—— 0″·5 (w 1)	—— 1838·58
—— AC 115°·0 (w 3)	—— 82″·7 (w 2)	
—— AB 227°·0 (w 5)	—— 0″·9 (w 1)	—— 1842·52

A binary star (*vicinissima*) with a companion in the *sf* quadrant, on the Serpent-bearer's left hand; it is 15° to the north-east of the bright star, η Ophiuchi, on the line towards Altair. A 5, and B 6, both pale white; C 10, light blue; two other stars in the field. This most difficult object is 88 ⚹. ɪ., and when discovered in April, 1783, it was merely wedge-shaped, and esteemed by Sir William "the closest of all his double stars." The more we study this active astronomer's labours, the more we are lost in admiration of his zeal and power.

Having noticed the "jam simplex" of Σ.'s Catalogue of 1827, and finding that it had defied the powers of the Dorpat telescope even to elongate it, I gave it a rather hopeless scrutiny in 1832; and making nothing of it, I noted the star C as a future reference. But on taking some examining gazes in 1838, I was surprised and pleased to find A B was measureable; and on repairing to Hartwell in the summer of 1842, I found it was quite elongated. The following are the results of the other observations on this very interesting system:

	Pos.	Dist.		Ep.
⚹.	331°·6	*oblong*		1783·37
	...	*simplex*		1825·67
Σ.	146°·0	*oblong*		1827·28
	192°·9	0″·35		1835·68
	199°·9	0″·44		1836·62

DCXXXI. 95 HERCULIS.

Æ 17ʰ 54ᵐ 43ˢ   Prec. + 2ˢ·54

Dec. N 21° 36′·1   —— S 0″·46

Position 261°·8 (*w* 9)   Distance 6″·1 (*w* 9)   Epoch 1833·78

A neat double star, between the Theban's head and the Eagle's tail, in the spot where Hevelius placed his Cerberus; it is 10° distant from *a* Ophiuchi on a north-north-east line, which leads upon β Lyræ. A 5½, light apple-green; B 6, cherry-red,—besides which there are two small stars in the *sp* quadrant, and a seventh-magnitude in the *np*. This beautiful object is 26 ⚹. ɪɪɪ., and presents a curious instance of difference in colour between components so nearly equal in brightness. Indeed, it was only on rigid comparison, that I was induced to mark the following star half a magnitude smaller than the preceding one, for the other observers note them as of the same size. Piazzi says, "Duplex. Comes ejusdem magnitudin. parumper ad boream sequitur." A slight movement in space has been assigned to A, but it is not sufficiently established yet for entering into the argument of B's being in physical connexion with it or not; the best values are:

P....Æ 0″·00   Dec. + 0″·05
Br... + 0″·04   + 0″·06
B.... + 0″·02   + 0″·06

From a comparison of the epochs of ⚹. and H. and S., there was a hope of 95 Herculis being a binary system, but the later measures go far to prove the object to be optical, and that the proximity of the stars is merely apparent, no connexion existing between them. These are the results for comparison with mine, on which the conclusion is founded:

	Pos.	Dist.	Ep.
⚹.	265° 51′	6″·10	1780·69
H. and S.	261° 52′	6″·23	1821·97
Σ.	261° 45′	6″·06	1829·90

DCXXXII.   21 M. SAGITTARII.

Æ 17ʰ 55ᵐ 01ˢ   Prec. + 3ˢ·62

Dec. S 22° 30'·6   —— S 0"·44

Position 317°·0 (*w* 1)   Difference Æ = 0ˢ·8 (*w* 1)   Epoch 1835·55

A coarse cluster of telescopic stars, in a rich gathering galaxy region, near the upper part of the Archer's bow; and about the middle is the conspicuous pair above registered,—A being 9, yellowish, and B 10, ash coloured. This was discovered by Messier in 1764, who seems to have included some bright outliers in his description, and what he mentions as nebulosity, must have been the grouping of the minute stars in view. Though this was in the power of the meridian instruments, its mean apparent place was obtained by differentiation from μ Sagittarii, the bright star about 2°½ to the north-east of it.

DCXXXIII.   70 OPHIUCHI.

Æ 17ʰ 57ᵐ 22ˢ   Prec. + 3ˢ·01

Dec. N   2° 32'·6   —— S 0"·23

Position	Distance	Epoch
Position 136°·4 (*w* 3)	Distance 5"·43 (*w* 4)	Epoch 1830·76
—— 132°·5 (*w* 6)	—— 5"·98 (*w* 4)	—— 1833·59
—— 130°·6 (*w* 8)	—— 5"·97 (*w* 6)	—— 1835·56
—— 128°·6 (*w* 6)	—— 6"·19 (*w* 6)	—— 1836·81
—— 127°·5 (*w* 6)	—— 6"·26 (*w* 5)	—— 1837·64
—— 126°·5 (*w* 8)	—— 6"·25 (*w* 6)	—— 1838·51
—— 122°·4 (*w* 9)	—— 6"·64 (*w* 6)	—— 1842·55

A binary star in the space between the left shoulder of Ophiuchus and the Serpent's tail, in a rich vicinity; and about 6° to the east-south-east of the bright star β Ophiuchi. A 4½, pale topaz colour; B 7, violet,—and these two point upon a third star, of the 12th magnitude, in the *sf* quadrant. There is also a little *comes* in the *sp*, preceding A by 5ˢ·5. A sensible proper motion influences the primary—and consequently the physically connected acolyte also—to the following values:

P.... Æ + 0"·30   Dec. − 1"·17
B....   + 0"·26   − 1"·09
A....   + 0"·22   − 1"·10

This very interesting object has so remarkable an angular velocity, together with such an appreciable alteration of distance, that I strongly urge such as investigate those extraordinary systems, to peruse the elaborate and able discussions of its orbit by Professor Encke, Sir John Herschel, and Herr Hofrath Mädler. To these a few words may be added, in illustration.

70 Ophiuchi was designated by the letter p, in the British Catalogue; but, as there is no such letter in Bayer's Map, Mr. Baily has properly rejected it in his late edition of Flamsteed. It is 4 ⩑. ɪɪ., and was thus mentioned in its discoverer's paper upon stellar changes: "The alteration of the angle of position that has taken place in the situation of this double star, is very remarkable. October 7, 1779, the stars were exactly in the parallel, the preceding star being the largest; the position, therefore, was 0° 0′ following. September 24, 1781, it was 9° 14′ nf; and May 29, 1804, it was 48° 01′ np; which gives a change of 131° 59′ in 24 years and 234 days. This cannot be owing to the effect of systematical parallax, which could never bring the small star to the preceding side of the large one." To this important passage, H. stamps additional value, by telling us it is particularly written, in his father's MS. observations, that the two stars run together along the equatoreal hair. With such a starting point, it has been tolerably easy to watch the general relations of this system, however involved the computations have proved, from their extreme delicacy. These are the selected epochs with which I compared my own:

		Pos.	Dist.	Ep.
⩑.	{	90° 00′	3″·59	1779·77
		318° 48′	2″·56	1804·41
Σ.		168° 42′	4″·66	1819·63
H. and S.		155° 42′	4″·27	1822·45
S.		148° 18′	4″·76	1825·56
Σ.		140° 14′	4″·78	1828·71
H.		138° 54′	5″·95	1830·36
D.		132° 49′	6″·14	1833·42

From these epochs, the annual retrograde march will be thus:

Herschel I.	5° 32′	1804·41	Myself	1° 37′	1833·59
Struve	9° 86′	1819·63	——	1° 00′	1835·56
H. II. and South	4° 61′	1822·45	——	1° 60′	1836·81
South	2° 58′	1825·56	——	1° 32′	1837·64
Struve	2° 57′	1828·71	——	1° 15′	1838·51
Dawes	2° 00′	1833·42	——	1° 01′	1842·55

which angular velocities indicate that the periastre passed unobserved; and it may be stated, in round numbers, that 70 Ophiuchi describes its ellipse in a period of about eighty years. To arrive at this conclusion has required the arduous application of able mathematicians; but the process by which the elliptical elements of the orbits of binary stars may be obtained with comparative facility, is clearly described by Sir John Herschel in a paper on the investigation of those orbits, which I quoted at No. CCCCLVI. These are his elements of the ellipse in question:

		Apparent Orbit.	Real Orbit.
Major semi-axis	=	4″·331	a = 4″·392
Minor semi-axis	=	2″·717	e = 0″·4667
Position of major-axis	=	310°	π = 294° 04′
Greatest apparent distance	=	6″·152	Ω = 183° 31′
Least apparent distance	=	1″·807	γ = 48° 05′
Position of maximum distance	=	123°	λ = 145° 46′
Position of minimum distance	=	245°	n = −4°·4812
			P = 80·34
			τ = 1807·06 A.D.

where a symbolizes the major semi-axis, e the excentricity, π the inclination of the projected perihelion, Ω the angle of position, γ the inclination of the plane of the orbit, λ the angle between major axis and line of nodes, n the mean annual angular motion, P the periodic time, and τ the epoch of the perihelion, or rather periastral passage. M. Mädler selected some normal places for stated epochs, and arrived at the conclusion that the motions of 70 Ophiuchi do not follow the Newtonian law. But he has not shown how, or where, the point of gravity of the measures can be placed. On the contrary, the phenomena of some of these wonderful binary compounds show, most unequivocally, that there are at least some of them subject to the same dynamic laws, and obedient to the same power of gravitation as governs our own system.

The investigation of this rapid revolver's path, occasioned both trouble and disquietude to H. In comparing the formulæ with observations, he found the only irreconcileable contradiction to the curve was offered by M. Struve's measures in 1818, 1819, 1825, 1826, 1827, and 1828. On which he says, "I have already had occasion to observe on the smallness of some of this eminent observer's distances as compared with my own. Whatever be the cause, whether accidental and limited to the particular cases, or general and extending over masses of observation, I trust he will pardon me for noticing it (in no spirit of evil cavil), as deserving careful examination on both our parts." Such was the temperate yet strong comment of one who justly takes the highest rank in practical and theoretic astronomy; and who endeavoured to reap the utmost value of all the observations which presented themselves, during his analytical treatment, to disentangle the quæsita from the data. To facilitate the numerical calculations of this and other orbits, for which Sir John, *mirabile dictu*, professes a great inaptitude, he invented and adopted a mechanical contrivance, which gives, by simple inspection and reading off, the solution of the transcendental equation,

$$u - e. \ \sin u = n \, (t - \tau)$$

for any given value of its right hand member. I was greatly amused by an examination of this ingenious machine at Slough, shortly after it was made; for notwithstanding the sublimity of its purpose, it was both simple and rude, apparently constructed from the ruins of a Dutch clock and a kitchen jack. With proper practical modifications, it applies to a large class of transcendental equations, of which this is only a particular, and the simplest, case.

It must, however, be confessed that M. Struve has displayed a very laudable anxiety to ascertain the exact value due to the results of his measures; and whether any constant error pervaded his observations. The perfect accordance as to angular position, whether measured on the continent or in England, satisfied that indefatigable astronomer that not a doubt can be entertained of their general correctness, for, in a letter to M. D'Ouvaroff, he says they prove "clairement qu'il ne se trouve dans ces directions aucune source d'erreur de cette nature et de quelque importance." It was otherwise with respect to distances. On comparing the results obtained by the Fraunhofer refractor, with his own former observations, and those of Sir John Herschel, Sir James South, and the Rev. Mr. Dawes, he found the great telescope made the measures smaller than the others. He therefore insisted, that the wire micrometer, by an optical perturbation, must necessarily give all the distances too great, especially with regard to the closest stars; and the evil, he holds, is aggravated according to the comparative weakness of the telescope. But experience has shown, that the Professor's argument cannot be wholly admitted; and that the *disturbing cause* is not yet shown.

Not satisfied with careful reductions, comparison, and reasoning, M. Struve resolved upon the most rigorous test yet devised. "In 1830," he says, "I agreed with M. Bessel, of Königsberg, to observe several stars in common. He has a magnificent heliometer, which forms the ornament of that observatory. The principle on which measures are taken with that instrument is totally different from that of the wire micrometer. An identity of distance given by the two observers would have been an irrefragable proof of their correctness; but a comparison of the distances of thirty-nine stars, taken by both, shows that those of Dorpat are, on an average, $0''\cdot19$ smaller than those of Königsberg. If I can assert with certainty that the Dorpat telescope is so superior in measuring distances that the non-accordance of results indicates an imperfection in the anterior measures; we must also grant that the Königsberg heliometer, though of considerably less optical power than the Dorpat tube, must still be placed in the same rank as a means of measuring. Observations made with such an instrument, and by such an astronomer as M. Bessel, are of the greatest weight." Consequently, in order to detect the hidden source of the error, M. Struve afterwards made numerous researches, and arrived at satisfactory results. There are those, however, who may regard the employing of some months' time in scrutinizing experiments, in order to clear off a quantity, amounting, perhaps, but to the fraction of a second of space, to be supererogatory; yet such an assumption will disappear on recollecting the extreme closeness of the first-class double stars. Many of these are considerably within one second, and if the parallax of the stars, as is most probable, be only $\frac{1}{10}$th of a second, then the sought error becomes of consequence, since that $\frac{1}{10}$th corresponds to a linear distance greater than the diameter of the Earth's orbit. Hence Sir J. Herschel, smarting under a load of the discordant measures of 70 Ophiuchi, soundly rates their "extravagant errors." But he adds, "I would not be misunderstood, as intending by this expression any reflection on the pains and diligence of the various observers, myself among the number, who have occupied themselves with measures of the distances of the double stars. When I speak of extravagant errors, of course I mean only extravagant with reference to the delicacy of the question. An error of half a second in the distance of a moderately close double star might be absolutely fatal in the computation of its orbit, if used in conjunction with others not affected by a proportional one, or erring the contrary way; but where is the double star, the history of whose measurements does not offer such an amount of palpable error?"

Sir John Herschel having pointed out, in 1825, a great diminution of angular velocity in this star, and it being inconsistent with the laws of central forces that this should take place without a corresponding increase of distance, an unusual weight has since been given to that condition; for the angular velocity is inversely as the square of the distance in the *apparent* as well as the *real* orbit, whatever be its position with regard to the line of sight. Under these circumstances, the star was well watched, and the labours of MM. Bessel, Struve, Mädler, Kaiser, Dawes, Schlüter, Galle, and other astrometers, attest the interest it inspired. Among the rest, I paid the greatest attention to my measures, and as I occasionally used a Wollaston diagonal-prism to place the stars either vertically or horizontally, as convenience demanded, I feel that my results merit considerable confidence. Still, being aware how greatly the inflection of light interferes in the exact contact between the wire and the luminous body, and that the parallax of the same lines, or what Σ. terms the optical illusion, opposes a difficulty proportionate to the deficiency of power, in accurately crossing such unequal stars, I was hardly prepared to enter the lists in presence of such instruments as those of Dorpat and Königsberg. I have, however, been highly gratified with a letter from M. Bessel, dated 14th November, 1842, of which the following is an extract:

I am greatly obliged to you for the communication of your observations of the double star *p* Ophiuchi, which appear to be as nearly agreeing to mine, as may be expected. I shall write both in chronological order:

1830·50	...	5″·474 B.	1836·52	...	6″·344 B.
1830·76	...	5″·43 S.	1836·81	...	6″·19 S.
1831·53	...	5″·679 B.	1837·52	...	6″·439 B.
1832·69	...	5″·794 B.	1837·64	...	6″·26 S.
1833·59	...	5″·98 S.	1837·69	...	6″·474 B.
1834·61	...	6″·127 B.	1838·51	...	6″·25 S.
1835·56	...	5″·97 S.	1841·74	...	6″·849 B.

On the whole your measured distances appear to be somewhat smaller than mine, but the difference is not very great, especially at the beginning. Till the year 1835, both series are almost perfectly agreeing one with the other; afterwards their difference becomes more sensible; but with respect to both latest observations, yours is, perhaps, by a trifling quantity too small, while mine is probably too great. At least, the observations of 1842 seem to indicate a small decrease in the distance, which, I believe, ought to be attributed to the observations either of 1841 or of 1842. I regret not having duly reduced the latter, so that I am not able, at this moment, to communicate to you their precise result.

Such is the comparison between the Königsberg heliometer and my refractor; between the mighty detector of parallax, and the more humble labourer in the stellar regions. " The instrument with which Bessel made these most remarkable observations," said Sir John Herschel, " is a heliometer of large dimensions, and with an exquisite object glass by Fraunhofer. I well remember to have seen this object-glass at Munich before it was cut, and to have been not a little amazed at the boldness of the maker who would devote a glass, which at that time would have been considered in England almost invaluable, to so hazardous an operation. Little did I then imagine the noble purpose it was destined to accomplish. By the nature and construction of this instrument, especially when driven by clock-work, almost every conceivable error which can affect a micrometrical measure is destroyed, when properly used; and the precautions taken by M. Bessel in its use, are such as might be expected from his consummate skill." The hazardous operation thus alluded to, was the act of cutting the object glass in two by a plane passing through the diameters of the lenses, and perpendicular to their surfaces.

Having communicated Professor Bessel's letter to my friend the Rev. W. R. Dawes, that zealous astronomer favoured me with the following tabulated scale of comparisons.   In a letter dated 7th March, 1843, he says:

I return Bessel's letter with many thanks, and I have the pleasure of handing you at the same time, a comparison of the measured distances of 70 Ophiuchi by four different observers, arranged in order of time.   The comparison is interesting, as three of the observers used the wire micrometer*.

Epoch.	BESSEL.	STRUVE.	SMYTH.	DAWES.	Epoch.	BESSEL.	STRUVE.	SMYTH.	DAWES.
1830·50	5″·474	..	..	..	1836·52	6″·344	..	..	..
·57	..	..	..	5″·530	·66	..	6″·137	..	..
·76	..	..	5″·43	..	·71	..	..	..	6″·472
·84	..	5″·310	..	..	·81	..	..	6″·19	..
1831·53	5″·679	..	..	..	1837·52	6″·439	..	..	..
·68	..	5″·410	..	..	·64	..	..	6″·26	..
1832·55	..	..	..	5″·71	·69	6″·474	..	..	..
·69	5″·794	..	..	..	·72	..	6″·152	..	..
·75	..	5″·553	..	..	·73	..	..	..	6″·478
1833·42	..	..	..	6″·14	1838·51	..	..	6″·25	..
·59	..	..	5″·98	..	1839·65	..	..	..	6″·534
1834·47	..	5″·852	..	..	1840·59	...	..	..	6″·628
·57	..	..	..	6″·13	1841·71	...	..	..	6″·635
·61	6″·127	..	..	..	·74	6″·849	..	..	..
1835·56	..	..	5″·97	..	1842·53	..	..	..	6″·724
·60	..	6″·108	..	..					

DCXXXIV.   362 P. TAURI PONIATOVII.

Æ 17ʰ 58ᵐ 17ˢ   PREC. + 2ˢ·78

DEC. N 11° 59′·8   —— S 0″·15

POSITION 257°·4 (w 7)   DISTANCE 6″·7 (w 5)   EPOCH 1831·67

—— 257°·9 (w 9)   —— 6″·9 (w 9)   —— 1838·56

A neat double star, in the space between the Polish bull and the Eagle's wing, being 8° to the east of α Ophiuchi, in the line towards Altair. A 8, straw-colour; B 8½, sapphire blue. This is 56 ℍ. III., the preceding component of which, is certainly the smallest; but Piazzi, who registered A of the 8th degree of brightness, says, " Duplex. Comes 7·8ᵉ magnitudinis 0″·6 præcedit paulisper ad austrum," which is not quite in the usual order of these matters. It has been thus measured:

ℍ.	Pos. 260° 18′	Dist. 7″·62	Ep. 1782·46
H. and S.	257° 39′	6″·75	1823·45
Σ.	257° 56′	6″·84	1830·09

Taurus Poniatowskii is a small asterism placed in the heavens, in 1777, by the Abbé Poczobut, of Wilna, in honour of Stanislaus Ponia-

---

* My results for 1842 had not been communicated to Professor Bessel, and consequently are not included in Mr. Dawes's tabular view.

towski, king of Poland; a formal permission to that effect having been obtained from the French Academy. It is between the shoulder of Ophiuchus and the Eagle, where some stars form the letter V, and from a fancied resemblance to the zodiac-bull and the Hyades, became another Taurus. Poczobut was content with seven component stars, but Bode has scraped together no fewer than eighty.

DCXXXV. 37 ♅. IV. DRACONIS.

ℛ 17ʰ 58ᵐ 39ˢ   Prec. − 0ˢ·03
Dec. N 66° 38'·1   —— S 0"·12

Mean Epoch of the Observation . . . . 1837·33

A planetary nebula, between the first twist in the Dragon's body and his head; a fancied line from Polaris to γ Draconis, both of which are well known to the practical astronomer, passes through it in nearly mid-distance, and it makes a triangle, rectangular and isosceles, with the Pole-star and β Ursæ Minoris, the northernmost of the two Guards,

the right angle being at β. This is a remarkably bright and pale blue object, which was discovered by ♅. in February, 1786, and described as having a disc about 35" in diameter, but with very ill-defined edges. There are several telescopic stars in the field, and the annexed diagram affords a notion of its aspect.

The nebula before us is situated in the pole of the ecliptic, whence, being easily found, and always above the horizon, it becomes an object of much interest and utility. The poles of the ecliptic, it will be remembered, are those points in the heavens which are farthest distant from the plane of the Earth's orbit; and as the ecliptic holds so determinate a position in Uranography, the situation of those poles is of paramount theoretic importance. But, for several reasons, the use of the equinoctial poles has been preferred in every-day practice.

DCXXXVI. 100 HERCULIS.

ℛ 18ʰ 01ᵐ 22ˢ   Prec. + 2ˢ·41
Dec. N 26° 04'·7   —— N 0"·12

Position 3°·0 (w 5)   Distance 13"·6 (w 3)   Epoch 1830·69
——— 2°·8 (w 9)   ——— 14"·1 (w 9)   ——— 1836·52

A neat double star, south of the hero's right hand, where some place the bunch of snakes. A and B, both 7, and both pale white, the preceding star being made the primary. This object is 41 Ħ. III., who has erroneously designated it 43 Herculis; and it is formed of Nos. 389 and 390 of Piazzi's Hora XVII. The other measures stand thus:

Ħ.	Pos. 1° 37′	Dist. 11″·71	Ep. 1781·78
H. and S.	2° 25′	14″·28	1823·46
Σ.	2° 54′	13″·85	1831·72

This object lies in a pretty open space midway between Wega and β Ophiuchi; and it is 11° from β Lyræ, on the line towards α Herculis.

DCXXXVII.   73 OPHIUCHI.

Æ 18ʰ 01ᵐ 37ˢ		Prec. + 2ˢ·98	
Dec. N 3° 58′·3		—— N 0″·14	

Position 260°·5 (w 8)	Distance 1″·7 (w 5)	Epoch 1834·60
—— 259°·9 (w 9)	—— 1″·5 (w 5)	—— 1838·74
—— 255°·0 (w 9)	—— 1″·4 (w 6)	—— 1842·39

A close double star, in the space between the left shoulder of Ophiuchus and the Serpent's tail; and 6°½ nearly east of Celbalrai, or β Ophiuchi. A 6, silvery white; B 7½, pale white; and the two point nearly upon a dusky telescopic star in the *sp* quadrant. This lovely object is 87 Ħ. I., of which Sir William says, that a power of 277 only shows it oval, and that it requires 460 to divide it. It certainly must have opened since then, for I see it plainly under very moderate magnifying powers. A comparison of the former measures with mine, under allowances for the discordancies of a difficult star, would imply the same.

Ħ.	Pos. 267° 12′	Dist. 0″·90±	Ep. 1783·32
H. and S.	257° 37′	1″·98	1822·46
Σ.	259° 44′	1″·54	1831·05

DCXXXVIII.   30 Ħ. VII. SAGITTARII.

Æ 18ʰ 03ᵐ 08ˢ		Prec. + 3ˢ·60
Dec. S 21° 36′·1		—— N 0″·27

Mean Epoch of the Observation . . . . 1836·68

A large and coarse cluster of minute stars, principally from 10th to 13th magnitudes, close to the upper end of the Archer's bow, and in the Galaxy. This was registered by its discoverer in May, 1786, and is No. 1998 of his son's Catalogue of 1830. It forms a rich field, but without any disposition to particular form. One bright yellow star has a surrounding galaxy, somewhat more clustered than the other portions.

The mean apparent place of the centre of this object, was obtained by differentiation from the præses of the following compound star, with the alignment of which it is identified.

DCXXXIX. μ^1 SAGITTARII.

Æ 18ʰ 04ᵐ 11ˢ	Prec. + 3ˢ·50
Dec. S 21° 05′·7	—— N 0″·36

Position AB 260°·0 (w 1)	Distance 10″·0 (w 1)	
—— AC 315°·0 (w 2)	—— 40″·0 (w 1)	Epoch 1835·58
—— AD 114°·5 (w 2)	—— 45″·0 (w 1)	

A multiple star on the north end of the Archer's bow, which has recently acquired the honours of the Greenwich list. A 3½, pale yellow; B 16, blue; C 9¼ and D 10, both reddish. This fine object, of which C and D point to a coarse double star in the *np* quadrant, may be found by the out-door gazer about 25° to the east-north-east of Antares, where it is crossed by a line dropped into the south, from *a* Ophiuchi through β. It was registered 7 H. v. in September, 1779, and described as "treble;" but Sir John Herschel, by the discovery of the minute point of light, B, made it quadruple. As this most delicate acolyte can only be caught after intense gazing under high powers, and a train of favouring circumstances, it may be needless to add, that its position and distance from the primary here given, are merely vague estimations, upon a datum afforded by A and C.

Τοξότης, Sagittifer, Sagittipotens, Sagittarius, or Arcitenens, is the ninth of the zodiacal and the third of the southern signs; and it is well marked by eight stars, which form two quadrangles resembling each other, four being in the Milky Way, and four of them out of it. In the days of Eratosthenes, this asterism was figured as a satyr; and so it appears on the Farnese Globe. Manilius describes the *mixtus equo* with a look "scowling and threatening as Hannibal's when commencing the battles of Trebia, Thrasymene, and Cannæ." It is now usually figured in the form of that imaginary animal called the centaur, with a bow drawn to the full extent, as if in the act of shooting off an arrow; and this, we are taught, is to personify Chiron, the schoolmaster, or Crotus, the huntsman. But though Jove seems to have nominated both those gentlemen to the berth, he himself held the tiller; for in the illuminated manuscript Almanack of 1386 it is written, "The Schoter es ye pncipal howce of Jupit." The chaste Diana, however, disputed her father's right to this "howce;" and by a coin of Gallienus, representing Sagittarius under the legend *Apollini Conservatori*, it seems that the tutelage was extended. Nor have the classical ancients had the field to themselves. Novidius asserts that the figure originally represented Joash, the son of Ahaziah, shooting off the arrows of Elisha; "but," observes Hood, "let them believe it that will."

These *niaiseries* are merely here resuscitated to advance the ends of justice; for the meagre mythological story of this creature may be placed among the many proofs before us, that the Greeks could not have been the first to give names to the constellations. And there are vestiges of other constellations on this spot. Sir William Jones traced the Hindú zodiac to have existed more than 3000 years ago, when the Sun entered Aswin; and among the early Orientals the stars γ, ϵ, δ, ζ, ϕ, τ, σ, ν, o, ξ and π, constituted the fan of lions' tails, which, according to the *Brahmánda Paráan*, was flirted by Múla, the wife of Chandra. Then again the Arabians had once very different figures from the classical ones, for their XXth Lunar Mansion, as may be gathered from their *al-na'ám al-wáridah*, and *al-na'ám al-sádirah*, the ostriches going out to water, or returning; but the ostrich being a non-drinking bird, it is read as *na'aïm*, camels or cattle. Their later terms *al-rámi*, the archer; *ḳaus*, the bow; *zujj-al-nushábah*, or *naṣl*, the point of the arrow,—are probably derived from Ptolemy. The vacant space between the Archer's shoulders and Capricorn's horns, was their XXIst Mansion of the Moon, and called *al-beldah*, "quod urbem oppidumve denotat," says Dr. Hyde; but why so blank a region should be designated *city*, the learned commentator sayeth not. Ḳázwíní, Fírúzábádí, and Tízíní, mention *al-beldah* and its boundary of six dark or small stars, called *el-ḳeládah*, or necklace. All this clashes with the claims of Astræus, Chiron, Atlas and Cleostratus: still the similitude of design is indicative of a common parental point between the classic centaur and the image recognised by the middle Arabs.

Although the genealogy of this "igneal trigon" is not pure, it has attracted much attention among occult sages, who of erst ruled the roast, insomuch that no undertaking was plotted without a preliminary consultation of the stars. According to the renowned astrologian doctor, Arcandum, whose book, printed in 1542, is "ryghte pleasaunte to reade," a man born in Sagittarius is to be thrice wedded, to be very fond of vegetables, to become a matchless tailor, and to have three special illnesses; but as the last attack of sickness is to befal the patient at eighty years of age, it is not of paramount moment. Such was the *science* in which potentates delighted! Well might "Veritas in profundo" be the astrologer's motto.

A line from the head of Andromeda, carried over a Pegasi, points on the north to Perseus, and on the south through Capricorn to the Archer; the which, as Sherburne archly tells us, *aims* at the meridian at midnight about the end of June, "two months after the Centaur has *galloped* over it." The galley-rhymester has given a general rule for picking up the constellation on a fine night:

> From Deneb in the stately Swan   describe a line south-west,
> Through bright Altair in Aquila   'twill strike the Archer's breast.

Sagittarius seems to have been much attended to on its enrolment among the old forty-eight constellations, and has thus varied the number of components, in its descent to our times:

Ptolemy	31 stars	Hevelius	26 stars
Tycho Brahé	16	Flamsteed	69
Halley	21	Bode	339

DCXL.   6 Σ. N. TAURI PONIATOVII.

Æ 18ʰ 04ᵐ 21ˢ   PREC. + 2ˢ·91
DEC. N   6° 49′·2   —— N 0″·38

MEAN EPOCH OF THE OBSERVATION . . . . 1835·55

A fine planetary nebula, in a rich vicinity, on the shoulder of the
Polish bull; with many telescopic stars in the field. It is small but
very bright, and is thought by Professor Struve, its discoverer, to be one
of the most curious objects in the heavens. Sir John Herschel, who
gave it a careful scrutiny, considered it as "something between a planetary nebula and a bright round nebula." It lies 7° from β Ophiuchi in
the direction of the Dolphin's lucida, that is, about east-by-north.

DCXLI.   2002 H. CLYPEI SOBIESKII.

Æ 18ʰ 07ᵐ 37ˢ   PREC. + 3ˢ·55
DEC. S 19° 55′·5   —— N 0″·67

POSITION 250°·0 (w 1)   DISTANCE 20″·0 (w 1)   EPOCH 1836·55

A telescopic double star, in Sobieski's shield, preceding a loose but
bright group in the *nf* quadrant. A 8½, and B 10, both grey. This
object is described by its discoverer, Sir John Herschel, as being placed
in a faint nebula, and he has so figured it; but to my instrument the
stars have only a glow, and some gleamy points of light around them.
The nebulosity is represented as being elliptical, and 50″ in diameter.
Having found μ Sagittarii by the alignment already given, this object
must be sought at the distance of 1°½ to the north-east; it is close upon
26 P. XVIII.

DCXLII.   24 M. CLYPEI SOBIESKII.

Æ 18ʰ 08ᵐ 49ˢ   PREC. + 3ˢ·51
DEC. S 18° 27′·5   —— N 0″·77

MEAN EPOCH OF THE OBSERVATION . . . . 1835·56

A beautiful field of stars, below the sinister base of the Polish shield,
and in a richly clustering portion of the Milky Way. This object was
discovered by Messier in 1764, and described as a mass of stars—a great
nebulosity of which the light is divided into several parts. This was
probably owing to want of power in the instrument used, as the whole is
fairly resolvable, though there is a gathering spot with much star dust.

A double star, H. and S., No. 264, follows in the *sf* quadrant, and a wider one *sp*, which is their No. 263; these must be the objects alluded to by Piazzi, Nota 25, Hora XVIII., " Quatuor sequuntur ad austrum, quorum una duplex." A line led from *a* Aquilæ to the south-west over λ Antinoi, and continued as far again, reaches 24 Messier.

DCXLIII. 16 M. CLYPEI SOBIESKII.

Æ 18^h 09^m 44^s	Prec. + 3^s·40
Dec. S 13° 50'·5	—— N 0"·85

MEAN EPOCH OF THE OBSERVATION . . . . 1835·56

A scattered but fine large stellar cluster, on the nombril of Sobieski's shield, in the Galaxy, discovered by Messier in 1764, and registered as a mass of small stars in the midst of a faint light. As the stars are disposed in numerous pairs among the evanescent points of more minute components, it forms a very pretty object in a telescope of tolerable capacity. Its mean apparent place was obtained by differentiation with the equatoreal* instrument upon *μ* Sagittarii, from which it lies north-half-east, at 7° distance; where it is on the line produced between *θ* Ophiuchi and *δ* Aquilæ, which also reaches Altair.

DCXLIV. 18 M. CLYPEI SOBIESKII.

Æ 18^h 10^m 36^s	Prec. + 3^s·48
Dec. S 17° 11'·7	—— N 0"·93

POSITION 322°·0 (*w* 1)   DISTANCE 35"·0 (*w* 1)   EPOCH 1835·56

A neat double star, in a long and straggling assemblage of stars, below the Polish shield. A 9, and B 11, both bluish. This cluster was discovered by Messier in 1764, and registered as a mass of small stars appearing like a nebula in a 3½-foot telescope; which affords another instance that the means of that very zealous observer did not quadrate with his diligence. The whole vicinity is, however, very rich, and there are several splendid fields about a degree to the south of this object. It was also differentiated from *μ* Sagittarii, for which it lies 4° to the north-by-east, in the direction of *ε* in the Eagle's tail.

---

* Though I have written equatoreal as a difference between the instrument and its equatorial relations, I have only custom to plead in justification. Thus the rare tract which Ramsden printed in 1774, is intituled, a *Description of a new universal Equatoreal*. A copy of this is preserved in the British Museum, in the *Opuscula Philosophica*. (486.)

DCXLV.   17 M.   CLYPEI SOBIESKII.

Æ 18ʰ 11ᵐ 23ˢ   Prec. + 3ˢ·46
Dec. S 16° 15'·8   —— N 0"·99

Mean Epoch of the Observation . . . . 1836·68

The horse-shoe, or Greek-*omega* Ω, nebula, just below Sobieski's
shield; discovered by M. Messier in 1764, and registered as a train of
light without stars, about 5' or 6' in extent. As with the two preceding
objects, its place hangs upon μ Sagittarii, from which it bears north-by-
east, about 5° distant, in the line

towards ε Aquilæ. In my telescope
charged with a moderate eye-piece,
this curious nebula is well seen,
though not to the extent of convo-
lution figured by Sir John Herschel.
A magnificent, arched, and irresolvable luminosity occupies more
than one third of the area, in a
splendid group of stars, among
which are Nos. 38 and 43, Hora
XVIII. of Piazzi; they are principally from the 9th to the 12th
magnitudes, reaching more or less all over the field, somewhat in the
accompanying form.

In describing this object, Sir W. Herschel tells us it is " a wonderful
extensive nebulosity of the milky kind. There are several stars visible
in it, but they can have no connection with that nebulosity, and are,
doubtless, belonging to our own system scattered before it." It was
also rigidly examined by his son, who says, " The chief peculiarities
which I have observed in it are, first, the resolvable knot in the following portion of the bright branch, which is in a considerable degree
insulated from the surrounding nebula, strongly suggesting the idea of
an absorption of the nebulous matter; and secondly, the much smaller
and feebler knot at the *np* end of the same branch, where the nebula
makes a sudden bend at an acute angle."

This vicinity, as must have been already inferred, is particularly rich
in clustering portions of the Milky Way. Indeed, the wonderful quantity of suns profusely scattered about here would be confounding, but
for their increasing our reverence of the Omnipotent Creator, by reveal-
ing to us the immensity of the creation. Space, the grand theatre of
astronomical meditation, is here illimitable; and so great is the number
of stars in some parts of this Via Lactea, that ḤℲ. observed 588 of them
in his telescope at the same time; and they continued equally numerous
for a quarter of an hour. In a space about 10° long, and 2°½ wide, he
computed that there were no fewer than 258,000 stars. Carrying this
view into adjoining regions, words and figures necessarily fail, for the

powers of mind falter in such vast and awful conceptions. Yet, despite of the anger of the Zoilus before alluded to, the inquiry deserves to be persevered in; in the words of poor Recorde:

> If Reason's reache transcende the skye,
> Why shoulde it then to Earth be bounde?
> The witte is wronged and leadde awrye,
> If mynde be maried to the grounde.

Warmed with the magnificence and boundless nature of his theme, Recorde proceeds:

> When Scipio behelde out of the high heavens the smallenes of the earth with the kingdomes in it, he coulde no lesse but esteeme the travaile of men most vaine, which sustaine soe muche grief with infinite daungers to get so small a corner of that lyttle balle,—so that it yrked him (as he then declared) to considere the smalnes of that their kingdome, whiche men so muche did magnifie. Whosoever therefore (by Scipion's good admonishment) doth minde to avoide the name of vanitie, and wyshe to attayne the name of a man, lette him contemne those trifelinge triumphes, and little esteeme that little lumpe of claye: but rather looke upwarde to the heavens, as nature hath taught him, and not like a beaste go poringe on the grounde, and like a scathen swine runne rootinge in the earthe.

Ovid had, however, been before him in appreciating the distinctive faculty of man, in the beautiful lines:

> Pronaque cum spectent animalia cætera terram;
> Os homini sublime dedit, cœlumque tueri
> Jussit, et erectos ad sidera tollere vultus.

But the efforts of Sir William Herschel upon that astral feature, the Galaxy, must ever command admiration, and will speak through successive ages to the advantage of science, in a silent but undying tongue. In 1784, having made some skilful gauges, he remarks: "By these observations it appears that the utmost stretch of the space-penetrating power of the 20-feet telescope could not fathom the profundity of the Milky Way, and that the stars which were beyond its reach, must have been further from us than the 900th order of distances." * * * "From the great diameter of the mirror of the 40-feet telescope we have reason to believe, that a review of the Milky Way with this instrument would carry the extent of this brilliant arrangement of stars as far into space as its penetrating power could reach, which would be to the 2300th order of distances, and that it would then probably leave us again in the same uncertainty as the 20-feet telescope." Such is the wondrous arrangement of the celestial bodies in space! Suns upon suns, scattered over the regions of immensity, which stretch towards infinity on every side; yet by the fiat of the UNFATHOMABLE ETERNAL, all is properly arranged and compensated for, so as to ensure unbroken harmony and stability, amid the apparent complexities of the action of an ever evolving and transmuting cause.

A writer of the day speaks of the dazzling constellations which " crowd the boundaries of space," but what he means by boundary is not told. To be sure, so finite a term is less bewildering than the above view of the immensity of the Universe, especially where it is thought

> The radiant orbs
> That more than deck, that animate the sky,
> Are life-infusing suns of other worlds.

DCXLVI.   40 DRACONIS.

$\cancel{R}$ 18^h 12^m 00^s Prec. $-$ 4^s·48

Dec. N 79° 58 ·3   —— N 1″·05

Position 235°·8 (w 9)   Distance 20″·5 (w 9)   Epoch 1836·37

——— 235°·3 (w 9)   ——— 19″·9 (w 9)   —— 1839·81

A fine double star, in the neutral ground between the left foot of Cepheus, and the Lesser Bear's flank, whence it not unfrequently appears as 40 Cephei vel Draconis. A 5½, and B 6, both white. This object is Flamsteed's Nos. 40 and 41 of the constellation; Piazzi's 62 and 63, Hora XVIII.; and 67 ♓. iv.; it has been thus micrometrically and accurately measured:

	Pos.	Dist.	Ep.
♓.	235° 33′	20 ·65	1782·78
H. and S.	235° 04′	21″·36	1822·29
Σ.	235° 38′	20″·62	1832·95

A confusion in the British Catalogue has created the suspicion of a great proper motion in one of these stars, since Flamsteed's time; but the various measures disprove its existence. My meridional observations, as I have elsewhere said, are not meant to grapple with the accurate determination of so nice an element; yet they would have detected the amount imputed. But the question has been satisfactorily settled by Mr. Baily, who says, that the $\cancel{R}$ of 40 Draconis was deduced by Flamsteed from that of 28 Draconis, and would require to be increased 6′ 16″·2 in order to correspond with Bradley's observations. It is to be found at nearly one quarter of the distance from Polaris to γ Draconis, to the south-east of ε Ursæ Minoris.

DCXLVII.   η SERPENTIS.

$\cancel{R}$ 18^h 13^m 02^s Prec. $+$ 3^s·14

Dec. S 2° 56′·0   —— N 1″·14

Position 76°·9 (w 5)   Distance 110″·9 (w 3)   Epoch 1835·61

A star with a very minute companion, in the caudine portion of the Serpent's body, nearly midway between η Ophiuchi and α Aquilæ. A 4, golden yellow; B 13, pale lilac; two other telescopic stars in the north portion of the field. This most delicate and difficult object was measured with an excellent annular micrometer, presented to me by Mr. Francis Baily; and the reductions were made according to his formulæ, preparatory to using it in an attack upon Halley's comet. It is 14 ♓. vi.; but there is an unaccountable difference between the details first registered and mine:

Pos. 99° 07′   Dist. 81″·04   Ep. 1780·47

which would lead us to suppose, that for ♓.'s 9° 07′ sf, we ought to read *nf*. But though this would somewhat adjust the position of the stars, the disparity in the distance between them would remain inexplicable. Probably the micrometer was to blame, in 1780.

This star is suspected of being variable; and it is charged with proper motions in space, to the following values, which are involved in the above precessions:

$$
\begin{array}{lll}
P.\ldots\; & \mathcal{R}\; - 0''{\cdot}67 & \text{Dec.} - 0''{\cdot}68 \\
B.\ldots\; & - 0''{\cdot}56 & - 0''{\cdot}65 \\
A.\ldots\; & - 0''{\cdot}59 & - 0''{\cdot}65 \\
\end{array}
$$

DCXLVIII. 28 M. SAGITTARII.

$$\mathcal{R}\;\; 18^{\text{h}}\; 14^{\text{m}}\; 41^{\text{s}} \qquad \text{Prec.} + 3^{\text{s}}{\cdot}69$$
$$\text{Dec. S}\;\; 24°\; 56'{\cdot}9 \qquad\qquad \text{——} \; \text{N}\; 1''{\cdot}28$$

MEAN EPOCH OF THE OBSERVATION . . . . 1836·70

A compact globular cluster of very minute stars, between the Archer's head and his bow. It is not very bright; and is preceded by two telescopic stars in a vertical line. Messier, who enrolled it in 1764, describes it as a nebula without a star, and seen with difficulty in his 3½ foot telescope. But Sir William Herschel resolved it, and placed it among the stellar clusters: his son has recommended it as a testing-object for trying the space-penetrating powers of telescopes. It lies nearly midway between β Ophiuchi and β Lyræ, and about half a degree to the north-east of 105 Herculis.

DCXLIX. 59 SERPENTIS.

$$\mathcal{R}\;\; 18^{\text{h}}\; 19^{\text{m}}\; 01^{\text{s}} \qquad \text{Prec.} + 3^{\text{s}}{\cdot}07$$
$$\text{Dec. N}\;\; 0°\; 06'{\cdot}3 \qquad\qquad \text{——} \; \text{N}\; 1''{\cdot}66$$

POSITION 313°·9 (w 5)	DISTANCE 4″·2 (w 7)	EPOCH 1831·58
—— 314°·7 (w 6)	—— 4″·4 (w 6)	—— 1835·49
—— 314°·2 (w 9)	—— 3″·9 (w 9)	—— 1842·53

A very neat double star, in the Serpent's tail. A 5½, yellow; B 8, indigo blue. This object is 12 Ӈ. I., and from observations made by Sir William in 1781 and 1802, he thought the position had retrograded about 2°; and that there was an increase of distance from two diameters to four or five of interval between the stars. He considered the latter alteration as well ascertained, and ascribable to a real motion of 59 Serpentis. This assurance was considerably strengthened twenty years afterwards, when H. and S. took it in hand, and concluded there was a rapid rotation of one star about the other, in a plane nearly passing through the eye. Recent observations, however, controvert these conjectures; and a comparison of the principal results, albeit they are not quite so coincident as might be expected, will stamp the relative fixity of the components:

	Pos.	Dist.	Ep.
Ḥ.	314° 33′	3″·50±	1781·79
H. and S.	318° 05′	4″·15	1822·95
Σ.	314° 02′	3″·95	1828·62
D.	314° 40′	4″·40	1830·62

To identify 59 Serpentis let an east-south-east ray be shot from β Herculis through a, which will be found to lie two-fifths of the way; and it is 38° south-half-west of Wega. It is closely followed by the little star, 77 P. XVIII.

DCL. 39 DRACONIS.

Ꞃ 18ʰ 21ᵐ 36ˢ   Prec. + 0ˢ·88

Dec. N 58° 42′·5   —— N 1″·89

Position AB 5°·5 (w 4)   Distance 3″·3 (w 4) ⎫
—————— AC 21°·7 (w 9)   —————— 89″·2 (w 7) ⎬ Epoch 1836·39
                                             ⎭

A triple star in the first inflection of the monster's neck, and lying in mid-distance between γ and δ Draconis. A 5, pale white; B 8½, light blue; C 7, ruddy. This rather difficult object is 7 Ḥ. I., and its suspected retrograde angular motion seems confirmed, if we may depend on the Slough data, though not to the mean velocity assigned. These are the registered results:

	Ḥ. 1780·78			H. and S. 1823·54	
Pos. A B	12° 52′	Dist. 2″·50±	Pos. A B	3° 55′	Dist. 3″·59
A C	26° 38′	caret	A C	21° 55′	90″·20

Σ. 1831·87
Pos. A B   6° 00′   Dist. 3″·12
     A C 21° 36′   88″·89

Though the primary has been overlooked by some of the authorities, it certainly seems to have a proper motion through space; these being the values assigned by Piazzi and Baily:

P.... Ꞃ − 0″·09   Dec. + 0″·02
B....   − 0″·06   + 0″·04

DCLI. 25 M. SAGITTARII.

Ꞃ 18ʰ 22ᵐ 14ˢ   Prec. + 3ˢ·53

Dec. S 19° 10′·2   —— N 1″·94

Position 149°·5 (w 1)   Distance 27‴·0 (w 1)   Epoch 1836·60

A loose cluster of large and small stars in the Galaxy, between the Archer's head and Sobieski's shield; of which a pair of 8th magnitudes, the principal of a set something in the form of a jew's harp, are above registered. The gathering portion of the group assumes an arched form, and is thickly strewn in the south, or upper part, where a pretty knot of minute glimmerers occupies the centre, with much star-dust around. It was discovered in 1764 by Messier, and estimated by him at 10′ in extent: it is 5° to the north-east of μ Sagittarii, and nearly on the parallel of β Scorpii, which glimmers far away in the west.

DCLII.   χ DRACONIS.

R 18^h 23^m 55^s Prec. — 1^s·07
Dec. N 72° 39′·8   —— N 2″·09

Position 54°·3 (w 1)   Difference R = 14^s·6 (w 1)   Epoch 1834·61

A star with a minute companion, on the main flexure of the creature's body; it is on the same parallel with γ Ursæ Minoris, and forms the south-west vertex of a nearly equilateral triangle with β Cephei and δ Ursæ Minoris. A 4½, pale yellow; B 13, bluish; and there is a third star, telescopic, seen under an averted eye in the *sf* quadrant. The primary is bright for the 4½ rank, but in this I have followed Piazzi; yet from Ptolemy's δ, almost all the observers down to Groombridge have rated it of the 4th magnitude. Its extraordinary proper motion through space, thus valued, brought it upon my working list:

P.... R + 1″·72   Dec. — 0″·33
B.... + 1″·76   — 0″·35
A.... + 1″·75   — 0″·36

The adjacent star to the south, φ Draconis, has been lately pronounced to be double by Σ., who designates the constituents to be of the 4½ and 6¼ magnitudes, 0″·6 apart. But this is in the *monstre* refractor of the new and grand observatory of Poulkova.

DCLIII.   δ URSÆ MINORIS.

R 18^h 23^m 56^s Prec. — 19^s·23
Dec. N 86° 35′·4   —— N 2″·05

Position 149°·0 (w 1)   Difference R = 53^s·7 (w 1)   Epoch 1834·74

A standard Greenwich star with a very distant telescopic companion, in the middle of the Lesser Bear's tail. A 3, greenish tinge; B 12, grey; and a minute star in the *np* quadrant. This star has an extraordinary movement in consequence of its situation so near the Pole, and the earlier observations are not sufficiently accurate to build upon. The declination seems steady enough; but Piazzi lumps all the motions in R together, and gives as "motus proprius annuus in arcu" a quantity

$= - 284''{\cdot}68$; while Mr. Baily assigns the value of its proper motions only, on the following scale:

$$\text{Æ} + 0''{\cdot}47 \qquad \text{Dec.} + 0''{\cdot}02$$

Flamsteed, in whose day it followed ν Aquilæ to the meridian, seems to have observed it only once, viz. on June 3, 1691; "and then," says Mr. Baily, "at the special request of Sir Christopher Wren." Such a slight has, however, been more than made up of late by the attentions of Messrs. Knorre, Encke, and Stratford, who have given its mean apparent place for every day in the year, in their respective ephemerides. This, however, could only be an accommodation to fixed observatories and grand instruments.

In sharing the somewhat unmerited honour of daily duty with Polaris, in the computations of the *Nautical Almanac*, it will be recollected that this is not the only encroachment which δ has made upon a. This star appears on the Catalogues as Yildun; but the epithet happens to be miscopied from Hyde's notes on Ulugh Beigh for *Yilduz*, the "star," in Turkish, and signifying *Yilduz Shemáli*, the North Star, *i. e.* a Ursæ Minoris.

DCLIV.   22 M. SAGITTARII.

$$\text{Æ}\ 18^{\text{h}}\ 26^{\text{m}}\ 25^{\text{s}} \qquad \text{Prec.} + 3^{\text{s}}{\cdot}66$$
$$\text{Dec. S}\ 24°\ 01'{\cdot}4 \qquad\text{——}\quad \text{N}\ 2''{\cdot}31$$

Mean Epoch of the Observation . . . . 1835·57

A fine globular cluster, outlying that astral stream, the Via Lactea, in the space between the Archer's head and bow, not far from the 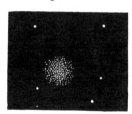point of the winter solstice, and midway between μ and σ Sagittarii. It consists of very minute and thickly condensed particles of light, with a group of small stars preceding by 3^{m}, somewhat in a crucial form. Halley ascribes the discovery of this in 1665, to Abraham Ihle, the German; but it has been thought this name should have been Abraham Hill, who was one of the first council of the Royal Society, and was wont to dabble with astronomy. Hevelius, however, appears to have noticed it previous to 1665, so that neither Ihle nor Hill can be supported.

In August, 1747, it was carefully drawn by Le Gentil, as seen with an 18-foot telescope, which drawing appears in the *Mémoires de l'Académie* for 1759. In this figure three stars accompany the cluster, and he remarks that two years afterwards he did not see the preceding and central one: I, however, saw it very plainly in 1835. In the description he says, "Elle m'a toujours parue tres-irrégulière dans sa figure, chevelue, et repandant des espèces de rayons de lumière tout autour de son diamètre." This passage I quote, "as in duty bound;" but from familiarity with the object itself, I cannot say that I clearly understand how or why his telescope exhibited these "espèces de rayons." Messier, who registered it in 1764, says nothing about them, merely observing that it is a nebula without a star, of a round form; and Sir William Herschel, who first resolved it, merely describes it as a circular cluster, with an estimated profundity of the 344th order. Sir John Herschel recommends it as a capital test for trying the space-penetrating power of a telescope.

This object is a fine specimen of the compression on which the nebula-theory is built. The globular systems of stars appear thicker in the middle than they would do if these stars were all at equal distances from each other; they must, therefore, be condensed towards the centre. That the stars should be thus accidentally disposed is too improbable a supposition to be admitted; whence Sir William Herschel supposes that they are thus brought together by their mutual attractions, and that the gradual condensation towards the centre must be received as proof of a central power of such a kind.

DCLV. α LYRÆ.

Æ	18ʰ 31ᵐ 30ˢ	PREC.	+ 2ˢ·01
Dec.	N 38° 38'·1	—	N 2"·75

POSITION 135°·2 (w 6)	DISTANCE 43"·1 (w 4)	EPOCH 1830·84
—— 137°·9 (w 8)	—— 42"·7 (w 4)	—— 1837·51
—— 140°·3 (w 9)	—— 43"·4 (w 9)	—— 1843·34

A standard Greenwich star, with a little companion in the *sf*, on the preceding *yoke* of the Lyre. A 1, pale sapphire; B 11, smalt blue; a third star in the *nf* quadrant, and various glimpse ones in the field of view: the whole preceded at a distance by a coarse triple star, on the same parallel of declination.

α Lyræ is one of the insulated bodies, and is worthy of ranking with Sirius, Canopus, and Capella. Yet, by the experiments of Dr. Wollaston, it appeared that the light afforded us by this star is about $\frac{1}{180,000,000,000}$th part of the Sun's light, or only about ⅓th part of the light of Sirius, but still it offers a glorious blaze. From its peculiar circumstances, it was selected by Dr. Brinkley for investigating the parallax, and after much toil he came to the conclusion, that the angle which the radius of the Earth's orbit subtends at that star, is not less than 2", nor much more than 2"·52. From this the distance of α Lyræ was estimated as being at least 400,000 times ninety-five millions of miles, or semi-diameters of our orbit. Piazzi also assigned it a similar value in parallax, and Calandrelli even more. To a measurement made by Sir William Herschel, under a magnifying power of 6450, its diameter subtended an angle of 0"·3553. But the case was fraught with difficulties apparently insurmountable; for the parallaxes hitherto assigned become proportionally less as the means of observation improve. Thus, after the closest and most skilful application to this star, Airy has pronounced that its annual parallax is too small to be sensible to our best instruments. M. Struve, finding this brilliant object the best adapted for his high northern station, observed it very closely during the apparitions of four successive years; of which the seventeen days of July, 1836, by means of thirty-four equations on the least squares, yield a resulting parallax of $0''\cdot125$, subject to the probable error of $0''\cdot055$. These computations yield to a Lyræ the distance of one and a half million times our own distance from the Sun; while those of the Bishop of Cloyne estimate the same space only at twenty billions of miles. This brought forward Struve's ingenious analogy into the vastness of space, by which it is inferred that the stars of the 1st magnitude are, in general, about two millions of our solar distance from us; those of the 6th magnitude, sixteen millions; those of the 8th magnitude, sixty millions; and those of the 12th or smallest visible in the Frauenhofer tube, 640 millions, or 60,800,000,000,000,000 of miles. This is a grand conception, of which the interest is not destroyed even by the subsequent splendid labours of Bessel, on 61 Cygni *.

It is impossible to gaze on a Lyræ and its minute companion, without recalling the Apostle's words, " one star differeth from another star in glory;" yet the little one bears illumination very fairly. The measures previous to my own, were:

H̶.	Pos. 116° 14′	Dist. 42″·99	Ep. 1792·32
H. and S.	132° 07′	42″·11	1822·87
D.	134° 58′	42″·52	1830·42

A discussion of all the results affords testimony that the deductions of Piazzi, Baily, and Argelander, as to the proper motions of the primary, are pretty exact, assuming the small star at rest; and that the connexion between the two is merely optical and accidental. The values assigned for these movements in space, are these:

P....	Æ + 0″·28	Dec. + 0″·25
B....	+ 0″·30	+ 0″·28
A....	+ 0″·29	+ 0″·29

a Lyræ is called Wega in the Catalogues, from *Wáki*, in the compound name of *al-nesr-al-wáki*, the falling eagle; a part being put for the whole in the Alphonsine Tables. It is also termed Lucida Lyræ, and is greatly in favour with navigators; but not so much so as it is to be, for, from the effects of the periodic revolution of the celestial equinoctial pole around the pole of the ecliptic, it will be within 5° of the North Pole about 10,000 years hence, and will be gazed at as the polar gem of the northern hemisphere. Instead, therefore, of passing the meridian within a few degrees of the zenith, as it now does in these

---

* This will in some measure recal Kepler's ingenious suggestion on the same point. He held that, on the surface of a sphere, there could be only thirteen points equally remote from each other and from the centre; and supposing the nearest fixed stars to be as distant from each other as they are from the Sun, he concludes that there are but thirteen stars of the 1st magnitude. At twice the distance from the Sun there may be four times as many, and so on.

latitudes, it will then remain nearly stationary with respect to the horizon. In the mean time its present place is readily found by running a ray from Arcturus through Alfecca; and this ray prolonged, leads to the Swan. Wega, Arcturus, and Polaris, form, very nearly, a large right-angled triangle. From the southward, a line from Antares, through Al Rá'i, in the head of Ophiuchus, carried a similar distance beyond, leads upon Lyra; and the three well known stars in the neck of Aquila point nearly upon it. To lisp it in numbers:

Altair in Aquila that flames,	and Wega's lucid light,
To Rasalague westward join'd,	form a triangle bright.
Of which the apex to the north,	the Lyre pertains unto,
A truly noble point it forms,	a gem of sapphire blue.

The Lyre is one of the old forty-eight constellations, and has been called Nablon, Λύρα, Cythara, Sulhafa, Testudo, Fides, Fidis, and Fidicula, from the Hebrew, Greek, Latin, and Arabic terms for a stringed instrument. But there is no end of its names. We are told, that the design originated from musical sounds struck out of the shell of a tortoise, which had become putrid during the inundation of the Nile; but the tenacious nature of intestines preserved them till the subsidence of the waters, when they dried into chromatic strings. Hence the ambiguous verse in the beautiful ode of Horace, wherein the mute fishes having the melodious voice of the swan was a sore puzzle, till Molyneux rubbed off the obscurity by thus explaining the matter:

> O mutis quoque piscibus
> Donatura cycni, si libeat, sonum !

The predecessors of Ptolemy called this asterism χέλυς, from which the Italians made their *Sheliak*, from *Shelyâk*, a name now improperly applied to β Lyræ. It has also obtained the designation of *Alohor* in the Catalogues, or *Al-lúrá*, the Greek word λύρα, in the Arabic character. For ages it was simply delineated as a cithara. In the celebrated MS. of Cicero's *Aratus*, now in the British Museum, which is pronounced to be of the second or third century, there is a large coloured drawing of the truly classical lyre; but in after times, somebody or other thought proper to place it in the claws of a descending bird, ycleped Vultur cadens. The Arabs are blamed for this, but it does not appear on their Borgian and Dresden Globes, while their Al Nesr, which Hyde renders *Vulture*, really means the large stone-eagle. Ideler, on Kazwíni, rubs the transcribers of the Alphonsine Tables, and by the known cucumbergerkin etymological process, converts the puzzling phrase, *pupilla deferens,* into *vultur cadens;* and from *vultur cadens* came our old Falling Grype. The dumpy base which the tortoise constituted, was drawn into beauty and utility by adding the horns of Apollo's bull; but while some ascribe this invention to Mercury, or to Orpheus, there are others who deem it to be an allusion to the harmony of the spheres. It was well known as *Jugum;* but Schiller's *Cœlum stellatum Christianum*, stamps it as the manger wherein our Saviour was placed; thereby destroying the dream of Novidius, that it symbolized the harp of David. The constituents have been thus numbered:

Ptolemy	. . . . 10 stars	Hevelius	. . . 17 stars
Tycho Brahé	. . . 11	Flamsteed	. . 21
Bayer	 13	Bode	 166

The elegant simplicity of form in the early Greek asterism, as contrasted with its modern successor, may be seen by any one who will compare the drawing above cited, or even that on the Farnese Globe, with the one which Hevelius produced 1000 years afterwards, where the awkward bird holds an instrument which neither in ancient nor modern times, ever had existence. In the *Astronomia Reformata*, Riccioli designates it *Albegala*, and its lucida *Brinek*.

DCLVI.   151 P. XVIII. LYRÆ.

$\mathbb{R}$ 18^h 32^m 47^s Prec. + 2^s·11
Dec. N 35° 55'·0   —— N 2"·85

Position 180°·2 (w 7)   Distance 3"·8 (w 4)   Epoch 1834·69

A very neat double star, on the preceding horn of the Lyre, and 3° to the south of Wega, its lucida. A 8, pale white; B 9, lilac, a distant telescopic companion in the following part of the field. This elegant object was discovered by Σ., at Dorpat, but appears to have been first measured by H. The best previous results stand thus:

H. Pos. 178° 50'   Dist. 3"·85   Ep. 1828·65
Σ.   180° 15'   3"·96   1830·95

DCLVII.   2 AQUILÆ.

$\mathbb{R}$ 18^h 33^m 31^s Prec. + 3^s·28
Dec. S 9° 12'·0   —— N 2"·92

Position 133°·8 (w 2)   Distance 55"·0 (w 1)   Epoch 1831·58

A brightish star with a minute companion, in the space between Antinous and the Polish shield, 20° to the south-south-east of β Ophiuchi. A 5, yellowish; B 11, purple; several minute stars in the field, which is in a condensed part of the Via Lactea. This object was first classed by Ħ. in 1781, under a distance of 42"·74, but no angle of position. He showed it to Dr. Watson, under very high magnifying powers, a practice which some hypercritics were then arraigning. Sir William, however, was not to be diverted from an object by groundless argument. "The naturalist," says he, "does not think himself obliged to account for all the phenomena he may observe; the astronomer and optician may claim the same privilege. When we increase the power we lessen the light in the inverse ratio of the square of the power; and telescopes will, in general, discover more small stars the more light they collect; yet, with a power of 227 I cannot see the small star near the star following o Aquilæ, when, by the same telescope, it appears very plainly with the power of 460: now, in the latter case, the power being more than double, the light is less than the fourth part of the former."

DCLVIII. 26 M. CLYPEI SOBIESKII.

Æ 18ʰ 36ᵐ 27ˢ   Prec. + 3ˢ·29
Dec. S 9° 33′·3   —— N 3″·18

Mean Epoch of the Observation .... 1835·26

A small and coarse, but bright, cluster of stars, preceding the left foot of Antinous, in a fine condensed part of the Milky Way; and it follows 2 Aquilæ by about half a degree. The principal members of this group lie nearly in a vertical position with the equatorial line, and the place is that of a small pair in the south, or upper portion of the field. This neat double star is of the 9th and 10th magnitudes, with an angle = 48°, and it follows an 8th, the largest in the assemblage, by 4ˢ. Altogether the object is pretty, and must, from all analogy, possess affinity among its various components; but the collocations and adjustments of these wondrous firmamental clusters, and their probable distances, almost stun our present faculties. There are many astral splashes in this crowded district of the Galaxy, among which some fine specimens of what may be termed *luminiferous ether*, are met with.

DCLIX. 5 AQUILÆ.

Æ 18ʰ 38ᵐ 13ˢ   Prec. + 3ˢ·09
Dec. S 1° 07′·5   —— N 3″·33

Position AB 121°·5 (w 9)   Distance 13″·3 (w 7) ⎫
—— AC 145°·0 (w 1)   —— 30″·0 (w 1) ⎭ Epoch 1838·60

A delicate triple star, between the Serpent's tail and Antinous, in a wavy nook of the Eagle's boundary, which the map-makers have carried to 10° south; whence this star sometimes appears as η Serpentis. A 7, white; B 8, lilac; C 14, blue. The object forms a curve into the *sf* quadrant, but C is so minute as to have escaped former observers except Ħ., who, however, did not measure it, and its place is here merely estimated for finding. Piazzi, indeed, saw a star which possibly may have been C, for he says, Nota 176, Hora XVIII., "Duplex. Comes 1″·4 temporis sequitur 30″ circ. ad austrum." This gives an angle of 144°·5, and 36″·6 for the distance. The former measures of A B, compared with mine, indicate permanence both in angle and distance; they are:

	Pos.	Dist.	Ep.
Ħ.	121° 37′	11″·90	1796·60
H. and S.	122° 42′	14″·46	1823·45
D.	121° 10′	13″·36	1830·64
Σ.	121° 29′	13″·22	1832·45

5 Aquilæ is about 17° to the south-west of Altair, in a line towards η Ophiuchi; and its place is indicated by a ray carried from Gemma through β and α Herculis.

DCLX. 110 HERCULIS.

Æ 18ʰ 38ᵐ 46ˢ   Prec. + 2ˢ·58
Dec. N 20° 23′·9   —— N 3″·37

Position 110°·0 (w 1)   Distance 55″·0 (w 1)   Epoch 1831·68

A most delicate though wide double star, in the space over the Eagle's tail. A 5, pale yellow; B 16, dusky. This object was among the tests sent to me by Sir John Herschel, for trial; and though the acolyte was caught by gleams on steady gazing in a darkened field, it is certainly a *minimum visibile* of my telescope. The details, of course, are mere estimations, but carefully taken. It lies nearly midway between β Ophiuchi and β Cygni. It is also 18° south-half-east of Wega, and exactly halfway between β Herculis and a Delphini.

DCLXI. ε LYRÆ.

Æ 18ʰ 39ᵐ 02ˢ   Prec. + 1ˢ·96
Dec. N 39° 30′·3   —— N 3″·40

Position		Distance		
AB	25°·3 (w 8)		3″·5 (w 4)	
—— AC	172°·5 (w 8)	——	207″·3 (w 5)	Epoch 1830·73
—— CD	157°·1 (w 7)	——	2″·8 (w 4)	
—— AB	23°·9 (w 8)	——	3″·2 (w 4)	
—— AC	172°·9 (w 9)	——	206″·8 (w 9)	—— 1836·45
—— CD	154°·6 (w 6)	——	2″·5 (w 5)	
—— AB	21°·9 (w 8)	——	3″·3 (w 5)	—— 1839·78
—— CD	152°·8 (w 7)	——	2″·5 (w 5)	
—— AB	20°·6 (w 6)	——	3″·2 (w 5)	—— 1842·59
—— CD	150°·9 (w 8)	——	2″·6 (w 5)	

A double-double, or rather a multiple star, on the frame of the Lyre, and only 1°½ to the north-east of Wega. A 5, yellow; B 6½,

ruddy; C (which is Flamsteed's 5 Lyræ), 5, and D 5½, both white. Between these pairs, and two-thirds over from ε towards 5, are three small stars forming a curve to the south, the two smallest being H.'s "debilissima" couple, of the 13th magnitude, having an angle = 220°, and about 45″ apart. Another star has, I understand, been added to these, by the Earl of Rosse's powerful telescope; but the annexed is the diagram as seen in my refractor, where A B are the lowest or northern pair, and C D the highest.

The proper motions in space assigned to this quadruple related system, form a link in the chain of evidence which proves the connexion; they are to these values:

ε Lyræ.			5 Lyræ.		
B.... Ⱅ	+ 0″·03	Dec. + 0″·07	B.... Ⱅ	+ 0″·03	Dec. + 0″·08
A.....	+ 0″·01	+ 0″·07	A....	+ 0″·22	+ 0″·09

and the following results, compared with my measures, will show the slow but certain change of the position angle of both the pairs, in a direction *np sf*, or retrograde:

ε Lyræ, A B.

Ⱨ.	Pos. 33° 55′	Dist. 3″·44	Ep. 1779·83
H. and S.	25° 53′	4″·01	1822·12
D.	23° 49′	3″·56	1830·53
Σ.	26° 06′	3″·03	1831·44

5 Lyræ, C D.

Ⱨ.	Pos. 173° 28′	Dist. 3″·50	Ep. 1779·83
H. and S.	159° 56′	3″·80	1822·42
D.	157° 30′	3″·01	1830·60
Σ.	155° 10′	2″·57	1831·44

This is a very elegant object, and merits the closest attention. The naked eye sees an irregular-looking star near Wega, which separates into two pretty wide ones under the slightest optical aid. Each of these two will be found to be a fine binary pair, and between the two sets are the minute acolytes above-mentioned. They present a vast field for contemplation: ε and 5 resemble each other so closely in magnitude, distance, orbital retrogradation, and proper motions, as to afford palpable evidence of their forming a twin system; and a combined rotation about a common centre of gravity, may be suspected. Though the resulting values of this retrocession vary, the absolute amount is not violently discordant when all the circumstances of the case are borne in mind; and they are unanimous in its *np sf* direction. Indeed, it may be roundly stated, that B will revolve round A in about a couple of thousand years; C will take a similar circuit around D, in perhaps half that time; and possibly both double systems may move about the central ones, in something less than a million of years. But what is this duration when compared to that astounding unit of time, the *annus magnus* of the whole creation! Imagination is lost in the conjecture.

DCLXII. ζ LYRÆ.

Ⱅ 18ʰ 39ᵐ 15ˢ   Prec. + 2ˢ·06
Dec. N 37° 26′·5   —— N 3″·42

Position 149°·6 (w 9)   Distance 43″·8 (w 9)   Epoch 1834·78

A fine double star, at the bottom of the preceding horn of the Lyre, about a couple of degrees due south of the preceding object, and the two, with Gemma, form a neat triangle. A 5, topaz; B 5½, greenish, being No. 189 P. Hora XVIII. This object is 2 Ⱨ. v., and its first register may be thus compared with results obtained by the data afforded in the Palermo Catalogue:

$$\text{Ⴙ. Pos. } 152° \ 18' \quad \text{Dist. } 41''\cdot99 \quad \text{Ep. } 1779\cdot66$$
$$\text{P. } \quad 148° \ 12' \quad \quad 44''\cdot20 \quad \quad 1800$$

Allowing for the components being charged with a small movement in space, the results of all the subsequent astrometers, show the relative constancy of these stars. The proper motions alluded to are thus stated for the star A, and there is every indication that B is also affected to a similar amount:

$$\text{P.... } \mathcal{R} + 0''\cdot04 \quad \text{Dec. } + 0''\cdot01$$
$$\text{B.... } \quad + 0''\cdot03 \quad \quad + 0''\cdot07$$

Bianchini thought he saw the southern of the two stars double, at Verona, in 1735, through telescopes of great focal length by Campini and Cellius. Short also saw it surrounded by five small stars; and several other reports obtained. But they are dismissed by Sir John Herschel, who observes, "Doubtless, in a part of the heavens so crowded with stars, numbers of minute stars may be seen near it in good telescopes; but the division of one of the large stars into two is a fact which we may be allowed to doubt." Among the *comites* most visible, are three of the 11th and 12th magnitudes near the northern verge of the field, and two in the *sp* quadrant, of which the largest is of the 10th lustre. ζ Lyræ is readily found.

DCLXIII. 197 P. XVIII. ANTINOI.

$$\mathcal{R} \ 18^\text{h} \ 41^\text{m} \ 07^\text{s} \quad \quad \text{Prec. } + 3^\text{s}\cdot21$$
$$\text{Dec. S } \ 6° \ 05'\cdot3 \quad \quad \text{—— N } 3''\cdot58$$

Position 168°·9 (*w* 3)　Distance 99''·0 (*w* 1)　Epoch 1833·66

A wide pair of stars, between the foot of Antinous and Sobieski's shield. A 7, orange tint; B 9, cerulean blue. This object is thus described by Piazzi: "1"·3 temporis, 1'½ ad austrum, sequitur alia 9ᵃ magnit." There are many telescopic stars in the field; and in the *sp* quadrant are two open double stars, each trending *np* and *sf*, and about 35" or 40" apart. Of these one pair is very faint, and the preceding and brightest constitutes Σ.'s No. 2391. It is posited about 4° to the west-by-south of the bright star λ Antinoi, in one of the strange contortions termed *hooks* by Mr. Baily, and in which the map-makers have specially delighted in this vicinity.

The asterism into which the sporades are here constellated, has ever been regarded as a needless memorial of a "sporco nume," and one which might now well be dispensed with. Ptolemy evidently eschews it, for after tabulating Aquila, he places half a dozen *amorphotæ*, from which, he says, if you like, you may form Antinous: wherefore Bailly's reprimanding him for the exaltation is not altogether in place, since it seems that the Bithynian was never regularly gazetted by the ancients. Among these, however, are the four fine stars, δ, λ, κ, and θ, which make the distinguishing rhombus, all of which are marked γ, or 3rd magnitude: on the Borgian globes this quadrangle appears as *el-dhalimaïn*, the two ostriches. Tycho brought the Bithynian forward in the *Progymnasmata*, as a separate constellation, and Hevelius followed him; but most astronomers lump his stars and those of the Eagle together. Thus Flamsteed omitted his figure on the 23rd plate of the great Atlas, whereon Aquila, Sagitta, Vulpecula, and Delphinus, are represented; but it appears in Bayer's *Uranometria*, 1603, and also on the neat plate in Kepler's *Stella Nova, in pede Serpentarii*, &c., 1606. Schickard, the "Astroscopius," designates the asterism *Antinous*, sive *Ganymedes;* but Thomas Hood, speaking of the Bithynian name, indignantly exclaims, "but whereupon that name should come I know not, except it were that some man devised it there, to currie favour with the Emperor Adrian."

DCLXIV. 11 M. ANTINOI.

Æ 18ʰ 42ᵐ 32ˢ   Prec. + 3ˢ·22
Dec. S 6° 27'·2   —— N 3"·69

Mean Epoch of the Observation . . . . 1835·57

A splendid cluster of stars, closely to the east-south-east of the above described object; it precedes the left foot of Antinous, and is on the dexter chief of Sobieski's shield. This object, which somewhat resembles

a flight of wild ducks in shape, is a gathering of minute stars, with a prominent 8th-magnitude in the middle, and two following; but by all analogy these are decidedly between us and the cluster. This, however, was not the opinion of Kirch, its discoverer, who, in 1681, described it as a small obscure spot, with a star shining through, and rendering it more luminous. Dr. Derham first resolved it into stars, with his 8-foot reflector, as shown in the *Philosophical Transactions* for 1733: "it is not,' said he, "a *nebulose*, but a cluster of stars, somewhat like that which is in the Milky Way." *That* in the Milky Way!

Dr. Halley drew up a description of the nebular wonders, in 1716. They then amounted to six; but he says, "there are undoubtedly more which have not yet come to our knowledge." He could little foresee the rich harvest which was soon to be reaped; but his reasoning was very fair for a commencement. "Though all these spots," he observes, "are in appearance but little, and most of them but a few minutes in diameter; yet since they are among the fixt stars, that is, since they have no annual parallax, they cannot fail to occupy spaces immensely great, and perhaps not less than our whole solar system. In all these so vast spaces it should seem that there is a perpetual uninterrupted day, which may furnish matter of speculation, as well to the curious naturalist as to the astronomer."

This fine object is on the shield by which Hevelius intended, FOR EVER, to honour the name of John III., king of Poland. In the *Prodromus Astronomiæ*, he appears to be uncommonly elated on having raised it to the perpetual memory of the glorious liberator of Vienna— "ob immensa ejus merita, heroicas animi dotes, magnanimitatem, et ob res strenuè, ac fortiter gestas." He was delighted in being able to place it in the happiest part of the firmament, where all the members and neighbours are significant. "I wish you to know, benevolent reader," he says, "that this shield consists of seven lucid stars, partly of the 4th magnitude; four of these are placed in the border of this shield, and designate the princes of our serene king, who at that time were all among the living. In the middle of the shield I have designed a cross, in eternal remembrance of the battles most happily fought by him for the Christian faith: three notable stars shine in this cross, of which one indicates his own royal person, another the queen's, and a third the princess's, his only daughter; so that these seven stars represent the whole reigning family." This, and much more, shows his anxiety and hope of its eternal duration; but, poor fellow, Mr. Baily has taken the field, and Sobieski is one of the first among the asterisms recently doomed to proscription. I hope his pruning-knife is to be applied to many other interlopers, most of whom are far more petty than this.

DCLXV. *ν*¹ LYRÆ.

Æ 18ʰ 43ᵐ 48ˢ		PREC. + 2ˢ·23	
DEC. N 32° 38'·0		—— N 3"·81	

POSITION A B 78°·0 (w 2)   DISTANCE 35"·0 (w 1)
—————— A C 124°·1 (w 8)   —————— 58"·6 (w 8)   } EPOCH 1836·81
—————— C D 208°·0 (w 1)   —————— 12"·0 (w 1)

A quadruple star on the cross-piece of the Lyre. A 6, pale yellow; B 13, bluish; C 11, pale blue; D 15, blue; and there are three other stars in the field. This object is 40 ʜ. v., who only measured A and C, though, perceiving B and not D, he records it as triple. Sir James South, who appears not to have seen B or D, also observed the stars A and C; and these are the results previous to mine:

ʜ.　Pos. 118° 27'　Dist. 56"·79±　Ep. 1781·73
S.　　　123° 58'　　59"·84　　1825·61

A comparison of these measures appeared to give an increase of angle = 5° 31' in forty-four years, and 3"·05 in the distance, a motion which is not confirmed by my observations. As B and D are too small to bear any illumination, the places above given are mere, though careful estimations, to point out their situation in the quadrangular figure.

This very delicate set is followed by ν^2 Lyræ, a whitish star of the 6th magnitude, at an angle of 175°, and a distance of about 15′ of space; and they are both just to the south of β, the following object.

DCLXVI. β LYRÆ.

Æ. 18^h 44^m 09^s PREC. + 2^s·21

DEC. N 33° 10′·8   —— N 3″·84

POSITION AB 150°·1 $(w\ 9)$ DISTANCE 45″·8 $(w\ 9)$ ⎫

—— AC 319°·5 $(w\ 2)$ —— 60″·0 $(w\ 1)$ ⎬ EPOCH 1834·73

—— AD 25°·0 $(w\ 1)$ —— 71″·0 $(w\ 1)$ ⎭

A standard Greenwich star, which, with its companions, forms a quadruple system on the frame of the Lyre. A 3, very white and splendid; B 8, pale grey; C 8½, faint yellow; D 9, light lilac; and there is a neat but minute double star in the south vertical. The two principal components of this object, as has been remarked by H. and S., are nearly coincident with the neighbouring ζ Lyræ, both in angle and distance; but the comparison cannot be applied to the relative brightness. Piazzi noticed the acolyte B, but considered it hardly visible; "3″ temporis," he says, "1′ circiter ad austrum sequitur alia vix pene visibilis:" yet he must also have observed it as one of his meridian objects, and accurately observed it too, for I find, by a reduction of the mean apparent places of Nos. 215 and 216, Hora XVIII., the following admirable results:

Pos. 150°·0 Dist. 47″·3 Ep. 1800

β Lyræ is called Sheliak, but improperly, since the word is corrupted from the Greek χέλυς, by the Arabians, a name intended to designate the whole asterism. A line drawn from Wega to Altair, about one quarter the distance, passes between β and γ Lyræ, which are very readily distinguishable in the lower, or southern part of the instrument; or, as our galley poet expresses it,

Along the line from Wega down,   near six degrees in space,
Tow'rds Altair, in the Eagle's neck,   you'll pitch on Sheliak's place.

β Lyræ has a small proper motion in space, of which the value has been thus assigned:

P.... Æ - 0″·13   Dec. - 0″·25
B....   + 0″·03   - 0″·02
A....   - 0″·03   - 0″·02

This star has recently been added to the *stellæ versatiles*. Ptolemy rated it γ in brightness, and most of his successors also ranked it a 3rd magnitude, except Hevelius, who classed it 3½. It is now less bright than γ, its neighbour, which we may presume was not the case when Bayer lettered them. It was, however, pronounced to be variable by Goodricke, from 3rd to 5th magnitudes, in a period of 6 days 9 hours; but neither the variation nor the period were considered as accurately ascertained. Lately, a comparison of his observations with

those of Westphal and Argelander, affords 6^d 10^h 34^m 56^s for its mutations from the maximum to the minimum of light. Cassini, when he
observed the new star of 1670, in Collo Cygni, compared it frequently
with β and γ Lyræ, without perceiving or suspecting that β was
variable. A standard scale of magnitudes is a desideratum.

DCLXVII.   226 P. XVIII. DRACONIS.

Æ 18^h 44^m 37^s PREC. + 0^s·87
DEC. N 59° 09'·1   —— N 3"·88

POSITION 98°·1 (*w* 6)   DISTANCE 1"·7 (*w* 3)   EPOCH 1833·97

A close double star, in the middle of Draco's neck, and nearly half-
way between γ Ursæ Minoris and δ Cygni. A 8 and B 9, both silvery
white, with a neat pair of small stars in the field, and followed on the
parallel at about 4^m by the bright binary system, *o* Draconis. No. 226
is a beautiful object, but, from the inflection of light, rather difficult to
manage. It was discovered and classed by Σ., and is No. 2410 of the
great Dorpat Catalogue. My measures were made before the arrival
of that active astronomer's in this country, and the reference for comparison was unfortunately later than it ought to have been, for there
seems strong presumption of a binary system; the results being:

Pos. 96°·2   Dist. 1"·67   Ep. 1831·87
97°·4   1"·49   1832·79
99°·0   1"·31   1834·91

DCLXVIII. σ SAGITTARII.

Æ 18^h 45^m 20^s PREC. + 3^s·72
DEC. S 26° 29'·3   —— N 3"·94

POSITION 243°·5 (*w* 2)   DISTANCE 309"·0 (*w* 1)   EPOCH 1837·62

A star with a distant companion on the Archer's right shoulder; it
is the eastern vertex of a splendid triangle formed with μ and δ, and is
on the parallel of Antares. A 3, ruddy; B 9½, ash-coloured; a small
star on a similar rhumb to the *sf* makes the whole form a triangle with
the apex to the north. A has a slight proper motion in space attributed
to it, of this amount:

P.... Æ − 0"·06   Dec. − 0"·11
B....   + 0"·09   − 0"·08

This star has been placed among the variable ones, under a probability of its varying from the 2nd to the 4th magnitudes; but its
low altitude might occasion apparent changes. Ptolemy, Ulugh Beigh,
Bradley, De Zach, and Mayer, have classed it 3; Flamsteed, 3½; Tycho
Brahé and Hevelius, 4; but Bode makes it 2, and Lacaille and Pigott, 2½.

Piazzi, however, is unusually severe upon the last-named gentleman, saying, that he cannot see with what reason he has considered σ Sagittarii as variable, nor whence he gathered that astronomers did not agree in its position : "Non video," ait, "quo fundamento Cl. Pigott stellam hanc inter variabiles recenseat, nec unde collegerit astronomos de ejus positione non convenire." (*Philosophical Trans.*, vol. LXXVI.) Now on referring to the passage in question, we find no mention of the *position*, and the remarks, which relate wholly to the comparative brilliance, must be deemed creditable to the zeal and ability of Mr. Pigott. He says, "Mr. Herschel, with great reason, has placed this star among those which probably have changed their magnitudes. I had long since remarked the singular disagreement in all the Catalogues, which induced me to observe it frequently, particularly in 1783, 1784, and 1785, when it appeared of the 2·3rd magnitude, and brighter than π Sagittarii."

This spot was of some note among the Arabians, and was probably then a relic of some earlier system of astronomy. There are eight bright stars in two divisions, the preceding of which is headed by γ Sagittarii, and the other by σ. The former is named *Min al-na'áim al-Wáridah*, and the latter, *Min al-na'áim al-Sádirah*, the going and returning camels, or *na'ám*, ostriches. The first set lie in the Milky Way, at the point of the Archer's arrow; and the two together form the XXth Mansion of the Moon. See No. DCXXXIX.

DCLXIX.   57 M. LYRÆ.

Æ 18h 47m 37s	Prec. + 2s·22
Dec. N 32° 50'·1	—— N 4"·13

Mean Epoch of the Observation . . . . 1835·57

This annular nebula, between β and γ on the cross-piece of the Lyre, forms the apex of a triangle which it makes with two stars of the 9th magnitude; and its form is that of an elliptic ring, the major axis of which trends *sp* and *nf*. This wonderful object seems to have been noted by Darquier, in 1779; but neither he nor his contemporaries, Messier and Méchain, discerned its real form, seeing in this aureola of glory only "a mass of light in the form of a planetary disc, very dingy in colour." Sir W. Herschel called it a perforated resolvable nebula, and justly ranked it among the curiosities of the heavens. He considered the vertices of the longer axis less bright and not so well defined as the rest; and he afterwards added, "By the observations of the 20-feet telescope, the profundity of the stars, of which it probably consists, must be of a higher than the 900th order, perhaps 950." This is a vast view of the ample and inconceivable dimensions of the spaces of the Universe; and if the oft-cited cannon-ball, flying with the uniform velocity of 500 miles an hour, would require millions of years to reach Sirius, what an incomprehensible time it would require to pass so overwhelming an interval as 950 times the distance! And yet, could we arrive

2 F 2

there, by all analogy, no boundary would meet the eye, but thousands and tens of thousands of other remote and crowded systems would still bewilder the imagination.

In my refractor this nebula has a most singular appearance, the central vacuity being black, so as to countenance the trite remark of

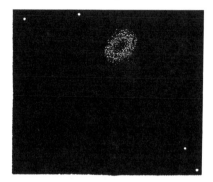

its having a hole through it. Under favourable circumstances, when the instrument obeys the smooth motion of the equatoreal clock, it offers the curious phenomenon of a solid ring of light in the profundity of space. The annexed sketch affords a notion of it.

Sir John Herschel, however, with the superior light of his instrument, found that the interior is far from absolutely dark. "It is filled," he says, "with a feeble but very evident nebulous light, which I do not remember to have seen noticed by former observers." Since Sir John's observation, the powerful telescope of Lord Rosse has been directed to this subject, and under powers 600, 800, and 1000, it displayed very evident symptoms of resolvability at its minor axis. The fainter nebulous matter which fills it, was found to be irregularly distributed, having several stripes or wisps in it, and the regularity of the outline was broken by appendages branching into space, of which prolongations the brightest was in the direction of the major axis.

DCLXX. θ^1 SERPENTIS.

Æ 18ʰ 48ᵐ 16ˢ   Prec. + 2ˢ·98

Dec. N 3° 59′·8   —— N 4″·19

Position 103°·9 (w 9)   Distance 21″·6 (w 9)   Epoch 1834·78

A neat double star on the tip of the Serpent's tail, lying about 14° to the east-south-east of Altair. A 4½, pale yellow; B 5, golden yellow, being θ^2, or Piazzi's No. 237; and there are other stars in the field. This fair and easy object for a moderate telescope is 6 H. iv., whose distance in 1779·80 was 19″·37; but no angle of position was then taken. By deductions from the mean apparent places of the Palermo Catalogue, Struve's early registers in the *Observationes Astronomicæ*, and the operations of Herschel and South, the exact registered measures, previous to my own, are:

P.	Pos. 98° 30′	Dist. 21″·80	Ep. 1800
H. and S.	104° 26′	21″·68	1822·11
Σ.	104° 00′	21″·59	1822·73

from which there appears to have been no appreciable alteration in a

lapse of thirty-five years, but what may be charged to the proper
motions in space, A having them to the following amount:

$$P.... \text{Æ} + 0'''.04 \quad \text{Dec.} + 0''.32$$
$$B.... \quad + 0''.05 \quad\quad + 0''.10$$
$$A.... \quad + 0''.01 \quad\quad + 0''.11$$

On the strength of a great number of observations, Piazzi also
assigns a spacial movement to B; but subsequent experience has not
confirmed his conclusion.

This star appears in Piazzi and other Catalogues as Alya, supposed
to be the Arabic *alyah*, a broad sheep's tail; but no such name is known
to the Arabian astronomers. It was substituted for *Alioth* found in the
Alphonsine Tables, by an ingenious but erroneous conjecture of Joseph
Scaliger's. Nor is it the name only which has given rise to discussion,
for there is much uncertainty as to its magnitude. Ptolemy ranked it δ
in brightness, and was followed by Ulugh Beigh, La Caille, and Pigott;
Tycho Brahé, Hevelius, Bayer, Flamsteed, Bradley, and De Zach, made
it of the 3rd magnitude; Piazzi and myself saw it constantly 4½; and
Montanari found it of the 5th. It must, therefore, be variable, and
should be carefully watched.

DCLXXI. 2024 H. AQUILÆ.

Æ 18ʰ 48ᵐ 25ˢ   Prec. + 2ˢ·83
Dec. N 10° 09′·6   —— N 4″·20

Position 220°·0 (*w* 1)   Distance 6″·5 (*w* 1)   Epoch 1835·57

A double star on the verge of a bright coarse cluster, on the bird's
southern wing. A 7½, yellow; B 11, dusky; and there is another neat
pair preceding these, having a distance of 25″ between them. This
object was discovered by H., with the 7-foot equatoreal, in 1828. The
observations here recorded, were made during occasional bright corus-
cations of an Aurora Borealis.

This object is placed in a barren vicinity between the two astral
streams of the Milky Way. It lies 6° to the north of θ Serpentis,
above described; and about 12° to the east of γ Aquilæ, on its parallel.

DCLXXII. *o* DRACONIS.

Æ 18ʰ 48ᵐ 50ˢ   Prec. + 0ˢ·88
Dec. N 59° 11′·7   —— N 4″·24

Position 347°·6 (*w* 7)   Distance 30″·4 (*w* 5)   Epoch 1830·78
—— 345°·5 (*w* 9)   —— 30″·3 (*w* 9)   —— 1837·89

A neat double star, in a fine field, on Draco's neck, and midway
between γ Ursæ Minoris and δ Cygni. A 5, orange yellow; B 9, lilac,

being Piazzi's No. 248. The results of the epochs previous to the above, including deductions from the mean apparent places of the Palermo Catalogue, are:

	Pos.	Dist.	Ep.
⛢.	0° 00′	26″·65	1780·76
P.	355° 00′	31″·00	1800·00
H. and S.	349° 11′	29″·95	1822·14

which, compared with my observations, show that the angle of position has a *np sf*, or retrograde, motion; and the distance may have increased. The measures of Σ. between 1832 and 1835, also countenance the conclusion, respecting an orbital change.

DCLXXIII.   11 AQUILÆ.

Æ 18ʰ 51ᵐ 44ˢ	Prec. + 2ˢ·76
Dec. N 13° 25′·0	—— N 4″·49

Position 240°·9 (*w 7*)   Distance 19″·1 (*w 5*)   Epoch 1832·61

A delicate double star on the Eagle's tail, where it forms the south-west vertex of an equilateral triangle with ε and ζ. A 7, pale white; B 10, smalt blue, accompanied by a wide pair of stars in the *nf* quadrant. This object is rather difficult to measure, from the feebleness of the *comes;* it is 32 ⛢. III., who merely estimated its distance at about 7″, in 1781. My *departure* is therefore taken from the following observation of Sir James South:

Pos. 240° 32′   Dist. 19″·66   Ep. 1825·02

The magnitude of the primary is assumed from the Palermo Catalogue, which is the plan throughout this cycle; but it certainly appeared bright enough to be rated among the 6th, on careful comparison.

DCLXXIV.   263 P. XVIII. AQUILÆ.

Æ 18ʰ 52ᵐ 40ˢ	Prec. + 2ˢ·73
Dec. N 14° 41′·7	—— N 4″·57

Position 289°·1 (*w 9*)   Distance 6″·5 (*w 5*)   Epoch 1836·55

A fine double star on the Eagle's tail, closely following the bright star, ε. A 8½, pale yellow; B 10½, sapphire blue. This handsome test object was discovered by Sir W. Herschel, and seems to have undergone no change, either in angle or distance, in an interval of upwards of half a century. These are the registered results for comparison:

	Pos.	Dist.	Ep.
⛢.	286° 00′	5″·00±	1783·40
S.	288° 19′	6″·68	1825·27
Σ.	288° 39′	6″·45	1830·96

DCLXXV. γ LYRÆ.

Æ 18ʰ 52ᵐ 57ˢ Prec. + 2ˢ·24

Dec. N 32° 28'·5 —— N 4"·59

Position 23°·5 (*w* 1) Difference Æ = 3ˢ·7 (*w* 1) Epoch 1834·59

A lustrous star with a minute distant companion, on the cross-piece of the Lyre; and 7° to the south-south-east of Wega. A 3, bright yellow; B 11, blue; other telescopic stars in the field. In Piazzi's and other Catalogues, γ is designated Sulaphat, from the Arabian *al-sulḥafát*, the tortoise, showing the part of the asterism in which it was placed. It has this amount of proper motion in space attributed to it, viz.:

$$P.... Æ + 0''·07 \quad Dec. - 0''·09$$
$$B.... \quad + 0''·03 \quad - 0''·02$$

This star appears as γ relative to brightness in Ptolemy, and he is followed by all the tabulating magnates, who rate it of the 3rd magnitude. But it has exhibited symptoms of variability, in being now brighter than β, which we may presume was not the case when Bayer lettered them. The subject demands close attention.

DCLXXVI. 274 P. XVIII. ANTINOI.

Æ 18ʰ 54ᵐ 31ˢ Prec. + 3ˢ·09

Dec. S 0° 55'·9 —— N 4"·72

Position AB 148°·6 (*w* 7) Distance 25"·8 (*w* 5) Epoch 1831·48

—— AB 146°·8 (*w* 9) —— 25"·6 (*w* 9) ⎫
—— BC 85°·0 (*w* 1) —— 2"·0 (*w* 1) ⎭ —— 1838·59

A delicate triple star, between the Eagle's wing and the left heel of Antinous; and 15° to the south of the two bright stars in the Eagle's tail. A and B, both 9, and both white; C 16, blue; many other minute stars in the field. Here the principal pair is compounded of Piazzi's Nos. 274 and 275, Hora XVIII., but Σ. discovered C, and made it a triple object. The *nova*, however, is so exceedingly small, that I only caught it by occasional glimpses; and then flitting it recalled Burns's flake of snow on a running stream. The estimation was, however, facilitated by assuming a distant star near the following parallel, as a station-pointer. The preceding micrometric measures of A and B, with a reduction of the mean apparent places in the Palermo Catalogue, are:

	Pos.	Dist.	Ep. 1800
P.	147° 48'	26"·40	
Σ.	149° 06'	26"·09	1822·67
H. and S.	148° 49'	26"·02	1823·48

DCLXXVII.   287 P. XVIII. DRACONIS.

Æ   18ʰ 54ᵐ 50ˢ   Prec. + 0ˢ·99
Dec. N 58° 00'·4   —— N 4"·74

Position 341°·0 (w 2)   Distance 0"·7 (w 1)   Epoch 1834·53

A close double star, in the back of Draco's neck, and about a degree south of ο Draconis, which has been already aligned. A 7, white; B 8, pale red. This was discovered by Sir W. Herschel in 1782, when it was wedge-shaped, with a fine black division between the components. Σ. designates it "pervicina." To me it is only elongated sufficiently to allow of my catching a measure at the angle, though notches appear at intervals. As to distance, when it happens to be below 1", all measurement, properly so called, ceases, and gives place to simple estimation. The difference between my position of the axis major, and that given under 43 Ḥ. i., is too small to be fairly imputed to the effect of rotation.

DCLXXVIII.   15 AQUILÆ.

Æ   18ʰ 56ᵐ 31ˢ   Prec. + 3ˢ·17
Dec. S   4° 15'·8   —— N 4"·89

Position 206°·6 (w 8)   Distance 34"·5 (w 6)   Epoch 1831·58

A fine double star, on the left heel of Antinous; it is midway between β Ophiuchi and α Capricorni, and rather more than a degree north-half-west of the bright star λ Antinoi. A 6, white; B 7½, lilac tint; other small stars in the field. Ḥ. saw this object in July, 1781, and classed it 13 v., but he does not appear to have measured it. Piazzi noted it thus: "Duplex. Comes 7·8ᵃ magnitudin. 1" temporis præcedit 30" ad austrum;" affording a comparison. The first micrometric results I find, with reductions from Piazzi, are thus:

	Pos.	Dist.	Ep.
P.	206° 00'	33"·30	1800
H. and S.	206° 44'	35"·61	1823·52

which shows that no sensible alteration has occurred in the interval.

DCLXXIX.   299 P. XVIII. LYRÆ.

Æ   18ʰ 56ᵐ 55ˢ   Prec. + 1ˢ·69
Dec. N 46° 42'·7   —— N 4"·93

Position 265°·0 (w 1)   Distance 50"·0 (w 1)   Epoch 1838·55

A wide double star, *hooked* to the boundary in the space north of the instrument; where a line from δ Cygni towards γ Draconis, nearly passes over it within half the distance. A 5½, white; B 14, blue. This delicate, though open, object, is No. 1362 of H.'s Fourth Series. The acolyte is of course too minute to admit of illumination, and its place is therefore merely estimated; but it is well seen, and points a little south of a line which would touch a distant 9th-magnitude preceding it.

DCLXXX. 302 P. XVIII. AQUILÆ.

Æ 18^h 57^m 59^s Prec. + 2^s·93

Dec. N 6° 18'·8   —— N 5''·02

Position 153°·9 (w 8)   Distance 10''·3 (w 3)   Epoch 1838·75

A very neat double star, on the margin of the Eagle's lower wing. A 7½, lucid white; B 9, cerulean blue; and the two point to a 10th-magnitude star in the *np* quadrant, at about three times the distance between A and B. This fine object is No. 2446 of the Dorpat Catalogue, where Σ.'s epoch is:

Pos. 154° 38'   Dist. 10''·133   Ep. 1831·70

302 P. xviii. lies on the preceding border of the following branch of the Milky Way; it is 10° due north of λ Antinoi, a star of the 3rd magnitude, and on the western parallel of β Aquilæ, from which it is about 13° distant. It is not difficult to pick up, being the brightest of its immediate neighbourhood.

DCLXXXI. ζ AQUILÆ.

Æ 18^h 58^m 02^s Prec. + 2^s·75

Dec. N 13° 37'·8   —— N 5''·02

Position 84°·0 (w 1)   Difference Æ = 8^s·4 (w 1)   Epoch 1832·47

A standard Greenwich star, with a distant companion, at the root of the Eagle's tail; and about 12° west-north-west of Altair, on a line towards Alphecca. A 3, greenish tint; B 11, livid; two other stars in the field, one following exactly on the equatoreal line, and the other near the *sf* vertical. As the *comes* defies the lamp illumination, the position is inferred by pointing the double-image of A, in the rock-crystal micrometer, to the direction of B. This star, as well as ε, has been designated *Dheneb el 'Okáb*, the Eagle's tail, from its situation; and the two are not without a suspicion of being connected, from the identity in signs and values of their proper motions, notwithstanding they are so far apart; those for ζ have been thus registered:

P.... Æ − 0''·14   Dec. − 0''·11

B....   − 0''·03   − 0''·06

DCLXXXII.   17 LYRÆ.

Æ 19ʰ 01ᵐ 23ˢ   Prec. + 2ˢ·26
Dec. N 35° 15′·2   —— N 5″·30

Position 331°·2 (w 6)   Distance 2″·9 (w 3)   Epoch 1834·79
——— 329°·9 (w 7)   ——— 3″·6 (w 5)   ——— 1838·82

A double star on the following edge of the Lyre's cross-piece; where it is about 7°½ south-south-east of Wega, and pretty closely following γ. A 6, light yellow; B 11, cerulean blue; a third star at a distance in the *sf* quadrant. This very beautiful and delicate object was discovered by Σ. with the Dorpat refractor, and forms No. 2461 of his great Catalogue, where these results are registered:

Pos. 330°·63   Dist. 3″·723   Ep. 1830·72

DCLXXXIII.   8 P. XIX. LYRÆ.

Æ 19ʰ 02ᵐ 21ˢ   Prec. + 2ˢ·04
Dec. N 38° 40′·7   —— N 5″·38

Position 122°·1 (w 5)   Distance 1″·4 (w 2)   Epoch 1834·86

A close double star, on the frame-base of the Lyre, following Wega on the eastern parallel at 5° distance. A 7½, yellowish; B 9½, pale white, and so impracticable under illumination, that the spherical crystal micrometer was applied. This is one of the first-class "vicinæ" of Σ.'s Catalogue of 1827, and in the great one of 1837, is registered thus:

Pos. 120°·95   Dist. 1″·267   Ep. 1831·05

DCLXXXIV.   13 P. XIX. LYRÆ.

Æ 19ʰ 03ᵐ 01ˢ   Prec. + 2ˢ·08
Dec. N 37° 39′·5   —— N 5″·44

Position AB 337°·0 (w 2)   Distance 18″·5 (w 1) ⎫
——— AC 350°·0 (w 3)   ——— 74″·8 (w 2) ⎬ Epoch 1835·73
——— CD 294°·0 (w 1)   ——— 5″·0 (w 1) ⎭

A double-double star on the base of the Lyre, and about 5° east-half-south from Wega. A 8, bright yellow; B 11, pale grey; C 9½, greenish; D 12, dusky. This very difficult compound object was discovered by Σ. with the powerful Dorpat refractor, and forms Nos. 2472 and 2473 of his great Catalogue.

After it had been duly examined in September, 1835, the telescope was turned upon 56 Draconis, a star of the 6½ magnitude, forming with another of nearly the same brightness, 31 ḤI. II., having a distance of upwards of 5″ between them. But as I only saw it single, there is either a mistake in the entry, a very extraordinary angular motion, or one of the components must be variable to invisibility. It therefore was not retained on my meridian list.

DCLXXXV. η LYRÆ.

Æ 19ʰ 08ᵐ 18ˢ   Prec. + 2ˢ·04
Dec. N 38° 52′·5   —— N 5″·89

Position 84°·8 (w 8)   Distance 28″·3 (w 5)   Epoch 1834·74

A neat double star on the following part of the frame, nearly on the parallel and 6° east of Wega. A 5, sky-blue; B 9, violet tint. This is a fair object for a moderate telescope, as it is sharply defined and bears illumination capitally: it is followed in the *sf* quadrant, about 90ˢ, by an open double star, which is H. and S., No. 289.

Christian Mayer observed η and "its satellite," as recorded in *De Novis in Cœlo Sidereo Phœnomenis*, 1779; and in August of the same year ḤI. picked it up, and registered it No. 2, Class IV. But as he placed it in Pos. 31° 51′ *sp*, with a distance of 26″, there was evidently some error, so that the only comparison available at my epoch was with H. and S., whose results are:

Pos. 84° 02′   Dist. 29″·34   Ep. 1823·46

DCLXXXVI. 2035 H. AQUILÆ.

Æ 19ʰ 08ᵐ 36ˢ   Prec. + 3ˢ·10
Dec. S 1° 11′·9   —— N 5″·91

Mean Epoch of the Observation . . . . 1835·67

A loose cluster between the lower wing of the Eagle and the thigh of Antinous, about 13° south-west of Altair, in the direction of a line from Wega passed through ε Aquilæ. It consists of a splashy group of stars from the 9th to the 12th magnitude, trending *np* and *sf*, with three brighter ones, about 7½, in the *sf* portion. Attentive gazing, under the smooth motion of the equatoreal clock, brings up the points of light, and reveals an indication of star dust. The new *monstre* telescopes will make easy work of it. It is on the eastern margin of the Milky Way, but there are few neighbours of note near it.

DCLXXXVII. 43 P. XIX. SAGITTARII.

Ɀ 19ʰ 09ᵐ 09ˢ Prec. + 3ˢ·51
Dec. S 18° 58'·8 —— N 5"·96

Position 170°·2 (w 4) Distance 38"·6 (w 2) Epoch 1835·57

A wide double star following the Archer's head, and terminating an outcrop of the Galaxy: it is 30° to the south-south-west of Altair, where it is the preceding of a group of small stars. A 8, white; B 11½, pale grey; in a field of stragglers, of which the largest, in the *np* quadrant, is of an orange tinge. This object was discovered by Ḫ̣I., who registered it under these measures:

Pos. 168° 45' Dist. 36"·05 Ep. 1782·60

DCLXXXVIII. 56 M. LYRÆ.

Ɀ 19ʰ 10ᵐ 19ˢ Prec. + 2ˢ·34
Dec. N 29° 54'·2 —— N 6"·05

Mean Epoch of the Observation . . . . 1835·66

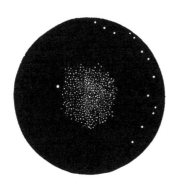

A globular cluster, in a splendid field, between the eastern yoke of Lyra's frame and the Swan's head: it is 5°½ distant from β Lyræ, on a south-east line leading to β Cygni, which is about 3°½ further. This object was first registered by M. Messier in 1778, and, from his imperfect means, described as a nebula of feeble light, without a star. In 1784, it was resolved by Sir William Herschel, who, on gauging, considered its profundity to be of the 344th order.

DCLXXXIX. 23 AQUILÆ.

Ɀ 19ʰ 10ᵐ 24ˢ Prec. + 3ˢ·05
Dec. N 0° 48'·0 —— N 6"·06

Position 12°·6 (w 7) Distance 3"·1 (w 3) Epoch 1833·68

A close double star, under the Eagle's southern wing. A 6, light orange; B 10, grey. This elegant object must be 14 Ḫ̣I. ɪ., instead of

24 Aquilæ, its neighbour in the *sf*, which is a single star. Sir William's position, however, is at variance with this conclusion, being

Pos. 162°·0   Dist. 3″·50±   Ep. 1781·58

but if a mistake of noting *s* for *n* in the quadrant has occurred, it will only differ 5° 21′ from my results. Indeed, in his mentioning the improvement of the *comes* under magnifying powers, the description answers so well for Flamsteed's 23, that I conclude there can be no question about it. Struve enrolled it No. 2492 of the Dorpat Catalogue; and it may be found to the south-south-west of Altair, at 12° distance, where a line through the heads of Herculis and Ophiuchus reaches it. The proper motion in space attributed to this star by Piazzi has been greatly reduced by Mr. Baily, and will probably disappear under future rigorous observation.

DCXC.   28 AQUILÆ.

Æ 19ʰ 12ᵐ 12ˢ   PREC. + 2ˢ·80

DEC. N 12° 05′·1   —— N 6″·21

POSITION 175°·7 (*w* 8)   DISTANCE 59″·8 (*w* 5)   EPOCH 1831·42

A wide and unequal double star, on the Eagle's back; it is 8° distant from Altair, on a west-north-west line leading upon ε, in the tail. A 6, pale white; B 10, deep blue. This is 34 Ħ. v., but though it was enrolled in 1781, there is only a random remark that the distance was about 35″. It was afterwards measured by Sir James South, from whose results, with my own, little, if any, appreciable motion is discernible; his measures yield:

Pos. 175°·06   Dist. 59″·28   Ep. 1825·04

DCXCI.   δ DRACONIS.

Æ 19ʰ 12ᵐ 30ˢ   PREC. + 0ˢ·02

DEC. N 67° 22′·8   —— N 6″·24

POSITION 19°·7 (*w* 1)   DIFFERENCE Æ = 2ˢ·7 (*w* 1)   EPOCH 1834·78

A bright star with a distant companion, in the first flexure of Draco. A 3, deep yellow; B 9½, pale red; other small stars in the field. A small amount of proper motion is assigned to A in the Palermo Catalogue, which has been largely increased by recent comparisons; but they all agree in the direction:

P.... Æ + 0″·07   Dec. + 0″·02
B....   + 0″·33   + 0″·07
A....   + 0″·33   + 0″·07

δ Draconis is the Nodus Secundus of various catalogues, and though rejected from the Greenwich list in 1830, is a notable star. Ptolemy marked it δ in lustre, and designated it ὁ βορειότερος. Flamsteed, however, registered it *austrina*, but erroneously; and Mr. Baily thinks he was "probably led into this mistake from considering its position with respect to the pole of the equator; whereas all Ptolemy's descriptions refer to the pole of the ecliptic." It is the most northern star on the western side of the square formed by δ, π, ρ, and ε; and is 18° to the north-north-east of the well-known star γ Draconis. Tízíní called it *al-taïs*, the goat; and it is just below the equilateral triangle formed by σ, τ, and υ, called *al-athâfî*, the tripod, which strictly means the three stones on which the Nomade Arab places his kettle.

DCXCII. 108 P. XIX. DRACONIS.

Æ 19^h 15^m 18^s Prec. + 0^s·59

Dec. N 62° 55'·1   —— N 6"·47

Position 349°·0 (*w* 3)   Distance 0"·5 (*w* 1)   Epoch 1833·78

A close double star on Draco's neck, 4°½ distant to the south of δ Draconis. A 8, and B 9, both white. This object was discovered by Σ., and is his No. 2509, being marked in the Catalogue of 1827, "cuneus, ni fallor." But though a very difficult object, I found that it gave, at intervals of gazing, a very appreciable axis major; and when the great work of the Dorpat astronomer arrived in 1837, I was not surprised to find these final results:

Pos. 345°·83   Dist. 0"·573   Ep. 1836·93

DCXCIII. δ AQUILÆ.

Æ 19^h 17^m 25^s Prec. + 3^s·01

Dec. N 2° 48'·0   —— N 6"·64

Position 259°·3 (*w* 1)   Distance 96"·5 (*w* 1)   Epoch 1833·62

A standard Greenwich star, with a distant and minute *comes*, on the southern or lower wing; and south-west-by-south, distant 9° from its lucida. A 3½, white; B 12, livid; other stars in the field, of which a pair in the north form a coarse double object. The position and distance of the nearest of these was obtained by estimations made in a dark field for the angle, and the Δ Æ by a bar for the distance. The proper motions of A are stated so similarly in amount and direction, that little doubt can exist of its being actually under weigh; these are the best ascertained values:

P.... Æ + 0"·18   Dec. + 0"·10

B....   + 0"·27   + 0"·11

A....   + 0"·24   + 0"·12

DCXCIV. 128 P. XIX. ANSERIS.

Æ 19ʰ 19ᵐ 29ˢ   Prec. + 2ˢ·62
Dec. N 19° 34'·7   —— N 6"·81

Position 44°·8 (w 3)   Distance 25"·0 (w 1)   Epoch 1833·58

A very delicate double star on the Goose's foot, in a small group to the north-north-west of Altair, and at one-third the distance between that star and Wega. A 6½, topaz yellow; B 13, deep blue; and A is the apex of a scalene triangle with two other stars to the north of it. As illumination is precluded to so small an object as B, the position was obtained with the ocular crystal micrometer, and the distance carefully estimated. It is Σ.'s No. 2521, with these measures:

Pos. 43°·58   Dist. 22"·647   Ep. 1829·40

DCXCV. 21 ♄. VIII. ANSERIS.

Æ 19ʰ 20ᵐ 30ˢ   Prec. + 2ˢ·49
Dec. N 24° 49'·3   —— N 6"·86

Mean Epoch of the Observation . . . . 1837·34

A large straggling cluster, on the neck of the Goose, and about 3° distant from β Cygni. It is preceded by a couple of 7th magnitude stars, and is remarkable for its brightest members assuming the form of a Greek Ω, even more strongly than 17 Messier. See No. DCXLV. There is much star dust in glimpses, at the lower opening of the *omega*, and also at the top of the field; but the distinctness is greatly interfered with by the brightness of the leaders.

DCXCVI. 149 P. XIX. CYGNI.

Æ 19ʰ 21ᵐ 56ˢ   Prec. + 2ˢ·15
Dec. N 36° 12'·5   —— N 7"·01

Position 61°·9 (w 8)   Distance 7"·0 (w 5)   Epoch 1833·79

A very elegant double star between the Lyre and the Swan's neck, about 8° to the north of β Cygni, and on the *nf* of a 4th-magnitude star. A 8½, white; B 9, pale blue. This is 69 ♄. II., and when H. and S. re-measured it, they concluded upon an angular change of + 5°·56 in an interim of forty-one years. This was not confirmed by my results, nor, as was afterwards found, by those of Σ. The error was grounded on a comparison of these data:

$$\begin{array}{llll}
\text{H.} & \text{Pos. } 60^\circ\ 48' & \text{Dist. } 6''{\cdot}0\ \pm & \text{Ep. } 1782{\cdot}82 \\
\text{H. and S.} & 66^\circ\ 44' & 7''{\cdot}43 & 1823{\cdot}57
\end{array}$$

DCXCVII.   6 VULPECULÆ.

$$\begin{array}{ll}
\text{Æ } 19^\text{h}\ 22^\text{m}\ 03^\text{s} & \text{Prec. } +\ 2^\text{s}{\cdot}50 \\
\text{Dec. N } 24^\circ\ 20'{\cdot}7 & \underline{\qquad}\ \text{N } 6''{\cdot}99
\end{array}$$

Position $146^\circ{\cdot}7$ (w 1)   Difference $\text{Æ} = 3^\text{s}{\cdot}3$ (w 1)   Epoch $1834{\cdot}70$

A star with a distant telescopic *comes*, close to Reynard's nose. A 4, deep yellow; B 10, pale grey; and it is followed in the *sf* by a curve of four little stars, in a rich field. This is to the south of β Cygni, $3^\circ\frac{1}{2}$ distant, and shows how even modern asterisms become distorted for the want of well-defined boundaries; for though designated in Vulpecula, it is on the neck of the Goose, and actually the Lucida Anseris. All the investigators of its proper motions, agree in the course it is taking:

$$\begin{array}{llll}
P.... & \text{Æ} - 0''{\cdot}30 & \text{Dec.} & -\ 0''{\cdot}11 \\
B.... & -\ 0''{\cdot}14 & & -\ 0''{\cdot}09 \\
A.... & -\ 0''{\cdot}17 & & -\ 0''{\cdot}09
\end{array}$$

Vulpecula et Anser is a modern constellation, crowded in by Hevelius to occupy a space between the Arrow and the Swan, where the Via Lactea divides into two branches. For this purpose he ransacked the *informes* of this bifurcation, and was so satisfied with the result, that the effigies figure in the elaborate print of his offerings to Urania. He selected it on account of the Eagle, Cerberus, and Vultur cadens. " I wished," said he, " to place a fox with a goose in the space of sky well fitted to it; because such an animal is very cunning, voracious, and fierce. Aquila and Vultur are of the same nature, rapacious and greedy:" and therefore he deems the intrusion to be conformable to the fables of the poets, and the rules of the astrologers. In 1672, while scrutinizing his asterism, he saw a new star in the head of the Fox; but it was visible for about two years only, nor has it been since identified. Vulpecula et Anser has been *almost* acknowledged by astronomers, and the two are often treated of distinctly and separately; collectively the constituents have thus gradually increased:

$$\begin{array}{llll}
\text{Hevelius} & . \ . \ . \ 27 \text{ stars} & \text{Piazzi} & . \ . \ . \ . \ 88 \text{ stars} \\
\text{Flamsteed} & . \ . \ . \ 35 & \text{Bode} & . \ . \ . \ . \ 126
\end{array}$$

Much licence ought to be granted to Hevelius, as he placed most of his constellations where they seemed to be wanted; and they have become almost of standard authority. But all astronomers must be thankful that the rage for intruding new asterisms into the heavens, has been checked. It is hoped that he who treats—*De Cœli Imaginibus*—in future, will find few or none of the recent mongrels.

DCXCVIII. 144 P. XIX. AQUILÆ.

Æ 19ʰ 22ᵐ 09ˢ   PREC. + 3ˢ·01
DEC. N 2° 34'·7   —— N 7"·03

POSITION 4°·9 (w 5)   DISTANCE 37"·0 (w 2)   EPOCH 1838·63

A very delicate though wide double star, on the border of the
Eagle's lower wing, where it follows δ on the eastern parallel, at about
a degree's distance. A 7, deep yellow; B 11, pale green; several other
minute companions are visible, and A and B point nearly upon a 9th-
magnitude star in the *nf* quadrant. This pretty object is one of Professor Struve's discoveries, and No. 2532 of the Dorpat Catalogue, under
the following results:

Pos. 5°·02 Dist. 34"·90 Ep. 1829·0

DCXCIX. 38 ♓. VI. AQUILÆ.

Æ 19ʰ 23ᵐ 55ˢ   PREC. + 2ˢ·87
DEC. N 8° 54'·1   —— N 7"·18

MEAN EPOCH OF THE OBSERVATION . . . . 1835·71

A stellar nebula on the Eagle's back, preceding Altair on the western
parallel by about 5°. This is a minute object in a rich Milky-Way field
of stars, of 8th to 15th magnitudes; and in the most powerful telescopes
it has a fan-shaped appearance. Sir William Herschel discovered it with
his 20-foot reflector in 1791; he saw it considerably bright and easily
resolvable, estimating its profundity as of the 900th order, indicating
a distance altogether overwhelming to the mind. In his review of this,
also with a 20-foot reflector, Sir John Herschel says, "It is like a nebula
well resolved, and is a curious object."

DCC. β CYGNI.

Æ 19ʰ 24ᵐ 16ˢ   PREC. + 2ˢ·42
DEC. N 27° 37'·7   —— N 7"·19

POSITION 55°·4 (w 5)   DISTANCE 34"·2 (w 5)   EPOCH 1830·81
—— 55°·6 (w 9)   —— 34"·4 (w 9)   —— 1837·58

A bright double star on the Swan's bill, about 13°½ to the south-
south-east of Wega. A 3, topaz yellow; B 7, sapphire blue; the colours
in brilliant contrast, by which term I do not mean the mere optical

complementary tints, but relating to these bodies as radiating their own coloured lights. The various measures of this fine object exhibit a singular coincidence, and by the impressment of the observations of Bradley and Piazzi, we gain a pretty long period before my first epoch. The results appear thus:

Bradley	Pos. 57° 34'	Dist. 34″·20	Ep. 1755·00
Herschel I.	54° 52'	34″·83	1782·45
Piazzi	54° 31'	34″·28	1800·00
Struve	54° 30'	34″·29	1821·76
Herschel II. and South	54° 45'	34″·38	1822·98
Dawes	55° 32'	34″·51	1830·54

These conclusions display a very remarkable relative constancy both in angle and distance, especially as both the components appear to be affected with proper motions, the amount of which does not differ so much in the several reductions, as in the course or direction of the march. These are the present values and signs, and the next rigorous comparison will decide between them:

Piazzi:		Baily:	
Star A.... Æ − 0″·07 Dec. + 0″·05		Star A.... Æ + 0″·03 Dec. + 0″·02	
Star B.... − 0″·13 + 0″·04		Star B.... + 0″·05 + 0″·04	

Ulugh Beigh and Tízíní call β Cygni, Minkár el dejájeh, the hen's beak; but its more usual appellation is Albireo. This name is of doubtful origin, introduced, as it appears, by Bayer, and ingeniously conjectured by Ideler to be a corruption of " ab ireo" in the Latin version of the *Almagest*, where the translator from the Arabic *euris* for *ornis* made *eurisim*, which he supposed to be the Greek 'Ερύσιμον, in Latin *Irio*. Hence the *Hierezim* of Riccioli, whose *Astronomia Riformata* exhibits some curious blunders of black-letter transcription. A search among the archives of Spain might probably bring the original MS. of the Alphonsine Tables to light; and such a discovery would well reward any pains that might be taken.

DCCI. 169 P. XIX. CYGNI.

Æ 19ʰ 25ᵐ 40ˢ   Prec. + 2ˢ·41
Dec. N 27° 55'·8   —— N 7″·32

Position 5°·2 (*w* 5)   Distance 5″·8 (*w* 3)   Epoch 1835·78

A delicate double star in the Swan's mouth, a little *nf* the last-described object. A 9, white; B 11, pale blue. This object is 99 ⨂.ɪɪ., and had been thus measured before my attack:

⨂.	Pos. 2° 12'	Dist. 4″·50±	Ep. 1783·71
S.	4° 22'	6″·84	1825·36
Σ.	5° 15'	5″·36	1830·69

whence, allowing for probable errors on so extremely difficult a star, it may have a slow direct orbital motion; but the fixity of its distance, during an interval of fifty-two years, seems to be established, by the coincidence of the obtained results.

DCCII. 185 P. XIX. ANTINOI.

Æ 19^h 28^m 01^s Prec. + 3^s·30

Dec. S 10° 46'·8   —— N 7"·51

Position AB 338°·4 (w 4)   Distance 3"·2 (w 2) }

—— AC 153°·5 (w 2)   —— 8"·0 w }   Epoch 1835·58

A delicate triple star, on the right knee of Antinous, and about 20° south-by-west from Altair, where it is closely following 37 Antinoi. A 9, light yellow; B 10, cerulean blue; C 12, violet tint; and the three lie nearly in line pointing to a distant small star in the *sf* quadrant. This was discovered at Slough about sixty years ago, and classed 13 ḤI. I., "a curious treble star;" but no measures were printed, Sir William merely stating that the two nearest were apart half the diameter of the large one, and the furthest about 7" or 8". He also well describes it as "the last star of a telescopic trifolium, similar to that in the hand of Aquarius."

If estimating the interval between the stars by the apparent diameter, is at all to be trusted for a distance, A and B must have widened since 1781; while, from a manuscript angle of position furnished from his father's papers by Sir John Herschel, there is presumptive evidence of a slow *sp nf*, or direct motion. But large allowances for errors of observation, are due to so difficult an object. The registered measures are:

	Pos.	Dist.	Ep.
ḤI.	307° 15'	0"·60+	1782·77
S.	316° 03'	4"·49	1825·59
Σ.	339° 34'	2"·84	1831·01

It is singular that Professor Struve, who measured this star in 1828, 1831, and 1832, should say, "Dubitari potest utrum hæc stella an 2545 sit Herscheli I. 13." Is it possible that he overlooked C, which at once identifies it? Sir James South, who observed it with his 7-foot equatoreal at Passy, near Paris, also says, "Sir William Herschel having described it as a triple star, I applied a power of 413, but no third star could even be suspected." It is plain enough in my refractor; I therefore conclude Sir James must have measured the following, instead of 13 ḤI. I.; but I have inserted it agreeably to his own entry, No. 720.

DCCIII. 186 P. XIX. ANTINOI.

Æ 19^h 28^m 05^s Prec. + 3^s·30

Dec. S 10° 30'·4   —— N 7"·52

Position 315°·7 (w 6)   Distance 3"·8 (w 3)   Epoch 1835·58

A very neat double star on the right knee of Antinous, and closely following the above-described object. A 7½, pale white; B 9, sky-blue;

and they are followed by another double star nearly on the parallel. There can be little doubt of this object's having been measured for 185 P., for though it is considerably brighter and more open, yet there is certainly a great resemblance in the position and distance of A and B. This is the No. 2545 of the Dorpat Catalogue above alluded to, and the last edition of that work contains these measures:

Pos. 315°·18   Dist. 3″·53   Ep. 1829·11

DCCIV. ε SAGITTÆ.

$Æ$ 19^h 30^m 03^s Prec. + 2^s·71
Dec. N 16° 06′·5   —— N 7″·67

Position 80°·9 (w 9)   Distance 92″·2 (w 8)   Epoch 1833·78

A star with a distant companion, on the Eagle's northern or upper wing, about 8° to the north-north-west of Altair, on the line towards Wega. A 6, pale white; B 8, light blue. This object, 26 H. vi., was registered as being 8° 32′ in the *sf* quadrant in 1780, by "extremely exact" observation; this, however, must have been an erroneous entry of *s* for *n*. By making the due alteration, the results are so remarkably coincident, as to clear off all doubt:

H.   Pos. 81° 28′   Dist. 91″·89   Ep. 1780·64
S.   81° 03′   91″·91   1825·03

While measuring this star on the 5th of October, 1833, the east wind had an extraordinary effect. It often produces changes in the appearance of the spectra of heavenly bodies; but during the time of this observation the spectrum fluctuated, whirled and danced, collapsed and expanded, and when close to the wires, seemed to flit about them like a butterfly. Returning to the same star on the 9th, with a fine south-west breeze, it moved with all the steadiness of a little planet. I record the circumstance, because both sets of measures were coincident; observation therefore need not always be discontinued, when from anomalous circumstances it is apparently unsatisfactory.

DCCV. 54 SAGITTARII.

$Æ$ 19^h 31^m 33^s Prec. + 3^s·44
Dec. S 16° 39′·2   —— N 7″·80

Position AB 42°·8 (w 4)   Distance 28″·5 (w 3) $\Big\}$ Epoch 1837·58
——— AC 280°·0 (w 1)   ——— 20″·0 (w 1)

A very delicate triple star in the space between the heads of Sagittarius and Capricorn; it is about 25° south-by-west of Altair, and nearly

10° west of β Capricorni, being the preceding member of a little triangle. A 5½, yellow; B 8, violet; C 16, blue; there are several other small stars in the field, three of which, at nearly equal distances, form a line in the *sf* quadrant.

This fine object was discovered by Sir John Herschel, but so small is C, that, in his No. 1424, he termed it of the 18th magnitude. As, however, by gazing intently, in the best weather and under smooth clock motion, I caught it by glimpses, I have assigned it a magnitude which I deem the *minimum visibile* of my telescope. And it should be remembered, that such mere points of light were scrutinized to prove the power of the instrument for its general work, rather than to establish data for an epoch by estimation only. Thus the mark (w 1) serves only to show that the object was identified.

DCCVI. 51 ☿. V. SAGITTARII.

| Æ 19ʰ 34ᵐ 56ˢ | Prec. + 3ˢ·39 |
| Dec. S 14° 31′·6 | —— N 8″·07 |

Mean Epoch of the Observation . . . . 1837·65

A pale blue planetary nebula between the heads of Capricorn and the Archer; and about 2° distant from the above-described object, on a line north-by-east, towards the Dolphin. This fine object is accompanied by several small stars, four of which form a square about it; and it is preceded at a little distance, by a vertical row of telescopic stars. It was discovered on the 8th of August, 1787; and was shown by Sir William Herschel to M. De La Lande, at Slough.

DCCVII. 241 P. XIX. AQUILÆ.

| Æ 19ʰ 35ᵐ 02ˢ | Prec. + 2ˢ·89 |
| Dec. N 8° 00′·5 | —— N 8″·08 |

Position 253°·7 (w 6) Distance 26″·8 (w 4) Epoch 1833·78

A delicate double star, on the Eagle's back, preceding its lucida a little south of the western parallel, by about a couple of degrees. A 7½, pale topaz; B 9½, lilac. Piazzi first noticed its duplicity in these words, "Duplex. Comes 10ᵐ magnit. 5″ ad austrum, 1″·3 temporis præcedit:" and the description is as accurately expressed, as though compound stars were his object. It was afterwards measured by Σ., No. 2562 of the Dorpat Catalogue, with these results:

Pos. 252° 32′ Dist. 27″·22 Ep. 1825·52

DCCVIII.   250 P. XIX. AQUILÆ.

Æ   19^h 36^m 36^s PREC. + 2^s·81
DEC. N 11° 59'·8   —— N 8"·20

POSITION 312°·0 (*w* 2)   DISTANCE 20"·0 (*w* 1)   EPOCH 1838·63

A very delicate double star on the upper or northern wing of Aquila, and 4° north-by-west of its lucida, somewhat in the line formed from β to γ. A 8½, white; B 14, blue; other stars in the field. This excellent test object, of course, will not bear the slightest illumination; the position is therefore by the spherical crystal micrometer, and the distance is by an assumed scale of comparison, which, with so minute a *comes*, is necessarily precarious. The former measures were:

Ħ.   Pos. 311° 24'   Dist. 22"·75   Ep. 1783·60
Σ.   315° 40'   18"·07   1829·63

DCCIX.   257 P. XIX. AQUILÆ.

Æ   19^h 37^m 21^s PREC. + 2^s·85
DEC. N 10° 23'·6   —— N 8"·26

POSITION 276°·5 (*w* 6)   DISTANCE 4"·0 (*w* 4)   EPOCH 1836·74

A delicate double star on the Eagle's back, and slightly *np* the bright star γ Aquilæ. A 8, white; B 10, smalt blue. When I attacked this object, the only previous measures to which I had access, were:

Ħ.   Pos. 278° 18'   Dist. 3"·00±   Ep. 1783·60
S.   276° 27'   3"·99   1825·56

The difference here observable, was presumed to indicate a very slow orbital change in the direction *np sf*, or retrograde; which opinion my results do not confirm. The observations are of difficult achievement.

DCCX.   16 CYGNI.

Æ   19^h 37^m 34^s PREC. + 1^s·61
DEC. N 50° 09'·3   —— N 8"·28

POSITION 136°·4 (*w* 9)   DISTANCE 37"·3 (*w* 9)   EPOCH 1834·63

A fine double star on the tip of the left or preceding wing, 10° to the north-west of a Cygni, and within a degree to the east-north-east of θ. A 6½, and B 7, both pale fawn-colour. This object is 36 Ħ. v., who classed it in October, 1781; no angles were then observed, and the distance was assumed at 30″ by a "pretty exact estimation." Piazzi observed both components with his altitude and azimuth circle, B being his No. 262, Hora XIX. A reduction from his mean apparent places was made, for comparison with the micrometrical results of H. and S.

P.   Pos. 141° 00′   Dist. 37″·02   Ep. 1800
H. and S.   135° 13′   37″·50   1823·57

Now, though this was severe upon Piazzi's meridional deductions, I could not but think it indicated a sensible angular change; my observations, however, which were most satisfactory, confute this notion. Yet a proper motion lately ascribed to A, which escaped Piazzi's scrutiny, to these values, may reconcile the difference:

B.... Æ − 0″·23   Dec. − 0″·14
A....   − 0″·27   − 0″·15

DCCXI. γ AQUILÆ.

Æ 19ʰ 38ᵐ 38ˢ   PREC. + 2ˢ·85
DEC. N 10° 13′·6   —— N 8″·36

POSITION 255°·0 (w 1)   DISTANCE 136″·0 (w 1)   EPOCH 1836·74

A standard Greenwich star with a minute companion, on the bird's back, being the upper of the three well-known stars by which Aquila is distinguished, and which point towards Wega. A 3, pale orange; B 12, dusky; other small stars around. γ is usually designated Tarazed, from the Arabian *sháhin tárá-zed*, the soaring or star-striking falcon; the name commonly given to Aquila by the Persians, as we are told by Mohammed Nasír-ad-dín of Tús. (HYDE, *Syntagma*, i. 32.) By several direct comparisons in 1834 and 1836, I can pronounce that it is now actually brighter than β, which, of course, was not formerly the case; and there is a great coincidence in the amount and direction found by the scrutineers of its proper motions:

P.... Æ + 0″·06   Dec. + 0″·04
B....   + 0″·09   + 0″·01
A....   + 0″·06   + 0″·01

DCCXII. 276 P. XIX. CYGNI.

Æ 19ʰ 39ᵐ 47ˢ   PREC. + 2ˢ·20
DEC. N 35° 42′·3   —— N 8″·45

POSITION 126°·5 (w 9)   DISTANCE 15″·0 (w 5)   EPOCH 1835·88

A neat double star preceding the Swan's neck, in the outer or western stream of the Galaxy, and about 10° north-by-east from β Cygni. A 8, and B 8½, both white, and there is a third star in the *sp* quadrant, about 2′ distant. Here B is Piazzi's No. 277; and the two form No. 13 of Sir William Herschel's List of 145 New Double Stars. The first micrometrical measures I find are those of Σ.:

Pos. 125° 30′   Dist. 14″·16   Ep. 1823·01

DCCXIII. 278 P. XIX. CYGNI.

Æ 19ʰ 39ᵐ 53ˢ   Prec. + 2ˢ·23
Dec. N 34° 37′·5   —— N 8″·46

Position 28°·8 (w 8)   Distance 39″·0 (w 6)   Epoch 1834·68

A double star, close to the Swan's neck, in the preceding line of the Galaxy, and 9° to the north-north-east of β. A 6, straw colour; B 8, smalt blue. This fine though wide object is nearly midway between two telescopic stars, lying just across the parallel in an np and sf line, and forming a lozenge with A and B. Its former measures were:

Ⱨ.   Pos. 32° 57′   Dist. 35″·20   Ep. 1783·73
S.   29° 11′   38″·74   1825·22

From these results, compared with my own, a slight retrograde angular change, and an increase of distance may be inferred: but such a change is equally attributable to proper motion as to binary affection. Time will teach us; for splendid telescopes are coming into action.

DCCXIV. δ CYGNI.

Æ 19ʰ 39ᵐ 58ˢ   Prec. + 1ˢ·87
Dec. N 44° 44′·6   —— N 8″·47

Position 30°·9 (w 5)   Distance 1″·5 (w 2)   Epoch 1837·78
—— 25°·6 (w 8)   —— 1″·8 (w 3)   —— 1842·56

A most delicate double star in the middle of the left or preceding wing, 14° to the west of a Cygni, on its parallel, and forming, with it and γ, a well-marked and nearly equilateral triangle. A 3½, pale yellow; B 9, sea-green. This beautiful object was registered 94 Ⱨ. I., under the following conditions:

Pos. 71° 39′   Dist. 2″·50   Ep. 1783·72

but in 1802 and 1804, that zealous astronomer could no longer perceive the *comes*. In 1823, the star was perfectly round to H. and S., who saw no elongation, nor anything to give suspicion of a companion. In July, 1825, Sir James South again examined it with his 7-foot equatoreal, at Passy, near Paris, in company with my late friend M. Gambart, of Marseilles. Both observers agreed that it was perfectly round and sharply defined, appearing like a planetary disc, under a magnifying power of 787. Yet in the following year Σ. found it again double, and thus registered it:

Pos. 40° 39′ Dist. 1″·91 Ep. 1826·55

It was therefore presumed to have undergone an occultation in performing a periodic and elliptical revolution of about forty years. Though neither the occultation nor the period of rotation can now be admitted, this presumption had the effect of delaying my attack. In 1837 I saw B plainly, a small green spot just on the edge of the brilliance of A, near its north vertical, almost pointing in the direction of a little telescopic star in the *nf.* On my re-examination of it at Hartwell House, in 1842, it was certainly a much easier object to cope with.

After my first observations were made, Professor Struve favoured me with a copy of his grand Dorpat Catalogue, which is a double star "Thesaurus." Having obtained very satisfactory measures of δ in 1836, that skilful observer makes these important remarks:

The angular motion is undoubted, but unless I am mistaken, so small that it can in no way suit a revolution of forty years. The diminution of distance is probably to be ascribed to the imperfection of the two measures of 1826, which is easily to be excused in stars so difficult to observe, chiefly from the great brightness of the larger. Hence it was more probable that the whole motion from the first measure of ♄. in 1783 to 1836, through fifty-three years, was only 39° 43′. But if we accept this, it cannot be explained why ♄. did not see the companion in 1802, unless we believe either that the light of the smaller star is variable, or that in this interval of time it had passed its perihelion. The latter seems probable.

DCCXV. χ CYGNI.

Æ 19ʰ 40ᵐ 21ˢ   Prec. + 2ˢ·27
Dec. N 33° 21′·7   —— N 8″·50

Position 72°·9 (w 9) Distance 25″·7 (w 7) Epoch 1835·67

A fine double star on the Swan's neck, in the Via Lactea, and $7°\frac{1}{2}$ distant from β in the Swan's bill, on a bearing north-north-east. A 5, golden yellow; B 9, pale blue; and there are numerous small stars in the field. It was classed 11 ♄. iv. in 1779, the distance then being 24″·83; but no measures appear till those of H. and S., which show:

Pos. 73° 18′ Dist. 25″·50 Ep. 1822·49

The results are, therefore, singularly accordant; and it now remains to watch the effect of proper motions, those for A being thus valued:

P.... Æ − 0″·10   Dec. − 0″·40
B....   + 0″·03   − 0″·41
A....   − 0″·01   − 0″·43

From these conclusions, it is probable that, ere long, it will be found that the proper motion in Æ is inappreciable; but the agreement of Piazzi, Baily, and Argelander, as to its amount and direction in declination, must be received as *proven.* But the amount of our knowledge respecting that mysterious energy, is as yet but small.

DCCXVI.　73 ♅. IV. CYGNI.

Æ 19ʰ 40ᵐ 35ˢ　　Prec. + 1ˢ·62
Dec. N 50° 07′·6　　—— N 8″·51

Mean Epoch of the Observation . . . . 1835·67

A very singular nebula on the tip of the Swan's preceding wing, closely following the 16 Cygni already treated of, and 5°½ north of δ. There are several telescopic stars about it, and it is preceded by two which are double. In my telescope it is small, and somewhat resembles a star out of focus; but both the Herschels agree, on viewing this object through their powerful instruments, that it appears to constitute a connecting link between the planetary nebulæ and the nebulous stars. It was discovered and registered in September, 1793.

DCCXVII.　π AQUILÆ.

Æ 19ʰ 41ᵐ 10ˢ　　Prec. + 2ˢ·82
Dec. N 11° 25′·4　　—— N 8″·56

Position 122°·0 (w 5)　Distance 1″·5 (w 3)　Epoch 1831·70
——— 121°·3 (w 8)　——— 1″·7 (w 5)　——— 1836·81

A close double star on the Eagle's northern wing, about 3° north-half-west from its lucida. A 6, pale white; B 7, greenish. This beautiful object, which may be termed a miniature of Castor, is 92 ♅. i., the earliest measures of which were:

♅.	Pos. 124° 24′	Dist. 1″·40+	Ep. 1783·65
H. and S.	135° 27′	1″·95	1823·70

from which data the angle was considered as having varied in an annual average motion, on the hypothesis of equal errors, of + 0°·314. But as Sir James South's re-examination of it, two years afterwards, showed a misreading of 10° in the last epoch, there appeared sufficient evidence of its fixity when I first attacked it. The following results, however, still exhibited such a discordancy from mine, that I determined on taking advantage of a fine opportunity of again working upon it:

S.	Pos. 123° 27′	Dist. 1″·54	Ep. 1825·61
Σ.	125° 51′	1″·56	1825·80
D.	124° 00′	1″·80	1830·54
H.	127° 24′	1″·68	1832·56

The disagreements of these conclusions are imputable to the delicacy and consequent difficulty of the object. But subsequent observations by both Σ. and D., coincide with my own in establishing the optical condition of the components.

DCCXVIII.   ζ SAGITTÆ.

Æ 19ʰ 41ᵐ 53ˢ   Prec. + 2ˢ·66
Dec. N 18° 44′·8   —— N 8″·62

Position 313°·4 (w 7)   Distance 8″·9 (w 7)   Epoch 1831·59
—— 312°·3 (w 9)   —— 8″·6 (w 9)   —— 1838·67

A neat double star above the reed of the arrow, 9° south-by-east from β Cygni, and 10° north of Altair.   A 5, silvery white; B 9, blue, improving greatly under the red illumination.   When I first took this object in hand, it was under marked circumstances, there being such grounds for supposing it a binary system, that it was distinguished by an asterisk, as a remarkable star, in the index to the Catalogue of H. and S.   But my measures, taken under every advantage at both epochs, are decisive as to its fixity.

ζ Sagittæ, erroneously designated ξ in the Palermo Catalogue, is 30 Ⓗ. ii.; and the measures of the discoverer having been corrected by his son, from the original papers, the following were the results which supported the presumed motion previous to my first epoch:

	Pos.	Dist.	Ep.
Ⓗ.	304° 10′	8‴·83	1781·88
Σ.	309° 30′	8″·77	1819·74
H. and S.	314° 32′	8″·82	1823·69
H.	318° 26′	9″·81	1829·63

Insignificant as this little asterism undoubtedly is, and there are few of even the modern ones more so, it is actually one of the old forty-eight constellated groups, and has held its station through the classic ages as 'Οϊστὸς, Τόξον, Sagitta, Telum, Jaculum, Arundo, Canna, and other names to the same purport.   It is the barbarous Istusc and Alahance of the Alphonsine Tables; and El-sahm, the arrow, on the Dresden Globes, whence the Sham of the Palermo Catalogue.   This constellation is placed between the bill of the Swan, and the stars called the Family of Aquila, with its point towards the east, and its notch in the west.   It was formerly about 4° in length, but the moderns have stretched it to upwards of 10°.   Aratus is commonly said to have made Sagitta a part of Aquila; but Grotius shows this to be erroneous, and that the mistake is traceable to the Latin version of Germanicus.   Certainly in the large celestial map which illustrates Morel's Aratus, (1559,) the bird and the arrow are distinctly separated.   The components have been thus numbered:

Hyginus	. . .	4 stars	Kepler	. . . .	8 stars
Ptolemy	. . .	5	Flamsteed	. . .	18
Griemberger	. .	7	Bode	. . . . .	34

ζ Sagittæ is closely preceded on the west, by No. 274 P. Hora XX., a star of the 8¼ magnitude.   This was considered also to be double, by one of my astronomical friends; but, after a very patient examination of it, I can pronounce it to be single.

DCCXIX.　295 P. XIX. CYGNI.

Æ　$19^h\ 42^m\ 44^s$　　PREC. $+\ 2^s\cdot29$

DEC. N $33°\ 02'\cdot4$　　——　N $8''\cdot69$

POSITION A B $221°\cdot0$ (w 2)　DISTANCE $165''\cdot0$ (w 1) $\Big\}$ EPOCH 1836·60

———— A C $172°\cdot0$ (w 2)　————　$238''\cdot0$ (w 1)

A variable star, with distant companions on the Swan's neck, and nearly 8° to the north-north-east of β Cygni. A, adopted from the Palermo Catalogue as 6½; B 10, reddish; C 9, livid.

This extraordinary star has been mistaken for χ, a 5th-magnitude star, which precedes. Kirch, after watching it from 1686 to 1690, detected a regular variation from the 5th magnitude to invisibility, and made its period 404½ days; and in this conclusion he was joined by Maraldi, Cassini, and Le Gentil. Pigott thinks from this, that the period is variable, as he found the star going through all its changes in an average of 396 days 21 hours. This object takes about 3½ months in increasing from its minimum to its maximum brightness, and in decreasing again; whence it may be regarded as invisible for six months; which tallies with Halley's observations with a "6-foot tube, that bearing a very great aperture, discovers most minute stars." From these premises we see that it must rotate on its axis; for as its position in the heavens remains unaltered during the recurring changes, the varying brightness can be attributed neither to orbital motion nor an altering distance.

This star has been greatly confounded with the neighbouring one, discovered by Jansen in 1600, in *pectore*, or rather in *eductione colli*, near γ, and considered as varying from the 3rd to the 10th magnitude. It will be recollected that this was the "nova" which called forth such remark about two centuries ago, when the astrologers easily accounted for its production by $\mathstrut♂\ ♂\ ♀$. These changes, however, are truly wonderful, "where," as Sir John Herschel has well expressed it, "but for such evidences, we might conclude all to be lifeless."

DCCXX.　α AQUILÆ.

Æ　$19^h\ 42^m\ 58^s$　　PREC. $+\ 2^s\cdot89$

DEC. N $8°\ 26'\cdot9$　　——　N $8''\cdot71$

POSITION $323°\cdot1$ (w 8)　DISTANCE $152''\cdot6$ (w 5)　EPOCH 1834·81

A standard Greenwich star with a distant companion, at the root of the Eagle's neck. A 1½, pale yellow; B 10, violet tint; and the measures of the several astrometers stood thus, at the time of my observation:

⩊.	Pos. 334° 44'	Dist. 143''·40	Ep. 1781·56
Σ.	326° 06'	153''·71	1821·85
H. and S.	325° 48	153''·37	1823·11

These results indicated a retrograde angular movement, which is corroborated by my operations, but not altogether the surmised increase of distance. The orbital change here shown, is amenable to the occult movement in space, termed proper motion, the detected value of which, in the large star, is so sensible, that the rigorous investigations of Piazzi, Baily, and Argelander, yield results which are almost identical, both in amount and direction. They are:

$$P.... \; Æ + 0''\cdot 51 \qquad Dec. + 0''\cdot 38$$
$$B.... \qquad + 0''\cdot 59 \qquad\qquad + 0''\cdot 38$$
$$A... \qquad + 0''\cdot 56 \qquad\qquad + 0''\cdot 39$$

and it will be for time to show whether the *comes* is also following in the same direction with equal, or even more rapidity than A. It should, therefore, be sedulously watched, although its minuteness renders it of difficult vision under illumination. The discrepancy indicated, nearly forty years before ♅.'s epoch, by Bradley's observations, is:

$$a \; Aquilæ \;\; . \;\; . \quad Æ + 21''\cdot 60 \qquad Dec. + 17''\cdot 4$$
$$The \; Comes . \;\; . \qquad + 35''\cdot 85 \qquad\qquad + 12''\cdot 7$$

which, as well as the recent measures, shows the movement of both to be in the same direction; but we must pause for further observations before a binary or physical connexion can be positively pronounced, because in the meantime, trying back from my own epoch upon ♅. and Σ., with Piazzi's value of proper motion for A, these results are afforded for contemplation:

$$♅. \quad Pos. \; 335°\cdot 4 \quad Dist. \; 153''\cdot 5 \quad Ep. \; 1781\cdot 56$$
$$\Sigma. \qquad 326°\cdot 0 \qquad\quad 152''\cdot 8 \qquad\quad 1821\cdot 85$$

Aquila is in that part of the Milky Way which is just below Lyra, and a line drawn from Arcturus, through the head of Hercules, reaches three stars in a diagonal line,—β, a, and γ,—pointing north-north-west towards Wega and the Dragon's head beyond the Lyre: these three are designated the Family of Aquila, and have been (*mirabile dictu*) mistaken, by rather green hands, for Orion's belt. The brackish rhymes further tell us:

In Via Lactea's beauteous stream | beneath the Swan and Lyre,
See where Jove's Eagle soars on high, | the type of strength and fire;
And mark the triangle in which | his lucida partakes,
Which form, if join'd with Deneb's beams | and Wega bright, it makes.

'Aeτός, Aquila, Jovis Ales, Armiger Jovis, Vultur volans, Eagle and Flying Grype, have been the several designations of this asterism with our astronomers; by the Arabians it was called *El-'okáb*, the eagle, and *El-nesr el-táir*, the flying eagle, whence a Aquilæ is generally known as Altair. It is one of the old Catasterisms of Hipparchus; and M. Dupuis very fancifully thought the name was given when the constellation was near the summer solstice, and that the bird of highest flight was chosen to express the greatest elevation of the Sun. Some of our astrologers ascribed mighty virtue to the Grype. The components of Aquila and Antinous are usually thrown together in catalogues; and they have been thus numbered:

Ptolemy . . . . 15 stars | Hevelius . . . . 42 stars
Tycho Brahé . . . 19 | Flamsteed . . . 71
Bayer . . . . . 32 | Bode 276

DCCXXI.   307 P. XIX. AQUILÆ.

Æ   19ʰ 44ᵐ 38ˢ   Prec. + 2ˢ·86
Dec. N   9° 56'·7   —— N 8"·84

Position 307°·8 (w 5)   Distance 15"·0 (w 1)   Epoch 1834·63

A very delicate double star on the upper or northern shoulder of the bird, and about 1°¼ to the north of Altair. A 7, lucid white; B 13, blue; there is also a star of the 10th magnitude at a distance in the *sp* quadrant, and another still smaller follows nearly on the parallel, the whole forming a pretty object. It is 28 Ꮮ. II., and when discovered, had been thus registered:

Ꮮ. Pos. 306° 28'   Dist. 12"·00 ±   Ep. 1781·56

As this double star was well seen under an interposed achromatic lens, then just completed for me by Mr. Dollond, I applied that simple increase of power to its measures. The adaptation consists of a concave lens which is fixed to an adapter tube, and introduced into the ocular focus of the telescope, where it adds little to the weight, and but four inches to the length of the instrument. This doubles the value of the magnifying power used, diminishes the magnitude of the spider lines, and seems to flatten the whole field of view; and these advantages are gained without any serious loss of light. In the present instance, the object is so delicate, that without such aid, it could have been merely estimated; but with an eye-piece magnifying 118 times, = 236, and the telescope driven by the equatoreal clock, the position was of comparatively easy measurement.

DCCXXII.   56 AQUILÆ.

Æ   19ʰ 45ᵐ 27ˢ   Prec. + 3ˢ·26
Dec. S   8° 59'·1   —— N 8"·90

Position 72°·5 (w 4)   Distance 43"·0 (w 1)   Epoch 1834·63

A delicate double star between the head of Capricorn and Antinous's leg, in an absurd boundary hook 17° due south of Altair. A 6, deep yellow; B 12, pale blue; and the two point very nearly to a brightish star in the *nf* quadrant, just below, or north of which, is a fine pair of 6th-magnitude stars, lying *np* and *sf* of each other, and both of a blue tint. They may prove to be connected.

This open test was examined after the preceding object, and also under the achromatic lens. The angle of position was therefore readily obtained, but the distance is a mere estimation, by △ Æ over a small bar in the eye-piece, expressly fitted for such cases.

DCCXXIII. 57 AQUILÆ.

Æ 19^h 45^m 58^s PREC. + 3^s·25
DEC. S 8° 38'·3   —— N 8"·94

POSITION 171°·5 (w 9)   DISTANCE 35"·4 (w 7)   EPOCH 1834·63

A fine double star in the bow of Antinous, and in the hook with the object above-mentioned. A 6½, and B 7, both blue; being Nos. 313 and 314, P. XIX. This being close to 56 Aquilæ was brought into the field under the achromatic lens, with a very powerful advantage; and the angular measures seemed to be really so perfect as to decide the disputed question of the star's fixity. The object is 14 ♅. IV., and from a comparison of his results with those of Σ. and H. and S. there appeared a retrograde orbital movement of − 0°·41 per annum, with a considerable increase of distance, a conclusion countenanced in some measure by deductions which I drew from Piazzi's meridional observations. The registered data stood thus, when my attack commenced:

	Pos.	Dist.	Ep.
♅.	188° 05'	29"·47	1780·59
P.	178° 30'	34"·6	1800·00
Σ.	171° 48'	36"·20	1821·79
H. and S.	171° 08'	36"·16	1823·60

There is therefore little doubt that Sir William Herschel inserted the wrong quadrant to his position, since 81°·55 sf instead of sp = 171°·55, would be satisfactory; but still there exists an error in the distance, which, in so easy and wide an object, is somewhat inexplicable. The proper motions assigned to A are disappearing under recent scrutiny; the statement now is:

		Dec.
P....	Æ + 0"·13	− 0"·07
A....	− 0"·02	+ 0"·02
B....	0"·00	+ 0"·04

DCCXXIV. 320 P. XIX. VULPECULÆ.

Æ 19^h 46^m 20^s PREC. + 2^s·64
DEC. N 19° 55'·5   —— N 8"·97

POSITION 147°·6 (w 9)   DISTANCE 42"·7 (w 9)   EPOCH 1834·72

A wide double star between the Fox and the Arrow, in the following stream of the Milky Way, and 11°¼ to the north of Altair. A and B, both 7, and both white, the latter being 321 P. XIX. This object is 106 ♅. V., all the measures of which, including deductions from Piazzi's admirable mean places, indicate relative fixity, being:

	Pos.	Dist.	Ep.
♅.	150° 42'	38"·90	1782·85
P.	147° 30'	43"·00	1800·00
H. and S.	148° 30'	42"·43	1823·63

Away in the sp quadrant is a triple star, the members of which are nearly in a line *sp* and *nf* across the parallel, of the 10th, 11th, and 13th magnitudes. These form 105 Ḥ. III., but the discoverer, as well as *Σ.*, only noted two of them; and in the great Dorpat Catalogue of 1837, it is marked No. 2595, *rej*.

DCCXXV.   71 M. SAGITTÆ.

$\mathbb{R}$ 19^h 46^m 36^s Prec. + 2^s·67
Dec. N 18° 22'·1 —— N 8"·99
Mean Epoch of the Observation . . . . 1836·65

A rich compressed Milky-Way cluster on the shaft of the arrow, and 10° north-a quarter-east from Altair. It was discovered by Mechain in 1781, and described by Messier as a nebula unaccompanied by stars, and of a very feeble light. Piazzi seems to have observed it meridionally as a star of the 8th magnitude, by admitting the light of a lamp upon it (312 P. XIX.), but his darkened field ought to have shown that it is flanked with four telescopic stars, besides other larger companions in view. It was first resolved into stars by Sir William Herschel, in 1783, who esteemed its profundity to be of the 243rd order.

DCCXXVI.   β AQUILÆ.

$\mathbb{R}$ 19^h 47^m 26^s Prec. + 2^s·94
Dec. N 6° 00'·7 —— N 9"·06
Position 343°·0 (*w* 1)   Distance 175"·0 (*w* 1)   Epoch 1834·50

A standard Greenwich star, with a telescopic *comes*, in the Eagle's neck, 2°½ to the south-south-east of Altair, and the lowest of the Family of Aquila. A 3½, pale orange; B 10, pale grey; with two other small stars following, besides the four preceding ones mentioned by Piazzi, Note 324, Hora XIX. A is ascertained to be more sensibly affected by proper motions than it was held to be; and the following are the results of the best scrutineers:

$$P.... \mathbb{R} - 0''·03 \quad \text{Dec.} - 0''·54$$
$$A.... + 0''·08 \quad\quad - 0''·47$$
$$B.... + 0''·11 \quad\quad - 0''·44$$

This star appears in many catalogues as Alshain, *i. e.*, *al-shâhin*, the falcon, which, though used as an Arabic word, is Persian: with many others, it was transplanted under the Khalifs. It is not now so bright as γ or δ, which could not have been the case in Mayer's time. Ptolemy marked it γ in brightness, as did also Ulugh Beigh, Tycho Brahé, Bradley, and La Caille; Flamsteed, Piazzi, and Zach, rate it 3½, but Hevelius could enter it no larger than the 4th magnitude.

DCCXXVII. ε DRACONIS.

Æ 19ʰ 48ᵐ 41ˢ   PREC. − 0ˢ·17
DEC. N 69° 51′·6   —— N 9″·15
POSITION 354°·6 (w 8)   DISTANCE 3″·1 (w 5)   EPOCH 1833·68

A fine double star in the *nf* flexure of Draco's back. A 5½, light yellow; B 9½, blue; and there is a third star at a little distance to the north of A. This elegant object is 8 Ḥ. ɪ., and these were the measures previous to mine:

Ḥ.   Pos. 333° 14′   Dist. 2″·50±   Ep. 1780·76
H. and S.   355° 21′   2″·59   1823·58
D.   354° 00′   3″·10   1830·60

These appeared to show a great increase of angle; but as Ḥ. had found it 354°·29 in 1804·39, there was still a doubt over the whole. My observations, however, followed since by the corroboration of the Dorpat measures, establish the relative fixity of the components; though they seem to have a small proper motion, of which that to A has been thus sifted:

P.... Æ − 0″·39   Dec. − 0″·30
B....   + 0″·18   − 0″·01

Though H. and S. rate the *comes* of this star as only of the 10th magnitude, it seems to have given them no small trouble, for they remark that it was "even uncertain whether the small star had really been seen at all." In my instrument it bears pretty full illumination, yet the inflection, from disparity of brilliance, rendered the measures rather ticklish. ε Draconis is on the preceding parallel of β Cephei; and it constitutes the north-west angle of the square formed by δ, π, ρ, and ε. σ Draconis, a star of the same magnitude with ε, and preceding it to the west-by-south, is *Al-atháfi*, the tripod; and it is remarkable for having a large proper motion in declination, amounting to about − 2″·0.

DCCXXVIII. ψ CYGNI.

Æ 19ʰ 51ᵐ 30ˢ   PREC. + 1ˢ·57
DEC. N 52° 01′·0   —— N 9″·36
POSITION 184°·7 (w 6)   DISTANCE 3″·4 (w 5)   EPOCH 1830·78
——— 184°·2 (w 9)   ——— 3″·5 (w 5)   ——— 1837·53

A fine double star, in the space between the tip of the Swan's preceding wing and the tail, being the middle one of three stars about 10° north-north-west of Deneb. A 5½, bright white; B 8, lilac. From Ḥ. having registered the acolyte in a wrong quadrant, there appeared to have been a great change in the angle of this pretty object, but H. on reference to his father's MS. altered the *np* to *sp*, which made an alteration of 180° by a *coup de plume.* The results of 15 ♅. II. are therefore:

	Pos.	Dist.	Ep.
♅.	180° 28′	4″·00±	1779·84
H. and S.	182° 00′	4″·32	1823·65
Σ.	184° 35′	3″·32	1831·39

DCCXXIX.   27 M. VULPECULÆ.

Æ   19ʰ 52ᵐ 39ˢ   Prec. + 2ˢ·59

Dec. N 22° 17′·1   —— N 9″·46

Mean Epoch of the Observation . . . . 1834·62

This is the double-headed shot or dumb-bell nebula, on the Fox's breast, close to 14 Vulpeculæ; it is about 7° to the south-east of β Cygni, and nearly half-way between it and the Dolphin. This magnificent and singular object is situated in a crowded vicinity, where field after field is very rich. It was discovered in 1764, and described by Messier as an oval nebula without a star. My instrument, however, shows several,

 of which three, lying *sp* and *nf,* nearly in the same direction with the dumb-bell, and 26ˢ distant, following, are sufficiently remarkable: the two largest are of the 8 and 9¼ magnitudes, and 6ˢ apart, on an ∠ = 73°, the latter having a small reddish telescopic companion *nf.*

27 Messier is truly one of those splendid enigmas, which, according to Ricciolus, are proposed by God, but never to be subject to human solution. As the axis of symmetry, or line through the centres of the principal masses, is not less in apparent diameter than 5′, the vastness of its extent is as utterly inconceivable as the dynamical maintenance of its form. Although ♅. could only bring it to a mottled nebulosity, yet, from some accidental stars, he considered it resolvable: he also concluded it to be a double nebula, each with a seeming nucleus, with their apparent luminiferous matter running into each other. But Sir John Herschel saw and figured a feature which had been overlooked, and which entirely altered the views of its physical constitution. This is "the faint luminosity which fills in the lateral concavities of the body, and converts them, in fact, into protuberances, so as to render the general outline of the whole nebula a regular ellipse, having for its shorter axis the common axis of

the two bright masses of which the body consists, that is to say, the *longer* axis of the oval form under which it was imperfectly seen by Messier." The subjoined sketch, showing it as seen in my telescope, is of course without this "filling in."

The Earl of Rosse has lately applied his powerful reflector upon this wondrous object. It bore very high magnifying powers, and though his lordship would not assert that it was actually resolvable, "in the absence of that complete re-solution which leaves no room for error," very evident symptoms of resolvability were shown. The direction of the axis of symmetry through this dense agglomeration of stars, as measured with the flat bar-eyepiece of my micrometer, is at an angle with the meridian of the world = 31°·8.

DCCXXX. 75 M. SAGITTARII.

Æ 19ʰ 56ᵐ 38ˢ Prec. + 3ˢ·55
Dec. S 22° 22'·0 —— N 9"·77

Mean Epoch of the Observation . . . . 1834·62

A globular cluster in the space between the left arm of Sagittarius and the head of Capricorn, and 7°½ degrees to the south-south-west of β Capricorni. It is a lucid white mass among some glimpse stars, with a large one in the *nf* verge of the field. It was discovered by Mechain in 1780, who considered it as a nebula without a star; but Messier viewed it as a mass of very small stars, which opinion, on an object which at best is rather faint, was bold. In 1784, it was resolved in Sir William Herschel's 20-foot Newtonian, and, on being gauged, was assigned a profundity of the 734th order. No wonder that this miniature of 3 Messier (No. CCCCXCII.) should be pale to the gaze.

DCCXXXI. 396 P. XIX. CAPRICORNI.

Æ 19ʰ 57ᵐ 51ˢ Prec. + 3ˢ·35
Dec. S 13° 22'·8 —— N 9"·86

Position 13°·0 (w 2) Distance 15"·0 (w 1) Epoch 1838·56

A most delicate double star near the tip of Capricorn's left horn, which precedes α Capricorni by about 3° on the western parallel. A 8½, pale yellow; B 14, blue; and there are many glimpse companions in the field. This object is 63 ʜ. ɪɪɪ., and though seen very distinctly in my instrument, is so troubled with variable refractions, that even the estimation I attempted was an affair of difficulty. The other registered details are:

ʜ. Pos. 16° 12' Dist. 14"·12 Ep. 1782·68
Σ. 11° 20' 13"·32 1827·67

2 H 2

DCCXXXII.   415 P. XIX. VULPECULÆ.

Æ 20ʰ 00ᵐ 12ˢ   Prec. + 2ˢ·63
Dec. N 20° 38'·7   —— N 10"·04

Position 340°·5 (ᵂ 8)   Distance 4"·5 (ᵂ 5)   Epoch 1838·70

A delicate double star, close to the arrow under the Fox's shoulder. A 8, pale white; B 10, sky-blue. This was discovered by Sir James South, at Passy, near Paris; and is one of those classed by Struve as "acervi," but it is in rather a fine galaxy splash of stars than a cluster. It is to be found midway between β Cygni and α Delphini, and nearly a degree west of Ϩ Sagittæ, a star of the 7th magnitude. The earliest results obtained and registered are:

S.   Pos. 340°·00   Dist. 4"'·518   Ep. 1825·63
Σ.   340°·42   4"'·450   1832·95

DCCXXXIII.   2 P. XX. AQUILÆ.

Æ 20ʰ 01ᵐ 51ˢ   Prec. + 2ˢ·73
Dec. N 16° 26'·8   —— N 10"·16

Position 13°·2 (ᵂ 8)   Distance 5"·9 (ᵂ 5)   Epoch 1838·58

A delicate double star, hooked into the Eagle's boundary near the shaft of the arrow, and situated 10° to the north-north-east of Altair. A 7, pale topaz; B 10, lucid blue; they are in a splendid vicinity, and in the *sf* is the star mentioned by Piazzi, "25" temporis aliæ 8ᵃ magnitud. sequitur 5' circiter ad austrum." The object before us is 70 Ḥ. II., and was thus registered previous to my epoch:

Ḥ.   Pos. 17° 03'   Dist. 4"'·50±   Ep. 1782·85
S.   14° 36'   6"·61   1825·64
Σ.   13° 40'   6"·43   1830·12

These results would imply orbital motion, but that the star is of too difficult measurement for an absolute opinion yet. The colours here assigned are under high magnifying powers, fine atmosphere, and easy motion; yet Ḥ. has said, "large, red; small, a deeper red." Now, whether Sir William's bias for red tints was as much owing to his mirror-metal, as to a peculiarity of vision, his observations afford proof that the singularity was not at all accompanied by any defect of vision. Indeed sharp-sightedness might be termed an attribute of Ḥ. These seemingly opposed qualities, however, as I have already shown, are not incompatible with each other.

DCCXXXIV. ϑ SAGITTÆ.

Æ 20^h 02^m 53^s Prec. +   2^s·64
Dec. N 20° 26′·6   —— N 10″·24

Position AB 327°·1 (w 6) Distance 11″·4 (w 6)
—————— AC 226°·6 (w 9)   ———— 70″·1 (w 9)   } Epoch 1834·77

A triple star near the head of the Arrow, and nearly in mid-distance between β Cygni and α Delphini. A 7, pale topaz; B 9, grey; C 8, pearly yellow. This object was discovered in August, 1780, and registered 24 ♄. iii.; but Sir William took no measures of position, giving the distance of A B = 11″ 4‴, and A C = 59″ 49‴, another proof of the existence of some cause of error in his micrometer when opened to considerable distances. Piazzi saw the duplicity of A, but the *comes* must have appeared smaller than of the 9th magnitude, for he says, "Duplex. Comes vix pene visibilis, 0^s·2 temporis, 5″ ad boream præcedit." By admitting this slight notice, and his observation of C (13 P. xx.), the epochs stand thus:

Piazzi, 1800:
A B Pos. 329°·2 Dist. 6″·0±
A C   225°·7   64″·5

South, 1824·98:
A B Pos. 327° 22′ Dist. 11″·777 | A B Pos. 326° 43′ Dist. 11″·405
A C   226° 49′   70″·088 | A C   326° 40′   70″·699

Struve, 1832·82:

A and C point in the direction of another star in the *sp* quadrant, at about the same distance; and A is followed on the parallel by a wide double star, in a fine field. The proper motions hitherto registered for θ are indecisive both in value and direction; the best are:

P.... Æ − 0″·01   Dec. + 0″·17
B....   + 0″·12   + 0″·09

DCCXXXV. 12 P. XX. ANTINOI.

Æ 20^h 03^m 06^s Prec. +   3^s·08
Dec. S 0° 35′·7   —— N 10″·26

Position 202°·2 (w 9) Distance 54″·2 (w 9) Epoch 1835·58

A wide double star on the youth's arm, 7°½ to the south-south-east of β Aquilæ, and just to the north of θ Antinoi, a star of the 3½ magnitude. A 8, and B 9, both pale grey, the latter being 13 P. xx.; there are several small stars in the field. This object is 136 ♄. v., and from a comparison of all the measures, including reductions from Piazzi's mean apparent places, it must be merely optical. The results stand thus:

♄. Pos. 204° 12′ Dist. 47″·09 Ep. 1783·70
P.   202° 30′   56″·00   1800
S.   202° 07′   54″·67   1825·01

DCCXXXVI.   30 P. XX. DRACONIS.

Æ   20ʰ 03ᵐ 24ˢ   Prec. +   0ˢ·79
Dec. N 63° 14′·1   —— N 10″·28

Position 163°·7 (*w* 5)   Distance 3″·0 (*w* 3)   Epoch 1832·91

A close double star between the Dragon's back and the elbow of Cepheus; it is 16° west-by-north of α Cephei, and about one-third of the distance between 64 and 66 Draconis. A and B, both 9, and both pale white. It was discovered by Professor Struve, and entered No. 2642 of the Dorpat Catalogue, with the remark, "australis duarum æqualium." In the larger work which followed, are these data:

Pos. 165°·20   Dist. 2″·447   Ep. 1832·51

DCCXXXVII.   26 P. XX. ANTINOI.

Æ   20ʰ 04ᵐ 25ˢ   Prec. +   3ˢ·06
Dec. N 0° 23′·6   —— N 10″·35

Position 207°·9 (*w* 9)   Distance 3″·5 (*w* 5)   Epoch 1832·80

A fine double star on the left arm of Antinous, and nearly 2° north-half-east of θ, and south-east of β Aquilæ. A 6½, and B 7, both white. This pretty object is 96 Ḥ. ɪɪ., and from the measures given in the Catalogue of Messrs. Herschel and South, it was considered as having a sensible orbital retrogression, which my observations did not confirm. These are the previous results:

	Pos.	Dist.	Ep.
Ḥ.	213° 48′	3″·00 ±	1783·70
Σ.	208° 09′	3″·86	1821·82
H. and S.	208° 12′	4″·10	1823·65

DCCXXXVIII.   43 P. XX. AQUILÆ.

Æ   20ʰ 06ᵐ 19ˢ   Prec. +   2ˢ·95
Dec. N 6° 06′·1   —— N 10″·50

Position 11°·7 (*w* 7)   Distance 43″·6 (*w* 5)   Epoch 1833·70

An open double star on the Eagle's beak, 4°¼ due east of β Aquilæ. A and B, both 8½, and both lucid white, the latter being 44 P. xx.; and there is a coarse pair in the *sf* quadrant, somewhat resembling this, but the components are smaller. The relative circumstances of this object will be seen on comparing the following deductions:

	Pos.	Dist.	Ep.
P.	14° 20′	45″·50	1800
S.	12° 49′	43″·89	1824·67

DCCXXXIX. *o*¹ CYGNI.

ℛ 20ʰ 08ᵐ 36ˢ   Prec. + 1ˢ·89

Dec. N 46° 15'·6   —— N 10"·67

Position AB 330°·0 (*w* 2) Distance 15"·0 (*w* 1)

————— AC 174°·1 (*w* 9)   ————— 106"·6 (*w* 9) ⎫

————— AD 333°·8 (*w* 9)   ————— 338"·0 (*w* 9,) ⎭ Epoch 1838·67

A wide quadruple object in a rich field on the Swan's left thigh, about 8° west-by-north of Deneb, and forming the vertex of a low isosceles triangle with that lucida and δ Cygni. A 4, orange; B 16, livid; C 7½, and D 5½, both cerulean blue; these colours were closely attended to, and the latter noted when the large star was hidden by the flat bar of the eyepiece magnifying 157 times, so that Struve's idea, that they are not the effect of contrast, is supported. Here A and C form 10 ♅. vi. B is a most minute point of light, to which Sir John Herschel drew my attention as a severe test which had escaped his father's gaze: it is, of course, far too delicate for metrical observation with my means, but its angle is not difficult to estimate, as it lies nearly on the line to D, which is *o*¹ Cygni. These are the recorded data, including reductions from the Palermo Catalogue, a means, however little intended, by which several important epochs are obtained. from the results of the altitude and azimuth circle:

	A and B	B not seen by ♅. P. S. or Σ.		
♅.	A and C	Pos. 182° 46'	Dist. 99"·98	Ep. 1779·84
P.	...	175° 20'	106"·50	1800
S.	...	173° 53'	106"·39	1824·66
Σ.	...	173° 59'·7	106"·85	1836·18
P.	A and D	324° 30'	338"·00	1800
Σ.	...	333° 41'·9	337"·33	1835·95

From these premises, it would certainly seem that ♅. had read off 10° wrong in the angle, and that his micrometer showed the defect at large distances already alluded to. The spacial movement, though small, must be admitted into the argument of probable changes, and the most recent values are:

P.... ℛ − 0"·03   Dec. + 0"·18

B.... + 0"·03   + 0"·05

Piazzi assigns a directly opposite proper motion to A and D, which tends to approximate them, and to diminish their angle a little: but it will require the utmost precision of modern observation, under the nicest manipulation of the best instruments, to assign the proper value and direction to those delicate quantities.

No such movement being assigned to C, its relation to A would be merely changed by that of the large star in an increase of distance, which, however, is not *proven* by subsequent measures; wherefore it will be prudent to suspend judgment, and wait twenty or thirty years longer, as H. wisely advises.

DCCXL. α^2 CAPRICORNI.

℞ 20ʰ 09ᵐ 10ˢ   Prec. + 3ˢ·33
Dec. S 13° 02′·1   —— N 10″·71

Position A a 145°·0 (w 1)   Distance   5″·0 (w 1)
——— A a 155°·7 (w 2)   ——— 198″·0 (w 1)   Epoch 1838·72
——— A B 291°·4 (w 9)   ——— 373″·4 (w 9)
——— B b 221°·8 (w 3)   ——— 43″·0 (w 1)

A standard Greenwich star and quintuple object, at the tip of Capricorn's right horn. A 3, pale yellow; B, which is a^1, 4, yellow; a 16, blue; a 9, ash-coloured; b 9½, lilac tinge. Here the principal objects form Prima and Secun. Giedi of the Catalogues, from *al-jedí*, the goat; and they present a fine double star to the naked eye, if a very good one. The telescope, however, shows it to be multiple; but it requires an instrument of no small power to reveal that most minute point of light, a, which long baffled my attempts to catch it. Indeed, when Sir John Herschel first wrote to me about it, in a letter of the 7th of October, 1831, the case appeared hopeless; but under clock-work motion, on a singularly fine September evening, I caught it in little evanescent flashes, so transient as again to recall Burns's snow-flakes on a stream. As it was somewhat in the direction of the *sf* star *a*, its position was within estimation; but the distance is nothing more than a vague guess. However, I was well pleased with seeing it, because this object is one of those to which Sir John has, in the most express way, called the close attention of astronomers, as probably shining by reflected light. The determination of this interesting fact, may be reserved for one of the new grand telescopes, and will indeed be a triumphant achievement.

The micrometrical measures of the two principal stars, as well as reductions from Piazzi's ℞s and Decs., are in such admirable accordance, during a period embracing thirty-nine years, as to attest relative fixity. Those previous to my own are:

P.	Pos. 291° 42′	Dist. 371″·20	Ep. 1800
H. and S.	291° 26′	372″·99	1822·58
Σ.	291° 24′	374″·20	1835·70

A spacial movement is attributed to both components, which, as far as they can be yet valued, indicate something of a physical connexion:

a^2 Capricorni.		a^1 Capricorni.	
P.... ℞ + 0″·04 Dec. + 0″·25		P.... ℞ − 0″·05 Dec. − 0″·08	
B.... + 0″·12 + 0″·03		B.... + 0″·05 + 0″·02	
A.... + 0″·05 + 0″·01		A.... + 0″·01 + 0″·02	

Αἰγόκερως, Ægoceros, Πάν, Pan, Αἴγιπαν, Ægipan, and Capricornus, have been the most general names of Capricorn; but it appears in the Arabo-Latin *Almagest* as Alcaucurus, and elsewhere as *Al-jedi*, the goat. It is usually termed the Xth sign in zodiacal order, and the IVth of the southern signs; being one of the old forty-eight asterisms which the Greeks are supposed to have received from the Egyptians. As a mere quadruped, it was considered to be in honour of Amalthæa, though she was already deified as Capella; but its fanciful compound form was held to commemorate Pan's leap into the Nile, in a *pan*-ic at Typhon's approach, whereby his hind-quarters became fishified. The early Venetian illustrations of Hyginus represent this singular portion with an overhand knot in it; but a fine old geographical MS. on vellum, in the Archbishop of Canterbury's library at Lambeth, of an unknown date, but belonging to Robert Hare in 1564, figures it as a goat: as it also is in Albumazar's *Introductorium in Astronomiam.* Postellus seizes upon such representations, to claim the sign for the scape-goat in Leviticus; but he was also pleased to consider Pisces as representing the fishes with which the multitude were miraculously fed. Other wiseacres found in Giedi a corruption of Gad, and therefore an Israelitish emblem. Now Macrobius had long before advertised, that as the sun approached this sign, he quitted his lower course, and ascended more and more, wherefore the figure of a goat was chosen to represent it, because that animal is addicted to climbing the sides of mountains. Cancer and Capricorn form the boundaries of the solar course in the zodiac, the *portæ solis:* the Platonists held that souls descended from heaven into mortal bodies through the first, and when released from corporeality, reascended through that of Capricorn, which last was therefore called the Gate of the Gods, as the former that of men. (See No. CCCXXXV.) Under these truly sage traditions, Sir Edward Sherburne waggishly observes that "it *climbs* the mid-heaven at midnight, about the end of July."

Although Capricorn is not a striking object, it has been the very pet of all constellations with astrologers, having been the fortunate sign under which Augustus and Vespasian were born, who thereby were entitled to the tutelage of Vesta: and this Sabæan superstition was honoured by medals, marbles, poems, and what not. It was not only of happy influence in classic times, but was also mightily looked to by the Arabians, who termed *a* and *β, Sa'd-adh-dhábih*, the lucky star of the slaughterer; and *γ* and *δ* were *Sa'dubnáshirah*, the fortunate star bringing good tidings. In this light the XXIInd Lunar Mansion was a popular one; and Kazwíní, Tízíní, Ferghání, and Fírúzabádí*, mention its happy tendency. Ulugh Beigh designates it *Fortuna mactantis;* but my friend Mr. Rothman found the curious bronze Arabian globe belonging to the Royal Astronomical Society, inscribed with "E Fortunâ debilis." As to the *muḥibbain*, or two friends, Beigel dismisses the name altogether; "moral beings," he observes, "being strangers to the Nomade Arab's heaven."

With such high authority before them, the astrologers of the Middle Ages of course adopted this sign in all its importance; nor were they long in discovering, that as the deluge was owing to a conjunction of all the planets in Capricorn, so would the great conflagration be occasioned by their conjunction in Cancer. We learn from the notes to the

---

* Fírúzabádí, of Khorásán, is the author of the *Ḵámús, i. e.,* Ocean, the most famous of all Arabic Lexicons.

Ysagogicus of Alchabitus, published in 1485, that when Saturn is in this earthly Trigon he exercises great influence over man,—"Caput et pedes habet;" Jupiter in the same sign rules "genua et oculos," and so on. The MS. Almanac for 1386, assures us that "whoso is borne in Capcorn schal be ryche and wel lufyd;" while the DOCTOR, as Arcandam is *par excellence* styled, shows that a man so born shall be addicted to all kind of ladies, shall have eight special illnesses, and shall die at sixty. He moreover adds, rather unpolitely to Augustus and Vespasian, that the sign is prosperous in dull and heavy beasts. It was the most frequent of the Uranographic amulets so long used, and may have been among the rubbish found in Sidrophel's pocket, wherein were

> A moon dial, with Napier's bones,
> And several constellation-stones,
> Engrav'd in planetary hours,
> That over mortals had strange powers.

As a counterpoise to the general impression in favour of Capricorn, Ovid, in making Acætes qualify for a sailor, is supposed by some to give it an unfavourable aspect. But though Ovid had no great cause to honour the nativity-star of Augustus, he certainly meant the Hædi (See No. CLXXXVI.) in the following lines:

> *Oleniæ* sidus pluviale *Capellæ*,
> *Taygetem*que, *Hyadas*que oculis, *Arctum*que notavi.

The ancients accounted Capricorn the Xth sign, and when the Sun arrived thereat, it made the winter solstice, with regard to our hemisphere; but the stars having advanced a whole sign towards the east, Capricorn is now rather the XIth sign; and it is at the Sun's entry into Sagittarius, that the solstice occurs, though the ancient manner of speaking is still retained. To identify the asterism, direct a line from Wega to Altair, and continue it 23° into the south-south-east, and it will pass between a and β in the head of Capricorn:

> A startling monster's hybrid form   your eyes will there assail,
> That sign so often dubb'd the Goat,   yet with a fish's tail;
> And though its figure is not large,   it brightly still doth glow,
> Its stars within the outline placed,   no *amorphotæ* know.

Sir William Herschel has tabulated the lustre and magnitude of the stars in Capricorn, in the *Philosophical Transactions* for 1796; and they have been thus successively numbered:

Ptolemy	28 stars	Hevelius	30 stars
Tycho Brahé	28	Flamsteed	51
Copernicus	29	Bode	154

DCCXLI. σ CAPRICORNI.

Æ   20ʰ 10ᵐ 10ˢ   PREC. +   3ˢ·47
DEC. S 19° 36′·8   —— N 10″·78

POSITION 176°·8 (*w* 8)   DISTANCE 54″·1 (*w* 4)   EPOCH 1837·61

A wide double star on the animal's *os frontis*, and rather more than 4° to the south of β. A 5½, yellow; B 10, violet; there are several distant stars in the *nf*. This object is 87 ☿. v., and it seems to have been relatively stationary from its first enrolment in 1782.

Hevelius, in his *Prodromus*, Maupertuis, in his *Figures des Astres*, and Dr. Derham, in his *Letter to the Royal Society*, in 1733, mention the positions of four nebulæ in the head of Capricorn, for which Messier searched in vain. Concluding that his search might have been with his 3½-foot telescope, I again ransacked the vicinity, and am pretty well satisfied of the non-existence of any nebula or cluster which the instrument of Hevelius could show, and yet be hidden from mine. As to the nebulous wisp (No. 2073), discovered by Sir John Herschel to the *sf* of σ, it was far beyond the reach of former observers.

DCCXLII. β² CAPRICORNI.

℞ 20ʰ 12ᵐ 01ˢ	Prec. + 3ˢ·38
Dec. S 15° 16′·9	—— N 10‴·92

Position 267°·2 (w 9)   Distance 204″·8 (w 9)   Epoch 1832·68

A wide pair of stars in the middle of Capricorn's right horn, and 2°½ distant from *a*, to the south-half-east. A 3½, orange yellow; B 7, sky-blue; and there are several telescopic companions, to be presently spoken of. The two principal form the Dabih major and Dabih minor of the Catalogues, a name derived from the *Sa'd-adh-dhábiḥ* already mentioned under *a* Capricorni. The object forms 28 ☿. vi., and was registered in 1780, but without other remarks than that B precedes, and is about 3′ distant from A. A reduction of the excellently determined ℞s and Decs. of Piazzi, however, afforded an epoch of comparison with the micrometrical measures of Sir James South; and these were the data when my attack was made:

P.	Pos. 263° 12′	Dist. 205″·50	Ep. 1800
S.	267° 06′	203″·71	1824·69

When I commenced testing the properties of my telescope in 1830, Sir John Herschel sent me the place of a double star which he had discovered, forming the vertex of an obtuse angle, and nearly isosceles triangle, with β¹ and β² Capricorni. This he pronounced to be one of

the most minute and delicate of such objects, being of the 17th and 18th magnitudes, and 3° apart; and he moreover added, that the telescope which was not competent to divide this pair, could not have the remotest chance of seeing the satellites of Uranus. Under this friendly warning, I took advantage of the very finest weather in the beginning of September, and it was truly fine, to try my power upon it. I gazed in profound quiet and darkness, and readily made out the object (x), from its situation near the following line of a trapezium formed by two telescopic stars to the *np* of β^1 and β^2; but I was utterly unable to split it, although I made every attempt that my eye-piece battery would admit of. The largest member of the group, β^2, is followed by a faint coarse double star, which is splendor itself as compared with the test just spoken of.

As this whole group, from its easy situation, forms an admirable criterion for proving the performance of telescopes in light, colour, penetration, and definition, I have frequently reverted to it; and am bound to admit, that the vision of the Herschelian test as a single point of light, is the maximum power of my instrument.

DCCXLIII. $\varkappa$ CEPHEI.

Æ 20ʰ 14ᵐ 08ˢ	Prec. − 1ˢ·78
Dec. N 77° 13'·6	—— N 11"·10

Position 124°·7 (*w* 8)	Distance 7"·5 (*w* 5)	Epoch 1833·75
—— 123°·8 (*w* 9)	—— 7"·5 (*w* 8)	—— 1838·83

A neat double star on the instep of the left foot, and about halfway from β Cephei to ϵ Ursæ Minoris, the inner star of the tail. A 4½, bright white; B 8½, smalt blue, colours beautifully shown. This fine object is 70 Ḥ. iii., and exhibited this promise of binarity, when I took it in hand:

Ḥ.	Pos. 122° 30'	Dist. 5"·79	Ep. 1782·74
Σ.	126° 12'	7"·08	1820·18
H. and S.	128° 04'	8"·14	1823·70

I was, therefore, disappointed on finding my results were altogether in favour of the fixity of the components; and this is since confirmed by Σ.'s later measures.

My attention was drawn to the mean apparent place of this star by Mr. Baily, who found a serious difference in the Æ of Piazzi's first and second Catalogues; my transit instrument soon showed that the value in the first was the correct one. This star and γ, though Ptolemy expresses ποδὸς to both, are very absurdly placed upon the knees of Cepheus in some recent maps and globes, and one foot upon the Little Bear's back. The figure is properly drawn upon the large Maps published by the Society for the Diffusion of Useful Knowledge; but though the best which have appeared, they have not been accepted of as a standard, as may be instanced in the large Globes just published, and supposed to be also under their patronage. Those maps, with Mr. Baily's reformed boundaries, ought to be made the authority.

DCCXLIV. 16 ⩗. IV. DELPHINI.

Æ 20ʰ 15ᵐ 15ˢ   PREC. + 2ˢ·67
DEC. N 19° 35′·6   —— N 11″·15

MEAN EPOCH OF THE OBSERVATION . . . . 1837·65

A fine though small planetary nebula, between the Dolphin's pectoral fin and the Arrow's head: it is nearly 6° to the north-north-west of *a* Delphini, exactly on the line from that star to Wega. It is in a coarse cluster, in the centre of which are four stars lying *sp* and *nf;* and the nebula follows them, between two stars of the 10th magnitude, and some glimpse companions. It was discovered in September, 1782.

Sir John Herschel, who has figured this object, suggests that the minute stars in close proximity may possibly prove to be satellites. "The enormous magnitude of these bodies," he remarks, "and consequent probable mass (if they be not hollow shells), may give them a gravitating energy, which, however rare we may conceive them to be, may yet be capable of retaining in orbits, three or four times their own diameter, and in periods of great length, small bodies of a stellar character."

DCCXLV. 113 P. XX. VULPECULÆ.

Æ 20ʰ 15ᵐ 47ˢ   PREC. + 2ˢ·58
DEC. N 23° 34′·2   —— N 11″·19

POSITION 222°·4 (*w* 2)   DISTANCE 45″·0 (*w* 1)   EPOCH 1837·88

A delicate double star on the Fox's loins, in the midst of a little stellar group closely preceding 118 P., and about 9° to the north-north-west of *a* Delphini. A 8, bluish white; B 14, indigo blue; and the two point to a dusky 11th-magnitude star in the *nf* quadrant, over a sort of crescent of little companions. The intense colour of B is well shown to the averted eye, in a darkened field. About a minute of time preceding this object, and 20′ south of it, is a minute close double star, whose components lie nearly on the parallel; this is Σ. No. 2672.

DCCXLVI. 116 P. XX. ANTINOI.

Æ 20ʰ 16ᵐ 28ˢ   PREC. + 3ˢ·06
DEC. N 0° 33′·3   —— N 11″·24

POSITION 28°·0 (*w* 3)   DISTANCE 30″·0 (*w* 1)   EPOCH 1834·81

A very delicate double star between the head and bow of Antinous. A 7½, white; B 12, grey. This object is in a poor field, and the *comes*

so minute that its distance is only an estimation, though its angle could
be caught by the spherical crystal micrometer.   It lies about 10° south-
east of β Aquilæ, where it is crossed by a line from a Delphini to
a Capricorni. Σ. first enrolled it, No. 2677, and IVth Class; and the
arrival of the Great Dorpat Catalogue brought these measures:

Pos. 28°·76   Dist. 33″·18   Ep. 1828·47

DCCXLVII.   29 M. CYGNI.

Æ 20ʰ 18ᵐ 17ˢ   Prec. +   2ˢ·21
Dec. N 37° 59′·9   —— N 11″·37

Position 299°·5 (w 1)   Distance 55″·0 (w 1)   Epoch 1835·48

A neat but small cluster of stars at the root of the Swan's neck, and
in the preceding branch of the Milky Way, not quite 2° south of γ;
and preceding 40 Cygni, a star of the 6th magnitude, by one degree just
on the parallel.   In the *sp* portion are the two stars here estimated as
double, of which A is 8, yellow; B 11, dusky.   Messier discovered this
in 1764; and though his description of it is very fair, his declination is
very much out: worked up for my epoch it would be north 37° 26′ 15″.
But one is only surprised that, with his confined methods and means,
so much was accomplished.

DCCXLVIII.   140 P. XX. ANTINOI.

Æ 20ʰ 19ᵐ 10ˢ   Prec. +   3ˢ·12
Dec. S 2° 37′·9   —— N 11″·44

Position 191°·0 (w 9)   Distance 59″·6 (w 9)   Epoch 1833·58

A wide double star, a little below the left arm of Antinous, and 5° east-
south-east of θ, a star of the 3½ magnitude.   A 7½, and B 8, both white,
and they point to a star of similar lustre in the lower verge of the field.
As B is Piazzi's No. 139 Hora XX., though it is rather sharp upon
his circle, I reduced his mean apparent places in order to gain a
comparison with Sir James South's micrometrical measures; and they
stood thus at the time of my taking the field:

P. Pos. 192° 30′   Dist. 61″·50   Ep. 1800
S.   189° 25′   59″·87   1825·00

This object derives interest in having, to each of its components, a *comes*
to the *np*, so that the figure forms a trapezium: intense gazing detects
a glimpse star to the south, which throws the diagram into a pentagon.
Of these, one of the 12th magnitude is nearest to A, bearing from it
311°, and about 40″ distant.

DCCXLIX. ρ CAPRICORNI.

Æ 20ʰ 19ᵐ 44ˢ   PREC. + 3ˢ·43
DEC. S 18° 20'·2   —— N 11"·48

POSITION A B 176°·7 (w 6)   DISTANCE   3"·8 (w 4) ⎫
———— A C 150°·7 (w 9)   ———— 236"·0 (w 5) ⎬ EPOCH 1830·73
⎭

A close double star with a distant companion in the *sf* quadrant, on the animal's right or following ear; being the middle star of three in a line 4° to the south-south-east of β. A 5, white; B 9, pale blue; C 7½, yellow, being Piazzi's No. 144, Hora XX. The two first of these constitute 51 Ƕ. II., discovered in 1782, and re-measured in 1802; but as from the defects of the old notation by quadrants one was noted *sf* and the other *sp*, there would have been inextricable confusion but for the later measures. The standard results previous to mine are:

Ƕ.   Pos. 174° 00'   Dist. 3"·00 ±   Ep. 1782·68
H. and S.   177° 17'   4"·03   1823·78

A and C form 29 Ƕ. VI., but it is registered, in August, 1780, without any measures of the angle, and noting the distance "about 2¼ min." But by deductions from Piazzi's Æs and Decs., and the measures of H. and S., we gain two epochs previous to my own:

P.   Pos. 149° 30'   Dist. 228"·20   Ep. 1800
H. and S.   150° 45'   238"·02   1823·78

DCCL. ο² CAPRICORNI.

Æ 20ʰ 20ᵐ 43ˢ   PREC. + 3"·45
DEC. S 19° 06'·4   —— N 11"·55

POSITION 239°·9 (w 9)   DISTANCE 21"·8 (w 9)   EPOCH 1832·59

A fine double star between the right ear and the eye, to the *sf* of the above object, at less than a degree's distance. A 6 and B 7, both bluish; and they point nearly upon a distant small star in the *nf* quadrant. Though the great southern declination of this object renders the measures liable to be affected by variable refractions, there is great coincidence in the results. The components being Nos. 153 and 154 of Piazzi's Hora XX., we may include deductions from his mean apparent places in the comparison:

Ƕ.   Pos. 239° 15'   Dist. 23" 50   Ep. 1782·68
P.   241° 15'   21"·00   1800
H. and S.   239° 43'   22"·06   1823·73

It may therefore be assumed, upon the evidence before us, that ο² Capricorni is an optical object.

DCCLI. 178 P. XX. DELPHINI.

Æ 20ʰ 23ᵐ 35ˢ   Prec. $+$ 2ˢ·86

Dec. N 10° 43'·6   —— N 11"·75

Position A B 256°·1 (*w* 9)   Distance 14"·3 (*w* 7) ⎫
—— A C 125°·0 (*w* 1)   —— 20"·0 (*w* 1) ⎬ Epoch 1835·91 ⎭
—— B b 210°·5 (*w* 3)   —— 0"·7 (*w* 1)   —— 1842·58

A delicate quadruple star in the preceding fin of the tail, and closely *sp* the bright star ε. A 7½ and B 8, both white; C 16, blue; b 9, yellowish; several small stars in the field. This was long viewed as a double star only, 16 Ḥ. iii., and as both the components were meridionally observed by Piazzi, the available data stood thus:

Ḥ.   Pos. 260° 18'   Dist. 12"·50   Ep. 1779·88
P.   253° 00'   15"·00   1800
S.   256° 25'   14"·69   1825·76

Soon after the last epoch, Sir John Herschel caught up the minute point of light C, which is only to be seen in my instrument by evanescent glimpses, under smooth equatoreal clock movement; but by means of a star in the *sf* quadrant, at about 30° from the vertical, the position admits of an approximate estimation. When, therefore, I examined this object at the close of 1835, it was registered as a triple star; how it became quadruple will be best seen in the following extract of a letter addressed to me by the Rev. W. R. Dawes, 27th October, 1840:

While observing 16 Ḥ. iii. last night, I met with an interesting circumstance. I felt dissatisfied with the definition of the smaller star, as it would appear oblong. Having measured it as a double star, I applied power 600. The night did not bear it at all, yet the elongated form of the smaller star was confirmed. I got two measures of the direction of the axis. I then altered the power to 420; vision much better; patiently adjusted the focus to the larger star A, which was perfectly round. On re-examining B with the same focus, the elongation of its disc was more obvious than ever. Struve measured this double star four times, once with power 480, and thrice with 320; and if it had been then of the shape it bears now, his scrutinizing eye would have detected it. Of course my observation requires confirmation; but at present I am persuaded this will turn out to be a new binary system.

Circumstances prevented my examining the case, until 1842, when it was one of the *agenda* which I took with me to Hartwell House, where, in the self-same telescope with which I had before measured A, B, and C, I saw that B was quite elliptical, with a very sensible major-axis. By the way, in scrutinizing such objects, pushing in the eye-tube to procure the expanded spectrum called the spurious disc, will, under delicate management, assist in showing whether the image be round, or otherwise; and the light of bright objects thus blunted, is less dazzling than when we let the rays corradiate at the focus. Another useful *ruse* to help the vision, is to withdraw the eye for a few seconds, and direct it towards the darkest recess of the observatory; after which, the field will appear to be comparatively bright.

DCCLII. 199 P. XX. CYGNI.

Æ 20ʰ 25ᵐ 57ˢ   PREC. + 1ˢ·85
DEC. N 48° 40'·6   —— N 11"·92

POSITION 278°·9 (w 6)   DISTANCE 60"·7   EPOCH 1833·66

A star with a telescopic companion, in a group on the Swan's tail, 4° north-by-west of Deneb. A 7, white; B 9½, pale blue. This is one of the group which Hevelius terms a nebula; it is 23 ♄. IV., mistakenly registered as ω² Cygni, which star is in the *sp*, and is single. The following are the previous data, in which it will be seen, that Sir William's distance must have been erroneously entered:

♄. Pos. 277° 23'   Dist. 30"·00   Ep. 1780·82
S.   278° 49'   61"·38   1825·14

DCCLIII. 103 ♄. I. DELPHINI.

Æ 20ʰ 26ᵐ 21ˢ   PREC. + 2ˢ·94
DEC. N 6° 53'·2   —— N 11"·95

MEAN EPOCH OF THE OBSERVATION . . . . 1837·59

A small but bright globular cluster, below the Dolphin's tail, and 3°¼ south of the star ε, where it is crossed by a line which passes from Wega below β Cygni. It is immediately preceded by a star of the 9th magnitude, and there is a coarse telescopic pair at a distance, near the preceding parallel; with several minute stars in the field. This object is a mass of very small stars, and is therefore installed among Sir John Herschel's test objects for trying the space-penetrating powers of telescopes: it was discovered in September, 1785.

DCCLIV. ω³ CYGNI.

Æ 20ʰ 26ᵐ 23ˢ   PREC. + 1ˢ·85
DEC. N 48° 41'·0   —— N 11"·95

POSITION 318°·9 (w 7)   DISTANCE 55"·2 (w 4)   EPOCH 1833·68

A wide double star in the Swan's left leg, in the group 4° north-by-west of Deneb, called *Rukbat al-dajájáh*, the hen's knee. A 5, pale red; B 10, grey; and these with two minute companions form a trapezium, so that the whole might be termed quadruple, of the VIth Class.

2 I

This object is without doubt 24 ℍ. iv., though both he and his son thought otherwise. I am led to this conclusion from the position of A and B as measured in October, 1780, it being 44° 19′ *np*, as well as from its being termed "treble," since with the 12th-magnitude star preceding them, a nearly equilateral triangle is formed. With this admission the whole answers very well to Sir William Herschel's description: 23 ℍ. iv. immediately precedes it.

DCCLV.   42 ℍ. VI. CEPHEI.

Æ 20ʰ 28ᵐ 17ˢ   Prec. + 1ˢ·21
Dec. N 60° 06′·2   —— N 12″·08

Mean Epoch of the Observation . . . . 1835·71

A large and rich cluster of minute stars on the left elbow of Cepheus: it is 12° due north of the above ω³ Cygni, and 3° west-south-west of η Cephei. The preceding portion of the most gathering part of the cluster is formed by a regular angle, or fan-shape figure; and the whole exhibits a grand but distant collocation of suns, which are evidently bound together by mutual relations, under the energy of a force which, though reason asserts its existence, imagination fails in conceiving.

It may assist him who fishes for this cluster to state, that an 8th-magnitude star precedes it by 35ˢ nearly on the parallel; and it is followed at twice that distance by a 7th-magnitude high up in the *sf*, which is itself preceded by a delicate telescopic pair. And there are other stars in the field.

DCCLVI.   β DELPHINI.

Æ 20ʰ 30ᵐ 03ˢ   Prec. + 2ˢ·80
Dec. N 14° 02′·6   —— N 12″·21

Position A B 105°·0 (*w* 1)   Distance 15″·0 (*w* 1) ⎫
—— A C 341°·8 (*w* 3)   —— 30″·0 (*w* 1) ⎭ Epoch 1834·79

A most delicate triple star on the Dolphin's body; it is 1°½ south-by-west of *a* Delphini, and in a line with β Cygni and γ Lyræ. A 4, greenish tinge; B 15, and C 12, both dusky. A and C constitute 35 ℍ. v., which was thus registered:

Pos. 348°·0 Dist. 25″·90 Ep. 1781·59

the distance, however, is marked "narrow." But neither Sir William nor Struve, whose No. 2704 it is, saw that it was a triple star, as found by Sir John Herschel in his IVth Series; and in truth C is a most flitting evanescent glimpse-object. The principal of the trio appears on

Piazzi's Catalogue under the barbarous term Rotanev, the which putteth derivation and etymology at defiance. See *a* Delphini. Rotanev has, however, been well marked down, and appears to have a slight spacial movement, of which these are the values:

$$P.... \; R + 0''·12 \qquad Dec. + 0''·08$$
$$B.... \qquad + 0''·12 \qquad \qquad - 0''·01$$

DCCLVII. 1 AQUARII.

$$R \quad 20^h \; 31^m \; 13^s \qquad Prec. + \quad 3^s·07$$
$$Dec. \; S \quad 0° \; 04'·3 \qquad —— \; N \; 12''·29$$

Position AB 220°·0 (*w* 1) Distance 20''·0 ⎫
—————— AC 45°·0 (*w* 1) ————— 35''·0 ⎬ Epoch 1838·69
⎭

A very delicate triple star between the bow of Antinous and the head of the Horse; it is 16° south of *a* Delphini, and as many east-south-east of Altair. A 5½, topaz tint; B 13, bluish; C 14, ruddy, A and B point to another minute distant *comes* in the *nf*, and there are other telescopic stars in the field. This is No. 2986 of H.'s Vth Series of Observations with the 20-foot telescope.

DCCLVIII. α DELPHINI.

$$R \quad 20^h \; 32^m \; 12^s \qquad Prec. + \quad 2^s·78$$
$$Dec. \; N \; 15° \; 21'·1 \qquad —— \; N \; 12''·35$$

Position 117°·8 (*w* 1) Distance 105''·0 (*w* 1) Epoch 1835·72

A bright star with a distant telescopic companion in the *sf* quadrant, on the middle of the Dolphin's body. A 3½, pale white; B 13, blue; and there are two small stars nearly in a line with A, away in the *sp*. A rigorous scrutiny of the observations has detected a slight spacial movement in A, but the doctors differ as to its amount and direction: these are the best values:

$$P.... \; R - 0''·09 \qquad Dec. + 0''·10$$
$$B.... \qquad + 0''·15 \qquad \qquad - 0''·02$$

a Delphini appears under the cacophonous and barbaric epithet Svalocin, on the Palermo Catalogue. But no poring into the black-letter versions of the Almagest, El Battáni, Ibn Yúnis, and other authorities, enables one to form any rational conjecture as to the mis-reading, mis-writing, or mis-application, in which so strange a meta-morphosis could have originated.

Δελφῖνος, Δελφὶς, Delphinus, Delphis, Curvus, Vector Arionis, Hermippis, Sinon or Simon nautis, are among the names by which the

Dolphin has been known, since its enrolment as one of the old forty-eight. Ḳazwíní says, "It contains ten stars near to each other. The bright star in the tail is called *Dheneb el dulfin*, the Dolphin's tail. The four in the middle, *a*, *β*, *γ*, *δ*, are called by the Arabs *el 'akúd*, the necklace, but by the vulgar *el ṣalib*, the cross; and that in the tail *'amúd el ṣalib*, the stem of the cross." Some of the classical ancients made out that it represented Apollo piloting Castalius from Crete; others that it was the Dolphin which carried Arion, for those fishes were then as notoriously attached to music as seals are now. Novidius, however, sees in it the whale which swallowed Jonah. But the usual representations of the asterism, bear little resemblance to any of the cetaceous tribe, it being mostly figured as a huge periwinkle pulled out of its shell; and certainly not "very like a whale." In the MS. of Cicero's translation of Aratus's astronomical poem* in the British Museum, (*Harleian*, No. 647,) there are coloured figures of the constellations, containing within their outlines, the prose descriptions of Hyginus: now these very early drawings exhibit a better dolphin than any of the later maps. So also in the early Venetian editions of Hyginus, the creature though armed with a horny nose, is not unlike a fish, and appears without that caudal periwinkle kink which Flamsteed, Hevelius, and Bode give it. But the learned in Oriental lore, assure us that we are all mistaken about it; in fact, that it is no fish at all, for that the Hindús, from whom the Greeks borrowed it, called it by the Sanskrit word *Zizumara*, sea-hog: they ought, however, to remember that the sea-hog is a porpoise. Much influence was ascribed to it in meteorology, and de Rheita says, "Sol exoriens cum Delphino ventos effert."

Although this asterism is small, and deficient in first or second rate constituents, it is bright and remarkable from a rhomboid formed of 3½ and 4th-magnitude stars. It lies nearly 15° north-east of *a* Aquila; where a line from *β* Lyræ through *β* Cygni, continued twice as far into the south-east will meet Svalocin, the lucida of the lozenge. The situation of the constellation is also pointed out in the brackish rhymes:

To heaven's grand arch from deepest seas, behold the Dolphin rise,
The grace, as old Manilius saith, of ocean and the skies:
'Tis placed between that space wherein the Eagle's wings are spread,
And those few stars unto the east which mark the Horse's head.

The ancients allotted ten stars to this fish, but Eratosthenes and Ovid only nine, because that was the number of the Muses; and the MS. of Cicero's Aratus just mentioned, asserts "sunt stellæ VIIII." There seems to have been less attention in drawing this creature, than usual. The artists should have taken advantage of the pliant form they adopted for a Dolphin, to turn his head down to the south-east, include half a dozen more tolerable stars in the figure, and still leave *a*, *β*, *γ*, and *δ*, in their relative places. The Catalogues have thus successively numbered the components:

Ptolemy	10 stars	Flamsteed	18 stars
Hevelius	14	Bode	51

---

* This poem had a great *run*. It was translated by Cicero, Hyginus, and Germanicus; was commented on by Grotius and other great men; and had the high honour of being quoted by St. Paul, also a Cilician, in Acts xvii. 28.

DCCLIX. 49 CYGNI.

Æ 20ʰ 34ᵐ 34ˢ   Prec. + 2ˢ·42
Dec. N 31° 44′·5   —— S 12″·52

Position 49°·2 (*w* 6)   Distance 2″·9 (*w* 3)   Epoch 1832·65
—— 48°·8 (*w* 8)   —— 3″·2 (*w* 5)   —— 1838·86

A close double star a little below the Swan's lower or right wing; it is 8° south-south-east of γ, and 2° south-south-west of ε Cygni, being the middle star of three nearly in a line. A 6, golden yellow; B 9, blue. This fine object was discovered by Sir William Herschel, and classed No. 98, of II. Class, under these data:

Pos. 58° 12′   Dist. 4″·00 ±   Ep. 1783·71

This gave a strong indication of an angular motion, in a direction *np sf*, or retrograde: but as the star is difficult, the discrepancies must be imputed to errors of observation. In 1837, the measures of Professor Struve arrived and confirmed this conclusion; yet he also had considered the star in motion, saying, " Fortasse in hac differentia motus manifestatur."

DCCLX. α CYGNI.

Æ 20ʰ 35ᵐ 57ˢ   Prec. + 2ˢ·04
Dec. N 44° 42′·7   —— N 12″·61

Position 102°·5 (*w* 2)   Distance 108″·5 (*w* 1)   Epoch 1837·65

A standard Greenwich star with a telescopic companion, at the root of the Swan's tail. A 1, brilliant white; B 12½, pale blue, the latter well seen when the large star was placed behind a central bar in the eye-piece. From the excellent series of meridional observations which have been made of α Cygni, its fixity *per se* seems doubtless: indeed, the results yielded to severe scrutiny, are interesting:

P.... Æ − 0″·08   Dec. 0″·00
B....   + 0″·03   0″·00
A....   + 0″·001   + 0″·005

The most usual name of α Cygni is Deneb, a word familiarized from the *dheneb-ed-dajájeh*, hen's tail, of the Arabians. It is also *Al-ridf*, the hindmost or pursuivant: and Arided, taken by Scaliger for *el-rided*, which an old Span.-Lat.-Arab. Dictionary showed him was a flower, whence the " quasi rosa redolens" in Riccioli's *Astronomia Riformata*. It is situated 25° east-by-north of Lyra, and therefore about a similar distance to the eye, as that star is from Polaris.

Cygnus is a well-known northern constellation between Lyra and Cepheus, and one of the old forty-eight. It was anciently called Ὄρνις, the bird, and Eratosthenes seems to have been the first and only one of that school who dubbed it Κύκνος. It was, however, in no want of names, as Ales, Milvus, Volucris, Olor, and Avis testify. Of Ornis, Robert Recorde angrily says, " Some men of too much oversight do translate it Gallina, a hen;" and it was also Kætha (*Kata?*), the water-fowl. Five of its largest stars form a very regular elongated cross, of which four, δ, γ, ε, and ζ, in a line across the Galaxy, were called by the Arabs, *el-fawaris*, the riders; the star opposite Deneb is not so perceptible as the others, but the cross is plainly marked without it. This was assumed as *Christi crux* by Schickard, the " astroscopium," in 1705; but the crude notions of Schickard, Novidius, Schiller, and other self-styled reformers of the sphere, were disregarded by astronomers, as only indicating derangement. In this, of course, nobody cared a straw for the absurd fables of the Greek mythology; they only adhered to the configurations as an established standard scale of astronomical recognition. The rhymester advises the star-gazer to look for the square of Pegasus, which having easily found,

From the wing's tip, Alpheras through,   now skim aslant the skies,
And lo! bedeck'd with glorious stars,   the soaring Cygnus flies:
Or, from the westward should you wish   the same to gaze upon,
Arcturus, Gemma, Wega, join   to lead you to the Swan.

The comparative brightness of the stars in Cygnus, was carefully registered by Sir William Herschel; whose observations will be found in the *Philosophical Transactions* for 1796 and 1797. It contains double, multiple, and variable stars, clusters, and nebulæ; and its components have been thus progressively shown:

Ptolemy	. . . 19 stars	Hevelius	. . . . 47 stars
Tycho Brahé	. . 27	Flamsteed	. . . 83
Bayer	 35	Bode	 360

DCCLXI. 52 CYGNI.

Æ 20ʰ 39ᵐ 04ˢ   Prec. + 2ˢ·47
Dec. N 30° 08′·3   —— N 12″·82

Position 56°·9 (*w* 8)   Distance 7″·0 (*w* 4)   Epoch 1835·73

A neat double star a little below the Swan's right wing. A 5½, orange; B 9½, blue; with a small double star following it. This object is 25 ♅. ɪɪ., and it is located 3° due south of ε Cygni, on a line pointing to the following component of the Dolphin's rhombus. The preceding measures were:

	Pos.	Dist.	Ep.
♅.	58° 57′	6″·00±	1780·69
S.	57° 20′	7″·20	1825·19
Σ.	57° 12′	6″·62	1830·18

whence its apparent fixity is established. Having completed the micrometric operations the illumination was removed; and after considerable attention a peculiar glow indicated the presence in the field of the extraordinary branched nebulosity 15 ♅. v., which H. has figured in the *Philosophical Transactions* for 1833.

DCCLXII. γ DELPHINI.

Æ 20ʰ 39ᵐ 15ˢ   Prec. +   2ˢ·78
Dec. N 15° 33′·2   —— N 12‴·83

Position 273°·6 (w 6)   Distance 12″·1 (w 5)   Epoch 1831·60
—— 273°·4 (w 9)   —— 12″·3 (w 5)   —— 1834·52
—— 273°·3 (w 8)   —— 11″·8 (w 7)   —— 1839·71

A beautiful double star on the Dolphin's head, and 2° due east of *a*. A 4, yellow; B 7, light emerald; with a third star in the *nf* quadrant, about 2′¼ distant. Sir William Herschel supposed that there was a considerable proper motion in one of these stars, but all the micrometrical measures coincide in establishing its fixity, as well as in countenancing the accuracy of the meridional observations. Bradley makes the Δ Æ = − 12″·5, and Δ Dec. + 0″·2, in 1755; and by deductions from Piazzi's places of Nos. 303 and 304, Hora XX., the following scale of comparison, from results previous to mine, is obtained:

	Pos.	Dist.	Ep.
H.	274° 09′	11″·82	1779·74
P.	273° 00′	13″·50	1800
H. and S.	273° 43′	12″·31	1823·68
D.	273° 20′	12″·02	1830·57

DCCLXIII. λ CYGNI.

Æ 20ʰ 41ᵐ 11ˢ   Prec. +   2ˢ·33
Dec. N 35° 54′·3   —— N 12‴·96

Position AB 104°·3 (w 8)   Distance 84″·9 (w 5)   Epoch 1834·81
—— Aa 130°·0 (w 1)   —— 0″·7 (w 1)   —— 1843·74

A close double star, with a distant companion, on the Swan's lower, or right wing: it is 5° south-east of γ, and is the northern vertex of a neat triangle with it and ε. A 5, B 10, and a 6, all bluish. The two wide stars constitute 32 H. vi., registered in September, 1780; and a comparison of H.'s measures, Sir James South's, and my own, making due allowances for errors of observation and instruments, imply fixity. But after I had considered the object as quite disposed of, I received intelligence of A's being double, as detected by Σ.'s gigantic telescope at Poulkova. It must either have come out from behind the other, or it ought to have been seen while under the action of micrometrical measurement, and magnifying power. Time will shew.

After this observation was made, the telescope was turned upon λ Cassiopeæ, another of Professor Struve's new tests; but after using every means to elongate it, there was no possibility of pronouncing that it was otherwise than single. The star, however, was not in its best position for accurate and effective gazing.

DCCLXIV.   η CEPHEI.

Æ   20ʰ 42ᵐ 01ˢ   Prec. + 1ˢ·22
Dec. N 61° 13′·2   —— N 13″·03

Position 29°·0 (w 1)   Distance 45″·0 (w 1)   Epoch 1834·74

A bright star, with a distant telescopic companion, on the bend of
the Æthiopian's left arm, and nearly midway on a line from Polaris to
Deneb. A 3½, pale yellow; B 13, dusky; several minute stars in the
field. The principal scrutineers of proper motion are agreed in the
amount and direction assignable to A:

P.... Æ + 0″·20   Dec. + 0″·81
B....   + 0″·20   + 0″·82
A....   + 0″·20   + 0″·80

DCCLXV.   4 AQUARII.

Æ   20ʰ 42ᵐ 57ˢ   Prec. + 3ˢ·18
Dec. S   6° 13′·2   —— N 13″·08

Position 45°·0 (w 1)   Distance 0″·5 (w 1)   Epoch 1834·69

A binary star, between Aquarius and Equuleus, being the middle
one of three stars pretty close together, 22° south-east of Altair, and
11° north-east of β Capricorni. A 6, pale yellow; B 8, purple. This is
44 Ḥ. I., and it certainly reflects no little credit upon Sir William in
having detected so very difficult an object. When I attacked it, the
companion appeared beyond my power; but after succeeding in making
it wedge-shaped in a direction towards a 14th-magnitude star in the
nf quadrant, long gazing brought up a bright point of light in the same
direction. This I estimated to the best of my judgment, and an inspection of all the data leads to the conclusion of a very considerable
angular motion. These are the previous conditions:

Ḥ. Pos. 351° 30′   Dist. 0″·30   Ep. 1782·68
Σ.   25° 00′   0″·81   1825·59

DCCLXVI.   72 M. CAPRICORNI.

Æ   20ʰ 44ᵐ 39ˢ   Prec. + 3ˢ·30
Dec. S 13° 07′·6   —— N 13″·19

Mean Epoch of the Observation .... 1836·72

A globular cluster of minute stars between Aquarius and the neck of
Capricorn; being 9° due east of *a* Capricorni, where it follows, at about

half a degree, 325 P. xx., a star of the 6½ magnitude. There are many telescopic stars in the field, a small pair of which closely follow the cluster. This object was discovered by the astronomical ferret M. Messier in 1780, and registered as a nebula: three years afterwards H. resolved it into stars, with his 20-foot reflector, and, on gauging, he pronounced its profundity to be of the 243rd order. It will show the reader the care and attention of Sir William, to give his next observation of this cluster, with the giant-reflector:

October 4, 1810. 40-feet telescope. Space-penetrating power 191·68. Magnifying power 280. Having been a sufficient time at the telescope to prepare the eye properly for seeing minute objects, the 72d of the *Connoissance des Temps* came into the field. It is a very bright object.

It is a cluster of stars of a round figure, but the very faint stars on the outside of globular clusters are generally a little dispersed, so as to deviate from a perfectly circular form. The telescopes which have the greatest light show this best. It is very gradually extremely condensed in the centre, but with much attention. even there, the stars may be distinguished.

There are many stars in the field of view with it, but they are of several magnitudes, totally different from the excessively small ones which compose the cluster. It is not possible to form an idea of the number of stars that may be in such a cluster: but I think we cannot estimate them by hundreds. The diameter of the cluster is about one-fifth of the field, which gives 1' 53"·6.

This cluster is followed, at about 5ᵐ △ Æ and 7' to the southward, by a trio of 10th-magnitude stars in a poor field: this is No. 73 of Messier's list, also registered in 1780.

DCCLXVII.   355 P. XX. EQUULEI.

Æ   20ʰ 44ᵐ 51ˢ   Prec. +   2ˢ·95
Dec. N   6° 44'·0   —— N 13"·21

Position 145°·0 (w 9)   Distance 39"·9 (w 9)   Epoch 1836·68

A wide double star in the space between the Horse's head and the Dolphin's tail: it is 9° south-by-east of γ Delphini, on a line led by it from γ Cygni. A and B, both 8½, and white, the latter being Piazzi's No. 356. The first micrometrical measures I find are those of Sir James South, with these results:

Pos. 144° 44'   Dist. 40"·60   Ep. 1824·54

This star affords one of the rare instances of error in Piazzi's Catalogue, and should be looked to. It has also been noticed by Professor Struve, No. 2733, who says: " In Piazzi vero Catologo novo invenies △ Æ = + 59"·4 arcus, △ Decl. = − 32"·5. In Æ errorum 30" suspicari licet." Reducing the places in the Palermo Catalogue, and then with Struve's supposed amount of error, the angle and distance will appear thus:

By the Catalogue . . Pos. 119°·0 Dist. 67"·7 ⎫
By Σ.'s cited error . .   137°·6   43"·3 ⎭ Ep. 1800

But even the latter position is too far out for Piazzi's accustomed accuracy. Trying back from my own measures, the △ Æ = + 23"·0 in arc, △ Decl. = − 33"·0.

DCCLXVIII.   376 P. XX. EQUULEI.

Æ 20ʰ 47ᵐ 40ˢ   Prec. + 3ˢ·00
Dec. N 3° 55'·6   —— N 13"·39

Position 286°·8 (w 4)   Distance 1"·8 (w 2)   Epoch 1833·65

A close double star between the Horse's head and the bow of
Antinous. A 6, orange tint; B 8, purple. To pick up this exquisite
pair, continue the alignment of the preceding object 2°½ further, and it
will find it the leader of three small stars beyond the Dolphin's tail,
which are pointed at by a line from Wega through β Cygni. It was
discovered by Professor Struve at Dorpat, and thus registered:

Pos. 289°·67   Dist. 2"·133   Ep. 1829·48

DCCLXIX.   χ DELPHINI.

Æ 20ʰ 48ᵐ 01ˢ   Prec. + 2ˢ·86
Dec. N 11° 57'·6   —— N 13"·41

Position 21°·0 (w 1)   Distance 55"·0 (w 1)   Epoch 1838·84

A delicate but wide double star, in the space between the Dolphin's
head and that of the Horse; being nearly 6° south-east of α Delphini.
A 6, white; B 14, pale lilac; A being the apex of a triangle formed
with two stars in the following field, and it is preceded by a distant star
near the parallel. This is No. 1592 of H.'s 20-foot Sweeps; and as
the *comes* defies the lamp, its bearings are only the result of comparative
estimation, though done with great care.

DCCLXX.   ε EQUULEI (1 Fl.)

Æ 20ʰ 51ᵐ 05ˢ   Prec. + 3ˢ·01
Dec. N 3° 41'·1   —— N 13"·61

Position A C 77°·6 (w 8)   Distance 10"·7 (w 6)   Epoch 1833·77
———— A B 290°·0 (w 2)   —————— 0"·5 (w 1) ⎫
———— A C 78°·1 (w 9)   —————— 11"·2 (w 8) ⎬ —— 1838·83

A most delicate triple star preceding the Horse's *os frontis*, and 12°
south-south-east of α Delphini, being the following of the three mentioned at No. DCCLXVIII. A 5½, white; B 7½, lilac; with a bright
star following at a distance, and a small one in the *nf* quadrant. This

was entered as the double star 21 Ḥ. III.; and described by Piazzi, "Duplex. Comes 0″·7 temporis sequitur fere in eodem parallelo." Under this impression, the measures before me were:

| Ḥ. | Pos. 84° 21′ | Dist. 9″·37 | Ep. 1780·59 |
| H. and S. | 79° 21′ | 12″·37 | 1823·58 |

I also, therefore, measured it as double in 1833, without noticing the elongation of A. But on receiving Professor Struve's great Catalogue, and perceiving that he made the object to be triple, I attacked it again, with the above success. It is clear that A and B are binary, for they could not now be overlooked by a diligent astrometer: nor can there be any doubt that they are equally affected by proper motions, that for A being thus valued:

| P.... | $R - 0″·11$ | Dec. $- 0″·15$ |
| B.... | $- 0″·09$ | $- 0″·14$ |

DCCLXXI. 429 P. XX. CYGNI.

Æ 20ʰ 53ᵐ 23ˢ   PREC. + 1ˢ·92

DEC. N 49° 50′·5   —— N 13″·76

POSITION 34°·6 (w 8)   DISTANCE 2″·1 (w 3)   EPOCH 1833·69

A close double star on the tip of the Swan's tail, and 5°½ north-by-east from Deneb. A 6. silvery white; B 7½, pale grey; the vicinity rich. This beautiful object is 97 Ḥ. I., and its registered measures being

Pos. 43° 36′   Dist. 1″·15±   Ep. 1783·73

Sir John Herschel concluded, by the observations of 1823, that there existed a mean annual motion of − 0″·2564. The subsequent examinations, however, of S. Σ. H. D. and myself afford no confirmation of the assumption. The differences are, therefore, imputable to the probable errors of observing an object of considerable difficulty.

DCCLXXII. λ EQUULEI (2 Fl.)

Æ 20ʰ 54ᵐ 19ˢ   PREC. + 2ˢ·96

DEC. N 6° 33′·3   —— N 13″·79

POSITION 225°·6 (w 7)   DISTANCE 2″·6 (w 4)   EPOCH 1833·72

A very neat double star closely preceding the Horse's nose, being 10°½ from γ Delphini, the following component of the Dolphin's lozenge. A 6, and B 6½, both white. This lovely object was discovered by Σ., being No. 2742 of the Dorpat Catalogue; and when his measures arrived, I was glad to find them in close harmony with the above.

My late friend, the Baron de Zach, was inclined to attribute a very sensible proper motion in $\cancel{R}$ to this star, but I know not upon what comparison: the values assigned on apparently the best grounds are:

$$P_{,}... \; \cancel{R} + 0''\cdot07 \qquad \text{Dec.} \quad 0''\cdot00$$
$$B.... \qquad + 0''\cdot03 \qquad\qquad - 0''\cdot03$$

DCCLXXIII. 59 CYGNI.

$\cancel{R}$ 20ʰ 54ᵐ 23ˢ   Prec. + 2ˢ·03
Dec. N 46° 54'·0   —— N 13''·82

Position AB 351°·9 (w 5) Distance 20·''0 (w 3) ⎫
—— AC 142°·0 (w 2) —— 30''·0 (w 1) ⎭ Epoch 1836·72

A delicate triple star in the Swan's tail, and the following component of a neat triangle about 3°¼ to the north-east of Deneb. A 5½, orange tint; B 10 and C 13, both blue. There can be little doubt of this object's being 22 Ħ. iv., mistakenly registered as 63 Cygni; it is also Σ. No. 2743, both of whom treated it as a double star. H. first made it triple, No. 209 of the 20-foot sweeps. After the arrival of Σ.'s Catalogue of 1837, observing that he considered the large star to be "alba subviridis," I again examined it in a darkened field, and to my vision it was a decidedly deep yellow. In order to ascertain whether my eye was biassed, or whether the difference of aperture was a cause, I referred the case to the Rev. James Challis, who, in December, 1841, saw A orange coloured, B decidedly blue, and C blue, in the great Northumberland equatoreal at Cambridge.

DCCLXXIV. 1 Ħ. IV. AQUARII.

$\cancel{R}$ 20ʰ 55ᵐ 27ˢ   Prec. + 3ˢ·27
Dec. S 11° 59'·3   —— N 13''·89

Mean Epoch of the Observation . . . . 1836·76

A planetary nebula in the middle of the Water-bearer's scarf; it is 12° following α Capricorni slightly to the north of its eastern parallel, where a line from the Eagle's tail over θ Antinoi, and as far again, reaches it. This object is bright to the very disc, and but for its pale blue tint, would be a very miniature of Venus. It was discovered at Slough in September, 1782; and is one of Σ.'s 9 rare celestial objects, appended to the Dorpat Catalogue of 1827, where its form is pronounced to be elliptical. According to the already-cited theory of Sir John Herschel, who makes its apparent diameter about 20'', if this object be only equally distant from us with the stars, its real dimensions must be such as would fill the whole orbit of Uranus. Now a globular body of the magnitude of the orbit of Uranus, would contain within its periphery more than 68,000 millions of globes as large as our Sun! The works and design of the Omnipotent Creator are inscrutable to the most brilliant human intellect: yet enough is revealed, both with regard to the wondrous Universe and our own mental capacity, to convince the reflecting mind that it is a mark of devotion which we owe to our Maker, to study with earnestness the beautiful and harmonious works around us, however their immensity may at first bewilder us. He who zealously applies himself, will verify the sacred promise, "Those who seek shall find." In worldly pursuits a long noviciate is devoted to acquire the imperfect concoctions of man: how much more is due to catch a glimpse of the imperishable laws of the CREATOR!

DCCLXXV. 12 AQUARII.

Æ 20ʰ 55ᵐ 37ˢ   PREC. +   3ˢ·18
DEC. S   6° 27′·1   —— N 13″·90

POSITION 191°·0 (w 5)   DISTANCE 2″·8 (w 3)   EPOCH 1834·82

A close double star in the space between the Water-bearer's scarf and the Horse's head, and about 24° south-by-east of α Delphini, being the brightest of several near the spot. A 5½, creamy white; B 8½, light blue. This elegant object is a discovery of Struve's, being No. 2745 of the Dorpat Catalogue. The registered micrometrical measures previous to my epoch, are:

S.	Pos. 191° 40′	Dist. 3″·227	Ep. 1825·67
Σ.	189° 36′	2″·667	1831·31

DCCLXXVI. 452 P. XX. CYGNI.

Æ 20ʰ 56ᵐ 13ˢ   PREC. +   2ˢ·29
DEC. N 38° 52′·9   —— N 13″·94

POSITION 297°·0 (w 3)   DISTANCE 17″·0 (w 1)   EPOCH 1832·87

A neat double star, on the verge of the Swan's right wing, following γ just south of its parallel, and 6° south-by-east of Deneb. A 7, deep yellow; B 11, emerald hue; there are several telescopic stars in the field, of which one in the *sf*, of the 13th magnitude, is sufficiently near to form with A and B a triple object. This was detected by Σ., No. 2748, but afterwards branded with "rej.;" considering, perhaps, the *comes* too distant and too small a point to merit measurement.

DCCLXXVII.   61¹ CYGNI.

Æ 20ʰ 59ᵐ 43ˢ   Prec. + 2ˢ·33
Dec. N 37° 58'·0   —— N 14"·15

Position 90°·5 (w 5)   Distance 15"·6 (w 5)   Epoch 1830·81

—— 92°·3 (w 6)   —— 15"·4 (w 4)   —— 1832·65

—— 93°·2 (w 5)   —— 16"·2 (w 4)   —— 1834·76

—— 93°·6 (w 8)   —— 15"·8 (w 5)   —— 1835·59

—— 95°·1 (w 6)   —— 15"·9 (w 7)   —— 1837·65

—— 96°·3 (w 8)   —— 16"·3 (w 6)   —— 1839·69

A binary star which, from its extraordinary motions, was placed on the Greenwich List, being then the smallest body so honoured. A 5½ and B 6, both yellow, but the small one is of the deepest tint. This most interesting object is 18 ♄. iv., and it is formed by Piazzi's Nos. 475 and 476, Hora xx. It is situated on the inner tip of the Swan's right wing, 7°¼ south-by-east of Deneb, and nearly on the eastern parallel of Wega.

This star must be regarded as one of the nearest to us, from the great rapidity of its proper motions; and it affords a positive instance of a double star which, besides the individuals revolving round each other, or about their common centre of gravity, has a progressive uniform motion towards some determinate region. This path is relatively spiral, but still so vast as to appear rectilinear; but too little is yet known of its amount and direction to refer it to definite laws. The values, however, have been pretty exactly ascertained:

	61¹ Cygni.			61² Cygni.	
Piazzi . .	Æ + 5"·38	Dec. + 3"·30	...	Æ + 5"·30	Dec. + 3"·00
Baily . .	+ 5"·18	+ 3"·24	...	+ 5"·28	+ 3"·03
Argelander	+ 5"·11	+ 3"·23	...	+ 5"·19	+ 3"·02
Taylor . .	+ 5"·54	+ 2"·93	...	+ 5"·52	+ 3"·12
Myself .	+ 5"·10	+ 3"·31	...	+ 5"·21	+ 3"·12

and while the journey through interminable space is incessantly performed at this most prodigious rate, the revolving action around each other, at a mean annual rate of + 0°·73, had been thus traced previous to my operations, admitting deductions from the meridional observations of Bradley, Christian Mayer, and Piazzi:

Bradley	Pos. 35° 24'	Dist. 19"·63	Ep. 1753·80
Mayer	50° 58'	15"·24	1778
Herschel I.	53° 32'	16"·08	1780·72
Piazzi	69° 18'	18"·20	1800
Bessel	79° 07'	16"·74	1812·30
Struve	83° 02'	15"·20	1819·90
Herschel II. and South	84° 41'	15"·43	1822·90
Dawes	90° 20'	15"·70	1830·66

These data afford much room for reflection. The anomalies are much greater than two stars with 16" between them ought to exhibit, but still there is some allowance to be made in a branch of astronomy which must, even yet, be considered as new. The angles and distances deduced from Æs and Decs. are as accordant as can reasonably be expected under so critical an ordeal; and the infirmity which affects the wire-micrometer, from the inflection of light interfering in the precise contact between the thread and the luminous body, may be admitted in palliation. There is strong presumption that all this will be better managed in future: meantime the average annual motion from the results before us, in the direction *sp nf*, or direct, may be thus roundly tabulated:

Mayer	. . .	+ 0°·65 1778	Dawes	. . .	+ 0°·72 1830·66
Herschel I.	. .	+ 0°·83 1780·72	Myself	. . .	+ 0°·98 1832·65
Piazzi	. . .	+ 0°·83 1800	——	. . .	+ 0°·43 1834·76
Bessel	. . .	+ 0°·80 1812·30	——	. . .	+ 0°·48 1835·59
Struve	. . .	+ 0°·51 1819·90	——	. . .	+ 0°·73 1837·65
H. II. and South	+ 0°·57 1822·90		——	. . .	+ 0°·59 1839·69

We here gain a general mean of 60°·9 in eighty-six years, or 0°·71 per annum, whence a revolution of about five centuries may be concluded upon: and while one star is thus going round the other, the pair is journeying through the vastness of space with incomprehensible velocity. Such motions afford a wondrous proof of illimitable power, whether we consider its exertion in the original production of these motions, or in controlling them. So far the vast mechanism of Infinite Nature is before us: but human reason is utterly incapable of speculation upon the end and aim of the UNFATHOMABLE WILL. Vexed at the presumption of ignorance, Milton said:

> HE his fabric of the heavens
> Hath left to their disputes, perhaps to move
> His laughter at their quaint opinions wide.

Piazzi was the first who detected the extraordinary movements of 61 Cygni, as shown at p. 10, l. VI. of the book *Del Reale Osservatorio di Palermo*, published in 1806. Yet in 1812, the *Moniteur Universel*, No. 189, attributes the discovery to Bessel, who indeed had sent an interesting memoir to Baron de Zach's *Monatliche Correspondenz*. In his Catalogue of 1814 Piazzi makes a reclamation, and Bessel handsomely acknowledged the priority of the Italian astronomer. Such was the note of preparation which drew attention to the object, and led to that splendid result, SIDEREAL PARALLAX. That two stars should be moving in an apparently rectilinear path at so large a rate, while other stars in the neighbourhood did not appear to be affected with any proper motion, was an incident of promise. By the delicate observations of MM. Arago and Mathieu, in 1812, it appeared that the distance of 61 Cygni was not less than 412,000 times the diameter of the earth's orbit; now, under an annual proper motion of 5″, it was hence deduced that the two stars were whirled along at the astounding rate of sixty thousand times faster than Mercury, which is the swiftest moving body in the planetary system. This, however, was based on the supposition that the observations were perfectly correct; but the lapse of six months, and the change of season, might easily lead to some slight change in the telescope, and the important quantity of a single second would be but the breadth of a spider's thread on the graduated circle of any instrument. Though the argument was insecure, it was a welcome approximation: and thus it remained until Bessel placed a foundation for more precise and definite conceptions of the vast distances of stellar bodies. The means by which he made the signal discovery of parallax, cannot be better given than in his own words, from a letter which he addressed to Sir John Herschel, under date of the 23rd of October, 1838:

I selected among the small stars which surround 61 Cygni, two between the 9th and 10th magnitudes; of which one (a) is nearly perpendicular to the line of direction of the double star; the other (b) nearly in this direction. I have measured with the heliometer the distances of these stars from the point which bisects the distance between the two stars of 61 Cygni; as I considered this kind of observation the most correct that could be obtained, I have commonly repeated the observation sixteen times every night.

The result of his assiduous measurement is, that in summer 61 Cygni was further from a by $0''{\cdot}620$ than in winter, and further from b by $0''{\cdot}437$. These numbers I derive from a mean of the observations in May, June, and July; and those of November, December, and January. The difference in the amount yielded by the two stars, is accounted for by a being almost at half the distance of b, and in a better angle of position: thus

$$\begin{array}{llll} a & \text{Pos. } 201°\ 29'\ 24'' & \text{Dist. } 461''{\cdot}617 \\ b & 109°\ 22'\ 10'' & 706''{\cdot}279 \end{array}$$

Yet it is only an apparent disproportion, for, after due reduction, the probable value is brought down to $0''{\cdot}3136$, according to the following explanation given by the illustrious discoverer:

As the mean error of the annual parallax of 61 Cygni is only $\pm 0''{\cdot}0202$, and consequently not $\frac{1}{15}$ of its value computed; and as these comparisons show that the progress of the influence of the parallax, which the observations indicate, follows the theory as nearly as can be expected considering its smallness, we can no longer doubt that this parallax is sensible. Assuming it $0''{\cdot}3136$, we find the distance of the star 61 Cygni from the Sun 657,700 mean distances of the Earth from the Sun: light employs 10·3 years to traverse this distance. As the annual proper notion of (61) Cygni amounts to $5''{\cdot}123$ of a great circle, the *relative* motion of this star and the Sun must be considerably more than sixteen semi-diameters of the Earth's orbit, and the star must have a constant of aberration of more than $52''$. When we shall have succeeded in determining the elements of the motion of both the stars forming the double star, round their common centre of gravity, we shall be able also to determine the sum of their masses. I have attentively considered the preceding observations of the relative positions; but I consider them as yet very inadequate to afford the elements of the orbit. I consider them sufficient only to show that the annual angular motion is somewhere about $\frac{2}{3}$ of a degree; and that the distance at the beginning of this century had a minimum of about $15''$. We are enabled hence to conclude that the time of a revolution is more than 540 years, and that the semi-major axis of the orbit is seen under an angle of more than $15''$. If, however, we proceed from these numbers, which are merely *limits*, we find the sum of the masses of both stars less than half the Sun's mass.

For these very delicate researches, the gold medal of the Royal Astronomical Society of London was awarded to M. Bessel: "researches," said Sir John Herschel, "which have gone far to establish the existence and to measure the quantity of a periodical fluctuation, annual in its period, and identical in its law with parallax." When Pliny pronounced the search after solar distance "penè dementis otii est," his mind could not have grasped the conception of stellar parallax.

For further hopes of understanding more upon this point, and therefore of the universe in which we exist, see the note at page 96; and also the result of Mr. Henderson's successful exertions, page 162.

DCCLXXVIII. 1 P. XXI. CYGNI.

Æ 21ʰ 01ᵐ 52ˢ   PREC. + 2ˢ·53
DEC. N 29° 33'·7   —— N 14''·29

POSITION 316°·5 (w 8)   DISTANCE 3''·5 (w 5)   EPOCH 1833·92

A very neat double star, towards the tip of the Swan's right wing, directly preceding ζ Cygni, the next described object, by about 1°¼ due west. A 6½, dull white; B 9, pale lilac; and there is a third star in the *sp* quadrant. This is 97 Ḥ. II., and from the following comparisons with mine, appears to be an optical object:

	Pos.	Dist.	Ep.
Ḥ.	315° 15'	4''·00±	1783·70
S.	315° 12'	3''·58	1824·70
Σ.	315° 36'	3''·55	1829·75

DCCLXXIX. ζ CYGNI.

Æ 21ʰ 06ᵐ 07ˢ   PREC. + 2ˢ·55
DEC. N 29° 34'·5   —— N 14''·55

POSITION 51°·8 (w 1)   DISTANCE 105''·0 (w 1)   EPOCH 1835·55

A secondary Greenwich star with a distant companion, on the tip of the Swan's right wing, and the following star of the Cross. A 3, pale yellow; B 10, sky-blue. The field is rich in small stars, though none of the *comites* are very close, at least in my telescope. There appears to be but little doubt that ζ Cygni is influenced by proper motion, though at present there is a considerable difference of opinion both as to its amount and direction:

	Æ	Dec.
P....	− 0''·09	− 0''·08
Br....	− 0''·10	+ 0''·08
B...	+ 0''·15	− 0''·06

DCCLXXX. δ EQUULEI.

Æ 21ʰ 06ᵐ 42ˢ   PREC. + 2ˢ·92
DEC. N 9° 21'·7   —— N 14''·58

POSITION 38°·8 (w 4)   DISTANCE 27''·1 (w 3)   EPOCH 1830·67
———— 37°·6 (w 9)   ———— 27''·9 (w 7)   ———— 1836·78
———— 36°·8 (w 6)   ———— 28''·2 (w 4)   ———— 1838·59

A delicate double star in the Horse's mouth, preceding ε Pegasi by about 7°½ on its western parallel, where a line carried from β Cygni over a Delphini, and $12°$ further into the south-east, intersects it. A $4\frac{1}{2}$, topaz yellow; B 11, pale sapphire; and there are other stars in the following part of the field. The previous results which enter into my discussion of this object, are:

	Pos.	Dist.	Ep.
H.	78° 21′	19″·53	1781·80
S.	41° 57′	26″·24	1825·26
Σ.	40° 40′	26″·49	1828·82
D.	39° 03′	29″·73	1830·56

From the differences between the first two of these epochs, it was concluded that no less an arc than $-36° 24'$ had been described in $43\frac{1}{2}$ years, being at the rate of $-0°·838$ per annum; and a change of distance to the amount of $+6''·707$, or $+0''·154$ per annum. But this is possibly overcharged; and perhaps, making every due allowance for observational errors, the annual angular retrograde movement, on a general average, may be valued about $-0°·65$. Still the particular mean for the subsequent lapse of $13\frac{1}{4}$ years, gives only an annual retrocession of $0°·38$ in angle, and an increase of $-0''·14$ in distance; and this relative change can be accounted for by the proper motions of the large star, which have been thus ascertained:

P....	$\mathbb{R} + 0''·08$	Dec.	$-0''·29$
B....	$+0''·12$		$-0''·28$
A...	$+0''·03$		$-0''·30$

On these grounds, it may be safely concluded that this very interesting star is only optically double, and that it does not carry its acolyte through space: in other words, that the components lie in almost the same visual line, one being immeasurably behind the other. From its magnitude and situation in the heavens, it is remarkable, and not to be confounded with any other; reasons for which Sir John Herschel thinks, under all circumstances, it is a fit object for the investigation of parallax.

The little asterism Equuleus was unknown to Aratus and Eratosthenes, and consequently to Cicero, Germanicus, and Avienus. Hence Hood remarks, "This constellation was named of almost no writer, saving *Ptolomee*, and *Alfonsus* who followeth Ptolomee, and therefore no certain tail or historie is delivered thereof, by what means it came into heaven." Yet it is one of the old forty-eight, and has been widely known under the various names, Ἵππου προτομή, Sectio equi, Equus prior, Equus minor, Equiculus, Præsegmen, the little horse, and the horse's head, &c. It is distinguished by a trapezium of four stars of the 4th magnitude (a, β, γ, δ), called Kitalpha, a corruption of the Arabian 'Kiṭ 'aṭ al-faras', a portion of the horse, but the epithet is now confined to a: and it was also *Aswini*, the horse's head, and *Al-faras' al-tháni*, the second horse, among the Orientals. Its components have been thus numbered:

Ptolemy	. . . . 4 stars		Hevelius	. . . . 6 stars
Tycho Brahé	. . . 4		Flamsteed	. . . 10
Kepler	 4		Bode	 36

and its place in the firmament is pointed out by a couple of lines from the brackish lays of the rhymester:

When Pegasus within our view, his spacious square doth spread,
Midway from Markab to Altair you'll find the Horse's head.

DCCLXXXI. 51 P. XXI. CEPHEUS.

Æ 21ʰ 07ᵐ 43ˢ   Prec. + 1ˢ·53
Dec. N 59° 19'·8   —— N 14"·64

Position 229°·5 (w 8)   Distance 1"·3 (w 4)   Epoch 1834·78

A close double star in the space between the Æthiop's head and his left arm; and nearly 3° south-by-west of Alderamin. A 6½, and B 7½, both silvery white. This pretty object was discovered by Professor Struve, and is No. 2780 of the Dorpat Catalogue, the second edition of which contains these results:

Pos. 228°·77   Dist. 1"·123   Ep. 1831·82

DCCLXXXII. 1 PEGASI.

Æ 21ʰ 14ᵐ 41ˢ   Prec. + 2ˢ·76
Dec. N 19° 07'·4   —— N 15"·05

Position 310°·8 (w 8)   Distance 36"·4 (w 5)   Epoch 1833·95

A double star between the head of Pegasus and the hind legs of the Fox; and about 10°½ south-by-east of ζ Cygni. A 4, but considered variable, pale orange; B 9, purplish. This object was first classed 20 Ħ. v., under measures "pretty accurate;" and these are the results previous to my epoch:

Ħ.   Pos. 308° 19'   Dist. 37"·10   Ep. 1780·69
S.   310° 11'   36"·86   1825·22

This close agreement indicates no notable change; yet the proper motions of A are of a very sensible value; proving B to be affected with a similar movement, or a variation must have been shown:

P.... Æ + 0"·15   Dec. + 0"·05
B....   + 0"·18   + 0"·09
A....   + 0"·16   + 0"·09

DCCLXXXIII. α CEPHEI.

Æ 21ʰ 14ᵐ 43ˢ   Prec. + 1ˢ·42
Dec. N 61° 54'·6   —— N 15"·05

Position 36°·0 (w 1)   Distance 150"·0 (w 1)   Epoch 1832·74

A standard Greenwich star with a distant companion, on the left shoulder of Cepheus; it is nearly in mid-distance between Polaris and

2 K 2

Deneb, and $8°$ south-a-quarter-west from β Cephei. A 3, white; B 10, pale blue, with a companion of the same magnitude and colour. This star appears on the Catalogues as Alderamin, a corruption of the *Aldera-imin* of the Alphonsine Tables, which is from the Arabic *al-dhirá' al-yemín*, the right arm: it has a decided proper motion in Æ, but that attributed to it in declination is disappearing under the recent close investigations:

P....	Æ + 0″·27	Dec.	− 0″·07
B....	+ 0″·35		+ 0″·01
A....	+ 0″·34		0″·00

Κηφεὺς, Cepheus, Inflammatus, Flammiger, and Dominus Solis, are the names by which European astrophilæ have distinguished one of the old forty-eight constellations, situated between Cassiopeia and Draco; where, of course, in our latitude, it is circumpolar. But Columella, in his Agricultural Almanac, seems to borrow his celestial rules from the Alexandrian Greeks, without considering the different parallel of Rome, for he makes the Great Bear rise and Cepheus set. The *effigies* of Cepheus has been sadly pulled about, although Ptolemy has given the exact places of his several limbs. Grotius, in his notes to *Germanicus*, complains of the liberties taken with the text of *Aratus*, in order to make it accord with the drawings of the artists; and similar freedom is still exercised. He is, however, pretty fairly represented on the noted planisphere of Geruvigus, and in Morel's *Aratus*, 1559; but the early Venetian illustrations to Hyginus, and Julius Firmicus, gird a long Toledo blade to his loins, such as an Argonaut never saw. Hevelius even places the Pole-star on his kilt; and I have already had occasion to grumble about the pulling out of his feet, at No. DCCXLIII. Yet the true position may be well inferred from the galley-rhymes:

Near to his wife and daughter see	aloft where Cepheus shines.
That wife, the Little Bear, and Swan,	with Draco bound his lines;
Beneath Polaris, twelve degrees,	two stars the eye will meet,
Gamma, the nomade shepherd's gem,	and *kappa*, mark his feet;
Alphirk, the Hindú's Kalpeny,	points out the Monarch's waist,
While Alderamin, beaming bright,	is on the shoulder placed:
And where o'er regions rich and vast,	the Via Lactea's led,
Three stars, of magnitude the fourth,	adorn the Æthiop's head.

Cepheus was an asterism of note among the Arabians as *al-Multahab*, the flaming, and *al-Aghnán*, the sheep; while γ was *Ar-rá'i*, the shepherd; and ρ, *Kelb ar-rá'i*, the shepherd's dog. A later school used the Greek name under their own enunciation of it; but instead of their writing Keípheus by a mistake of the letter *káf* for *phe*, they wrote it *Kekáus*, which, to use Scaliger's expression, "ridiculè excarnificatum est."

Cepheus has been an object of great attention among astronomers, as it exhibits some choice and remarkable variable and double stars, nebulæ, and clusters. Sir William Herschel has tabulated the comparative brightness of its components, in the *Philosophical Transactions*, 1797, the numbers of which have thus progressively increased under the advance of optical means:

Ptolemy	. . . . 13 stars	Hevelius	. . . 51 stars
Tycho Brahé	. . 11	Flamsteed	. . 35
Bayer	 17	Bode	 294

DCCLXXXIV. β EQUULEI.

R̶ 21ʰ 14ᵐ 57ˢ   Prec. + 2ˢ·97
Dec. N 6° 07′·9   —— N 15″·07

Position AB 317°·0 (w 2)   Distance 35″·0 (w 1) ⎱
——— Bb 15°·0 (w 1)   ——— 3″·0 (w 1) ⎰ Epoch 1836·68
——— AC 275°·0 (w 1)   ——— 50″·0 (w 1)

A star with three very delicate *comites* forming a quadruple set on the Horse's cheek, at about 4° south-east of δ, on the line towards Fomalhaut. A 5½, lucid white; B 13, grey; b 16, dusky; C 14, blue; and there are two brightish stars in the following part of the field. This group is a severe trial upon vision and estimation; it was recommended to me by Sir John Herschel, in a letter of 11th June, 1831: "As you are testing your telescope, there is a pretty test object in β Equulei, which is a coarse triple, and one of the small stars itself is a pretty first-class double star." On receiving this notice I soon made out B and C; but it was not till long afterwards that, gazing while the telescope was driven by the equatoreal clock, I perceived the minute point of light close to and nearly north of B. The large star appears to be affected by a spacial movement to the amount perhaps of + 0″·11 in R̶, but that in declination must be slight indeed.

DCCLXXXV. 15 M. PEGASI.

R̶ 21ʰ 22ᵐ 13ˢ   Prec. + 2ˢ·90
Dec. N 11° 27′·4   —— N 15″·48

Mean Epoch of the Observation . . . . 1836·72

A globular cluster between the mouths of Pegasus and Equuleus, forming the northern vertex of a triangle, obtuse and nearly isosceles, of which the base is ε Pegasi and δ Equulei. This fine object was disco-

vered by Maraldi in 1745, and registered as "une étoile né-buleuse, assez claire, qui est composée de plusieurs étoiles." Messier could not quite make this out, but in 1764 described it as a nebula with a star, its form circular and centre bril-

liant; and the place he has assigned to it is very considerably in error. Thus it remained till 1783, when Sir William Herschel resolved it into stars, and found it a good object for proving the telescope's space-penetrating power; he estimated its profundity to be of the 243rd order.

Although this noble cluster is rated globular, it is not exactly round, and under the best circumstances is seen as in the diagram, with stragglers branching from a central blaze. Under a moderate magnifying power, there are many telescopic and several brightish stars in the field; but the accumulated mass is completely insulated, and forcibly strikes the *senses* as being almost infinitely beyond those apparent *comites*. Indeed, it may be said to appear evidently aggregated by mutual laws, and part of some stupendous and inscrutable scheme of involution; for there is nothing quiescent throughout the immensity of the vast creation.

DCCLXXXVI. β AQUARII.

Æ 21ʰ 23ᵐ 07ˢ   Prec. + 3ˢ·16
Dec. S 6° 16′·4   —— N 15″·53

Position 320°·0 (w 2)   Distance 25″·0 (w 1)   Epoch 1833·73

A standard Greenwich star, with a minute acolyte, on the Water-bearer's right shoulder; and in mid-distance between Fomalhaut and the Dolphin's rhombus. A 3, pale yellow; B 15, blue; and there is a 10th-magnitude star at a distance in the *nf* quadrant. This is a most difficult object, and one requiring the utmost delicacy of treatment to procure even an estimation of; it is 76 Ḫl. v., and was thus first registered, with the word "inaccurate" placed to the distance:

Pos. 325° 48′   Dist. 33″·27±   Ep. 1782·55

β Aquarii is the *Kalpeny* of the Hindús, and the Sadalsuud of the Palermo and other Catalogues. The latter is from the Arabian *sa'd-as-su'úd*, the luckiest of the lucky, being the XXIVth Lunar Mansion, embracing this star, ξ Aquarii, and β Capricorni. The preceding, or XXIIIrd Lunar Station, was *sa'd-al-bula'*, the fortunate swallower or absorber, because one of the two stars seems to swallow or absorb the light of the other; it contains ε, μ, and ν Aquarii. The distance of β from Regulus, by Ptolemy, differs from that of modern astronomers by + 19′·30.

DCCLXXXVII. 2 M. AQUARII.

Æ 21ʰ 25ᵐ 10ˢ   Prec. + 3ˢ·09
Dec. S 1° 32′·1   —— N 15″·64

Mean Epoch of the Observation . . . . 1836·72

A fine globular cluster preceding the Water-bearer's neck, and about 5° north-half-east from β, the above-described object. This appears to have been discovered by Maraldi in 1746, while hunting up M.

Cheseaux's comet. Some years afterwards, Messier described it as a nebula containing no star, centre brilliant and surrounded by a circular light, altogether resembling the nucleus of a comet. Maraldi shows that little was then understood about nebulæ, for after mentioning that he could make out no stars, he continues, " Ce qui me parut fort singulier; car la plupart des étoiles qu'on appelle nébuleuses sont environnées d'un grand nombre d'étoiles; ce qui a fait juger que la blancheure que l'on y découvre, est l'effet de la lumière d'un amas d'étoiles trop petites pour être aperçues par les plus grandes lunettes." Now it is well established that, even where a globular cluster may not appear insulated, the stars belonging to it may be easily distinguished from those which happen to be scattered about, or upon, it.

This magnificent ball of stars condenses to the centre, and presents so fine a spherical form, that imagination cannot but picture the inconceivable brilliance of their visible heavens, to its animated myriads. It was observed and figured by Sir John Herschel, No. 2125, who observes, that as the total light of the cluster does not exceed that of a star of the 6th magnitude, it follows that several thousands of the 15th magnitude must be required to equal one of the 6th. It was tested by Sir William Herschel with his 7, 10, and 20-foot reflectors; and  he pronounced it to be a cluster of very compressed exceedingly small stars. This result was splendidly proved when, in September, 1799, he showed it to Professor Vince in that wonderful effort of the day, the 40-foot telescope: " the scattered stars," he observes, " were brought to a good well-determined focus, from which it appears that the central condensed light is owing to a multitude of stars that appeared at various distances behind and near each other. I could actually see and distinguish the stars, even in the central mass." By submitting it to the same process which he had already applied to fathom the Milky Way, he estimated its profundity to be of the 243rd order.

In his remarks on 2 Messier, Sir John Herschel says, " It is like a heap of fine sand!" The expression is remarkable, inasmuch as Signor Cacciatore, showing me this object at Palermo, in 1814, observed that the components were about as difficult to enumerate as " l'arena delle spiaggie marittime." This, however, is a noted method of estimating the stellar host, having been resorted to in essays, sermons, lectures, and guides to knowledge. Thus Booker, a censor of the press in 1655, compliments Bagwell for making astronomical mysteries plain to the " meanest capacity," by arithmetic:

> I wax hoarse
> Already, as I view thy counting course,
> And thy ingenious fancy, to pourtray
> From sands to stars a plain and pleasant way.

DCCLXXXVIII.   39 M. CYGNI.

Æ   21ʰ 26ᵐ 29ˢ   Prec. +   2ˢ·12
Dec. N   47° 43′·8   —   N 15″·72

Mean Epoch of the Observation . . . . 1836·72

A loose cluster, or rather splashy galaxy field of stars, in a very rich vicinity between the Swan's tail and the Lizard, due south of β Cephei, and east-north-east of Deneb. This was picked up by Messier in 1764, with his 3½-foot telescope, and registered as being a degree in diameter. Among them there are several pairs, of which a couple were slightly estimated; the first being the brightest star (7ᵐ) and its *comes*, and the second a pretty pair of 10th-magnitudes:

First pair. Pos. 26° 0′ (*w* 1) Dist. 85″·0 (*w* 1)} Ep. 1836·72
Second pair.   12° 0′ (*w* 1)   8″·0 (*w* 1)}

DCCLXXXIX.   β CEPHEI.

Æ   21ʰ 26ᵐ 31ˢ   Prec. +   0ˢ·81
Dec. N   69° 51′·7   —   N 15″·72

Position 250°·2 (*w* 9) Distance 13″·5 (*w* 9) Epoch 1833·54
——   251°·0 (*w* 9)   ——   13″·7 (*w* 9)   ——   1843·16

A standard Greenwich star, double, on the left side of the monarch's girdle, and two-thirds of the distance from Polaris to Alderamin. A 3, white; B 8, blue, with a coarse but very minute double star preceding. This beautiful object is 6 Ḥ. iii., and was recorded double by Piazzi; but he made the companion too small; "1″·4 temporis, fere in eodem paralello, præcedit alia 11ᵉ magnitudinis." The published results of measures at my attack were:

Ḥ.   Pos. 254° 32′ Dist. 13″·12 Ep. 1779·67
H. and S.   250° 25′   13″·16   1823·58

whence a very slow retrogression of angle was surmised; but all the recent observations confirm its fixity; insomuch, that even the proper motions once attributed to A, are fast dropping upon a zero. At present they may be thus stated:

P.... Æ − 0″·17   Dec. − 0″·03
B....   + 0″·03   − 0″·04
A....   + 0″·02   − 0″·05

β Cephei is known as Alphirk, and Ficares, from the Arabian *kawákib-al-firk*, stars of the flock, which *a*, *β*, and *η* were supposed to represent. In Arabia they may be either sheep or antelopes. Herds and flocks constituted very natural imagery among the Nomades; and at one epoch the starry heavens appear to have been almost filled with them.

DCCXC. 3 PEGASI.

Æ 21ʰ 29ᵐ 45ˢ   Prec. + 2ˢ·98
Dec. N 5° 54′·2   —— N 15″·89

Position 349°·7 (ᵂ 6)   Distance 39″·1 (ᵂ 6)   Epoch 1830·73
—— 349°·5 (ᵂ 8)   —— 39″·3 (ᵂ 8)   —— 1837·81

A double star between the heads of Pegasus and Aquarius; the preceding of a trio about 4° to the south-south-west of ε Pegasi. A 6, white; B 8, pale blue, and there is a small double star in the field about 5′ *np* and 8″ apart. This object is 98 Ḥ. v., and Nos. 216 and 217 P. xxi. Previous to my first examination, I had reason to suppose a slight alteration of angle had taken place in an interval of upwards of forty years, together with a sensible increase of distance; but my measures, the last of which were made in the midst of incessant electric scintillations of Aurora Borealis, countenance relative fixity. These are the results on which my impression was based:

	Pos.	Dist.	Ep.
Ḥ.	352° 48′	34″·72	1782·76
P.	350° 48′	36″·90	1800·00
Σ.	350° 30′	39″·21	1821·54
H. and S.	348° 58′	39″·52	1823·79

DCCXCI. 30 M. CAPRICORNI.

Æ 21ʰ 31ᵐ 16ˢ   Prec. + 3ˢ·43
Dec. S 23° 52′·4   —— N 15″·97

Mean Epoch of the Observation . . . . 1836·70

A fine pale white cluster, under the creature's caudal fin, and about 20° west-north-west of Fomalhaut, where it precedes 41 Capricorni, a star of the 5th magnitude, within a degree. This object is bright, and from the straggling streams of stars on its northern verge, has an elliptical aspect, with a central blaze; and there are but few other stars, or outliers, in the field.

When Messier discovered this, in 1764, he remarked that it was easily seen with a 3½-foot telescope, that it was a nebula, unaccompanied by any star, and that its form was circular. But in 1783 it was attacked by Ḥ. with both his 20-foot Newtonians, and forthwith re-solved into a brilliant cluster, with two rows of stars, four or five in a line, which probably belong to it; and therefore he deemed it insulated.

Independently of this opinion, it is situated in a blankish space, one of those chasmata which Lalande termed *d'espaces vuides*, wherein he could not perceive a star of the 9th magnitude in the achromatic telescope of sixty-seven millimétres aperture. By a modification of his very ingenious gauging process, Sir William considered the profundity of this cluster to be of the 344th order.

Here are materials for thinking! What an immensity of space is indicated! Can such an arrangement be intended, as a bungling spouter of the hour insists, for a mere appendage to the speck of a world on which we dwell, to soften the darkness of its petty midnight? This is impeaching the intelligence of Infinite Wisdom and Power, in adapting such grand means to so disproportionate an end. No imagination can fill up the picture of which the visual organs afford the dim outline; and he who confidently probes the Eternal Designs cannot be many removes from lunacy. It was such a consideration that made the inspired writer exclaim, "How unsearchable are His operations, and His ways past finding out!"

DCCXCII.   248 P. XXI. CEPHEI.

Æ 21ʰ 34ᵐ 00ˢ   Prec. + 1ˢ·86
Dec. N 56° 46'·1   —— N 16"·11

Position AB 120°·3 (w 9)   Distance 11"·7 (w 5) }
—— AC 339°·5 (w 9)   —— 19"·7 (w 5) }   Epoch 1833·83

A neat triple star preceding the Æthiop's tiara or fillet, and 5°½ south-by-east of Alderamin. A 6, pale yellow; B and C both 8¼, and both grey. This is 71 Ḫ. iii., and is thus mentioned by Piazzi: "Triplex: sociarum alia 1" temporis parumper ad boream præcedit, et alia 1"·5 temporis sequitur parumper ad austrum: utraque 8·9ᵃᵉ magnitudinis." This note, though describing the *comites* a little more distant, still places them in the same relative position with their primary. These were the registered measures:

	Ḫ. 1782·74				S. 1824·81	
AB	Pos. 125° 24'	Dist. 11"·59		AB	Pos. 121° 38'	Dist. 11"·95
AC	343° 57'	18"·62		AC	339° 03'	19"·39

On the whole, the changes here observable, cannot be entirely attributed to orbital motions; nor, as H. remarked, can they be accounted for by supposing the stars B and C at rest, and the central star A only in motion; since it would require to have advanced eastward to suit C, but westward to suit B. Through all anomalies, here are palpable gleams of a force which penetrates through creation; and wondrous is the inference! Those mighty orbs move through space in a curve which it bewilders the mind to figure, obediently to a mysterious impulse which binds them in affinity with some vast central phenomenon.

DCCXCIII. 256 P. XXI. CEPHEI.

Æ 21ʰ 35ᵐ 24ˢ   Prec. + 1ˢ·86
Dec. N 56° 51'·5   —— N 16"·18

Position 57°·1 (w 9)   Distance 12"·5 (w 8)   Epoch 1836·79

A very neat double star closely following the above object. A 8, white; B 9, pale violet. This object is 72 Ḥ. iii., and was registered double by Piazzi, in these words, "Duplex: comes 9ᵃ magnit. 1"·5 sequitur." Considering instrumental and other imperfections, it may be said to show no change during an interval of fifty-four years; the following being the only published data, with which it was in my power to make comparisons:

Ḥ. Pos. 58° 00'   Dist. 13"·11   Ep. 1782·74
S.   56° 25'   12"·41   1825·08

DCCXCIV. ε PEGASI.

Æ 21ʰ 36ᵐ 19ˢ   Prec. + 2ˢ·97
Dec. N 9° 08'·7   —— N 16"·23

Position AB 327°·0 (w 1)   Distance 85"·0 (w 1) ⎫
—— AC 324°·3 (w 8)   —— 138"·1 (w 8) ⎭ Epoch 1833·67

A second-rate Greenwich star, with two distant *comites*, in the mouth of Pegasus. A 2½, yellow; B 14, blue; C 9, violet; and a 9th-magnitude star of a violet tinge, follows at a distance. The two first of this trio form 103 Ḥ. vi., "very unequal;" but as they lie in the direction of the third, Sir James South concluded A and C to have been the object enrolled by Sir William. The distance was, therefore, held to be greatly in error, since the first measures were:

Ḥ. Pos. 322° 45'   Dist. 90"·93   Ep. 1782·89

and those of Sir James, No. 798:

Pos. 322° 59'   Dist. 138"·51   Ep. 1825·15

A and C are thus mentioned by Piazzi, Nota 260, Hora XXI.: "8" temporis, 1' ad boream, alia 8ᵃ magnit. præcedit:" of course his instrument had no chance of B.

ε Pegasi is known as Enif on the Charts and Catalogues, from the Arabic word *enf*, the nose; it was also called *Fom*, or *Fam-al-faras*, the horse's lip. A line from β Cygni led over the Dolphin's rhombus, and carried as far again into the south-east, hits upon it; and the rhymester further advises:

Where Alpherat so brightly decks	the captive Lady's head,
Across yon square project a ray,	(south-west it must be led,)
And pass between the two first stars,	so nearly north and south;
Extend beyond, twice ten degrees,	and reach the Horse's mouth.

DCCXCV. μ CYGNI.

Æ 21ʰ 36ᵐ 59ˢ   Prec. + 2ˢ·65
Dec. N 28° 01′·4   —— N 16″·27

Position A B 113°·8 (*w* 8)   Distance   5″·6 (*w* 9) ⎫
——— A C 61°·5 (*w* 9)   ———   216″·0 (*w* 7) ⎬ Epoch 1832·79
                                                    ⎭
——— A B 114°·3 (*w* 9)   ———   5″·4 (*w* 9) ⎫
——— A C 61°·2 (*w* 9)   ———   216″·8 (*w* 9) ⎬ —— 1839·62
                                                    ⎭

A beautiful double star, with a distant companion in the *sf* quadrant, at the eastern extreme of the asterism and between the hoofs of Pegasus, 19° due north of the above-described ε Pegasi. A 5, white; B 6 and C 7½, both blue. The clear components of this object form 15 Ḥ. iii., and the *comes* was thus mentioned by Piazzi, Nota 266, Hora XXI.: " 0″·3 temporis in eodem paralello alia 8ᵃᵉ magnit. sequitur." At my first epoch, the registered measures were as follows:

Ḥ.   Pos. 110° 15′   Dist. 6″·92   Ep. 1779·80
H. and S.   113° 04′   5″·74   1823·69

An occult spacial movement may be accountable for the changes here shown, since proper motions have been detected in A, in which it is probable that B partly partakes. These are the assigned values:

P.... Æ + 0″·36   Dec. − 0″·27
B....   + 0″·27   − 0″·22
A....   + 0″·26   − 0″·24

The distant star C is 267 P. xxi., and by deductions from the Palermo Catalogue, and Sir James South's measures, we obtain these data for comparison:

P. Pos. 62° 18′   Dist. 216″·50   Ep. 1800
S.   61° 17′   217″·40   1823·69

DCCXCVI. ϰ PEGASI.

Æ 21ʰ 37ᵐ 24ˢ   Prec. + 2ˢ·71
Dec. N 24° 54′·7   —— N 16″·29

Position 310°·0 (*w* 2)   Distance 12″·0 (*w* 1)   Epoch 1836·66

A very delicate double star in the right fetlock of Pegasus, and 16° to the north of ε Pegasi, where it is closely followed by two small stars on the parallel; but the vicinity is otherwise barren. A 4, pale white; B 13, purple. This is No. 43 of Ḥ.'s last Catalogue, containing the 145 new double stars: and the *comes* is described as "about north, a little preceding." It is now, however, well up in the quadrant; but bearing no illumination its place is only by estimation, though carefully managed upon a vertical and parallel to which my eye was familiarized.

DCCXCVII. 285 P. XXI. CEPHEI.

Æ 21ʰ 38ᵐ 37ˢ   Prec. + 1ˢ·83
Dec. N 58° 02′·9   —— N 16″·35

Position 38°·0 (w 1)   Difference Æ = 5ˢ·0 (w 1)   Epoch 1833·88

A richly-coloured star with a distant companion, in the preceding part of the Monarch's anadema, and 5° to the south-south-east of Alderamin. A 6, deep orange tint; B 10, pale blue; many other small stars in the field. Piazzi has entered A on his Catalogue under the designation *Garnet Sidus*, saying, "Stellam hanc rubei sub obscuri coloris primum apparuisse supponitur circa annum 1782." This is derived from Sir William Herschel's luminous paper on the proper motion of the Sun and the Solar System, in the *Philosophical Transactions* for 1783, wherein he mentions that it was not marked by Flamsteed. He further observes, "It is of a very fine deep garnet colour, such as the periodical star *o* Ceti was formerly, and a most beautiful object, especially if we look for some time at a white star before we turn our telescope to it, such as *a* Cephei, which is near at hand."

DCCXCVIII. 312 P. XXI. PEGASI.

Æ 21ʰ 44ᵐ 04ˢ   Prec. + 2ˢ·81
Dec. N 19° 04′·8   —— N 16″·62

Position 95°·0 (w 1)   Distance 15″·0 (w 1)   Epoch 1838·66

A most delicate double star, in a barren field, between the head and shin of Pegasus, and about 10° north-by-east of ε Pegasi. A 6½, white; B 14, blue. This object is No. 947 of H.'s 20-foot Sweeps, and classed triple, having a minute *comes* in the *np*, of the 17th magnitude. But this I could not manage to get a sight of, notwithstanding the coaxings exerted. B, though best seen with an averted eye, was plain enough at times, even under a slight illumination.

DCCXCIX. 20 PEGASI.

Æ 21ʰ 53ᵐ 18ˢ   Prec. + 2ˢ·91
Dec. N 12° 21′·4   —— N 17″·06

Position 330°·0 (w 1)   Distance 35″·0 (w 1)   Epoch 1838·66

A most delicate double star, close to the animal's cheek, 5° east-north-east of ε Pegasi, and 15° west-by-south from Markab. A 5½, lucid white; B 14, blue; several glimpse stars in the field. This is a less difficult object than the one just described, which was turned upon after the data here given had been estimated, as a preparation. This is No. 289 of H.'s First Series.

DCCC.   29 AQUARII.

Æ   21ʰ 53ᵐ 41ˢ   Prec. + 3ˢ·29
Dec. S 17° 43'·9   —— N 17"·07

Position 242°·0 (w 8)   Distance 4"·5 (w 8)   Epoch 1830·78

A beautiful double star, placed on the tail of Capricorn, 4° east-by-south of the bright star δ Capricorni, and 17° south-a-quarter-west of α Aquarii. A 6 and B 8, both brilliant white; and they are followed at a distance by a star nearly on the parallel. The action of the telescope on these stars, was considerably improved by the application of a central paper disc upon the object-glass, of two inches in diameter, as the images came up *cleaner.* The former measures are:

Σ.   Pos. 243° 34'   Dist. 4"·50   Ep. 1823·19
S.   243° 22'   4"·37   1824·68

A movement in space is attributed to A, of which these are the most accurately investigated values, however much the results differ:

P.... Æ + 0"·27   Dec. + 0"·07
B....   + 0"·03   + 0"·08
A....   − 0"·04   + 0"·04

DCCCI.   α AQUARII.

Æ   21ʰ 57ᵐ 33ˢ   Prec. + 3ˢ·08
Dec. S   1° 05'·7   —— N 17"·25

Position 44°·0 (w 1)   Distance 116"·0 (w 1)   Epoch 1837·71

A standard Greenwich star, with a minute companion, on the Water-bearer's left shoulder. A 3, pale yellow; B 13, grey; and the two point nearly upon another telescopic star in the *sp* quadrant. α Aquarii is called Sadalmelik, properly *Sa'd-al-melik*, the king's lucky star; but others read *Sa'd-al-mulk*, the lucky star of the kingdom. A line from Alpherat led through Markab, and carried as far again into the south-west, hits it. The rhymester here observes:

From Scorpio, to where Aries shines,   you catch no brilliant ray,
Through twice two interjacent signs,   to mark your trackless way;
Yet would you know where, from his urn   Aquarius pours the stream,
From fair Andromeda descend,   o'er Markab's friendly beam.
Or from bright Wega cast your glance,   and through the Dolphin's space,
Then just as far again you'll find,   the Water-bearer's place.

The proper movements formerly given to a Aquarii, have been nearly annihilated under recent rigorous comparisons, and may be thus shown:

$$P.... \; \mathcal{R} - 0''\cdot12 \qquad \text{Dec.} - 0''\cdot05$$
$$B.... \qquad + 0''\cdot03 \qquad\qquad + 0''\cdot01$$
$$A.... \qquad - 0''\cdot01 \qquad\qquad + 0''\cdot01$$

῾Υδροχόος, Hydrochoos, Aquarius, Stellæ effusoris aquæ, Juvenis, and Ganymede, are among the names which the classical ancients gave to the Water-bearer. Although from its possessing no star larger than of the 3rd magnitude, this asterism is not a very conspicuous one, yet its double-stars, clusters, and nebulæ, together with its being the XIth sign in the zodiac, and also the XIth part of the ecliptic, give it great astronomical importance. The catalogues of its components have successively run thus:

Ptolemy	. . . 45 stars	Hevelius . . .	48 stars
Copernicus	. . 45	Flamsteed . .	108
Tycho Brahé	. . 41	Bode	343

In the Arabo-Latin Almagest, this constellation appears as Indrodurus; but the Arabians called it *Sákib el-má*, the water-drawer. It had immemorially been represented as a man pouring water from a vase; but as Moslems are forbidden to draw the human figure, the Arabs represented this sign by a saddled mule, carrying on his back two water-barrels. Among astrologers, it was of the aëry triplicity, and a sign of no small note, since there was no disputing that its stars possessed influence, virtue and efficacy, whereby they altered the air and seasons "in a wonderful, strange, and secret manner." Who could doubt this? Was not the direct power of the celestials upon elementary bodies fully proven, to the satisfaction of Lilly and Fisk, by Sir Christopher Haydon, the redoubtable defender of astrological faith against the "barbarous opinions" of M. Chambre? Let political economists, medical men, and mankind in general, look to this. Is it not written, in the before-cited Almanack of 1386, that in Aquarius "it es gode to byg castellis, and to wed, and lat blode?" Does not the *Ysagogicus* of Alchabitus, 1485, gravely assure us, that when Saturn is in Aquarius, he has man completely in his clutches, "caput et collū habet?" and Jupiter in the same sign "humeros, pect.' et pedes?" And all the authorities clearly show, that when this asterism appears in the horizon the weather invariably proves rainy. These *facts* being supported by reasoning deemed both logical and conclusive, let no one scepticize.

DCCCII. ξ CEPHEI.

$\mathcal{R}$ 21ʰ 59ᵐ 10ˢ	Prec. $+$ 1ˢ·70
Dec. N 63° 51'·0	―― N 17"·32

Position 289°·5 (w 7)	Distance 5"·6 (w 6)	Epoch 1830·91
―――― 288°·8 (w 9)	――― 5"·8 (w 9)	―― 1839·65

A splendid double star in the right axilla of Cepheus, being the upper member of a vertical line of small stars, east-north-east of Alderamin and south-south-east of β Cephei. A 5 and B 7, both bluish. A comparison of the various observations establishes the fixity of this fine object, which is 16 H. II.; and it had these registered data when I first took the field:

	Pos.	Dist.	Ep.
H.	290° 18′	5″·00	1779·85
H. and S.	293° 15′	5″·82	1823·62

DCCCIII. π^1 PEGASI.

Æ 22ʰ 02ᵐ 09ˢ	Prec. + 2ˢ·65
Dec. N 32° 23′·8	—— N 17″·45

Position AB 258°·7 (w 3)	Distance 74″·0 (w 2)	
—— AC 90°·2 (w 3)	—— 185″·0 (w 1)	Epoch 1835·56

A star with two small *comites* on the animal's left fore fetlock, where a line carried from Algenib across the square through β, points out a couple of stars not quite as far again to the north-west. A 5, bright yellow; B 10, blue; C 11, dusky; and on the parallel with A and C follows π^2 Pegasi, a star of the 4th magnitude, with Δ Æ $= 44^s$.

Besides the stars above-mentioned, I had registered a small star of the 12th magnitude, pretty close to A, and nearly on the line to B; but some doubts arising while the object was under reduction, I requested the Rev. James Challis to examine it with the large Northumberland equatoreal. My request was kindly acceded to in August, 1842; and as Mr. Challis did not perceive the supposed star marked on my diagram, I have omitted it.

DCCCIV.   11 P. XXII. CEPHEI.

Æ 22ʰ 03ᵐ 12ˢ	Prec. + 2ˢ·01
Dec. N 58° 30′·7	—— N 17″·50

Position AB 317°·1 (w 8)	Distance 20″·9 (w 8)	Epoch 1830·89
—— AB 316°·8 (w 9)	—— 21″·4 (w 9)	
—— B b 150°·0 (w 2)	—— 0″·5 (w 1)	—— 1839·77

A fine double star on the following part of the Æthiop's diadem, in a little group to the south-east of Alderamin, and on the parallel of β Cassiopeæ. A 6, and B 6½, both white; b 7, pale; and there are two very minute stars in the sf quadrant, the nearest being close to the vertical. From a reduction of Piazzi's mean apparent places, and the measures of H. and S., these were the data when I took them in hand:

	Pos.	Dist.	Ep.
P.	319° 30′	20″·80	1800
H. and S.	315° 13′	20″·09	1823·74

On comparing these results with those of my first epoch, I was satisfied that the anomalies were owing to instrumental errors, and therefore closed accounts with the object. But on the arrival of Professor Struve's great Catalogue, finding he had seen B double, I replaced it on the working list. Thus advised, it was readily enough seen to be *eggy* towards A; but my utmost efforts only made it slightly elongated, without any separation of discs. He who fishes for this test, must bear in mind that Σ.'s No. 2810 precedes it on the parallel by 30^m, and the two objects are so similar in lustre, colour, and mutual relation, that though they are wide apart, a mistake might occur.

DCCCV. 41 AQUARII.

ÆR. 22^h 05^m 27^s PREC. + 3^s·33
DEC. S 21° 52'·0   —— N 17"·59

POSITION 119°·4 (*w* 5)   DISTANCE 4"·8 (*w* 5)   EPOCH 1830·71

A beautiful double star between the Water-bearer and the Southern Fish; it is nearly in a line between Fomalhaut and δ Capricorni, and 1°¼ to the west-north-west of 47 Aquarii, a star of the 5th magnitude. A 6, topaz yellow; B 8½, cerulean blue. This object is No. 6 of H.'s 145 New Double Stars; it was registered in September, 1787, but without measures. Piazzi noted it thus: "Duplex; comes 8·9^æ magnitudinis sequitur 0"·2 temporis, 3" circiter ad austrum." It was first submitted to the micrometer by H. and S., who obtained these results:

Pos. 120° 42'   Dist. 5"·17   Ep. 1823·75

DCCCVI. 33 P. XXII. PEGASI.

ÆR. 22^h 06^m 37^s PREC. + 2^s·88
DEC. N 16° 24'·2   —— N 17"·64

POSITION 315°·4 (*w* 5)   DISTANCE 6"·5 (*w* 2)   EPOCH 1833·63

A very delicate double star in the space between the head and legs of Pegasus, 10° from ϵ towards β, or rather less than half way. A 7½, lucid yellow; B 10½, sea-green; and there are four small stars forming a curve below, or north of it. This beautiful object was discovered by Professor Struve, and is No. 2877 of the Dorpat Catalogue, in the second edition of which are these measures:

Pos. 316° 27'   Dist. 7"·63   Ep. 1828·95

Small as is the acolyte of this object, its emerald tinge is very marked; and, from snatches made while the larger one was behind a thick wire, it can hardly be entirely the effect of contrast.

DCCCVII. 75 Ⱨ. VIII. LACERTÆ.

Æ 22ʰ 08ᵐ 59ˢ   Prec. + 2ˢ·36
Dec. N 49° 05′·1   —— N 17″·74

Mean Epoch of the Observation . . . . 1834·74

A large and loose cluster of stars from the 9th to 14th magnitudes, on the Lizard's mouth; where a line carried from the Pole-star through the tiara of Cepheus, and led 8° further, strikes it. A neat double star forms the vertex of a telescopic triangle near the middle of the group. Ⱨ. registered this in September, 1786, and described it as 16′ in length from *sp* to *nf*, but the whole vicinity is very rich, especially to the north. The double star here mentioned is Σ.'s No. 2890, Class III., "in acervo."

DCCCVIII. 65 P. XXII. LACERTÆ.

Æ 22ʰ 11ᵐ 56ˢ   Prec. + 2ˢ·61
Dec. N 36° 58′·1   —— N 17″·86

Position 193°·4 (*w* 9)   Distance 15″·2 (*w* 9)   Epoch 1835·84

A neat double star on the tip of the creature's tail; a line led north-north-west from *a* Pegasi to the west of *η*, and about 10° beyond, will find it following 1 Lacertæ, a star of the 5th magnitude. A 6½, pale white; B 9, livid. This object is 17 Ⱨ. III., and No. 341 of H. and S., although by mistaking it for 1 Lacertæ they questioned its identity. The registered measures, when I took it in hand, were thus:

	Pos.	Dist.	Ep.
Ⱨ.	193° 44′	13″·70	1779·89
H. and S.	191° 17′	15″·62	1823·72

and as Sir William Herschel marked his distance "inaccurate," the *fixity* of this, as an optical object, may be concluded.

DCCCIX. γ AQUARII.

Æ 22ʰ 13ᵐ 23ˢ   Prec. + 3ˢ·09
Dec. S 2° 11′·5   —— N 17″·91

Position 130°·3 (*w* 3)   Distance 40″·0 (*w* 1)   Epoch 1834·70

A very delicate but wide double star on the Water-bearer's left arm, and 4° east-by-south from Sadalmelik. A 4, greenish tinge; B 14, purple, but though so minute it is fairly seen under the action of the rock-crystal micrometer. A has a spacial movement, the values of which are thus stated by the scrutators:

$$P.... \, \mathit{R} - 0''\cdot 11 \qquad \text{Dec.} + 0''\cdot 05$$
$$B.... \qquad + 0''\cdot 20 \qquad \qquad + 0''\cdot 04$$
$$A.... \qquad + 0''\cdot 16 \qquad \qquad + 0''\cdot 03$$

γ Aquarii is the Sadachbia of the Palermo and other Catalogues: the epithet is derived from the Arabian *sa'd-al akhbiyah*, the lucky star of hidden things, or hiding places, because, when it appears, the earthworms creep out of their holes. This is the XXVth Mansion of the Moon, composed of γ, ζ, η, and θ.

Padre de Rheita, an early improver of telescopes, thought he perceived five satellites attending upon Jupiter when in this vicinity, whereupon, in compliment to the reigning pontiff, he dubbed them "the stars of Urban Octavus;" and the fact was announced in a Latin treatise by C. Lobkowitz, in 1643, *De novem Stellæ circa Jovem*, with the appropriate motto, *Consideravi opera Tua, et expavi*. The planet, however, soon deserted his companions, and the stars proved to be the little group in front of the Urn. So poor De Rheita took nothing by his motion; but after thus complimenting the Pope, and discovering Sta. Veronica's *sudarium*, (see No. CCCLXXV.) though he was called to Rome by the Superior of his Order, he died a simple Capuchin, in 1660, at the age of sixty-three.

DCCCX. 2 LACERTÆ.

R 22h 14m 25s $\qquad$ Prec. + 2s·46
Dec. N 45° 43'·9 $\qquad$ —— N 17''·95

Position 6°·9 (*w* 1) $\qquad$ Distance 35''·0 (*w* 1) $\qquad$ Epoch 1835·84

A delicate but wide double star on the Lizard's shoulder, where a line from Polaris carried by the east of Cepheus's tiara, and 11° further, will find it the lucida of a fine galaxy field. A 5, pale yellow; B 13, orange tint; three other stars in the *sf.* As the *comes* defies illumination, the estimations were carefully made in a darkened field.

Lacerta, *vel* Stellio, is a little constellation formed by Hevelius at Dantzig, 160 years ago. As some of the *informes* between Andromeda and Cygnus were of the 5th magnitude, he thought they needed an intelligible designation, and therefore gathered ten of the brightest, and made his Lizard; the place being too narrow for any other description of animal. But he figured it as a strange weasel-built creature with a curly tail, the which was copied by Flamsteed. Very like a lizard! Seeing the kind of space it occupied, Royer wished to place the sceptre of his sovereign, the noted Louis XIV., in its stead, but the reptile prevailed. The components have been thus numbered:

Hevelius	. . . . 10 stars		Piazzi	. . . . 37 stars
Flamsteed	. . . 16		Bode	 60

2 L 2

DCCCXI. 33 PEGASI.

$\mathbb{R}$ 22^h 15^m 58^s Prec. + 2^s·88
Dec. N 20° 02′·5   —— N 18″·01

Position AB 181°·6 (w 5)	Distance 2″·7 (w 3)	Epoch 1831·74
—— AC 344°·0 (w 6)	—— 56″·9 (w 4)	
—— AB 180°·2 (w 6)	—— 2″·5 (w 3)	—— 1838·88
—— AC 341°·0 (w 9)	—— 56″·6 (w 7)	
—— AB 178°·9 (w 6)	—— 2″·7 (w 3)	—— 1839·69
—— AC 340°·8 (w 9)	—— 57″·9 (w 8)	

A triple star preceding the chest of Pegasus, and nearly midway between β and ϵ, where it is the leader of six stars of similar size, lying in a line. A 6½, yellowish; B 10, blue; C 8, pale grey. The first and third of these stars constitute 99 H. v., and were thus described by Piazzi, Nota 88, Hora XXII.: "Telescopica in eodem verticali, 50″ circiter ad boream." Taking this note into the account, the following data were obtained as points of departure:

H.	Pos. AC	0° 48′	Dist. 45″·05	Ep. 1782·76
P.		0° 00′	50″·+	1800
H. and S.		345° 45′	56″·04	1823·71

In reviewing this object, Professor Struve detected the close companion to A, and enrolled it in the Dorpat Catalogue. But my attention was first drawn to it by the Rev. W. R. Dawes, who had accomplished its measurement at Ormskirk, with a 5-foot telescope; which rendered its escaping the searching eye of H. the more remarkable. These are the results with which Mr. Dawes favoured me:

Pos. AB 182° 40′ Dist. 3″·05 Ep. 1830·77

This gentleman has since obtained measures in Regent's Park, in Mr. Bishop's observatory, and they will shortly appear in the series of double-star operations about to be published from thence. From my acquaintance with the whole *modus operandi*, I may pronounce, that it will be a valuable addition to our sidereal stock. The *prima facie* evidence of the apparent rotation of A and C was supported by my observations; but it was perceivable that the proper motions of the principal are equivalent to the changes of angle, and that the micrometric operations afford a confirmation of the fact of spacial movement in A to the eastward; also, that the values assigned are near the mark. All the scrutators agree in the course which it is taking, and a very few years will suffice to show, whether there is an optical or physical connection among the components. Piazzi remarks, "Motus annuus in $\mathbb{R}$ + 0″·80 ex Flamst. et ex nostris observationibus;" but he has registered the quantity obtained by comparison with Bradley:

P....	$\mathbb{R}$ + 0″·40	Dec. − 0″·01	
B....	+ 0″·41	− 0″·02	
A....	+ 0″·34	− 0″·01	

DCCCXII.   53 AQUARII.

Æ   22ʰ 17ᵐ 52ˢ   Prec. +   3ˢ·25

Dec. S 17° 33′·1   —— N 18″·09

Position 301°·5 (w 9)   Distance 9″·9 (w 9)   Epoch 1831·76

A neat double star on the Water-bearer's right thigh, and about 14° north-north-west of Fomalhaut. A and B, 6½, and both pale white. The components of this beautiful object are Piazzi's 93 and 94, xxii., from which we gain results to compare with the measures of H. and S.:

	Pos.	Dist.	Ep.
P.	290° 30′	12″·70	1800
H. and S.	303° 07′	10″·03	1823·26

The data are more discordant than would have been expected; but a slow change in the angle of position, and also in the distance, may be owing to proper motions, which, though they escaped the vigilance of Piazzi, are thus valued:

$$B. \quad Æ \begin{cases} A + 0''\cdot30 \\ B + 0''\cdot24 \end{cases} \quad Dec. \begin{cases} A + 0''\cdot05 \\ B + 0''\cdot04 \end{cases}$$

$$A. \quad Æ \begin{cases} A + 0''\cdot25 \\ B + 0''\cdot18 \end{cases} \quad Dec. \begin{cases} A + 0''\cdot03 \\ B + 0''\cdot03 \end{cases}$$

DCCCXIII.   ζ AQUARII.

Æ   22ʰ 20ᵐ 35ˢ   Prec. +   3ˢ·08

Dec. S   0° 50′·2   —— N 18″·19

Position	Distance	Epoch
356°·0 (w 7)	3″·6 (w 5)	1831·83
—— 355°·3 (w 8)	—— 4″·0 (w 5)	—— 1832·71
—— 353°·8 (w 7)	—— 3″·8 (w 4)	—— 1834·90
—— 352°·4 (w 7)	—— 3″·5 (w 4)	—— 1838·04
—— 348°·9 (w 9)	—— 2″·7 (w 6)	—— 1842·59

A binary star on the Water-bearer's left wrist, about 6° distant from Sadalmelik, on its eastern parallel; where it is in the middle of three other stars, which altogether form a figure resembling the letter Y. A 4, very white; B 4½, white. This fine object is 7 ℍ. ii., and was thus recorded by Piazzi: "Duplex; utraque ejusdem magnitudinis, et in eodem verticali, et altera ab altera 3″ distans." In 1804 ℍ. perceived that the components were physically connected; and these are the comparing data on which I proceeded to inquire into their conditions:

	Pos.	Dist.	Ep.
ℍ.	18° 21′	4″·56	1779·70
P.	0° 00′	3″·+	1800
Σ.	358° 18′	4″·40	1820·92
H. and S.	359° 29′	4″·99	1822·27
S.	1° 04′	4″·01	1825·73
D.	356° 12′	3″·95	1830·55

The discordances here observable, threw considerable doubt upon the rotation suspected by ♅.; and from the angle and distance being both discrepant, the case was considered to be beset with difficulties, of which the principal might arise from the inflection of light in so splendid and close a pair. But I was pleased to find ζ of such easy measurement, that my own epochs are sufficient to establish its *np sf*, or retrograde, motion, at an annual rate of − 0°·66 per annum. The half century between ♅. and D. yields an average of − 0°·43; so that by roundly assuming a mean of half a degree yearly, there may be a period of 750 years. From a hasty protraction it seems to have a very elongated orbit, with the axis major more than twice the minor, and inclined at about 15°; but the observations cover such a very small part, that all deductions must, for some years yet, be extremely uncertain.

The proper motions assigned to this star, have also exhibited such contradictions in amount and direction, as to add to the confusion of the question; but under these discordant values, one remarkable fact is established, namely, that while the movement in space is occurring, the two stars retain so relative a station with regard to each other, as to prove their mutual connexion. By a comparison from Bradley, Piazzi found the course was +, but in another from Mayer it was −; these, together with the results of Mr. Baily's investigation, are:

$$P....\begin{cases} Æ + 0''\!\cdot\!26 & \text{Dec.} + 0''\!\cdot\!07 & \text{(Nota 111, Hora XXII.)} \\ − 0''\!\cdot\!10 & − 0''\!\cdot\!14 & \text{(Palermo Catalogue.)} \end{cases}$$
$$B.... \quad + 0''\!\cdot\!21 \qquad + 0''\!\cdot\!05 \quad \text{(MS. Catalogue.)}$$

DCCCXIV.   37 PEGASI.

Æ 22ʰ 21ᵐ 53ˢ   Prec. + 3ˢ·03
Dec. N 3° 37'·3   —— N 18''·24

Position 116°·8 (*w* 5)   Distance 1''·3 (*w* 2)   Epoch 1835·81
——— 118°·9 (*w* 8)   ———— 1''·1 (*w* 3)   ——— 1839·66

A binary star on the mane, and near the animal's head: it is the following of three small stars which Hevelius termed a nebula, 7° to the east-north-east of Sadalmelik, where it is struck by a line from Alpherat through Markab. A 6 and B 7½, both white. This close and beautiful object was discovered by Σ., and is No. 2912 of the Dorpat Catalogue, where it was registered as one of the Ist Class "pervicinæ." Under such a notice from such an experienced astronomer, I was glad to find that I could manage to get a tolerable angle at my first epoch; and knowing of no other measures, I was satisfied in the mere record of its duplicity. But on the arrival of Σ.'s grand Catalogue, comparisons were afforded, by which its binarity was proved; and this was still further confirmed by my second series of observations. Assuming as the earliest epoch,

Σ. Pos. 112°·63 Dist. 1''·16 Ep. 1831·12

it is clear that the angle is undergoing a rapid change direct, already indicative of a period of about five centuries.

DCCCXV. δ CEPHEI.

Æ 22ʰ 23ᵐ 14ˢ	PREC. + 2ˢ·20
DEC. N 57° 35′·9	—— N 18″·28

POSITION 192°·2 (w 9)   DISTANCE 40″·9 (w 9)   EPOCH 1837·71

A fine though wide double star closely following the monarch's tiara, where a line led east-south-east from Alderamin, till it cuts the western or preceding parallel of β Cassiopeæ, strikes it. A 4½, orange tint; B 7, fine blue; the colours, under proportionately magnifying powers, being in fine contrast. This object is 4 I̶H̶. v., who gave it a distance of 38″·3, without any angle of position; and it constitutes Piazzi's Nos. 134 and 135, Hora XXII., which, in other words, are δ¹ and δ² Cephei. By computing the mean apparent places in the Palermo Catalogue, we obtain an epoch, thus:

P.	Pos. 195° 00′	Dist. 41″·70	Ep. 1800
H. and S.	191° 16′	41″·61	1822·87

At first sight, proper motion might be thought to interfere, but there is great reason to conclude that both components are similarly affected. These are the best values of the principal:

P....	Æ − 0″·10	Dec. − 0″·15
B....	+ 0″·06	− 0″·01
A....	+ 0″·05	− 0″·02

δ Cephei was discovered to be variable by Mr. Goodricke, who assigned it a periodical variation of 5 days 8 hours and 37½ minutes. See his paper in the *Philosophical Transactions* for 1786. It is so slightly variable that the changes are not easily seen, unless at its maximum and minimum brightness. A discussion of Goodricke, Pigott, Westphal, and Argelander, gives about 5 days, 8 hours, and 30 minutes, for its corrected period.

DCCCXVI. 8² LACERTÆ.

Æ 22ʰ 28ᵐ 46ˢ	PREC. + 2ˢ·65
DEC. N 38° 48′·5	—— N 18″·48

POSITION AB 185°·5 (w 9)	DISTANCE 22″·5 (w 9)	
—— Bb 155°·3 (w 4)	—— 27″·8 (w 3)	EPOCH 1834·59
—— AC 144°·8 (w 8)	—— 82″·2 (w 5)	

A quadruple star, forming, as I̶H̶. remarked, an arch, in the Lizard's tail; it is the preceding component of a neat triangle, upwards of 20° north-west of Alpherat. A and B 6½, and both white; b 11, greenish; C 10, blue; and there is a fifth star away in the *sp* quadrant. This is 86 I̶H̶. ɪv., and though registered by him as quadruple, had but its two principal stars measured, the 163 and 164 of Piazzi's Hora XXII.; and thus they stood:

$$\text{H.} \quad \text{Pos. } 185° \ 30' \quad \text{Dist. } 17''·22 \quad \text{Ep. } 1782·77$$
$$\text{P.} \qquad 218° \ 58' \ (!) \qquad 20''·60 \qquad 1800$$

The next astrometers who took the field, not perceiving the minute star b, examined the object as triple, as follows:

$$\text{H. and S.} \begin{cases} \text{Pos. A B } 184° \ 21' & \text{Dist. } 22''·67 \\ \text{A C } 145° \ 15' & 82''·52 \end{cases} \text{Ep. } 1823·74$$

Here the anomalies are palpably owing to error, and the fixity of the object appears unquestionable. The following of the two principal stars is designated A, for though Piazzi makes them both of equal magnitude, it is certainly brighter than the preceding one.

DCCCXVII.   10 LACERTÆ.

$$\text{Æ } 22^h \ 32^m \ 05^s \qquad \text{Prec. } + \ 2^s·67$$
$$\text{Dec. N } 38° \ 13'·2 \qquad \text{———} \ \text{N } 18''·58$$

Position 48°·2 (w 4)   Distance 58''·9 (w 2)   Epoch 1830·73

A wide double star at the root of the reptile's tail, and the south-east component of the triangle above described. A 6½, white; B 10, violet. This is 97 H. v., and appears to be stationary, the other measures being:

$$\text{H.} \quad \text{Pos. } 51° \ 15' \quad \text{Dist. } 52''·57 \quad \text{Ep. } 1782·76$$
$$\text{S.} \qquad 48° \ 41' \qquad 60''·44 \qquad 1825·21$$

Here Bode introduced a bit of flattery, in 1787, to immortalize the immortal Frederick II. of Prussia. He published a drawing of the *Honores Frederici* in his *Jahrbuch* for 1790; the crown, the laurel, the sword, the pen, as typifying the monarch, the hero, the sage, and the pacificator. It has already passed from the skies!

DCCCXVIII.   ζ PEGASI.

$$\text{Æ } 22^h \ 33^m \ 29^s \qquad \text{Prec. } + \ 2^s·98$$
$$\text{Dec. N } 9° \ 59'·9 \qquad \text{———} \ \text{N } 18''·63$$

Position 15°·5 (w 1)   Distance 65''·0 (w 1)   Epoch 1831·68

A secondary Greenwich star, with a minute *comes*, on the middle of the animal's neck; where a line from Alpherat passed over Markab, and led 7° further to the south-west, reaches it. A 3, light yellow; B 13, dusky; and there are three other telescopic stars in the field, under a moderate magnifying power. It appears to have a decided spacial movement in Æ, however ill-defined that in declination may be, from the following values of the investigators, viz.:

P.... Æ + 0″·08   Dec. − 0″·06
Br...   + 0″·10   − 0″·05
B....   + 0″·09   + 0″·02

Piazzi derived the amount here given, from comparison with Bradley; that from Flamsteed and La Caille being + 0″·58, and from Roemer + 0″·16. Hence, though the velocity differs, they all take the same course. But Mayer, in his *Commentatio de Motibus Propriis,* changes the signs, and gives − 0″·42 in Æ, and − 0″·29 in declination.

ζ Pegasi is designated Homam in the Palermo and other Catalogues, from *Sa'd al-homám,* the hero's happy star, of the Arabians. Tízíní called it *Sa'd al nu'ám,* the ostrich's lucky star; and it is included in the group known as *Su 'údu-l-nujúm,* the fortunate stars, so named because they appear to the Bedowín Arabs at the dawn of day, on the approach of spring.

DCCCXIX.   200 P. XXII. AQUARII.

Æ   22ʰ 34ᵐ 40ˢ   Prec. +   3ˢ·15
Dec. S   9° 08′·8   —— N 18″·67

Position 314°·4 (w 7)   Distance 2″·9 (w 4)   Epoch 1833·75
—— 313°·8 (w 9)   —— 2″·7 (w 6)   —— 1838·69

A very neat double star in the streamlet at the mouth of the Waterbearer's vase, and 19° south of ζ Pegasi, above described. A 7, and B 8½, both white. This object is 50 Ħ. i., and these were its registered data when I took it in hand:

Ħ.   Pos. 311° 12′   Dist. 3″·00±   Ep. 1782·74
Σ.   317° 22′   ...   1821·92
H. and S.   321° 19′   3″·40   1823·70

Here was certainly a hope that a very substantial change of angle had occurred in forty-one years, but it is probable that all discrepancies are imputable to the instrumental errors, to which so difficult an object is liable. Indeed, Sir William Herschel has expressly stated, that his angular measure was inaccurate on account of the low magnifying power used, "probably," he adds, "3° or 4° too small:" as to his distance I have only given an average estimate, the data being merely from the apparent diameter of the large star. Still it may turn out to be a physical object, though a long interval of time will be required to prove it such; meantime the aberrations of A from the common laws of precession, amount to:

P.... Æ − 0″·29   Dec. − 0″·18
B....   − 0″·11   − 0″·11

From the south-west movement which even Mr. Baily's reduced value would have given A, in the interval here embraced, the proper motions seem to apply to B also.

DCCCXX. 209 P. XXII. AQUARII.

Æ 22ʰ 36ᵐ 56ˢ   Prec. + 3ˢ·16
Dec. S 10° 29'·0   —— N 18"·74

Position 63°·5 (w 2)   Distance 10"·0 (w 1)   Epoch 1834·68

A very delicate double star between the streamlet and the Water-bearer's left thigh, and 1°½ south-half-east from the object just described. A 8, white; B 12½, violet tint. This was discovered by H̶., and registered 69 iii., under these results:

Pos. 60° 57'   Dist. 12"·77   Ep. 1782·74

that is, supposing Sir William made a mistake in noting it as being in the *sf* quadrant, instead of the *nf*. This must have been the case, since there can be no doubt of the star's identity, by his descriptive remarks: "In dextro femore. Full 1½ degree n. following the 64th ::, in a line parallel to λ and φ Aquarii; the largest of two that follow a very obscure triangle in the finder."

DCCCXXI. ξ PEGASI.

Æ 22ʰ 38ᵐ 42ˢ   Prec. + 2ˢ·98
Dec. N 11° 21'·4   —— N 18"·80

Position AB 120°·0 (w 1)   Distance 15"·0 (w 1)⎫
—— AC 32°·5 (w 2)   —— 110"·0 (w 1)⎭ Epoch 1834·79

A most delicate double star, with a distant companion, on the animal's neck; within a couple of degrees north-east-by-north of ζ, before aligned. A 5, pale yellow; B 15, blue; C 12, dusky. The two first components form No. 301 of H.'s First 20-foot Series; and as the *comes* is only to be caught under intense gazing, with the most favourable circumstances of atmosphere and instrument, I added C for identification. Sir John Herschel rated B of the 18th magnitude, but it certainly is not my *minimum visibile:* he observes, that it bears illumination well, and is *therefore* blue. The spacial movements of the *præses* have been registered under these values:

P.... Æ + 0"·32   Dec. — 0"·36
B....   + 0"·23   — 0"·43
A....   + 0"·21   — 0"·47

Here the amount of proper motion in declination given by Piazzi, is from comparisons with Bradley and Flamsteed, and is the smallest quantity of the three; but from his own observations he made it no less than — 0"·80. I regret the being unable to give further particulars, *in limine,* upon so interesting a point.

DCCCXXII. τ^1 AQUARII.

Æ 22ʰ 39ᵐ 13ˢ   Prec. + 3ˢ·19
Dec. S 14° 53′·9   —— N 18″·81

Position 112°·8 (w 8)   Distance 30″·4 (w 5)   Epoch 1830·79
—— 113°·5 (w 8)   —— 29″·7 (w 5)   —— 1838·67

A fine double star above the Water-bearer's left knee, and nearly midway of a line between Fomalhaut and ζ Pegasi. A 6, white; B 9½, pale garnet. This object is 80 Ħ. v., and these were its circumstances previous to my attack:

	Pos.	Dist.	Ep.
Ħ.	109° 54′	36″·78	1782·66
S.	112° 47′	30″·54	1825·21
D.	114° 19′	29″·46	1830·57

There appears to be evidence here of a progressive increase of angle, and a diminution in the distance. But the star is difficult, and all the recent observations, including those of Σ., indicate its fixity.

DCCCXXIII. 219 P. XXII. AQUARII.

Æ 22ʰ 39ᵐ 35ˢ   Prec. + 3ˢ·11
Dec. S 5° 03′·5   —— N 18″·82

Position AB 247°·4 (w 8)   Distance 4″·2 (w 5) ⎫
—— AC 158°·0 (w 8)   —— 55″·1 (w 9) ⎭ Epoch 1835·88

A triple star following Situla, at the mouth of the vase, by about 2°¼ on its eastern parallel; and it is two-thirds of the way from Fomalhaut towards ζ Pegasi. A 7½, yellow; B 8, and C 9, flushed white. This is 57 Ħ. ii., but though registered "treble" by Sir William, the third is only vaguely estimated. Piazzi said, "Duplex. Comes 7·8ᵃ magnitudinis, 0″·2 temporis sequitur, 3″ ad boream. Hæc cum sequenti [*meaning C, which is his* 220,] ut triplex spectari potest." My point of departure was:

		Pos.	Dist.	Ep.
H. and S.	AB	245° 36′	4″·35	1822·90
	AC	162° 33′	57″·38	

Although the differences here observable are not anomalous, they are sufficiently large, under such good and tolerably easy operations, to be remarkable. There appears to be a diminution both in the angle and distance of A and C; and that inference was countenanced on the arrival of the Dorpat measures. But the whole change may be ascribable to the proper motion of one or other of the stars,—that for A being considered as:

	Æ	Dec.
Br...	− 0″·17	− 0″·30
A....	− 0″·21	− 0″·29

DCCCXXIV. α PISCIS AUSTRALIS.

ℛ 22ʰ 48ᵐ 48ˢ   Prec. + 3ˢ·31
Dec. S 30° 28'·3   —— N 19''·08

Position 195°·0 (w1)   Difference ℛ = 4ˢ·8 (w1)   Epoch 1833·65

A standard Greenwich star with a very distant companion near the vertical of the *sp* quadrant, in the mouth of the Southern Fish. A 1, reddish; B 9½, dusky blue. This is the well-known Fomalhaut, a nautical star, the lunar distances of which are computed and given in the Ephemeris: it derives its name from the Arabian *fom-al-ḥút*, the fish's mouth; more fully written *fom-al-ḥút al-jenúbí*, or piscis australis. Strict investigation has decisively established a proper motion in this star, and the most esteemed values are these:

P.... ℛ + 0''·33   Dec. − 0''·26
B....   + 0''·39   − 0''·15

Fomalhaut is one of the best observed of the old Catalogue objects, there being only a difference of − 2'¼ in the distance between it and Regulus, as compared between Ptolemy and the moderns. Delambre thus drew a parallel:

REGULUS AND FOMALHAUT.

Difference of Longitude { *Syntaxis* . . .	6° 04' 30''		
{ *Moderns* . . .	6° 03' 59''·20		
Excess of Ptolemy	—— 30''·40		
Polar Distance . . . { *Syntaxis* . . .	110° 40' 00''		
{ *Moderns* . . .	111° 06' 16''		
Excess of Ptolemy	—— 26' 16''		
Distance . . . . . { *Syntaxis* . . .	200° 58' 00''		
{ *Moderns* . . .	201° 00' 37''		
Excess of Ptolemy	—— 2' 37''		

The colour of this star, and its apparent race after the Sun, made it a marked object, and the observer will find that the difference is not little between the slow march of Polaris over the meridian, and the gallop of Fomalhaut. But it affords a favourable means of experimenting on the atmospheric property which ℋ. calls the "prismatic power," and which he defines as "that refractive quality whereby the atmosphere disperses the rays of light, and gives a lengthened and coloured image of a lucid point." This power, which depends on the obliquity of the incident ray, diminishes with the increase of altitude, and therefore he recommends attention to the subject, since in delicate long polar-distant observations its effect ought not to be overlooked. But though I clearly saw the prismatic spectrum afforded by my lowest stars, I was unable to make any satisfactory experiments for correcting the measures thereby.

'Ιχθύς νότιος, 'Ιχθύς μέγας, Piscis notius, Piscis magnus, Piscis australis, Piscis austrinus, are names by which the Southern Fish was known to the classical ancients. It is a small asterism south of the Water-bearer, and is represented with its head turned towards the east, and its tail to the west. In the early Venetian editions of Hyginus, there is a smaller fish close under it, *remora* fashion, interfering with the *solitarius* by which that astronomer, from its insulated position, designated Piscis notius. In the celebrated MS. of Cicero's *Aratus*, before quoted, the figure is inscribed S̄. STELLAE XII., to which number six *amorphotæ* were added by Ptolemy, since whose time the constituents have thus progressively increased:

Ptolemy	. . . 18 stars	Brisbane	. . . 36 stars
Flamsteed	. . . 24	Bode	 77

The Arabs had drawn a figure in this portion of the heavens, probably before they were acquainted with the Greek constellations: of this, Fomalhaut was *Difda' al-auwel*, the first frog; and β Ceti was *Difda' al-tháni*, the second frog. The Mosaicists altered all this, for they held the asterism to represent the barrel of meal belonging to Sarephtha's widow; but Schickhard pronounces it to be the fish taken by St. Peter with a piece of money in his mouth. In either case it is easily found. A line carried from Aquila to the west-south-west, over the tail of Capricorn, hits it; as does another from the north-east by α and β Ceti. To the north, a line from Polaris, through the preceding components of the square of Pegasus, reaches it after passing over the group in the urn-stream of Aquarius. To those in the more southern latitudes, the rhymester thus advises:

He who would trace where dimly seen	the austral Fish doth glide,
Must start from Scheat, and through Markab	his occult line will guide:
When forty-five degrees thus pass'd,	conduct him to the south,
The dazzling Fomalhaut he'll find,	in Piscis notius' mouth.
When there, south-eastward cast the eye—	how glorious! how fine!
Achernar with Canopus bright	and Fom'lhaut form a line.

The Rev. Charles Turnor (*Newtonian Turnor*), has recently lent me a very valuable MS. Almanac, of 1340, which, by a note on the fly-leaf, appears to have belonged to the library of the convent "De Babewell," near St. Edmund's Bury: it is in excellent preservation, and in the original wooden binding. To the astronomical antiquary it is replete with interest, containing Solar, Lunar, and Planetary Tables, *cum canonibus;* the geographical positions of the then most important parts of the world, and "Tabulæ Magistri Johannis de Lineriis de quinque Planetis retrogradis," &c., also *cum canonibus*. Among its contents is a list of thirty-five principal stars, with their Arabian designations, and their constellation places, brought up from Ptolemy's Catalogue by applying 45″ for the annual change of longitude, and allowing the latitudes to remain unaltered. In this precursor of Maskelyne's Greenwich List, is our present star, Fomalhaut, distinguished as *Al-jenúbi*, southernmost, and as *Os piscis mĩdiani*, which is clear of the obscurity that prevailed from Ptolemy's having inserted it in both ʿΥδροχόος and ʾΙχθὺς νότιος. Mr. Baily, in his new edition of Ptolemy, No. 670, thinks that an error of nearly 3° may have arisen in the latitude of Fomalhaut, from some of the glossators' mistaking the numeral κγ[1] for κγ. To another star, λ Orionis, he has given latitude "*ʋοτ*. 16° 30′," appending the note—"L(*iechtenstein*) *lat.* 18°·50; but even 16° 30′ is nearly 3° too great, and there is probably some mistake in all the copies." Now the following extract from Master de Lineriis, will show that the supposition is well grounded, and that this early list of stars was copied from a text of Ptolemy now unknown to us:

Nomina Stellarum Fixarum.	Loca Stellarum.		Latitudies earum.		Prec.	Lat. Stellᵐ. ab Equint.		Prec.	Gdus equibus medent.Celum.	
	Signa.	gdus.	mnta	gdus.	mnta	Lat.	gdus.	mnta.	Lon.	gdus.	mnta.
Fom Alhout Algenubi *Os Piscis midiani* }	Aqrī'.	21	57	23	0	Merid.	35	49	Merid.	334	21
Raz Algauze *Caput Gemn* . . }	Gem.	11	55	13	50	Sept.	8	37	Sept.	74	25

The name and sign of the last star are evidently erroneously transcribed; but the error interferes very little with the identity stamped by the numerals. A strict examination of this rare work is very desirable, as much light may be thrown on the Solar and other Tables: and I am happy to add, that this is about to be undertaken, at my request, by Mr. Harris, the zealous Assistant-Secretary of the Royal Astronomical Society, who has had much acquaintance with early writings, and great practice in computation.

As to Johannes de Lineriis, my friend Mr. de Morgan, who may be termed the *furet* of scientific biography, tells me, that little is known of him, except that he was of Amiens, and has left some Tables in manuscript. Delambre merely says that he "fit quelques observations qu'on trouve dans les Œuvres de Gassendi. Tome vi., p. 502."

DCCCXXV.   16 LACERTÆ.

Æ 22ʰ 49ᵐ 06ˢ   PREC. + 2ˢ·72
DEC. N 40° 45'·1   —— N 19"·09

POSITION AB 345°·0 (*w* 1)   DISTANCE 25"·0 (*w* 1) }
—— AC 46°·7 (*w* 8)   —— 64"·0 (*w* 5) }   EPOCH 1836·58

A delicate triple star, in the space between the Lizard's back and the left hand of Andromeda. A 6, bright white; B 15, pale blue; C 9½, reddish; a fourth star at a distance in the *np* quadrant. This object is 85 IⱮ. IV., and one of singular difficulty, from the minuteness of the nearer component. The measures when I commenced were:

IⱮ.   { Pos. AB 349° 37'   Dist. 20"·45 }   Ep. 1782·76
{ AC 45° 36'   56"·62 }

H. and S. { AB *small star not seen.* }   1822·87
{ AC 45° 19'   64"·54 }

16 Lacertæ is claimed for Andromeda by some compilers, but is now usually regarded as the last in the Lizard. A line from Markab led north-half-west between β and η Pegasi, and carried as far again, finds it to the south-south-west of a pair of stars, of 4th and 5th magnitudes, north and south of each other.

DCCCXXVI. β PEGASI.

Æ 22ʰ 56ᵐ 01ˢ   Prec. + 2ˢ·88

Dec. N 27° 13′·0   —— N 19″·26

Position 199°·0 (w 1)   Distance 75″·0 (w 1)   Epoch 1836·88

A bright star with a distant minute companion, on the animal's left fore-leg. A 2, deep yellow; B 15, blue; there is a 10th-magnitude star in the *sp*, and a 9th one follows A on the parallel. This object is No. 1842 of H.'s IVth Series of 20-foot Sweeps; but as I see the *comes* steadily, I consider his estimation of its magnitude as too small. β is the Scheat of the Palermo and other Catalogues, probably a corruption of *sâ'id*, an arm, or cubit; it is generally called *Menkib al-feres*, the horse's shoulder, by the Arabs. Its spacial movements have been ably investigated, and the values are corroborative of each other, being as follows, including that from Mayer's *Commentatio de Motibus Propriis:*

	Æ	Dec.
M....	+ 0″·24	+ 0″·02
P....	+ 0″·24	+ 0″·20
B....	+ 0″·24	+ 0″·17
A....	+ 0″·22	+ 0″·15

Scheat precedes Alpherat, in the head of Andromeda, a little south of its western parallel, by 16°; where the line of the Pointers carried over the Pole, and 62° beyond, reaches it. We hear that

> The brilliant Scheat, some three degrees,   upon a south-west line,
> Points where the hero's lucky stars,   call'd *my* and *lambda*, shine.

DCCCXXVII. α PEGASI.

Æ 22ʰ 56ᵐ 47ˢ   Prec. + 2ˢ·98

Dec. N 14° 20′·8   —— N 19″·29

Position 305°·0 (w1)   Difference Æ = 17ˢ·2 (w 1)   Epoch 1833·78

A standard Greenwich star, with a very distant telescopic companion, at the junction of the animal's wing and shoulder, 13° south of Scheat, just described. A 2, white; B 11, pale-grey; and the two point to a third star about as far again in the *np* quadrant. The proper motions assigned to A are thus registered, again including Mayer:

	Æ	Dec.
M....	− 0″·16	+ 0″·04
P....	+ 0″·02	− 0″·07
B....	+ 0″·11	+ 0″·01
A....	+ 0″·08	− 0″·01

The discrepancy of these conclusions is the more surprising, as the star is a well-drilled one: Piazzi observed it no less than 162 times in Æ, and 55 times in declination.

α Pegasi is usually termed Markab, a thing ridden upon, a vehicle, a ship; here, perhaps, a saddle, as in Hebrew, from a version of the

Almagest, in which language the *Alphonsine Tables* were probably taken. It is called *Matn-al-faras*, the horse's withers, by the Arabs; being placed, as Kazwíní remarks, in that part of the body where the neck grows out of the shoulders. Sir William Herschel ranked it among the insulated stars.

Pegasus was simply called Ἵππος, the horse, by Aratus and the Greeks; Πήγασος, Equus alatus, sprung only from later poetic invention. Eratosthenes, it is true, mentions it, but see in what manner: "Some think," says he, "the horse is that Pegasus which, after the fall of Bellerophon, flew to the stars. But this cannot be, as it has no wings:" it therefore follows, that those appendages are a later addition. The Romans termed it Equus; but Germanicus uses the designation Pegasus, while Ovid calls it Equus Gorgoneus, in allusion to its fabled origin. Cornipes and Sonipes ales, were poetical terms for Bellerophon's horse; but certain "*konnyng squiers*" talk of a Jewish legend which carries it to greater antiquity, as being Nimrod's horse. It merits the name Equi-sectio more than Equuleus, since only the fore part of a horse, with a pair of wings, is figured; and that in an inverted position, the head being further from the north pole than the body. It occupies a considerable space, and is at once recognised by a large equilateral square, formed by the four fine bright stars, *α*, *β*, *γ* Pegasi, and *α* Andromedæ,—or Markab, Scheat, Algenib, and Alpherat. It is a *paranatellon* to Aquarius and Pisces, and has been thus registered:

Ptolemy	. . . . 20 stars	Hevelius .	. . 38 stars
Tycho Brahé .	. . 23	Flamsteed	. . 89
Bullialdus .	. . . 24	Bode .	. . . 393

Those who seek for more than Apollo's winged horse in this constellation, assert that it was emblematic of a ship in Phœnicia; while the still more erudite say that Pegasus is compounded from the Egyptian word *pag*, to cease, and *sus*, a ship, as symbolizing the cessation of navigation owing to the inundation of the Nile. Q. E. D. Among the neighbouring Arabs, the rectangular figure was called *Al-delw*, a water-bucket, and constituted the XXVIth and XXVIIth Lunar Mansions. The first was at *α* and *β*, called *Fargh al-delwi-l-muḳaddem*, the hither emptying place of the bucket, or its lip; the hindmost being *γ* Pegasi and *α* Andromedæ; while *τ* and *υ*, in the middle of the square, were recognised as *Al-kereb*, the joining of the two cross bars of wood placed diagonally over the well, to which the bucket-rope is fastened. The later Arabian astronomers, borrowing from the Greeks, called the constellation *Al-faras al-aʿdham*, the larger horse; and *Al-faras al-tháni*, the second horse.

From its situation relatively to Cepheus and Cassiopea, and being between Andromeda and Aquarius, Pegasus is of easy recognition. A line, or arc of a circle, from *β* and *α* Ursæ Majoris, crosses the pole and passes the leaders of the great square. To align the vicinity, observe the rhymester's rules:

A line athwart this spacious square	to north-west, shows the Swan,
Transverse to that a north-east ray,	points Perseus upon;
And on that way two beauteous stars	divide the space between,
In equal portions, from Mirak	to where Al'mak is seen.

DCCCXXVIII. 55 ☿. I. PEGASI.

Æ 22ʰ 56ᵐ 58ˢ   Prec. + 2ˢ·99
Dec. N 11° 27'·9   —— N 19"·29

MEAN EPOCH OF THE OBSERVATION . . . . 1830·91

An elongated nebula in the animal's mane, and about 3° due south of Markab, the above-described object. This is very faint; but after long gazing under clock-work motion, it comes up, trending very nearly north and south, having a telescopic star at each extreme. It was discovered by ☿. on the 19th of October, 1784, and described as being 4' long and 2' broad; not only is its aspect very different from the apparently spherical nucleated masses so often occurring, but it appeared a mere streak even in Sir John Herschel's 20-foot reflector, tapering at each end. Now, the sphere being the general figure assumed in consequence of the particles mutually attracting each other, we can only suppose that the lenticular appearance before us, is a vast ring lying obliquely to our line of vision. Observation and analogy unite in showing that this object is at an inconceivable distance beyond the stars. Yet there remains an interesting and important inquiry: I allude to the apparent connection of one or two most minute stellar points with some of the smaller nebulæ, as exemplified in several of these celestial gossamers. Sir John has noted several very remarkable instances in his great Catalogue of 1830; and they will form excellent objects for the new gigantic reflecting telescopes.

DCCCXXIX. 306 P. XXII. PEGASI.

Æ 22ʰ 59ᵐ 49ˢ   Prec. + 2ˢ·85
Dec. N 31° 57'·7   —— N 19"·36

POSITION 147°·3 (w 7)   DISTANCE 8"·3 (w 5)   EPOCH 1830·91
—— 146°·4 (w 9)   —— 8"·5 (w 6)   —— 1833·88

A fine double star, in the space north of the animal's chest, and in a blank vicinity 4¼ north-half-east from β Pegasi. A 7, bright white; B 8½, sapphire blue. It was discovered by ☿. in September, 1784; and both components were seen by Piazzi, who, however, only made observations on the leader, saying, "Duplex: præcedens observata." The first micrometric measures I find, are those of H. and S.:

Pos. 148° 19' Dist. 8"·716 Ep. 1823·75

Here the intervals of time are too short, and the operations too difficult, to pronounce upon the angular differences' being merely instrumental, or not. The distances, under due consideration of these conditions, may be said to be in perfect coincidence.

DCCCXXX.   57 PEGASI.

 Æ 23ʰ 01ᵐ 27ˢ   Prec. + 3ˢ·02
Dec. N 7° 48'·7   —— N 19"·39

Position 197°·0 (w 3)   Distance 35"·0 (w 1)   Epoch 1836·73

A very delicate double star, between the mane of Pegasus and the head of the preceding fish, about 6° south-half-east of Markab; and it is the middle one of three of similar magnitudes, within a degree of each other.   A 5½, orange; B 13, greenish; and these are followed nearly on the parallel by a third star.   This object being too difficult for exact metrical observations, the angle of position is obtained by the spherical crystal micrometer, and the distance carefully estimated.

This is No. 16 of ₩.'s List of 145 New Double Stars; and it was enrolled in November, 1784, with "position about 20° or 30° *sp*," and in distance rated of the IVth Class.   It then became No. 3173 of H.'s 20-foot Sweeps; and No. 2982 of the Dorpat Catalogue, the second edition of which exhibited these measures:

Σ. Pos. 198°·10   Dist. 32"·562   Ep. 1831·06

An almost imperceptible movement in space is attributed to the large star, of which the following are the best values:

P.... Æ + 0"·05   Dec. + 0"·08
Br...   + 0"·02   − 0"·01
B....   + 0"·03   + 0"·01

DCCCXXXI.   π CEPHEI.

Æ 23ʰ 02ᵐ 49ˢ   Prec. + 1ˢ·88
Dec. N 74° 31'·4   —— N 19"·42

Position AB 241°·5 (w 3)   Difference Æ = 11ˢ·8 (w 1)   Epoch 1838·75
——— Aa 330°·0 (w 1)   Distance . . . 1"·8 (w 1)   ——— 1843·77

A close double star with a distant minute companion, preceding the Æthiop's right leg, and the nearest star of note to the south-west of γ, in the curve towards β.   A 5, deep yellow; a 10, purple; B 12, blue; and A and B point nearly upon a distant small star in the *sp* quadrant. The nice precision of modern observation has detected a spacial movement in π, which, though slight, and variously valued, is decisively registered; here is the amount:

P.... Æ − 0"·09   Dec. − 0"·25
B....   + 0"·02   − 0"·05

π Cephei obtained a place on my working list, from having a *comes* marked closely to the north of it on Sir John Lubbock's map.   On a scrutiny which I gave it in 1838, I found nothing nearer than B, but was not satisfied with the definition of A.   Thus the matter would have rested, had not M. Struve, with the gigantic telescope at Poulkova, detected the little star a: thus apprised, it appeared, on steady gazing, sufficiently obvious at Hartwell to admit of an estimation.

Messrs. Lassel and Dawes saw this very difficult object at Liverpool, 7th of September, 1843, with Mr. Lassel's Newtonian reflector, of nine inches clear aperture, constructed by himself. The eye-piece magnified 400 times. This success, it is hoped, may elicit a valuable contribution of future observations from that quarter.

DCCCXXXII.   430 ♓. II. PISCIUM.

Æ   23ʰ 06ᵐ 36ˢ   Prec. + 3ˢ·05
Dec. N 3° 39'·7   —— N 19"·50

Mean Epoch of the Observation . . . . 1838·78

A faint nebula in the eye of the preceding or western Fish, and about 10° south-half-east of Markab; consequently, within a degree to the north-north-west of γ Piscium. This object is so dim as to be only perceptible under settled gazing, and clock-work motion, when it faintly gleams among the telescopic stars in the field. It was discovered by ♓. on the 30th of August, 1785, and described as being 4' long by 1' in breadth. H., whose No. 2216 it is—though entered there by mistake as 230 ♓. ii.—describes it as 80" in length; and both those astronomers mention its being preceded by a still fainter nebula in the *sp* quadrant, of which my instrument gave no trace.

After gazing upon this pale nebula, I turned the telescope upon that vaporous mass, Encke's comet, on the diaphanous night of October 17th. It was then in an irregular trapezium of Milky-way telescopic stars, in Æ 2ʰ 5ᵐ and Dec. + 48° 6'; but its glow was at least in the proportion of three to one over the nebula, faint as the wonderful and incoherent wanderer certainly was at that time.

DCCCXXXIII.   ψ¹ AQUARII.

Æ   23ʰ 07ᵐ 30ˢ   Prec. + 3ˢ·12
Dec. S 9° 57'·5   —— N 19"·52

Position 310°·5 (*w 7*)   Distance 49"·5 (*w 5*)   Epoch 1834·89

A double star in the centre of the upper part of the stream which Aquarius pours out, being the first of three similar stars, ψ¹, ψ², and ψ³, and rather more than one-third of the way from Fomalhaut to α Andromedæ. A 5½, orange tint; B 9, sky-blue. This is 12 ♓. iv., registered in November, 1779; but as Sir William gives a distance of 23"·10, and

2 M 2

says "pretty accurate," it would appear that ψ^1 cannot be the star. He adds, however, "it is the first of the three ψs," so there is no alternative. The departure, therefore, from which I infer the object to be merely optical, is Sir James South's observation:

Pos. 311° 08′ Dist. 49″·83 Ep. 1824·80

The three ψs have a decided spacial movement, which is treated of by Mayer, in his *Commentatio de Motibus Propriis;* that of ψ^1 is thus stated upon more recent investigation:

P....	Æ + 0″·36	Dec. − 0″·20
B....	+ 0″·42	0″·00
A....	+ 0″·39	− 0″·02

In the description of the Sphere, or Frame of the World, by Wyllyam Salysbury, in 1550, it is stated that, "the iiii. starres that be at the ryght hande's ende of Aquarius, are called Urna." This is, however, in the stream, where a principal star, κ, is called Situla, and by the Arabs *Al Delw*, both meaning the water-bucket. But Gassendi and others have derived Situla from *sitis*, thirst, it having been looked upon as a star of dryness by those Sidrophels who viewed the Water-bearer's urn as an oven.

DCCCXXXIV. 94 AQUARII.

Æ 23ʰ 10ᵐ 41ˢ   PREC. + 3ˢ·14
DEC. S 14° 19′·7   —— N 19″·58

POSITION 344°·9 (w 8)   DISTANCE 13″·5 (w 5)   EPOCH 1831·87
——— 345°·4 (w 9)   ——— 14″·0 (w 9)   —— 1838·91

A neat double star in the space between the stream and the left knee of Aquarius, and one-third of the distance from Fomalhaut to α Andromedæ, in a north-north-east direction. A 6, pale rose-tint; B 8½, light emerald. This pretty object is 34 ꟷ. III., and constitutes Nos. 41 and 42 of Piazzi's Hora XXIII., the places of which yield 13″·6 for the distance between them in 1800. By taking a mean epoch between ꟷ.'s observations in 1781 and 1802, we obtain the following register previous to my commencement:

ꟷ.	Pos. 342° 45′	Dist. 13″·750	Ep. 1792·16
Σ.	346° 36′	13″·991	1821·92
H. and S.	346° 41′	14″·998	1822·88

A consideration of the whole results implies fixity; and my last measures, were under such favouring circumstances, that I have very great confidence in their epoch: and this conclusion is borne out by the last Dorpat Catalogue.

Sir William Herschel, in 1781, familiarly described 94 Aquarii as being between ψ and ω towards δ. This is very correct, but it requires intimacy with the constellation to be available. As to Sir William himself, he could unhesitatingly call every star, down to the 6th magnitude, by its name, letter, or number.

DCCCXXXV. ο CEPHEI.

Æ 23ʰ 12ᵐ 04ˢ	Prec. + 2ˢ·40

Dec. N 67° 14'·2	—— N 19"·60

Position 173°·8 (w 5)	Distance 2"·5 (w 4)	Epoch 1834·95

An elegant double star in the space between the right arm and knee of Cepheus, and lying about 20° south-south-west of Polaris, with γ Cephei mid-way between them. A 7, orange yellow; B 9, deep blue; the colours in fine contrast. This fine object was discovered by M. Struve, and is No. 3001 of the Dorpat Catalogue; the last edition of which shows these measures as a mean:

Pos. 174°·97	Dist. 2"·353	Ep. 1832·84

Little can be said upon the data, until a longer lapse of time has intervened, when it may very probably prove to be a physical object. Meantime it must be noted that A has decided proper motions, although the amount has latterly been considerably lessened. By the most rigid scrutinies, they are thus valued:

P.... Æ + 0"·52	Dec. − 0"·16

B.... + 0"·24	 − 0"·04

DCCCXXXVI. 69 P. XXIII. AQUARII.

Æ 23ʰ 15ᵐ 29ˢ	Prec. + 3ˢ·11

Dec. S 9° 20'·2	—— N 19"·67

Position 272°·1 (w 9)	Distance 7"·5 (w 7)	Epoch 1834·79

A very neat double star in the middle of the upper part of the urn-stream, and 20° on the north-north-east line extending from Fomalhaut towards α Andromedæ. A 8, and B 8½, both flushed; but the smaller one has the reddest tint. This pretty object was discovered by Piazzi, and is thus entered in the Palermo Catalogue: "Duplex. Comes 8·9æ magnitud. præcedit in eodem fere parallelo." The first micrometric measures I meet with for a departure, are those taken by Sir James South at Passy, near Paris:

Pos. 274° 04'	Dist. 7"·981	Ep. 1824·80

As with the above-described ο Cephei, we must here also pause for a longer epoch; recollecting, moreover, that A has been convicted of spacial movement, to the following registered amount:

P.... Æ − 0"·06	Dec. − 0"·04

Br... − 0"·13	 − 0"·10

B.... − 0"·11	 − 0"·09

And these values, supposing B to remain stationary, would, in half a century, increase the angle to 340°, and diminish the distance to about 5".

DCCCXXXVII.   52 M. CEPHEI.

Æ   23ʰ 17ᵐ 10ˢ   Prec. +   2ˢ·63
Dec. N 60° 43′·1   —— N 19″·70

Mean Epoch of the Observation . . . . 1835·65

An irregular cluster of stars between the head of Cepheus and his daughter's throne; it lies north-west-by-west of β Cassiopeæ, and one third of the way towards a Cephei.   This object assumes somewhat

of a triangular form, with an orange-tinted 8th-magnitude star at its vertex, giving the resemblance of a bird with outspread wings.   It is preceded by two stars of the 7th and 8th magnitudes, and followed by another of similar brightness; and the field is one of singular beauty under a moderate magni-fying power.   While these were under examination, one of those bodies called falling stars passed through the outliers.   This phenomenon was so unexpected and sudden as to preclude attention to it; but it appeared to be followed by a train of glittering and very minute spangles.

This cluster was discovered by M. Messier, " le Préposé du Ciel," in 1774, at Paris, and was then described as " a mass of very small stars blended with nebulous matter, and requiring a good telescope to dis-tinguish them; it looks like a solid ball of stars, compressing into a blaze of light, with stragglers."

DCCCXXXVIII.   4 CASSIOPEÆ.

Æ   23ʰ 17ᵐ 45ˢ   Prec. +   2ˢ·62
Dec. N 64° 24′·3   —— N 19″·70

Position A B 225°·0 (w 6)   Distance   97″·5 (w 2)
——— A C 260°·0 (w 9)   ——— 218″·0 (w 1)   Epoch 1834·67
——— C D 256°·0 (w 1)   ——— 10″·0 (w 1)

A coarse quadruple object, closely south of Cepheus's right hand, and about 27° south-south-west of Polaris, on a line led over γ Cephei. A 5, pale yellow; B 9, yellowish; C 11, and D 13, both blue.   This was entered upon my working list as a triple star, 24 Ḥ. vi., which was thus described in 1780: " Treble.   Two are large.   Distance about 2′. A third is obscure.   Distance about 1¾ min.   They form almost a rectangle."   As this does not quadrate with my diagrams, when the great Northumberland equatoreal was mounted at Cambridge, I requested the Rev. James Challis to scrutinize the object, which he kindly did, with similar results to my own. The sensible amount of proper motion once attributed to the larger star, is disappearing under recent investigation.

Finding from Nota 81, Hora XXIII. of the Palermo Catalogue, that Piazzi had failed in finding the star 3 Cassiopeæ, I carefully fished for it, under the hope that if it were only variable, I might catch it up; but *non inventa* was also my note. Since I was thus foiled, Mr. Baily's edition of Flamsteed appeared, in which he observes, "There is no observation of this star in the 2nd volume of the *Historia Cœlestis*, the references given by Miss Herschel belonging to another and a different star, No. 3224 of this Catalogue. In MSS. Vol. 26 B, page 36, the position is said to have been obtained *per distantias;* probably from the star which is called *supra* τ, in Vol. 2, page 63. If so, I apprehend it must have been introduced through some mistake in the trigonometrical computation; as there is no star to be found corresponding with the position here given." It is plain, therefore, that 3 Cassiopeæ must not be included among the lost or missing stars, since there is no proof of its ever having been duly enrolled.

DCCCXXXIX. 101 P. XXIII. CASSIOPEÆ.

ℛ 23ʰ 22ᵐ 40ˢ   Prec. + 2ˢ·72

Dec. N 57° 40′·1   —— N 19″·78

Position A B 270°·5 (*w* 8)   Distance 74″·1 (*w* 5) ⎫
——— B C 345°·0 (*w* 2)   ——— 20″·0 (*w* 1) ⎭   Epoch 1830·81

A multiple star between the head of Cepheus and the throne of his daughter, where a line from Polaris, led through γ Cephei, and carried double that distance southwards, strikes it. A 5, light yellow; B 7½, white; C 14, blue. Of this fine group I diagrammed six components, intending at first only to measure Nos. 100 and 101 of Piazzi, whose mean apparent places I had reduced for obtaining this comparison:

P.   Pos. 269°·8   Dist. 75″·80   Ep. 1800
H. and S.   270°·0   73″·95   1822·87

To these I added the star which I have designated C, and in so doing, of course observed B very closely, but without noticing anything particular about it. On my requesting the Rev. W. R. Dawes to examine this object ten years after my measures were taken, he distinctly made B to be closely double, the primary having a small companion at an angle of about 222°, and 1½″ distance. This was very remarkable; inasmuch as I cannot but think, had it been *out*, that either Σ., H., S., or myself, must have seen it, especially as I made such careful estimations of C, that I might almost have registered it, *observatio egregie certa;* and moreover, it is so much smaller than the companion to B, that, to say the least, the latter ought to have been seen. At all events, there is every prospect that by this accident, a binary system has been added to our sidereal knowledge, and it is now in the charge of its able discoverer. I have since seen it, and with B and C a beautiful triple star is formed.

DCCCXL. ι PISCIUM.

Æ	23ʰ 31ᵐ 43ˢ	Prec. +	3ˢ·05
Dec.	N 4° 45'·6	——	N 19"·89

Position 39°·5 (w 1)   Distance 199"·0 (w 1)   Epoch 1835·83

A secondary Greenwich star with a distant companion, in the middle of the preceding Fish's back; lying about 14° to the south-east of α Pegasi, and the same distance south-west of γ Pegasi. A 4½, light yellow; B 12, pale blue; and there is another telescopic star to the *sf* of it. There is now no longer any doubt of the leader's being under the influence of a very sensible movement in space, since the rigorous comparisons of Messrs. Piazzi, Baily, and Argelander, agree in the direction, and almost in the amount. The registered values are these:

$$P.... \quad Æ + 0''·30 \qquad Dec. - 0''·55$$
$$B.... \qquad + 0''·44 \qquad\qquad - 0''·44$$
$$A.... \qquad + 0''·42 \qquad\qquad - 0''·44$$

DCCCXLI. χ ANDROMEDÆ.

Æ	23ʰ 32ᵐ 33ˢ	Prec. +	2ˢ·92
Dec.	N 43° 27'·0	——	N 19"·90

Position AB 186°·0 (w 1)   Distance 47"·0 (w 1)⎱
  —— AC 295°·5 (w 2)   —— 98"·5 (w 1)⎰ Epoch 1836·72

A wide but delicate triple star on the northern hand of the manacled lady; it lies midway between β Pegasi and α Cassiopeæ, and about 18° from each. A 5, brilliant white; B 14, dusky; C 12, ash-coloured; and there is a fourth star nearly on the preceding parallel at a distance. The two closest of these stars constitute No. 1898 of Sir John Herschel's 20-foot Sweeps; and the *comes* is so flitting and uncertain an object, that the estimations here given may at best be only ranked as comparative guesses. It forms, however, a severe test-object; but as those distant and minute companions do not probably belong to the primary, they constitute the best points with which to compare the deviations of the large star.

Here the improvement of vision by looking aslant at the *comites*, is another proof that the axis of the eye, that is, the direction in which we habitually look, is not always that in which we see objects best.

DCCCXLII. γ CEPHEI.

Æ. 23ʰ 32ᵐ 47ˢ   PREC. + 2″·40
DEC. N 76° 44′·7   —— N 19″·90

POSITION 270°·0 (ʷ 1)   DISTANCE, *not taken*   EPOCH 1832·77

A second-rate Greenwich star on the Æthiop's right foot, with a very distant telescopic companion on the preceding parallel. A 3, yellow; B 14, dusky; a coarse pair of 12th-magnitude stars in the *sp.* Though γ is not yet a standard of the Ephemeris, it must become one, for in about 2360 years it will be the polar star. Its motions then may puzzle the unborn, for it has now a decided movement through space to the following amount:

$$P....Æ - 0″·24 \quad Dec. - 0″·11$$
$$B.... \quad - 0″·24 \quad\quad + 0″·17$$
$$A.... \quad - 0″·29 \quad\quad + 0″·16$$

Here Piazzi's *minus* amount in declination, is derived from comparisons with Bradley; but he adds, Nota 155, Hora XXIII., "Motus proprius in declinatione ex La-Caille, et ex nostris observationibus + 0″·22." And this appears to be the best value.

γ Cephei is called Errai in the Palermo and other Catalogues, from the Arabian *ar-rá'i*, a shepherd. See p. 500. In Sir John Lubbock's Maps it is placed on the foot, as Ptolemy directs; but some still more recent publications figure it upon the knee. See No. DCCCXLIII. κ Cephei. It can be readily identified, since it forms an obtuse angle with Polaris to the north, and β Cephei to the south-west, at equal distances; and the galley-rhymester *sings:*

The Pointers of the Greater Bear   the Pole-star notify,
That barrier cross, and soon the stars   of Cepheus meet the eye;
Just twelve degrees prolong that line   tow'rds Via Lactea's host,
And there bright *Gamma* will be seen,   the Arab shepherd's boast.

DCCCXLIII. 171 P. XXIII. ANDROMEDÆ.

Æ. 23ʰ 36ᵐ 39ˢ   PREC. + 2ˢ·93
DEC. N 45° 29′·5   —— N 19″·94

POSITION AB 105°·0 (ʷ 3)   DISTANCE   5″·0 (ʷ 1) }
—— AC 155°·0 (ʷ 1)   —— 175″·0 (ʷ 1) } EPOCH 1833·75

A very delicate double star, north of Andromeda's left hand, and in the galaxy, about 12° south-south-west of α Cassiopeæ, where there is a little group, of which the principal is ψ. A 7½, topaz yellow; B 13, blue; C 10, plum-colour; and the three point to a triplet of small stars in the *sf* quadrant. A and B were first classed as double by M. Struve, as No. 3034, Class II., in the Dorpat Catalogue of 1827. When the volume with the measures afterwards arrived, I was gratified on finding his results thus stated:

Pos. 103°·77   Dist. 5″·353   Ep. 1831·85

DCCCXLIV.   107 AQUARII.

Æ   23ʰ 37ᵐ 42ˢ   Prec. +   3ˢ·12
Dec. S 19° 34′·1   —— N 19″·95

Position 141°·8 (*w* 7)   Distance 5″·5 (*w* 5)   Epoch 1832·80

A very neat double star, in the group near the centre of the urn-stream, 16° north-east of Fomalhaut, and 15° due west of β Ceti. A 6, bright white; B 7½, blue. This object is 24 ♅. II., discovered in August, 1780, but registered, without any angle of position, under an estimated distance of 2½ diameters. Piazzi, Nota 177, Hora XXIII., describes it thus: "Duplex. Comes 7·8ᵃᵉ magnitudinis sequitur 0‴·1 temporis, 3″ circiter ad austrum." Thus affording a point which, compared with the micrometrical measures of H. and S., gives some countenance to a retrograde motion and increasing distance; thus:

P.   Pos. 153° 30′   Dist. 3″·34   Ep. 1800
H. and S.   143° 30′   5″·06   1823·79

My own measures appear to continue the retrocession, though in a smaller ratio. It remains to be seen how they may be affected by proper motion, that for A having been valued in a manner which, alone, would have increased the angle:

P.... Æ + 0″·12   Dec. − 0″·03
B....   + 0″·12   + 0″·03

DCCCXLV.   179 P. XXIII. PISCIUM.

Æ   23ʰ 37ᵐ 49ˢ   Prec. +   3ˢ·07
Dec. S   0° 37′·4   —— N 19″·95

Position 230°·0 (*w* 1)   Distance 3″·0 (*w* 1)   Epoch 1833·79

A most delicate double star under the preceding Fish, and about midway between Fomalhaut and α Cassiopeæ. A 8½, pale white; B 15, blue; and there is a star of the 10th magnitude in the *nf* quadrant, not far from the vertical, which identifies the object. This was discovered by M. Struve, and is No. 3036 of the Dorpat Catalogue. It is a difficult test, and requires long attention to discern it clearly: when, therefore, Σ.'s measures arrived, I was glad to find the above estimation so near the mark; they gave:

Pos. 228°·23   Dist. 2″·417   Ep. 1832·50

DCCCXLVI. 216 P. XXIII. PEGASI.

Æ 23ʰ 44ᵐ 49ˢ   Prec. + 3ˢ·05
Dec. N 11° 02′·2   —— N 20″·00

Position 102°·0 (w 8)   Distance 18″·5 (w 6)   Epoch 1834·81

A neat double star on the animal's wing, lying about 6° to the south-west of γ Pegasi. A and B, both 8½, and both silvery white. B is Piazzi's No. 217, and a reduction of the mean apparent places, yields a means of comparison with ꙅ.'s 7-foot measures:

P. Pos. 108° 18′   Dist. 16″·60   Ep. 1800
H.   102° 28′   18″·07   1831·80

But this is so hard a trial upon the Altitude and Azimuth Circle, to be pitted against the Polar Axis and Micrometer, that nothing conclusive can as yet be pronounced upon it.

DCCCXLVII. 30 ꙅ. VI. CASSIOPEÆ.

Æ 23ʰ 49ᵐ 07ˢ   Prec. + 2ˢ·97
Dec. N 55° 49′·6   —— N 20″·02

Position 220°·0 (w 1)   Distance 5″·0 (w 1)   Epoch 1835·68

A fine galaxy cluster of minute stars, on a ground of star-dust, on the upper part of Cassiopea's chair or throne: it is 3° south-south-west of β, and not much further to the west of α, where it lies in mid-distance between ρ and σ, stars of the 5½ and 6th magnitudes, each of which has a companion of the like brilliance. It is, indeed, a very glorious assemblage, both in extent and richness, having spangly rays of stars which give it a remote resemblance to a crab, the claws reaching the confines of the space in view, under an eye-piece magnifying 185 times. With this form

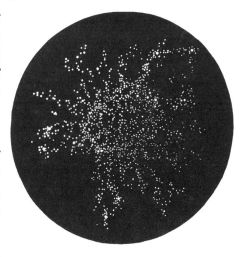

in the mind's eye, the imagined head will be in the *np*, the tail in the *sf*, and where the eyes would be, is the minute close double star of the

11th and 12th magnitudes, above estimated. There are several other pairs in the figure, especially towards the tail. The crab itself is but a mere condensed patch in a vast region of inexpressible splendour, spreading over many fields.

This beautiful object was first seen by the indefatigable Miss Herschel, whose labours are so intimately connected with those of her illustrious brother; and who, enrolling it in the autumn of 1783, realized the poet's description:

> Some sequestered star
> That rolls in its Creator's beams afar,
> Unseen by man; till telescopic eye,
> Sounding the blue abysses of the sky,
> Draws forth its hidden beauty into light,
> And adds a jewel to the crown of night.

It is with sincere pleasure that I am enabled to state, by a letter from Sir John Herschel, dated 21st November, 1843, that this lady is

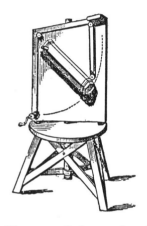

still living, at Hanover, nearly ninety-four years of age, and in the enjoyment of her mental faculties, after such persevering labour in the pursuit of science. And astronomy is very largely in her debt: for, besides her continuous task as the Slough amanuensis, she was assiduous as an observer; and with her little Newtonian Sweeper, of only twenty-seven inches focal length, and field of 2° 12', under a power of 20, she discovered various clusters and nebulæ, and no fewer than five comets, of which that of 1795, is the wonderful short-period one designated Encke's. This interesting instrument, of which the annexed is a sketch from memory, still exists, in the possession of its owner and former user, at Hanover. Owing to the admiration with which I have always viewed Miss Herschel's career, it was most gratifying to be one of the components of a council of the Royal Astronomical Society, which paid her a well-earned tribute of respect, in 1835. It cannot be better told, than in the words of the Report to the Fifteenth General Meeting of the Society. After noticing the narrow grounds and false principles which would interfere with the decision, it proceeds:

> But your Council has no fear that such a difference could now take place between any men whose opinion would avail to guide that of society at large; and, abandoning compliment on the one hand, and false delicacy on the other, submits, that while the tests of astronomical merit should in no case be applied to the works of a woman less severely than to those of a man, the sex of the former should no longer be an obstacle to her receiving any acknowledgment which might be held due to the latter. And your Council therefore recommends this Meeting to add to the list of honorary members the names of MISS CAROLINE HERSCHEL and MRS. SOMERVILLE, of whose astronomical knowledge, and of the utility of the ends to which it has been applied, it is not necessary to recount the proofs.

30 H. vi. would alone justify Miss Herschel's nomination. In gazing upon these myriads of universes, we cannot but be reminded of Gaspar Schott, the friend of Kircher and the precursor of Piazzi at Palermo;

who, writing so far back as 1667, congratulates man on being able to contemplate the multitude of heavenly bodies which the munificence of GOD has vouchsafed to manifest to us. But he predicts that many new stars, and even planets, will show themselves to those that search; for he thinks it " decidedly the part of intolerable arrogance to believe that our sight, lately strengthened by the invention of the telescope, though with the powers of a lynx, has surveyed all the stars; and it is the part of folly to wish to bind them by certain numbers and names: *stultitiæ, certis numeris et nominibus constringere velle."* There are brilliant proofs that Schott did not overshoot the mark in his prediction.

DCCCXLVIII. 240 P. XXIII. ANDROMEDÆ.

Æ 23ʰ 49ᵐ 56ˢ   PREC. + 3ˢ·04
DEC. N 23° 27'·6   —— N 20"·03

POSITION 313°·9 (κ 8)   DISTANCE 9"·4 (w 6)   EPOCH 1833·88

A fine double star, which, though designated in Andromeda, is on the body of Pegasus, and *hooked* into the precincts of Pisces; being only 5° south-south-west of a Andromedæ. A 8½, pale white; B 9, yellowish; and two of the same magnitude follow, as remarked by Piazzi, one to the north, and one to the south. This was discovered by Sir James South, at Passy, and his mean result is thus stated:

Pos. 44° 38' *np*   Dist. 9"·361   Ep. 1825·70

The interval between this epoch and mine is too brief, allowing for a moderate difficulty of observing the object, to pronounce a decisive opinion upon its evidence: still the data are so coincident, that it must, in our present state of knowledge, be deemed an optical pair, casually juxtaposed in the heavens.

DCCCXLIX. σ CASSIOPEÆ.

Æ 23ʰ 50ᵐ 55ˢ   PREC. + 2ˢ·99
DEC. N 54° 51'·8   —— N 20"·03

POSITION 325°·2 (κ 6)   DISTANCE 2"·9 (w 4)   EPOCH 1831·61
——— 323°·9 (κ 9)   ——— 3"·1 (w 5)   ——— 1834·88
——— 323°·7 (κ 8)   ——— 3"·0 (w 4)   ——— 1838·96

A beautiful double star on the lady's left elbow, and one degree south of No. DCCCXLVII., which lies, as already mentioned, between σ and ρ. A 6, flushed white; B 8, smalt blue; the colours are clear and distinct, though less fine than those of ε Boötis, of which this is a miniature. It is 5 ♄. I., described as being at the vertex of a telescopic isosceles triangle turned to the south. The astrometers attended it well, and the following were the known data when I first attacked it:

♄.	Pos. 330° 28′	Dist. 2″·50±	Ep. 1780·67
H. and S.	327° 21′	2″·92	1823·08
D.	324° 01′	2″·92	1830·60

From a re-examination of this object in 1804, by Sir William Herschel, he was led to infer that a very considerable orbital change had occurred in the interim; but the data here given are against such a surmise. My angles of position in 1834, appeared to be almost perfection: they were taken with a central paper disc, of two inches diameter, applied to the object-glass, which contrivance, though attended with a slight diminution of light, in this instance improved the images, by giving them sharper points.

DCCCL. 9 CASSIOPEÆ.

Æ 23ʰ 56ᵐ 02ˢ   Prec. + 3ˢ·03

Dec. N 61° 23′·8   —— N 20″·04

Position AB 330°·0 (w 2)   Distance 80″·0 (w 1) ⎫

——— AC 240°·0 (w 1)   ——— 150″·0 (w 1) ⎬ Epoch 1834·81

——— AD 194°·6 (w 5)   ——— 244″·5 (w 2) ⎭

A wide quadruple group closely following the right hand of Cepheus, and 3° north-half-west of β Cassiopeæ. A 6, white; B 11, and C 12, both dusky; D 8, deep yellow. The two first of these constitute 79 ♄. v., which was thus registered:

Pos. 320° 03′ Dist. 52″·67 Ep. 1782·66

A and D form a conspicuous pair, and are thus described by Piazzi, Nota 265, Hora XXIII.: "Alia 8ᵃ magnitud. 10″ circiter temporis præcedit, 3′ circiter ad austrum." These were micrometrically measured by Sir James South, at Passy, but without his perceiving either B or C; and the obtained results were:

Pos. 195° 37′ Dist. 245″·42 Ep. 1824·84

From the disparity of distance between A and B, as given by ♄. and that found by myself, I thought there might still be another *comes* so minute as to escape my instrument. I therefore requested the Rev. James Challis to scrutinize my diagram with the large Northumberland equatoreal, at Cambridge, the aperture of which is double that of my own instrument; but he found nothing within B. For identifying the nearest components, he kindly forwarded me their angular positions and apparent distances.

Pos. A B 331° 47′ Dist. 82″·27 ⎫ Ep. 1842·65
⠀⠀⠀ A C 239° 12′ ⠀⠀⠀151″·85 ⎭

The very sensible amount of spacial movement which Piazzi attributed to 9 Cassiopeæ, is disappearing before the strictness of recent investigation; and it will probably, under the standard observations now in hand, soon entirely vanish.

***

L'Envoy.

THUS, "gentle reader," have I conducted you through the Cycle of XXIV. Hours, by a course which, however new, is very practicable; and I hope that its interest is sufficiently seductive to wind you up, like a clock, to run the same round again. To me the task has been onerous, laborious, and expensive; it has occupied the vigils of two or three Olympiads in making many thousand observations, and has given daily employment for a similar period in the reductions and discussions. Still the subject is so fascinating as to retain its influence, and, *Deo volente*, I may yet return to it. Such study constitutes a sort of scrutiny into futurity, and truly elevates the mind above the mundane system. We cannot see, without peculiar emotion, that the great law of gravitation, so recently acknowledged and demonstrated in our own system, is decidedly extended throughout the vast heavens; and it is palpable, that all the perceptible universe is amenable to the operation of time and space, motion and force, in similar relations with our own, and equally open to mathematical disquisition. Thus the whole firmament, with its countless and glorious orbs,—which, though sustaining apparently independent positions, are but individual constituents of one Majesty of Creation,—in the absence of a larger comprehension, countenances the sagacity of the oft-cited ancient dogma, that

"GOD WORKS BY GEOMETRY."

INDEX.

2 N 2